CN00918483

Principles of Soilscape and Landscape Evolution

Computational models are invaluable in understanding the complex effects of physical processes and environmental factors which interact to influence landform evolution of geologic time scales. This book provides a holistic guide to the construction of numerical models to explain the co-evolution of landforms, soil, vegetation and tectonics, and describes how the geomorphology observable today has been formed. It explains the science of the physical processes and the mechanics of how to solve them, providing a useful resource for graduates studying geomorphology and sedimentary and erosion processes. It also emphasises the methods for assessing the relative importance of different factors at field sites, enabling researchers to select the appropriate processes to model. Integrating a discussion of the fundamental processes with mathematical formulations, it guides the reader in understanding which processes are important and why, and creates a framework through which to study the interaction of soils, vegetation and landforms over time.

GARRY WILLGOOSE is a professor in the Faculty of Engineering and Built Environment at the University of Newcastle, Australia. His research focuses on understanding the spatial and temporal dynamics of drivers of hydrology and erosion including landforms, soils, vegetation, soil moisture and fire. He developed the internationally utilised SIBERIA landscape evolution model, for which he was awarded the Lorenz Straub Award. He later customised SIBERIA to examine the long-term sustainability of rehabilitated mine sites, and developed further models for co-evolving hydrology, vegetation and soil for these sites.

This book was worth the wait! What started as a description of a pioneer modeling effort thirty years ago has ended up as a comprehensive treatise on soil and landscape evolution enriched by the experience of Willgoose. Hydrologists and geomorphologists interested in a quantitative understanding of what goes on in the critical surface zone of the geosphere must read this book.

Rafael L. Bras, Georgia Institute of Technology

If it moves, model it! There is no better synthesis of all the various elements in landscapes and soil than this lifetime compilation. Willgoose examines the many mechanisms operating in the landscape, at scales from continental tectonics down to the soil profile, demonstrating how he and others have built them into functional, mutually consistent and inter-connecting models. Its greatest strengths lie in the incorporation of soil processes – physical breakdown, mixing and weathering; and in how principles and models have been applied to the management of degrading spoil heaps.

Mike Kirkby, University of Leeds

An outstanding synthesis that thoroughly addresses both the theoretical basis and practical application of landscape evolution modelling — a benchmark of its kind.

Stuart Lane, University of Lausanne

Principles of Soilscape and Landscape Evolution

Garry Willgoose
The University of Newcastle, Australia

CAMBRIDGE
UNIVERSITY PRESS

University Printing House, Cambridge CB2 8BS, United Kingdom

One Liberty Plaza, 20th Floor, New York, NY 10006, USA

477 Williamstown Road, Port Melbourne, VIC 3207, Australia

314–321, 3rd Floor, Plot 3, Splendor Forum, Jasola District Centre, New Delhi – 110025, India

79 Anson Road, #06–04/06, Singapore 079906

Cambridge University Press is part of the University of Cambridge.

It furthers the University's mission by disseminating knowledge in the pursuit
of education, learning, and research at the highest international levels of excellence.

www.cambridge.org
Information on this title: www.cambridge.org/9780521858793
DOI: 10.1017/9781139029339

First published 2018

Printed in the United Kingdom by TJ International Ltd. Padstow Cornwall

A catalogue record for this publication is available from the British Library.

Library of Congress Cataloging-in-Publication Data
Names: Willgoose, Garry, author.
Title: Principles of soilscape and landscape evolution / Garry Willgoose, The University of Newcastle, Australia.
Description: Cambridge : Cambridge University Press, 2018. | Includes bibliographical references and index.
Identifiers: LCCN 2017035905 | ISBN 9780521858793 (hardback : alk. paper)
Subjects: LCSH: Soil structure. | Soil physics. | Soils. | Soil science. | Landscape changes.
Classification: LCC S593.2 .W55 2018 | DDC 631.4/3–dc23 LC record available at https://lccn.loc.gov/2017035905

ISBN 978-0-521-85879-3 Hardback

Additional resources for this publication at www.cambridge.org/willgoose.

Contents

Preface

This book has been a long time coming. The original idea, suggested by Peter Eagleson at my PhD defence in 1989, was to turn my PhD on landform evolution into a book co-authored with my PhD supervisors Rafael Bras and Ignacio Rodriguez-Iturbe. However, shortly thereafter I became involved in applying my landform evolution model, SIBERIA, to assess the long-term evolution (1,000–10,000 years in the future) of a rehabilitation design for a uranium mine in Australia, Ranger Uranium Mine (Willgoose and Riley, 1998a). It soon became clear to me that the characteristics of the eroding surface at Ranger evolved much faster than the landform, and the major uncertainty on the safety assessment of the containment structure was the evolving erodibility of the landform surface, not the landform evolution per se. Only millimetres of erosion significantly changed the erosion characteristics of landforms (which were many metres high) long before any significant landform evolution was apparent. Parallel projects at other mine sites confirmed that this behaviour was not unique to Ranger. Very quickly the science focus of my work moved away from landform evolution into what we now call soilscape evolution. The original 1989 book's focus was too narrow. Unfortunately our progress in soilscape evolution modelling proved slow and computationally difficult, even more so than the original development of SIBERIA, at least in part, because there was very little existing quantitative work to build upon. The original book went on the backburner despite significant progress during short sabbaticals at Lancaster University in 1995 and MIT in 2000. The final nail in the coffin was when I became seriously ill in 2001 as a result of involvement in a flood-forecasting project in China, and the book was abandoned until I recovered enough strength, some years later, to be able to revisit the project. In the meantime, landform evolution research had exploded worldwide.

The book you are holding is a very different book from the original plan. It is a more holistic discussion of how to model the key issues in the evolution of landscapes (landscapes being the combination of landform, soils, vegetation and so on). The intellectual core of this book is the evolution of soilscapes. Soils are central because they constrain the evolution of (1) the landform through the soil's interaction with the sediment transport processes and (2) the vegetation through vegetation's dependence on the water and nutrients stored in the soil. Moreover, soils are the single biggest store of carbon in the terrestrial environment and thus central to future trends in climate change. The various storage processes within the soil profile determine the timing, and lags between cause and effect, of almost all terrestrial processes. Soil's importance is reflected in the Critical Zone research agenda where the major component of the biogeochemical reactor that is the Critical Zone is the soil and its interface role in the water, geology and atmospheric cycles on Earth (NAS, 2010).

The focus of this book is on physically based modelling of processes, but I won't shy away from empirical and stochastic models if they are the best (current) fit. Yet the book is not just about the mathematics of the processes because we first need to know what processes to model in any particular situation. The book discusses the key concepts and the empirical evidence underpinning the mathematical representation of these processes, and studies of when and where these processes may be dominant or otherwise. The book is not a comprehensive review of the literature of landscape forming/evolving processes and their implications. That would require a much, much bigger book. Hopefully, however, even those readers whose primary interest is not *modelling* of landscape evolution may still find the discussions here worthwhile. Of course, where appropriate, I discuss, sometimes in gruesome detail, the numerical methods and the mathematics.

The range of processes that are potentially important in a landscape evolution model is quite broad. To keep the book treatment manageable, the book contains two types of chapters, process and synthesis. The process chapters describe a specific set of processes in detail in a self-contained fashion independent, as much as is possible, from the other process chapters. The variable length and content of these process chapters reflect the variation in *mathematical quantification* heritage and potential, not any assessment on my part of their importance. The synthesis chapters assemble those process chapters into what we recognise as a landscape evolution model. These latter chapters focus on how the individual processes interact and how the processes are assembled into a coherent model.

On the question of modelling philosophy my inclination is to favour simple models over complex ones. With all the nonlinear feedbacks in time and space even simple landscape evolution models can generate surprisingly complex behaviour. Furthermore, complex models have the endemic problem of equifinality, where many of the processes lead to the same or very similar behaviour, so it is difficult to use the models alone to infer what processes have caused the behaviour observed in the field. This complexity and equifinality tends to obscure the first order processes in the system. Some field workers have criticised my, and my modelling colleagues', naivety in ignoring some process or other in our modelling, but I have always been guided by this desire for simplicity over complexity. A colleague, Greg Tucker, once summed it up with a quote he had heard somewhere: 'If we include in our model every process that occurs in the field, then our model will be as difficult to understand as is the field'.

There are many people to thank. Rafael Bras and Ignacio Rodriguez-Iturbe started me on my landscape evolution journey, and I know they were disappointed that the original book plan did not come to fruition. The soils chapters in this book were mostly written while on sabbatical at the Community Surface Dynamics Modelling Systems Facility (CSDMS) at Colorado University, Boulder, in 2015, and I thank James Syvitski for supporting the quiet time I needed to write. I'd also like to thank Bob and Suzanne Anderson and Greg Tucker for interesting conversations on soils and rivers during that time.

During my career there have been a number of PhD students, post-docs and colleagues at Newcastle who have contributed to the landscape and soilscape research, including Min Chen, Sagy Cohen, Ken Evans, Yeboah Gyasi-Agyei, Greg Hancock, Eleanor Hobley, George Kuczera, Dene Moliere, Mariano Moreno de las Heras, Hemantha Perera, Patricia Saco, Saniya Sharmeen, Dimuth Welivitiya and Tony Wells. I particularly note the ongoing 25-year collaboration with Greg Hancock. I am especially indebted to Steven Riley (then Head of the Mining Impacts Section) and Arthur Johnson (then Institute Director) at the Environmental Research Institute of the Supervising Scientist (*eriss*) for supporting the original pioneering application of landform evolution modelling to the assessment of the long-term 1,000-year waste containment safety at the Ranger Uranium Mine back in 1990. This support is even more notable because they assessed the likelihood of success at less than 1%, though fortunately they didn't tell me that at the time. It's been a very productive 27-year collaboration, and counting. Among colleagues outside Newcastle there are a few that deserve particular thanks for their contribution and collegiality. First among equals is Mike Kirkby who has always had something interesting to say, a valued colleague when I was at Leeds University, and with whom I spent a very enjoyable week talking soils while writing this book. Others are Alex McBratney and Budi Minasny at Sydney University who have been a mine of information about mathematical soil science; the late Frank Henderson (Henderson, 1966) who, when I was an undergraduate, impressed on me the power of mathematics to understand how processes work in the field; Keith Stolzenbach at MIT who reinforced that same message during my PhD; and Bill Dietrich who after a very controversial AGU talk in 1987 (where I presented a mathematical analogy between leaf veins and river networks) stressed the need to explain in plain English what the mathematics meant. The list of individual research sponsors is too long to cite, but, aside from the ongoing collaboration with *eriss*, special thanks go to the US National Weather Service, which partly sponsored my SM and PhD at MIT; the Australian Water Research Advisory Council (AWRAC), which sponsored my post-doc fellowship; the Australian Research Council, which has been an ongoing sponsor (notably with an Australian Professorial Fellowship in 2006–2010); and the international mining and nuclear waste industry.

Some colleagues have been crazy enough to agree to review draft chapters, and the book is better for it. Thank you to (University of Newcastle unless otherwise noted) Pippa Chapman (University of Leeds), Anna Giacomini, Eleanor 'Nellie' Hobley (TU Berlin), George Kuczera, Budi Minasny (University of Sydney), Gary Sheridan (University of Melbourne), In-Young Yeo and Omer Yetemen. Greg Hancock and Tom Vanwalleghem (University of Córdoba) reviewed multiple chapters. Special thanks go to Natalie Lockart (and Linden in utero, a budding earth system scientist) for reviewing the entire text of the penultimate version. Thanks also to all those colleagues who answered my questions about their published work. Of course, any remaining stupidity is my responsibility alone.

Finally I would like to thank everybody who personally supported me through the dark years after I fell seriously ill in 2001. I have thanked you all privately. I just want to put that thanks on the public record. And while she doesn't feel she needs to be thanked publicly there has been my wife, Veronica Antcliff, who has put up with a lot during my career and this book project.

Before we get to the serious stuff there is one story that colleagues have insisted I recount. That is how my landform evolution model, SIBERIA, got its name. Before I give the answer I'll note that it's not an acronym, and herein lies an amusing story involving researchers who should remain nameless to protect their reputation. These researchers, on reading my early papers, were puzzled why I had called it SIBERIA. Thinking that the name was an acronym (and that perhaps the acronym had been explained in a previous paper that they had not read) they then spent several days over morning tea brainstorming what SIBERIA was an acronym for. They got to E and then gave up. The reality is somewhat different. On a return visit to MIT soon after my PhD I was in Rafael Bras's office with Ignacio and him. I was explaining my latest work where I had derived an analytic solution relating slope, area and elevation for landforms that were not in dynamic equilibrium (Willgoose, 1994a). Ignacio was increasingly pacing up and down in the office, puffing on his pipe faster and faster, the air getting thicker and thicker, as he is wont to do when he is excited. After what seemed like 30 minutes of pacing and puffing (but was probably much less) he burst out in his thick Spanish accent 'Garry, Garry, Garry ... Garry, Garry, Garry ... what on earth is this Siberia elevation stuff you are talking about?' My Australian accent had defeated him, yet again. And I thought he was excited about the mathematical result. I was deflated. That evening over a beer with Glenn Moglen (who was a PhD student at the time) I recounted this incident. Glenn burst out laughing and said that's a great name for a computer code. And so it was.

1 Introduction

> To look upon the landscape . . . without any recognition of the labour expended in producing it, or of the extraordinary adjustments of streams to structures and of waste to weather, is like visiting Rome in the ignorant belief that the Romans of today have had no ancestors.
>
> (Davis, 1899, p. 496)

There can be no better, nor more succinct, description of the motivation for, and content of, this book. For continental and global environmental modelling applications we now have high-quality global databases of topography (e.g. the Shuttle Radar Topography Mission [SRTM]), an emerging database for rates of change of topography (e.g. from the GPS, GLONASS and Galileo global positioning systems, and the GRACE gravity mission), and a global program to deliver high spatial resolution soil functional and chemistry data (the GlobalSoilMap program). For smaller site-specific applications LIDAR provides high-resolution and accuracy topography (Tarolli, 2014), ground-penetrating radar and other proximal remote sensing technologies (e.g. gamma spectroscopy) provide spatially distributed soils data, and sensors on drones can provide spatially distributed vegetation and soil chemistry data. While Davis could not foresee these datasets, his argument that for maximum benefit we need to know how the topography, soils and vegetation arose is as valid today as it was in 1899.

By understanding how the landscape system changes with time, then, we will be better able to diagnose why it is changing. This will allow us to better manage our landscapes, differentiate man-made impacts from natural evolution, and potentially predict (or at a minimum assess the risks of) future evolutionary scenarios and any anthropogenic interventions. While paleo-analogue and current-day analogue sites can indicate likely behaviour, their usefulness depends on how well we can generalise from their analogue behaviour to the site that we are interested in. That may be difficult. In some cases, such as the author's engineering applications of designing man-made structures for containing mining and nuclear waste, there may not be any analogue to guide our management decisions.

1.1 Why Mathematically Model

In the absence of exact physical analogues or scaled model experiments (e.g. using nondimensionalised parameters), if we wish to quantify landform evolution, then we are forced to rely on mathematical models to predict the future and guide decision-making. These models may range from empirical multiple regression correlations (e.g. gridded soil data obtained by correlating soil chemistry laboratory results with gridded landscape attributes such as topography; Holmes et al., 2004) through to graphical methods, or they may be computer models. Ideally the complexity of the model adopted for any particular study is a function of the domain knowledge, the computational tractability of the science and the project objectives (e.g. Larsen et al., 2016). Computer models can range from simple scripts in MATLAB or Python, to complex code with thousands of lines in Fortran or C. This book is about these models.

Many existing computer codes in landscape evolution modelling tend to be complex. In some cases that complexity is essential, in others not so essential. In some of the chapters in this book we will use quite short and simple computer models because these are best to display the implications of the physical principles. These codes can be downloaded from GitHub and other open source sites, or can be obtained from the author directly.

These models are only as good as the process representations incorporated into them. Typically the generic process representations are based on a synthesis of field, experimental and theoretical work. For quantitative application at any specific site there will almost certainly be some need for site-specific customisation. This customisation will involve the selection of what processes are dominant at that site and a calibration of the parameters of those processes at that site. Most of the chapters in this book are about the generic process representations that are accepted in the respective research and applications communities, and some characterisation of where and when these processes are important. As part of the presentation of these processes there will be some discussion of process rates and model parameters, and general information about how they have been derived. We will discuss calibration and validation briefly in a moment, but the topic of site-specific process calibration (i.e. experimental design, parameter calibration and model uncertainty analysis) is quite important for defensible quantitative landscape evolution applications and will be addressed in detail in Chapter 15.

One problem with complex models is that it can be hard to understand how they work and identify the cause-and-effect linkages. For researchers this makes them difficult to work with because there is a significant learning curve to climb before we can be completely confident that we can apply these codes to our (novel) sites. Moreover, in very complex models a single person cannot hope to be completely familiar with every aspect of the model, particularly when a model has been extended numerous times by a series of PhD students and post-docs who have subsequently moved on to other things.

In conclusion, if possible, simple models are best. They allow simple transparent explanations. That is the guiding principle throughout this book.

This book is about much more than landforms alone. A central rationale is to support the emerging field of soilscape evolution modelling. Many authors over many years have recognised the link between soils and the landforms over which they drape (e.g. Jenny, 1941, 1961; McBratney et al., 2000) and that soils evolve in response to the climate, geology and biosphere (e.g. Birkeland, 1990; McFadden and Knuepfer, 1990; Schaetzl, 2014). It seems only natural that these two views should be merged into the ongoing work with landform evolution models to create coupled models of soilscape and landform evolution. Many gaps remain to be filled (particularly with regard to quantification), but the field is now sufficiently mature that we can see a way forward and conceptualise what a coupled model should look like. This book pulls together the different threads of work in soilscape and landform evolution to provide this intellectual framework.

The abbreviation LEM has been used interchangeably in recent years to refer to either landform evolution models or landscape evolution models. Landforms are just one component of landscapes, along with soils, vegetation and soil flora and fauna. Much of this book is about the evolution of the soils, and how they interact with the evolving landforms and the environment. However, the terms landscape evolution model, soilscape evolution model and dynamic vegetation model are a bit of a mouthful, so we will abbreviate references to them throughout the book:

- LEM: Landform evolution model, modelling only the evolution of the landform/topography and nothing else. This is currently the common usage.
- SEM: Soilscape evolution model, modelling the evolution of the soils across the landscape and nothing else (i.e. assuming the landform is fixed). Soilscape is the term used in the soil science community to refer to a soil model (typically without any time-varying evolutionary component) applicable across a landscape (e.g. a gridded digital soil map), and the term soilscape distinguishes these types of models from soil profile/pedon models that model the vertical profile of the soil at a given point in isolation. A soil catena is a subset of soilscape modelling, typically applying to individual hillslopes rather than an entire catchment.
- SLEM: A coupled soilscape and landform evolution model, a model that links an LEM and SEM.
- BioSEM, BioSLEM: One of an SEM or SLEM that has a coupled biophysical model. A BioSEM may have an active biological weathering component to the soil profile development (e.g. organic acids from the breakdown of soil organic matter, fungal breakdown of rock fragments). A BioSLEM might feature the coupling of a vegetation model (typically something like the dynamic vegetation models [DVMs] that are typically coupled with ecohydrology and climate models) with an SLEM.

When I think of a landscape evolution model I think of something like a BioSLEM. This book will touch on the construction of all four types of models, with a heavy emphasis on SEM.

1.2 Modelling Philosophy: What Is a Model?

Before we get down to the nuts and bolts of model making, it is worthwhile to reflect on the applications for a (landscape evolution) model. These applications will influence the way we construct and use our models. It is often said, 'All models are wrong, it's just that some are useful.' While this is a gross overgeneralisation, this does capture the idea that models are approximations of the field, but that they capture to first order the characteristic(s) of interest. They do not replicate everything in the field because of the assumptions and approximations that are either implicit or explicit in the construction of the model. Models have two main uses: (1) numerical experimentation tools to improve our understanding of natural systems and/or (2) predictive tools to predict behaviour when the system is modified in some way.

1.2.1 Numerical Experimentation Mode

In this mode models are used in the same way as field and laboratory experiments. Experiments are done where components of the physics are changed to explore changes in behaviour. These changes may be changes in rate parameters or parameters that change the interactions and feedbacks, or may involve complete substitution of one set of physics for another. In this way we can quantitatively discern the impact of changes in process properties and how they change system response. The main problem with this is that, just as with field and laboratory experiments, it can be difficult to determine whether the observed changes in system response occur only for the parameters adopted for those experiments (or a particular site for field experiments). The main advantage that numerical experiments have is that it is possible to completely control the system. The user has the ability to hold some parameters steady while changing others, something that can be difficult to do for field catchments and even sometimes in laboratory experiments. By doing this it may be possible to ascertain how to design an experiment that can distinguish and test competing hypotheses for field behaviour. For instance, if changing a process (parameters, mathematical formulation and so on) does *not* change some characteristic we can observe and measure in the field, then, at least for this process, there is no value in performing such an experiment because the field experiment will also probably not show any significant differences. In this sense using models for numerical experimental laboratories is no different from the way science was done before computers, it is just that we can now do preliminary tests of what to expect in the field and hopefully design more robust experiments. This does not negate the value of purely curiosity-driven experiments to 'see what happens'. However, it does impose rigour into experiments where quantitative conclusions might be possible about cause and effect. Without the model it is not possible to know if the proposed processes, when quantified, actually lead to the quantitative effect that is observed. Quantification is key.

The type of model that is well suited to numerical experimentation is one where it is easy to change the construction and/or parameters of the model. Large complex models, while having the kitchen sink of all processes built into them, tend to be less well suited to this problem because it can be hard to be confident that you fully understand all the assumptions built into the model. Also, large complex models more commonly have problems with equifinality (e.g. Hancock et al., 2016). Equifinality (Beven, 1996) is where (1) two processes or (2) multiple combinations of parameters lead to similar responses so it is difficult to discern which process/parameters actually cause the observed behaviour.

1.2.2 Prediction Mode

In this mode models are used for making predictions of behaviour (1) at other sites or (2) for future situations. These predictions may then be compared with some societal, environmental or regulatory requirements to see if the modified system's behaviour is satisfactory. In principle our numerical experimentation models can be used for this problem, but there are a number of subtle differences. The most important difference is that the model must be capable of predicting system behaviour for the criteria that are used for assessment and ideally provide a measure of how reliable that prediction is. In some cases this acceptance criteria may be relative to unmodified behaviour (e.g. how much better will the system be after modification, expressed as, say, a percentage or absolute change relative to no modification).

It is common to observe that the exact deterministic results from a landform evolution model are sensitive to many model inputs. While small changes may change the average behaviour of the model only a small amount (e.g. average erosion rate across a catchment), these same changes may dramatically change the location of specific features (e.g. the two simulations may have a completely different location for a gully). Thus if the acceptance criterion is a limit on the maximum erosion at a specific location, it may be difficult to demonstrate that the modified behaviour at a specific location is better, even though the average for the entire landform is, in fact, better. Moreover, if small changes in model inputs have a dramatic effect on the location of a gully, it may be impossible to convince regulators that the model predictions are useful indicators of the performance of proposed field modifications.

Finally, if project proponents (e.g. a low-level nuclear waste repository that needs to be shown to be safe for 10,000 years – current requirements in the United States and Australia for radioactive mining waste and low-level nuclear waste) are to spend large amounts of money, they need to know that the quantitative predictions of the model are accurate, or at least justifiable to regulators and defensible in court. Regulators and courts, for instance, will often want to see short-term predictions of behaviour that can be tested quantitatively against observed behaviour to assess whether the model is 'correct' or not.

1.2.3 Model Calibration, Validation and Uncertainty Analysis

Finally, we need to be able to test our models. This is challenging because we are dealing with systems that evolve over thousands to millions of years, and because it is unusual to have situations where we can do independent replicates of landform evolution. Thus we are left with the difficult situation that our models will not match observed data perfectly (since they are approximations of the field) and we don't know how much difference between the models and the field is acceptable. Moreover, we may be able to calibrate our models to a specific field case (i.e. change the physics and its parameters until we get a satisfactory fit), but this demonstrates only that the model is feasible (i.e. with appropriate parameters the model can fit the data), not that the model is correct. Indeed, the hypothesis-testing literature (and a computer model is simply a hypothesis written in a computer language) indicates that we can never prove a model correct, we can only prove it incorrect. It may fit all the current data, but the data to disprove it may be just around the corner. This ability to reject a model is called falsifiability and is a key component of model testing (Popper, 1959). A related aspect of this is that a good model makes predictions that can potentially be falsified. If the model cannot make predictions that can be falsified, then it is not possible to test it (Pfister and Kirchner, 2017).

Nor can we blindly adopt similar physics from other types of models that have been tested for different applications. For instance, there are numerous well-validated and tested agricultural erosion models (e.g. RUSLE, CREAMS, WEPP). They are based on one of the broad range of shear stress–driven detachment and sediment transport models (see Chapter 4). They have all been empirically fitted to field data so that they all satisfactorily predict erosion over relatively short time scales of a few years for paddock-scale problems. However, when applied over long time scales of millennia or more, the specific parameters used in the different transport models yield qualitatively different landforms, even though over agricultural time scales of a few years the differences between the models are small (e.g. Willgoose and Gyasi-Agyei, 1995). The reason for this is the long-term feedbacks between changes in topography and erosion, something that is not modelled in the agricultural erosion models, so differences become apparent only over the longer term.

One unique aspect of landform evolution models (and it is not yet clear whether this is also true of soilscape evolution models; Phillips, 1993) is they show sensitivity to initial conditions and external inputs so that the deterministic output (e.g. the exact values of elevation at every node in the model) can vary dramatically with only small changes in inputs or parameters (e.g. Willgoose and Gyasi-Agyei, 1995; Willgoose et al., 2003). Channel and valley locations can move significant distances laterally for only small changes in model inputs. Thus it is not possible to make a direct comparison of the exact values at each location of the model and the field data. Rather it is necessary to identify statistics of the landforms that will be the same if the physics is modelled correctly and then compare these statistics. One might then ask, What is the set of essential statistics that is required to completely describe a landscape? This set of essential statistics can then form the focus of any model testing.

Key outputs needed for model testing are confidence limits on model predictions or confidence limits on field measurements. For instance, the *t*-test is commonly used for testing the differences between models and experiments. The *t*-statistic is the difference between the two experiments divided by the variability of the experiments, and this variability can be determined from the confidence limits. As we have mentioned, field replication of landscapes is difficult, so it is rarely possible to calculate confidence limits on field measurements. An alternative is to calculate confidence limits on model output. In principle this is simple if we have accuracy estimates on model parameters and exogenic inputs. We can use Monte Carlo simulation with the model. This process is to repeatedly run the model with different parameters, where the different parameters are determined by sampling the parameter values from the probability distribution of the model parameters. In this way we can generate a number of output simulations, and we can derive the probability distribution of the outputs for the model and thus determine confidence limits on the predictions (e.g. Willgoose et al., 2003). Likewise we can explore the effect of unknown climate by running simulations with randomly generated climate data to explore the effect of climate variability. The problem with this process is that it can be extremely computer intensive. Depending on how many parameters need to be varied, it is common for thousands of simulations to be necessary.

This problem of excessive computer time is even worse for satisfying many regulatory requirements in engineering applications because it is common that acceptance criteria are expressed as a small probability of some criterion or criteria being exceeded (e.g. one in one million chance of failure in any one year is common for high-consequence industrial applications). Thus not only is it necessary to do Monte Carlo simulation to obtain the probability distribution of the output, but the small probability required for acceptance means that the results we are trying to determine are on the tail of the probability distribution. The further out we are on the tail of the probability distribution, the greater is the number of simulations that are required to adequately sample enough rare events to define the tail of the probability distribution.

The key conclusion from this is that to be able to provide confidence limits on model simulations it is crucial to have fast models. Thus it is not simply enough to be able to simulate some processes, but it is also necessary to be able to simulate these processes efficiently so that confidence limits can be generated in a reasonable compute time. The book will return to this issue of algorithmic efficiency repeatedly.

1.3 Model Framework: The Model of Everything

The title of this section is a tongue-in-cheek description by a colleague of the vision of this book and what constitutes a 'complete' landscape evolution model. Given the complexity of the processes and interactions, it is worth taking a step back for a moment to present a big picture overview of the landform evolution system: What are the components of the modelling framework and how do these components interact? This framework informs the organisation of the chapters that follow. This overview will be presented in three stages starting at the largest length scale, the continental scale, zooming down to the catchment/river network scale, and then zooming in until we reach the smallest length scale, the hillslope scale.

1.3.1 Continental Scale

Figure 1.1 overviews the important processes at the continental scale. At this length scale (10–100 km) the weight of the land on the crust has significant interactions with the tectonics of mountain building. The loss of mass by erosion lightens the load on the crust and the crust rises in response by isostacy (Section 12.2). If mountains are being created by convergence of two plates, then this unloading by erosion also has significant feedbacks on the rate and nature of the interaction between the two plates underneath the emerging mountain range (Section 12.3). An important interaction is that the topography of the mountain range changes the spatial distribution of rainfall (Section 3.1.1) with higher rainfall on the windward side of mountains than the leeward side. However, not only does this increased rainfall increase runoff (and everything else being equal increases erosion), but it also increases the vegetation density (increased density decreases erosion), and it is the competing effects of increased runoff and increased vegetation that determines whether the erosion rate is higher on the rainy side or the drier side (Section 16.1.2).

The rivers (and the hillslopes linked to them) are the transport mechanism to remove sediment from the mountain ranges to the ocean (Chapter 4); these river network and hillslopes will be discussed below.

While this book will not cover marine processes, the fate of the eroded sediment is shown in Figure 1.1. Most of

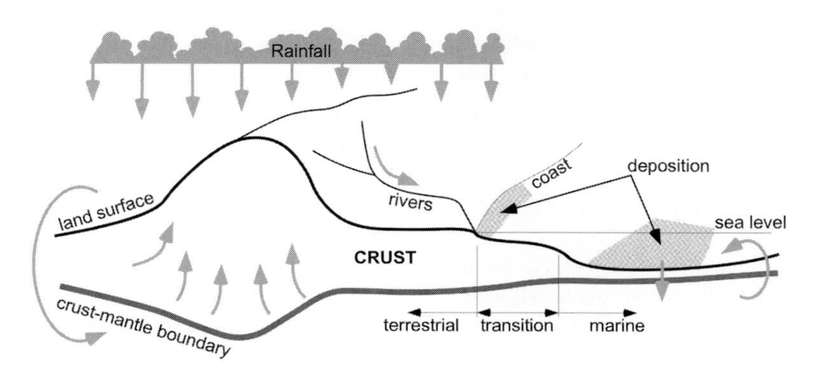

FIGURE 1.1: A schematic of the processes that occur at the continental scale including mountain building, isostatic rebound, river erosion, spatially variable rainfall and deep marine deposition. The heavy grey arrows are process fluxes and their predominant direction of action. The two circular arrows represent the feedbacks between (left) erosional reduction of elevation and rising in the crust, and (right) deep-sea deposition and depression of the crust.

the particulate and dissolved load is deposited either on the continental shelf (either in the delta at the river mouth or farther offshore) or in basins in the deep ocean beyond the shelf. The extra load from this deposited material results in the crust being depressed downward; that is, not only does erosion on the land interact with the crust and mantle so the crust rises, but the deposited sediment in the ocean also interacts with the crust and mantle, so the crust falls.

The continental shelf is an important transition region between terrestrial processes and marine processes. The depth below current sea level of the ocean edge of continental shelves worldwide is of the order of 140 m, which coincides with the level of the ocean during the ice ages of the last million years. Ice core data indicate that about 80% of the last million years has been ice age, while the current, interglacial, conditions (since about 10,000 years ago) have prevailed for less than 20% of this period. Thus the topography of the continental shelf regions around many continents will largely reflect terrestrial processes, with about 10,000 years of recent marine processes overprinted on them.

Some final points to note about Figure 1.1 are the following:

1. The arrows for uplift of the crust under the mountain range are not vertical, because mountain building involves convergence of the crust as well as vertical uplift. Moreover there is the possibility of relative lateral motion horizontally between the converging plates (e.g. the San Andreas fault in California).
2. The rainfall is spatially and temporally variable to reflect the effect of the changing topography on rainfall. Note that this spatial distribution of rainfall may vary seasonally, and year to year, as a result of changes in the prevailing wind directions. Some signature research sites have broadly consistent prevailing wind directions (e.g. Taiwan, New Zealand, Chile), while for other sites wind directions even today vary in direction seasonally as a result of seasonal monsoons (e.g. the Himalayas).

1.3.2 Catchment and River Network Scale

Figure 1.2 overviews the important processes at the catchment and river network scale. At the continental scale the main transport process for sediment is the channel network. Most of the sediment that is transported by the river network comes from the hillslopes, either by (1) mass movement (e.g. landslides) in parts of the river network that are steep or (2) by fluvial erosion and soil creep in flatter areas (Figure 1.2a).

Broadly speaking the river network consists of two types of channels (Sections 4.4 and 4.5): (1) *bedrock channels* where the river flows across bare exposed bedrock and (2) *alluvial channels* where the river flows through deposited sediment. These two channel types are end members in a transition from bedrock channels to alluvial channels as the sediment load in the river increases, but this distinction suffices for the moment. Alluvial channels occur in regions of deposited sediment called floodplains and typically meander from side to side within this floodplain. Bedrock channels are typically constrained laterally by the erosion-resistant bedrock channel sides and so tend to be straight rather than meandering. The floodplain is a result of net deposition of sediment on the floodplain by the alluvial channel (typically at high flows), and these sediments are remobilised as the river meanders from side to side (typically at lower flows). The residence time of the sediment within the floodplain is measured in many thousands of years. Recent work has also indicated that floodplains are a major storage site for organic carbon (Section 10.7).

Thus the process of erosion of the mountains is one of delivery of sediments from the hillslopes to the channel. This sediment (both mineral and organic matter) then passes through a mixture of bedrock and alluvial channels on its way to the ocean. In the alluvial channels there is an exchange of sediment between the channel and the floodplain deposits with some of the channel sediment being deposited in the floodplain, while some of the floodplain materials are remobilised from the floodplain deposits (Sections 4.5 and 10.7). Typically the finest, lightest and most mobile sediments will pass through the floodplain without being exchanged with the floodplain sediments (Section 4.2.1.3). This leads to what is commonly referred to as a 'strings and beads' plan of the river network where the strings are channel reaches without adjacent floodplains and the beads are river reaches with their adjacent floodplains.

Figure 1.2b shows a schematic of a river long profile showing the transition from a bedrock channel reach to an alluvial channel/floodplain reach and back again. For low sediment loads the bottom of the bedrock channel will be clear of sediment so bedrock incision will erode the channel. For high sediment loads the bottom of the channel in the bedrock reach will be periodically covered by sediment waves and pulses that move down the reach. Thus at high sediment loads the bottom of the bedrock channel will be partially protected by sediment. Incision can occur only when the bedrock channel base is not covered by sediment. Thus the bedrock channel will incise into the bedrock periodically, with the percentage of time in the

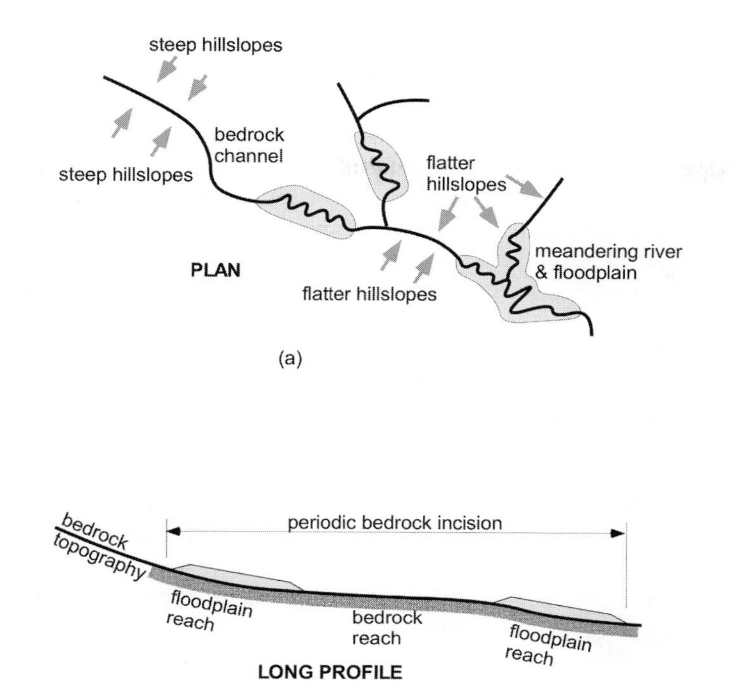

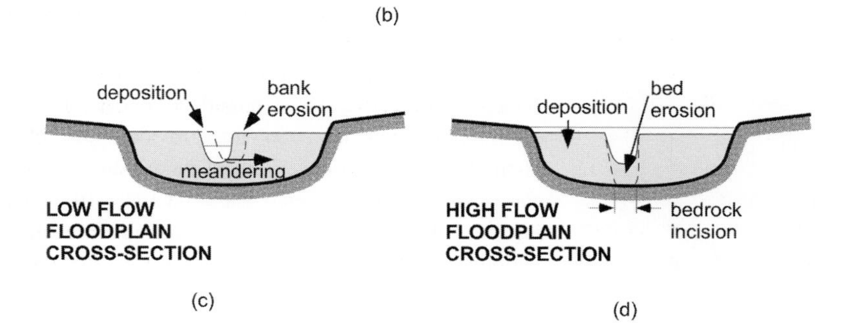

FIGURE 1.2: A schematic of the processes and interactions in channel networks at the catchment scale: (a) a plan of the river network showing the 'string and beads' arrangement of channels and floodplains, (b) a long section along a river showing the periodic change in channel type from bedrock to alluvial and back again, (c) a typical alluvial channel/floodplain cross section at low flow showing the lateral movement of the channel (left to right in this figure) within the floodplain as a result of meandering and (d) a typical alluvial channel/floodplain cross section at high flow showing the excavation of the channel cross section down to and exposing the underlying bedrock, and the deposition of particulate matter on the floodplain.

bedrock incision mode being a function of the percentage of time that the bedrock bed is not covered by sediment. Likewise for the alluvial channel there will localised stretches where the bedrock underneath the floodplain is exposed. Moreover, during high flows the channel is scoured and can be excavated down to bedrock (Figure 1.2d). Thus periodically an alluvial channel can be incising bedrock (and thus the bedrock under the floodplain is lowered), though for most of the time and for most locations the channel is largely in balance with the meandering channel excavating the floodplain sediment on the outside of the meander bend and depositing inside the bend, so that there is an exchange of sediment and organic matter between the channel and the floodplain (Figure 1.2c). At high flow, sediment and organic matter are deposited on the floodplain (Figure 1.2d). The residence time in the floodplain deposits is then a function of the rate of floodplain excavation by meandering relative to the width of the floodplain. The residence time is a key parameter for organic matter sequestration in the floodplain because it determines how much of the deposited organic matter decomposes in situ versus being remobilised by bank erosion at some time after its original deposition.

1.3.3 Hillslope Scale

Figure 1.3 overviews the important processes at the hillslope scale. The four panels overview different components of the hillslope processes: (a) mineral transport,

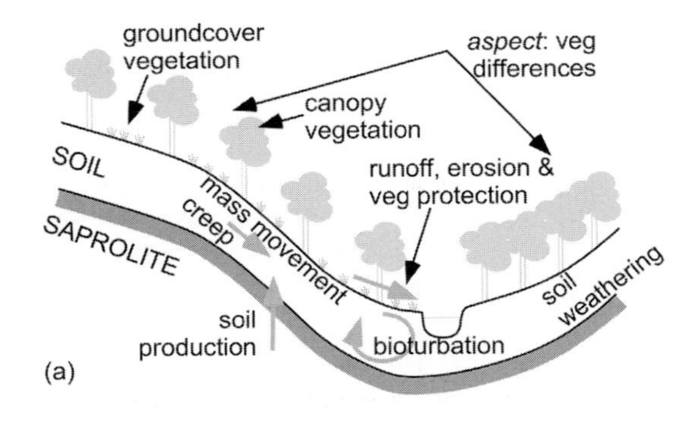

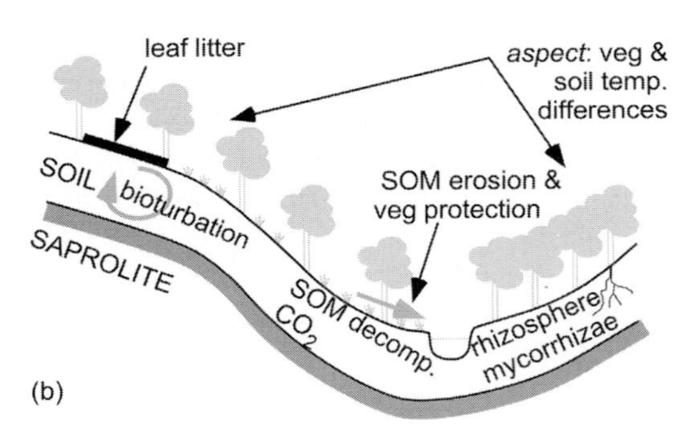

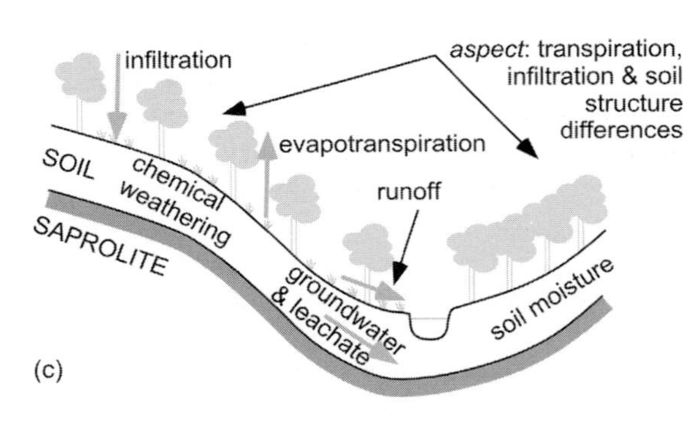

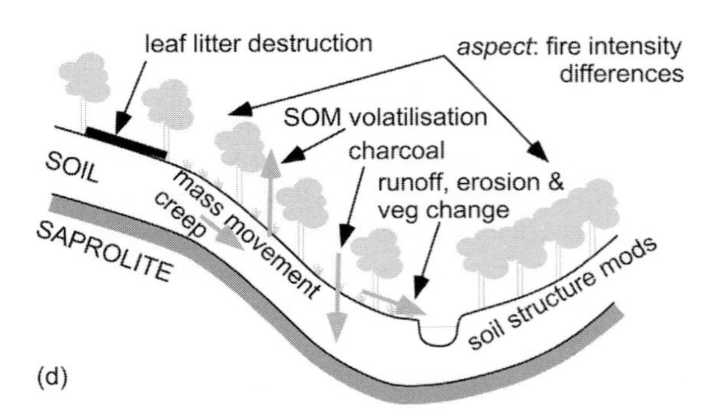

FIGURE 1.3: A schematic showing two hillslopes draining down to a central channel. The four panels each show the processes operating for the four main components of the hillslope system: (a) soil mineral matter, (b) soil organic matter, (c) the water balance

(b) soil organic matter and carbon transport, (c) water balance and (d) vegetation and fire. The hillslopes deliver the material that is then transported to the coast by the rivers. Most of the material transported by the rivers has been sourced in some fashion from the hillslopes, so the characteristics of the material eroded from the hillslope determine the material in the river and the rate of river transport.

Starting with the mineral matter (Figure 1.3a) the main transport mechanisms are fluvial erosion (Chapter 4), soil creep (Chapter 9) and mass movement (Chapter 13). The fluvial erosion is largely a function of runoff generation from rainfall (Chapter 3). The main modifier of the fluvial erosion rate is the groundcover provided by vegetation and its leaf litter (Chapter 14) though the canopy also provides some protection, but it is less important than groundcover. The dynamics of soil production and weathering influence the characteristics of the sediment on the surface of the soil. Soil production is the process of conversion of the underlying saprolite into soil, and its rate is a function of the thickness of the soil (Chapter 6). Within the soil profile a range of weathering processes are active that are (1) breaking the soil fragments into smaller and smaller fragments (Chapter 7) and (2) chemically transforming the soil rock fragments into secondary minerals (mostly clays) and leachate (that is transported out of the hillslope by water as dissolved load) (Chapter 8). The soil is vertically mixed by biological processes called bioturbation (Sections 7.3.5, 8.3.4). The roots of vegetation provide much of the strength of the hillslope to resist landsliding and other forms of mass movement (Chapters 13 and 14). Finally at mid-latitudes the aspect of the hillslopes can make a significant difference to vegetation cover, so many of the transport processes above vary depending on whether the hillslope is polar-facing or equatorial-facing.

Figure 1.3b shows the organic matter cycle on the hillslope. The main processes are vegetation providing organic matter, in the form of leaf litter, which is incorporated into the soil profile (Chapters 10 and 14). Much organic matter is generated within the profile from plants roots (the rhizosphere), mycorrhizae hyphae and microbiology. Mycorrhizae hyphae (i.e. fungi roots) are important in developing soil structure and increasing soil infiltration rates, and by mass are the most significant

CAPTION FOR FIGURE 1.3: (*cont.*) and (d) vegetation and fire. The heavy grey arrows show process fluxes and their predominant direction of action. The *aspect* label indicates those processes and properties that are different between polar- and equatorial-facing hillslopes for mid-latitude catchments.

component of soil organic matter (SOM). In the same way that bioturbation mixes the mineral matter from top to bottom in the soil profile, so bioturbation mixes the SOM. Within the profile the SOM decomposes over time, and the decomposition rate is initially fast for fresh SOM as the most reactive components decompose. The decomposition rate slows over time since the remaining SOM contains a greater percentage of components more resistant to decomposition (Chapter 10). SOM is preferentially eroded (i.e. sediment is enriched with SOM) at the surface relative to the mineral matter because it is less dense than the mineral matter. Finally the decomposition of the SOM (as well as respiration from vegetation roots; see below) generates carbon dioxide within the soil profile. Carbon dioxide levels within the soil profile can be very high (near 100%), and when this dissolves in water it generates acid that drives chemical weathering of the soil mineral matter (Chapter 8). The main impacts of aspect on the organic matter cycle are in changing vegetation density (and thus the generation rate for SOM source material) and soil temperature (the SOM decomposition rate increases strongly with soil temperature).

The water cycle is illustrated in Figure 1.3c. For most of the processes in this book the main characteristic of importance is the soil moisture because it influences (1) the growth rate for plants, (2) the amount of soil organic matter stored in the soil (wetter soils biodegrade faster), (3) the rate of chemical weathering in the soils and (4) the initial wetness of the soil when rainfall occurs so that wetter soil generates more runoff. The main sources that increase the soil moisture in the water cycle are rainfall and infiltration. The main sinks that decrease soil moisture are (1) bare soil evaporation, (2) plant transpiration (a function of plant growth rate, Chapter 14), (3) surface water runoff (a driver of fluvial erosion, Chapter 4) and (4) groundwater flow and leachate transport (Chapter 8). Higher soil moisture is associated with higher plant growth rate, surface runoff, creep and landsliding rates and lower fire occurrence. The main impacts of aspect are (1) the infiltration rate is changed by changes in soil structure due to SOM and soil grading and (2) higher transpiration from vegetation on equatorial-facing slopes.

Finally, the vegetation and fire cycles are summarised in Figure 1.3d. The main feedbacks from vegetation are in (1) plant roots, which increase the strength of the soil making it more resistant to landsliding, (2) groundcover and leaf litter, which provide soil surface protection against fluvial erosion, and (3) root respiration, which generates carbon dioxide within the soil profile, a source of acid active in chemical weathering. The main impacts of fire on landscape evolution are by modifying the impacts of vegetation. Fire removes groundcover vegetation and leaf litter layer and, if intense enough, the canopy vegetation as well. The loss of ground protection by the removal of groundcover and leaf litter results in fluvial erosion increasing dramatically until the vegetation recovers in a year or two. A further fire impact is that the SOM in the top of the soil is volatised so that the soil structure created by SOM (mostly mycorrhizae hyphae and polysaccharides) is destroyed. This results in reduced infiltration, increased runoff and increased erosion. For those trees killed by fire their root strength is reduced so that the resistance to mass movement and creep provided by plant roots is reduced, and mass movement rates increase for a decade or so after fire. The charcoal that is created by fire is incorporated into the soil, and since it is decomposition resistant, it is effectively sequestered forever within the soil. The main impact of aspect on vegetation is that the differences in vegetation and soil moisture lead to different fire recurrence rates and intensity. The denser the vegetation the higher the fire intensity, while the lower the soil moisture the dryer the fuel (e.g. leaf litter), leading to higher fire intensity.

What should be clear from the complexity of Figure 1.3 is that many factors on the hillslope interact and may result in either positive or negative feedbacks. Thus modelling the evolution of only the mineral matter processes (as was common for the early landform evolution models) independently of the soil organic matter, vegetation, fire and water is modelling only one part of a highly connected story. The challenge is that some parts of the processes in Figures 1.1–1.3 are quite difficult to quantify, so modelling them is difficult. This difficulty is no reason to ignore them, however, and the aim of this book is to provide an overarching framework within which we can quantify these processes and test their impacts on landscape evolution.

1.4 Scope of the Book

This book focuses on modelling principles, not modelling results. Of course, we are guided by what questions we wish to answer, but a book including extensive discussions of modelling results (and by necessity an extensive discussion of, and comparison with, field behaviour) is beyond the scope of this one. Likewise the book is not primarily about the underlying physics that we are modelling other than to focus on what questions we wish to answer, and this will determine what physical principles are important. Finally, all models are approximations, and we need to understand both our problem and its dominant physics to understand which approximations are appropriate. While we would like our models to be as realistic as possible, sometimes that is not feasible or practical. We may not fully understand the physics, in which case we may use the model as a numerical experiment as discussed in the previous sections. Even if we fully understand the physics we may not have the required computer power to fully solve the problem, so we are forced to simplify the model formulation to generate any solution whatsoever.

Accordingly, the focus here is to provide the reader with a broad holistic background of the processes that might need to be modelled in any particular situation, and how modellers currently solve these processes. The aim is to provide the reader with enough intuition to understand why modellers solve problems with the methods they do, and with enough background to launch into the gruesome details in the scientific literature when required.

The focus of this book is on temperate terrestrial processes. Unfortunately, that has meant leaving out specific discussion on marine and associated processes (e.g. subaerial, submarine and estuarine processes), the cryosphere (e.g. snow, glaciers and ice sheets), extra-terrestrial terrains (e.g. Mars, Pluto), aeolian landforms (e.g. dunes) and anthropogenic effects (e.g. farming, land degradation, forestry, waste repositories). Furthermore, other than highlighting links between vegetation and soil organic carbon, soil microbiology and its impacts on soil chemistry is also not discussed. This was a pragmatic decision to keep the scope of the book manageable, even though it is recognised that these processes have shaped some terrestrial locations.

The book is organised into a number of logical sections.

Chapter 2 discusses in general terms the principles underpinning a landscape evolution model, and how the mathematical components fit together in a mass and energy balance framework.

Chapter 3 presents a general climate, hydrology and geomorphology background that will be repeatedly used throughout the book.

Chapter 4 discusses fluvial erosion and deposition. It is discussed early in the book because it is central to both soilscape and landform evolution.

Chapters 5 through 11 discuss soilscape modelling. They discuss the various components of a soilscape evolution modelling framework and how it involves aspects of pedogenesis and soil transport. Each of these chapters is relatively self-contained, and the totality of the chapters describes how to model soilscape evolution on a landform that doesn't change significantly with time. Chapter 11 focuses on how the soilscape components fit together in an evolutionary framework and exemplifies some of the key work in the area using these models.

Chapters 12 through 14 examine the nonsoil landscape evolution processes. They discuss the various components of a landscape evolution model that characterise how landforms evolve. Some physical processes such as creep, typically discussed as parts of landform evolution models, appear in the soilscape section because they are truly soil processes rather than landform processes. Accordingly the observant reader will find that organisation of some topics within chapters purposely departs from the traditional organisation of discussions about landform evolution models.

Chapter 15 discusses how the landform evolution components fit together. This chapter covers the important topics of calibration and validation of soilscape and landscape evolution models, and uncertainty analysis of their output.

Chapter 16 concludes with some examples of coupled landscape evolution models. The linkage between landform, soilscape and vegetation evolution models is exemplified here. It ends with a little future-gazing at what I see as new and challenging applications, and some of the unresolved science challenges.

Readers can download the codes used in this book to exemplify the evolutionary processes discussed at Github (search for 'Willgoose soilscape') or from the Cambridge University Press webpage for this book (www.cambridge.org/willgoose).

1.5 Further Reading

For anybody interested in the history of landform evolution modelling there are two key references: Leopold et al. (1964) and Carson and Kirby (1972). While showing their age, these two books, through their focus on fundamental processes, still provide good overviews and insight. A more current overview, albeit not modelling specific, is provided by Anderson and Anderson (2010). An excellent overview of current challenges in modelling soils and processes within the soils is provided by Vereecken et al. (2016).

For the areas not covered in this book, suggested readings are anthropogenic interactions (Wainwright and Millington, 2010), engineering applications (Willgoose and Hancock, 2010), glaciers and cryosphere (LeB Hooke, 2005) and submarine processes (Peakall and Sumner, 2015).

2 Constructing a Landscape Evolution Model – Basic Concepts

2.1 Overview

In this book we will be discussing the components involved in the construction of a landscape evolution model, and the mathematical and numerical representation of those components. This landscape evolution model will simulate landforms, soils and vegetation. The details of the components will follow in Chapters 4 and beyond. This chapter aims to provide the overarching framework within which these components will fit, to provide a structure for the following chapters.

2.2 The Governing Equations

2.2.1 A Simple Landform and Soil Evolution Model Formulation

We start the discussion by overviewing how a landform evolution model is traditionally constructed. The landform is discretised in space (either on a regular grid of nodes in x and y, or on an irregularly spaced set of nodes with a Triangular Irregular Network [TIN]). Given some initial condition for this landform, the landform is subjected to a suite of environmental processes and evolved through time. This evolution through time requires a discretisation of the processes with time.

A commonly used model of landform evolution on its own, with no soils, is

$$\frac{\partial z}{\partial t} = U - \underline{\nabla} \cdot \underline{q}_{\text{sa}} = U - \left(\frac{\partial q_{\text{sa},x}}{\partial x} + \frac{\partial q_{\text{sa},y}}{\partial y} \right) = U + E \qquad (2.1)$$

where z is elevation (positive vertically upwards in metres), t is time, U is tectonic uplift (in metres/year) and q_{sa} is the sediment transport vector expressed as a volume flux (cubic metres of sediment/unit width/time) and E is the erosion (positive for deposition and negative for erosion, the sign convention defined to be consistent with z). The s subscript indicates that it is a flux of sediment, the x, y subscripts indicate the flux in the x and y directions, respectively, the a subscript indicates that it is the *actual* sediment transport to distinguish it from *potential* sediment transport to be used later in the book. The sediment transport is a vector quantity pointed in the direction of sediment transport and is in units of volume of sediment per unit width per unit time. The sediment transport in volume units can be converted to sediment transport in mass units by multiplying by the bulk density of sediment (i.e. $\underline{q}_{\text{sam}}$ (mass/unit width/time) $= \rho_b \underline{q}_{\text{sav}}$ (volume/unit width/time), where the subscripts m and v are used to differentiate the units). Note that Equation (2.1) is simply a statement of volume balance since the divergence term $\underline{\nabla} \cdot \underline{q}_{\text{sa}}$ is the difference between the sediment transport inflow and the sediment transport outflow at a point. Equation (2.1) can be formulated as a mass balance equation

$$\rho_b \frac{\partial z}{\partial t} = \rho_s U - \left(\frac{\partial q_{\text{sam},x}}{\partial x} + \frac{\partial q_{\text{sam},y}}{\partial y} \right) \qquad (2.2)$$

where ρ_s is the density of the saprolite/bedrock underlying the surface. The subscript m indicates that the sediment flux is mass units rather than the volume units used in Equation (2.1). Note that the difference between Equations (2.1) and (2.2) is that the bulk density of the sediment ρ_b is typically less than the density of the underlying rock ρ_s because sediment has voids between the sediment particles, while the bedrock/saprolite has less voids ($\rho_b = \rho_s(1 - \Delta n)$ where Δn is the change in porosity resulting from the conversion of saprolite into soil). For igneous and metamorphic bedrocks there will be virtually no voids, while for sedimentary bedrocks the porosity may be very similar to the sediment.

That Equation (2.2) is an equation for mass balance can be demonstrated if we look at the mass balance on a

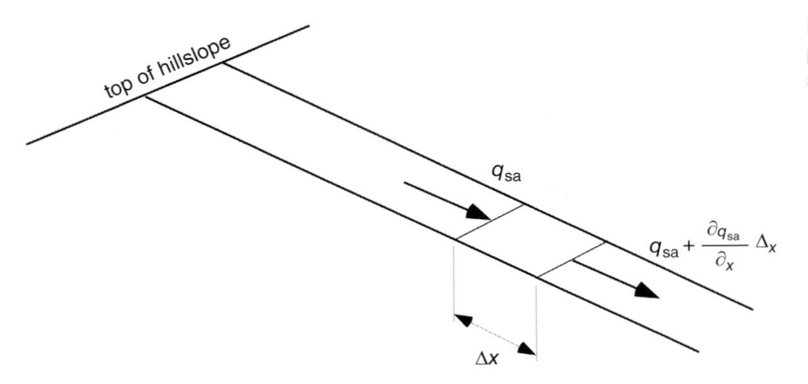

FIGURE 2.1: Schematic of the planar hillslope with unit width for surface erosion.

one-dimensional hillslope that is unit width (Figure 2.1) for a small section of the hillslope Δx long and for a short time period Δt

$$\begin{aligned}\rho_b \frac{\partial z}{\partial t}\Delta x \Delta t &= \rho_s U \Delta x \Delta t + (\text{inflow} - \text{outflow})\Delta t \\ &= \rho_s U \Delta x \Delta t + \left(q_{\text{sam},x} - \left(q_{\text{sam},x} + \frac{\partial q_{\text{sam},x}}{\partial x}\Delta x\right)\right)\Delta t\end{aligned} \tag{2.3}$$

where 'inflow' and 'outflow' are the sediment mass fluxes into the upslope boundary and out of the downslope boundary of the control volume, respectively. This simplifies to

$$\rho_b \frac{\partial z}{\partial t}\Delta x \Delta t = \rho_s U \Delta x \Delta t - \frac{\partial q_{\text{sam},x}}{\partial x}\Delta x \Delta t \tag{2.4}$$

and dividing through by $\Delta x \Delta t$ we obtain

$$\rho_b \frac{\partial z}{\partial t} = \rho_s U - \frac{\partial q_{\text{sam},x}}{\partial x} \tag{2.5}$$

which is the one-dimensional form of Equation (2.2).

Ahnert (1976) proposed a simple extension of Equation (2.1) to simultaneously model soil thickness, D:

$$\begin{aligned}\frac{\partial z}{\partial t} &= U - \left(\frac{\partial q_{\text{sa},x}}{\partial x} + \frac{\partial q_{\text{sa},y}}{\partial y}\right) = U - E \\ \frac{\partial D}{\partial t} &= P - E\end{aligned} \tag{2.6}$$

where E is the erosion rate (in metres/year) and P is the rate of conversion of saprolite to soil at the bottom of the soil profile (also in metres/year). Current coupled landform and soilscape evolution models typically use some variant of this formulation. Ahnert implicitly used a volume formulation for flux since he did not incorporate a sediment bulk density term.

An extension to Equation (2.6) recognises that in the conversion from saprolite to soil the density and porosity changes. In the absence of mass loss due to chemical dissolution there is a corresponding volume increase. The simplest way to extend the equations is to define the elevation of the saprolite-soil boundary as S, the bedrock conversion rate, P, as the depth of saprolite converted to soil per unit time, and the increase in porosity in the conversion from saprolite to soil is Δn so that

$$\begin{aligned}\frac{\partial z}{\partial t} &= \frac{\partial S}{\partial t} + \frac{\partial D}{\partial t} = U + \frac{P\Delta n}{1-\Delta n} + \frac{E}{\rho_b} \\ \frac{\partial D}{\partial t} &= \frac{P}{1-\Delta n} + \frac{E}{\rho_b} \\ \frac{\partial S}{\partial t} &= U - P \\ E &= -\left(\frac{\partial q_{\text{sam},x}}{\partial x} + \frac{\partial q_{\text{sam},y}}{\partial y}\right)\end{aligned} \tag{2.7}$$

We note that Equation (2.7) is consistent with Equation (2.2) when uplift equals soil production ($U = P$) so the soil depth is constant, and this can be seen by reformulating Equation (2.2) and the elevation equation in Equation (2.7), respectively

$$\begin{aligned}\rho_b \frac{\partial z}{\partial t} &= \rho_s U + E \Rightarrow \frac{\partial z}{\partial t} = U + \left(\frac{\rho_s}{\rho_b} - 1\right)U + \frac{E}{\rho_b} \\ \frac{\partial z}{\partial t} &= U + \frac{P\Delta n}{1-\Delta n} + \frac{E}{\rho_b} = U + \left(\frac{\rho_s}{\rho_b} - 1\right)P + \frac{E}{\rho_b}\end{aligned} \tag{2.8}$$

recalling that $\rho_b = \rho_s(1-\Delta n)$. This formulation makes several points explicit:

- The surface elevation is a result of the combination of the elevation of the saprolite boundary and the depth of soil so that rather than the elevation of the surface being the fundamental property to be modelled, it is the saprolite-soil interface that is the fundamental elevation property to be modelled.
- The elevation of the surface is a function of the conversion rate from saprolite to soil and the increase in the porosity that results from it. Consider a simple thought

experiment where uplift and erosion are zero but the soil is thickening due to bedrock weathering. The elevation of the surface increases as the lower porosity saprolite is converted to higher porosity soil and the soil layer thickens.

- This formulation is written in terms of porosity change in the absence of mass loss due to chemical dissolution. When the change in porosity is solely due to chemical dissolution it is possible for porosity to increase without a consequent increase in the volume of the soil so the depth of soil does not change.

However, this formulation for soil depth is still deficient because it only allows the soil thickness to vary as a result of vertical fluxes: (1) removal of soil at the surface due to erosion and (2) the soil production at the soil profile base. It does not allow the soil column to move down the hillslope. In a similar way to sediment transport at the surface, if more soil flows out downslope than flows in from upslope, then the soil will thin, so if we define a soil transport flux, q_{dm}, where the subscript d indicates that this is the flux of the soil, and m indicates that it is the mass flux (Figure 2.2), then Equation (2.7) becomes

$$\frac{\partial z}{\partial t} = U + \frac{E}{\rho_b} + \frac{P\Delta n}{1-\Delta n} - \frac{1}{\rho_b}\left(\frac{\partial q_{\text{dm},x}}{\partial x} + \frac{\partial q_{\text{dm},y}}{\partial y}\right)$$
$$\frac{\partial D}{\partial t} = \frac{P}{1-\Delta n} + \frac{E}{\rho_b} - \frac{1}{\rho_b}\left(\frac{\partial q_{\text{dm},x}}{\partial x} + \frac{\partial q_{\text{dm},y}}{\partial y}\right)$$
$$\frac{\partial S}{\partial t} = U - P \tag{2.9}$$

where we note that not only does the soil mass flux term potentially thin the soil but it also reduces the surface elevation. A number of processes cause soil to move downslope, and they will be discussed in detail in Chapters 9 and 13. One soil transport process is soil creep (Chapter 9), which is akin to a viscous flow of the soil downslope. The creep transport rate increases with increasing slope. So as the slope increases as you move downslope (e.g. from the top of the catchment divide), then the creep transport rate also increases (i.e. $\frac{\partial q_{\text{dm},x}}{\partial x} > 0$ if x is positive in the downslope direction), and from Equation (2.9) the soil thins and the elevations decreases by the same amount as the soil thins. Another soil transport process is shallow landsliding where the soil layer fails and flows/slides rapidly downslope (Chapter 13). While it is clear that after the landslide event the soil has thinned, the more interesting question is whether over the long term and averaged over many landslides (some of which occur upslope and deliver soil to the location, while others remove sediment from the location and deliver it downslope) the net effect of landslides is to thin the soils. It seems intuitively reasonable to suggest that at the tops of hillslopes soils will thin (more soil is removed than delivered), and at the bottom of the hillslopes soils will thicken and form colluvial deposits.

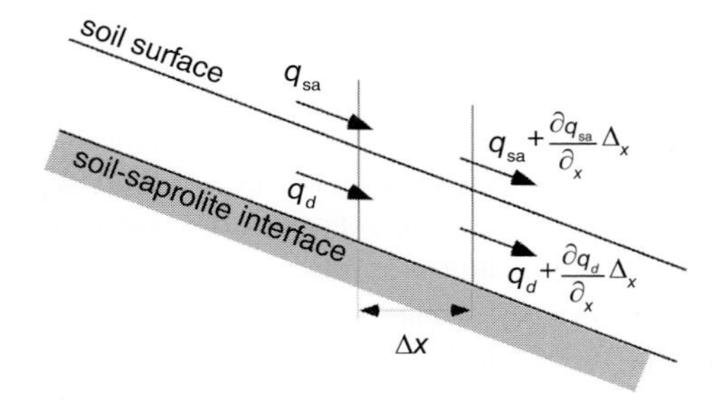

FIGURE 2.2: Schematic of the planar hillslope with unit width for surface erosion and downslope movement of the soil profile.

The formulations above are referred to as Eulerian equations because they solve the mass balance equations in time at a fixed location. In a following section (Section 2.2.3) we will introduce an alternative approach to model formulation that has some conceptual advantages over this approach, called a Lagrangian formulation. But, first, we will talk about numerically solving the Eulerian equations above.

2.2.2 Time Discretisation and Operator Splitting

It is rare that we can solve the equations in the previous section analytically, so we need to discretise these equations so that we can solve them numerically. We now discuss how the equations are solved with time. If we discretise Equation (2.9) in time, we can write a simple explicit Euler timestepping discretisation to this equation (the Euler timestepping algorithm should not be confused with Eulerian equation formulation; Euler was a man of many talents):

$$\frac{z_{t+1} - z_t}{\Delta t} = \left(U + \frac{E}{\rho_b} + \frac{P\Delta n}{1-\Delta n} - \frac{1}{\rho_b}\left(\frac{\partial q_{\text{dm},x}}{\partial x} + \frac{\partial q_{\text{dm},y}}{\partial y}\right)\right)\Bigg|_{t=t}$$
$$\frac{D_{t+1} - D_t}{\Delta t} = \left(\frac{P}{1-\Delta n} + \frac{E}{\rho_b} - \frac{1}{\rho_b}\left(\frac{\partial q_{\text{dm},x}}{\partial x} + \frac{\partial q_{\text{dm},y}}{\partial y}\right)\right)\Bigg|_{t=t}$$
$$\frac{S_{t+1} - S_t}{\Delta t} = (U - P)|_{t=t} \tag{2.10}$$

where the subscripts in t and $t+1$ are the times at which the values are defined, and the $()|_{t=t}$ indicates that the term inside the parentheses is evaluated at time t. After reorganisation, the equations solved numerically are

$$z_{t+1} = z_t + \left(U + \frac{E}{\rho_b} + \frac{P\Delta n}{1-\Delta n} - \frac{1}{\rho_b}\left(\frac{\partial q_{\text{dm},x}}{\partial x} + \frac{\partial q_{\text{dm},y}}{\partial y}\right)\right)\bigg|_{t=t} \Delta t$$

$$D_{t+1} = D_t + \left(\frac{P}{1-\Delta n} + \frac{E}{\rho_b} - \frac{1}{\rho_b}\left(\frac{\partial q_{\text{dm},x}}{\partial x} + \frac{\partial q_{\text{dm},y}}{\partial y}\right)\right)\bigg|_{t=t} \Delta t$$

$$S_{t+1} = S_t + (U-P)|_{t=t}\Delta t \qquad (2.11)$$

A number of problems exist with using this explicit Euler method to solve this set of equations in practice (mostly its poor numerical stability requiring small timesteps), but it will suffice for the current discussion, as it is simple and easy to explain. To solve Equation (2.11) we need to evaluate all the properties on the right-hand side of the equation at time t to determine the new value on the left-hand side at time $t+1$. Equation (2.11) is in the form of a state-space equation. The terms z, D and S are referred to as states, and their values at time t define everything we need to know about the system. If we know only z, D and S at time t, then we completely know the system at time t; this is the definition of a state. In state-space systems the behaviour of the system at time t (i.e. its rate of change) is completely defined by the 'states' (and properties that can be derived from the states). The flux terms (q_{sa} and q_d) change depending on the states, and they are determined by their 'constitutive equations'. For instance, in the previous section we talked about soil creep, q_d. The form of the q_d dependence on the states is determined by the constitutive equation for q_d (e.g. how it changes with slope, which itself is a function of the surface elevation state z). The constitutive equation in turn will contain a series of 'parameters' that determine the rates of the processes, and how they change with changes in the states.

The form of the constitutive equations and their parameters is the subject of the subsequent chapters in this book. However, one aspect of Equation (2.11) is worth discussing here, because it conceptually underlies the construction of the overarching landscape evolution framework. Equation (2.11) consists of three processes: (1) the evolution of the landform and soil due to erosion, (2) the soil production at the base of the soil profile and (3) the soil transport downslope. While it is convenient to consider them all together in one set of equations for the timestepping, it is not necessary to calculate them simultaneously as in Equation (2.11). Equation (2.11) can be split into three sets of equations that are solved sequentially at each timestep. This technique is called 'operator splitting' (Celia and Gray, 1991). Equation (2.11) can be rewritten as follows:

Step 1: Erosion modelling

$$z^{+}_{t+1} = z_t + \left(U - \frac{1}{\rho_b}\left(\frac{\partial q_{\text{sam},x}}{\partial x} + \frac{\partial q_{\text{sam},y}}{\partial y}\right)\right)\bigg|_{t=t} \Delta t$$

$$D^{+}_{t+1} = D_t + \left(-\frac{1}{\rho_b}\left(\frac{\partial q_{\text{sam},x}}{\partial x} + \frac{\partial q_{\text{sam},y}}{\partial y}\right)\right)\bigg|_{t=t} \Delta t$$

$$S^{+}_{t+1} = S_t + (U)|_{t=t}\Delta t \qquad (2.12)$$

Step 2: Soil Production

$$z^{++}_{t+1} = z^{+}_{t+1} + \left(\frac{P\Delta n}{1-\Delta n}\right)\bigg|_{t=t} \Delta t$$

$$D^{++}_{t+1} = D^{+}_{t+1} + \left(\frac{P}{1-\Delta n}\right)\bigg|_{t=t} \Delta t$$

$$S^{++}_{t+1} = S^{+}_{t+1} + (-P)|_{t=t}\Delta t \qquad (2.13)$$

Step 3: Transport Downslope

$$z_{t+1} = z^{++}_{t+1} + \left(-\frac{1}{\rho_b}\left(\frac{\partial q_{\text{dm},x}}{\partial x} + \frac{\partial q_{\text{dm},y}}{\partial y}\right)\right)\bigg|_{t=t} \Delta t$$

$$D_{t+1} = D^{++}_{t+1} + \left(-\frac{1}{\rho_b}\left(\frac{\partial q_{\text{dm},x}}{\partial x} + \frac{\partial q_{\text{dm},y}}{\partial y}\right)\right)\bigg|_{t=t} \Delta t$$

$$S_{t+1} = S^{++}_{t+1} \qquad (2.14)$$

where the + and ++ superscript notation is used to indicate the intermediate states produced by the operator splitting. It can be seen that by substituting Equations (2.12) and (2.13) into (2.14), we retrieve the original Equation (2.11). For this fairly trivial example there seems little advantage to this reformulation, but operator splitting is useful in controlling the complexity of the equations that we need to solve. One significant advantage is that it allows us to formulate optimal numerical solvers that are different for each of the different components of the physics. We will discuss this later. However, as an example, Equation (2.12) can be a very difficult equation to solve because of numerical instability and nonlinearities, and it typically requires very small timesteps. By operator splitting we can solve Equation (2.12) at very small timesteps but have large timesteps for (2.13) and (2.14) reducing the computations. Experience also shows that we can sometimes improve the performance per timestep if the $()|_{t=t}$ terms are recalculated for the new + and ++ intermediate states rather than for the original states at $t=t$ as written in Equations (2.12)–(2.14), but that may be offset by the extra effort that is required to recalculate the terms inside the parentheses at the new intermediate states rather than doing all the calculations for the original states at $t=t$. The operator splitting in

Equations (2.12)–(2.14) has used a very simple explicit Eulerian timestepping algorithm. This example was used because of the simplicity of the explanation above; operator splitting can be used for any timestepping algorithms. There are other conceptual advantages to operator splitting, but it is premature to discuss them now, and they will be discussed later.

2.2.3 The Lagrangian Conceptualisation and Process Locality

The formulation for transport continuity in the previous sections is based on applying a mass balance at a given location (e.g. a grid node point) with time. Because these equations track mass balance by balancing fluxes into and out of a *fixed location,* these equations are called Eulerian equations. An alternative approach tracks a unit of sediment as it moves downslope from its starting location to the end location where it stops moving. This specified volume or mass of soil/sediment has been given a number of names in the literature including package, cohort and precipiton; we will use the term 'package'. A method of tracking the dynamics of a package of sediment from source to sink is called a Lagrangian method.

How a Lagrangian formulation works is best shown with an example. Consider the example of a landslide (Figure 2.3). There are three components to a single landslide: (1) the probability that any given location will have a landslide at any given point in time (i.e. a package will start to move), (2) how much sediment will be moved by that landslide when it occurs (i.e. how big is the package) and (3) how far that landslide will travel once triggered (i.e. what is the path followed by the package). Given those three pieces of information you can track the sediment package as it travels downslope and determine the trajectory of that sediment package. If we did this for many landslide events over some period and summed up the net effect of all the package trajectories on the evolution of the landform, then this is a Lagrangian solution to

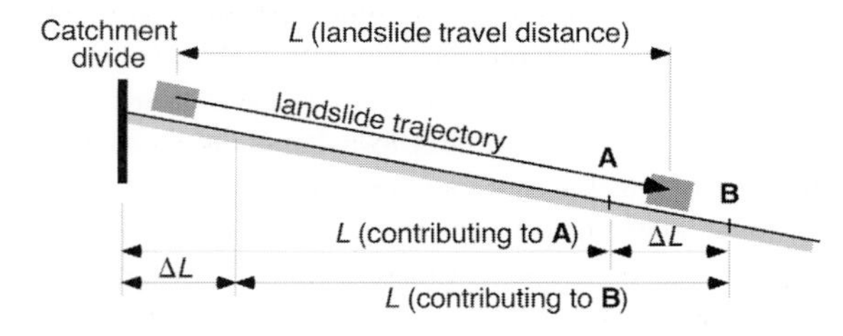

FIGURE 2.3: Schematic of the Lagrangian representation of a landslide moving downslope.

the landslide erosion problem. This has been used as the basis of landform evolution models (Chase, 1992; Crave and Davy, 2001). The conceptual appeal of this approach is obvious. The problem is, however, that we need to be able to characterise the three components above from field data. The first two can and have been characterised from remote sensing studies of landslide-prone areas (e.g. Hovius et al., 1997; Stark and Hovius, 2001). The third is more difficult because it relies on matching a specific landslide scar with a specific deposition area. Moreover the distance the package travels is a function of the exact details of the path that is taken and the energy dissipation along that path. We discuss this in more detail in Chapter 13.

Here we will show how the two model formulations, Eulerian and Lagrangian, are consistent, and by doing so highlight aspects of both approaches. Consider the landslide example again and assume a unit width slope. Assume that every landslide has the same probability per unit area per unit time of being triggered and this probability is the same at every location in the landform p. We will also assume the mass of the package of sediment transported by each landslide M, and the distance travelled by this package L are always the same. The probability p is defined as the probability that a landslide will be triggered in unit time within a unit area, so the probability of triggering in a region of length Δx on the unit width hillslope in a time period Δt is $p\Delta x\Delta t$. We start from the top of the catchment divide and consider the package passing a location A (Figure 2.3) that is distance L from the catchment divide. For every trigger location upslope of A the package of sediment triggered will pass A. If we consider location B that is a location $L + \Delta L$ from the divide, then it has the same mass of sediment pass it as location A. For location B those landslides that are triggered at a distance between 0 and ΔL from the divide deposit their sediment between A and B and do not pass location B. The key concept is that at any location only landslides within a distance L upslope of that location contribute to the sediment flux past that location. Accordingly we can calculate the average number of landslides in the region distance L upstream for the unit width of hillslope as $N = pL\Delta t$, and if each of these landslides contains mass M of sediment, then the flux past location A is

$$q_s = pLM \tag{2.15}$$

This is then the Eulerian equivalent of the Lagrangian representation of the landsliding process. The Eulerian flux equation in Equation (2.15) is equivalent to the average of the Lagrangian description. If we insert Equation (2.15) into Equation (2.1), then this example results in zero erosion anywhere since $\frac{\partial q_s}{\partial x} = 0$ with x positive in the downslope direction. For net erosion to occur one or more of p, L and M must increase downslope. Of course, this is a very simple example. and all three terms will vary in space. Moreover when the three terms change in space an equation equivalent to Equation (2.15) will not be as easy to derive. We will discuss these issues in Chapter 13 when landsliding is discussed.

The reason for going through the derivation leading to Equation (2.15) is that in later chapters we will discuss several processes that are more naturally understood and mathematically formulated in a Lagrangian framework. In some cases it will easier to numerically solve them by converting them to a Eulerian form, while in other cases it may be easier to numerically solve the Lagrangian form using cellular automata or particle-tracking approaches. The reader needs to remember that though the conceptualisation is different, Eulerian and Lagrangian representations are simply different ways of looking at the same problem and the reader should view them as fundamentally consistent methods. This will be particularly useful when we consider fluvial erosion in Chapter 4, where it will be convenient to switch between the formulations as each approach highlights different aspects of the physics.

The numerical methods for directly solving Lagrangian methods are generally lumped under the headings of particle tracking, cellular automata or agent-based methods. To demonstrate the principle for particle tracking we can think of the landslide problem again. At every timestep it is determined whether a landslide has been triggered at a node. If it is triggered, then the landslide package size is calculated (i.e. the mass). This package then moves downslope based on whatever dynamics for the transport has being adopted, thus determining the trajectory. For instance, Chase (1992) moved the mass in the steepest downslope direction until it reached a location where the slope gradient was below a threshold. Another way would be to model the kinetic energy of mass (e.g. gain of kinetic energy as it moves to lower elevation, loss of energy through frictional losses) until the kinetic energy is zero, at which location the mass stops moving and is deposited. Each landslide is modelled separately and the landform evolution is the net result of the many landslides that occur over the time modelled in the simulation.

Finally one thing that the Lagrangian formulation highlights is a property of transport processes called 'locality'. A process is referred to as 'local' if the process is completely determined by the properties of the process at the location at which the equations are being solved (e.g. at the node where we are calculating erosion/deposition). A local process can be written as an Eulerian differential equation at that location. A process is 'nonlocal' if it requires information from other parts of the landform to solve the equation at the node. Thus a process that requires only the local slope (e.g. diffusion, Chapter 9) to determine sediment flux (and thus elevation and soil depth change at that point) is local. A process that requires knowledge of what the catchment looks like upstream is nonlocal. A simple example to consider is erosion. Erosion (Chapter 4) requires knowledge of discharge at the node, which requires knowledge of the catchment area (and perhaps upstream soil, vegetation and so on), so erosion is strictly speaking nonlocal. However, we can convert erosion to a local process if we separately calculate the catchment area and discharge, and then use this in the erosion equations. This is a somewhat trivial example, but, on the other hand, considerable debate surrounds whether landsliding, with its strong dependence on energy dissipation along the landslide pathway, can be modelled as a local process, and we will discuss this further in Chapter 13. The reason for discussing locality here is that the derivation resulting in Equation (2.15) is for a very simple geometry. Consider now the case where the hillslope gradient varies downslope and where these variations are significant within the distance L. This means that the conditions that apply when the sediment package starts moving are different from the condition distance L downslope. Specifically L, the distance the package travels, may change as it moves downslope. Thus the flux at location A will depend on the way the hillslope interacts with the package as it moves from the point from which it started moving to the location A. This makes the flux process at location A nonlocal. However, it may be possible to reformulate the process as local by modelling not only the mass transport but also the momentum and/or energy. In this reformulation the energy dissipation rate would be calculated locally at A based on the energy dissipation processes at A.

2.2.4 Lagrangian Simulation Approaches

The discussion in the previous section leads to an obvious conceptualisation of how to model landslides (and perhaps other processes) within LEMs based on the explicit tracking of material as it moves downslope.

Essentially the idea is to track the movement of a particle that represents a property of interest based on some forcings. A simple example is the movement of a sediment particle as a result of a flow field. This example falls under the heading of 'particle tracking' in the

characterises the uplift term (i.e. $\frac{U\langle t\rangle}{\langle z\rangle}$) and the other the soil creep term (i.e. $\frac{D\langle t\rangle}{\langle x\rangle^2}$). In the spirit of nondimensionalisation nomenclature let's call these numbers **Nu** and **Nc**, respectively (**u** for uplift and **c** for creep). The most important thing to note is that provided that **Nu** and **Nc** do not change, the nondimensional solution to Equation (2.19) (i.e. $z^*(x^*, t^*)$ where z^* is a function of x^* and t^*) will also not change. So, for instance, if we change the vertical scale $\langle z\rangle$of the catchment (e.g. give the catchment higher relief), then provided U is changed so that **Nu** remains the same value, then the results from the equation (i.e. the change of z^* with x^* and t^*) will be the same. This shows that provided the physics of Equation (2.16) is an appropriate representation of the field processes, we can rescale different catchments having different areas and uplift and compare them.

If we rearrange Equation (2.19) we can extract even more information from the governing equation, and demonstrate one of the more powerful aspects of scaling theory, which is to characterise the relative dominance of different processes in the governing equation:

$$\frac{\partial z^*}{\partial t^*} = \frac{U\langle t\rangle}{\langle z\rangle}\left(1 + \frac{D\langle z\rangle}{U\langle x\rangle^2}\frac{\partial^2 z^*}{\partial x^{*2}}\right) \tag{2.20}$$

The nondimensional number inside the parentheses multiplying the soil creep is just **Nu/Nc**, but its importance is that it determines the relative importance of soil creep to uplift. If $\frac{D\langle z\rangle}{U\langle x\rangle^2} \gg 1$ then the equation is dominated by the soil creep term, while if $\frac{D\langle z\rangle}{U\langle x\rangle^2} \ll 1$ then the equation is dominated by the uplift term. This interpretation should be intuitively obvious from the ratio of D/U in the nondimensional number, but the nondimensional number indicates that the relative dominance is also a function of the vertical and horizontal scales.

Finally, the most subtle aspect, and the one that gives scaling analysis its name, is when the scales are adopted so that the nondimensional scaled x, z and t are of O(1). The terminology O(1) means that the value is of order 1, that is approximately equal to 1, as opposed to approximately equal to either 0.1 or 10. This can be easily achieved by selecting the scales $\langle z\rangle, \langle x\rangle$ and $\langle t\rangle$ appropriately. The implication of this is that $\frac{\partial z^*}{\partial t^*} = O(1)$ and $\frac{\partial^2 z^*}{\partial x^{*2}} = O(1)$, so that for Equation (2.20) to be internally consistent if both processes are important, then $\mathbf{Nu} = O(1)$ and $\mathbf{Nc} = O(1)$. This is consistent with $\mathbf{Nu}/\mathbf{Nc} \gtrless 1$ characterising the relative scale of processes. The reader may need some convincing on this point, but this conclusion is generally true if the nondimensionalisation scales are chosen carefully, and is a potentially important consideration in defining scales.

2.3.3 Definition of the Appropriate Scales

The most difficult aspect of nondimensionalisation, the one that everybody struggles with, is defining the scales in the nondimensionalisation exemplified by Equation (2.17).

The scaling analysis at the end of the section above suggests that we wish to define scales that result in the nondimensionalised variable being of O(1).

A subtler, and often ignored, issue is that the scale should be reasonably insensitive to random sampling error. All scales will have some sampling error in them since they are measured from a single realisation of an inherently random process (e.g. catchments will vary from site to site even if geology, climate and process are identical at each site). Scales that depend on the mean of what is being measuring rather than some extreme of the probability distribution are preferred because the sampling error is less for mean properties. Even better is something that is an integrated measure of the landscape (pedants will point out that the mean is in fact an integrated measure by the nature of how it is calculated). Integrated measures smooth out noise. A poor choice is something that is a derivative because derivatives accentuate noise (e.g. slope is inherently noisier than elevation, even if you take the means).

A good case study is how to define the vertical elevation scale of a landform $\langle z\rangle$. An obvious, but poor, choice is to use the difference between the minimum and maximum elevation of the landform, the 'relief'. Certainly, if the elevations are divided by this scale, the nondimensionalised elevations will be O(1). But the relief is an example of an extreme of a probability distribution. Its value can change dramatically simply if a single isolated mountain peak occurs in the catchment, even if the bulk of the landform's elevations are unaffected. A better choice for the vertical scale here would be the mean elevations above the minimum elevation, though this is still impacted by the minimum elevation, but in practice the minimum elevation is much less variable than the maximum. An obvious example of a nondimensionalised graph that suffers from this poor choice of vertical scale for nondimensionalisation is the hypsometric curve, where elevations are scaled between the minimum and maximum elevation (see Section 3.3.5). Moreover, the hypsometric integral, the area under the curve, proposed as an indicator of landscape development and process (Strahler, 1952; Willgoose and Hancock, 1998) can dramatically change if even a single isolated peak is added to the landform.

Even for the horizontal scales of a catchment there are better or worse choices. For the catchment scale $\langle x\rangle$ the square root of the drainage area appears to be a good choice. However, when we define Equation (2.16) in x and y rather just x, the opportunity to nondimensionalise differently in the two directions arises so that we have a horizontal scale in the x direction and a (potentially) different horizontal scale in the y direction. The most obvious case where we might to do this is to characterise catchment and/or channel network shape. One way of determining channel network shape is to calculate the ratio of network length to network width, where length is the length of the longest stream in the network and width is maximum width of the network width function. Alarm bells ought to be immediately ringing at the mention of 'maximum' because again this is an extreme of a probability distribution (in this case one maximum for length and another for width). It is better to use mean width and mean stream length. The shape of the catchment could then be characterised by the nondimensional number ${}^{\langle x\rangle}/_{\langle y\rangle}$ and the area nondimensionalised as ${}^{A}/_{\langle x\rangle\langle y\rangle}$.

Finally, there are scales other than time and length. Taking Equation (2.16) as an example, if the uplift is spatially varying (i.e. $U = U(x)$), then it might be appropriate to nondimensionalise the uplift using, for instance, the mean uplift $\overline{U}$ so that we can define a nondimensional uplift $U^*(x) = U(x)/\overline{U}$ (effectively in this case the spatial pattern of uplift). Equation (2.20) would then look like

$$\frac{\partial z^*}{\partial t^*} = \frac{\overline{U}\langle t\rangle}{\langle z\rangle}\left(U^*(x) + \frac{D\langle z\rangle}{\overline{U}\langle x\rangle^2}\frac{\partial^2 z^*}{\partial x^{*2}}\right) \tag{2.21}$$

This use of rescaling is quite general, can be applied to any property in an equation and allows spatially variable properties to be included in any nondimensional number developed from the governing equations. Of course, if the spatial pattern of $U^*(x)$ changes, then the deterministic solution of Equation (2.21) will change. However, in many cases (particularly if the spatial variability is random) we are interested in characterising statistical properties, and in these cases the exact spatial pattern (i.e. $U^*(x)$) may not be critical whereas the magnitude (i.e. $\overline{U}$) is.

2.4 Perturbation Analysis for Analysis of Stochastic Processes

Erosion time series vary randomly with time, yet in most case we are interested in the cumulative geomorphic effectiveness of these variable time series. In many cases what we need to know about the geomorphic effectiveness of the time series is simply the mean over time so that we can calculate the long-term average impact of the time-varying land-sculpting process. The sections below outline a simple technique that can be used to determine the averages of these time series. In many cases the analyses that follow give exact answers that allow us to exactly determine the mean of a process (i.e. this might be thought of as being the geomorphologically effective average of the process) in terms of the process parameters and the statistics of the forcing inputs. Some more complex forms of expression require approximations, but even in these cases we can obtain first-order approximations to the geomorphologically effective average of the process. This discussion provides enough detail to be able to understand the averaging analyses presented in this book.

2.4.1 Perturbation Analysis Principles

It is convenient to split a random process into its mean and a random perturbation around that mean so that a random process X can be written as

$$X = \overline{X} + X' \tag{2.22}$$

where $\overline{X}$ is the mean or expectation of X (i.e. $\mathrm{E}[X] = \overline{X}$) and X' is the random perturbation around that mean. We will see below that a useful consequence of this definition is that the expectation of the perturbation X' is $\mathrm{E}[X'] = 0$. All random variables are split into a mean and perturbation about the mean using Equation (2.22). The variance of X is then

$$\sigma_X^2 = \mathrm{E}\left[\left(X - \overline{X}\right)^2\right] = \mathrm{E}\left[\left(X'\right)^2\right] \tag{2.23}$$

and the covariance between two random processes X and Y is

$$\sigma_{XY}^2 = \mathrm{E}[(X'Y')] \tag{2.24}$$

It is also worth remembering some simple properties of the expectation process:

- The expectation of a nonrandom number is just that number (e.g. $\mathrm{E}\left[\overline{X}\right] = \overline{X}$; $\mathrm{E}[R] = R$ if R is a nonrandom number)
- A variable that is not random (i.e. a number) can be taken outside an expectation (i.e. $\mathrm{E}[RX] = R{\cdot}\mathrm{E}[X]$ if R is a nonrandom number)
- The expectation of the sum of two random variables is equal to the sum of the expectations of the variables taken separately (i.e. $\mathrm{E}[X + Y] = \mathrm{E}[X] + \mathrm{E}[Y]$) and
- The expectation of the multiplication of two independent random variables is the multiplication of the expectations (i.e. $\mathrm{E}[XY] = \mathrm{E}[X]{\cdot}\mathrm{E}[Y]$).

These properties will allow us to break up more complex problems, that cannot be directly solved using the techniques below, into a combination of problems that can be solved.

2.4.2 Examples

To demonstrate the principles involved in perturbation analysis, some simple examples are provided below.

2.4.2.1 Example 1: Linear Combination of Two Random Processes

Consider a random process Z that is the product of two random inputs X and Y so that $Z = X{\cdot}Y$. Expressing this as perturbations yields

$$\overline{Z} + Z' = \left(\overline{X} + X'\right)\left(\overline{Y} + Y'\right) \tag{2.25}$$

If we wish to know the mean of Z, we simply take the expectation of this equation after first expanding the right-hand side

$$\begin{aligned}\overline{Z} &= \mathrm{E}\left[\overline{Z} + Z'\right] \\ &= \mathrm{E}\left[\left(\overline{X} + X'\right)\left(\overline{Y} + Y'\right)\right] \\ &= \mathrm{E}\left[\overline{X}\,\overline{Y} + \overline{X}Y' + \overline{Y}X' + X'Y'\right] \\ &= \mathrm{E}\left[\overline{X}\,\overline{Y}\right] + \mathrm{E}\left[\overline{X}Y'\right] + \mathrm{E}\left[\overline{Y}X'\right] + \mathrm{E}[X'Y'] \\ &= \overline{X}\,\overline{Y} + \overline{X}{\cdot}\mathrm{E}[Y'] + \overline{Y}{\cdot}\mathrm{E}[X'] + \mathrm{E}[X'Y']\end{aligned} \tag{2.26}$$

We can further simplify this by recalling that the mean of the perturbation is zero (i.e. $E[X'] = 0$) and by using the definition of the covariance between X and Y so that

$$\overline{Z} = \overline{X}\,\overline{Y} + \sigma_{XY}^2 \tag{2.27}$$

The implications of Equation (2.27) are that (1) this equation is exact, (2) the mean of Z is not, in general, equal to the product of the means of the X and Y, (3) the difference from the product of the means is due solely to the covariance between X and Y (note that this equation is equal to the result for independent processes when the covariance is zero) and (4) while Equation (2.27) appears to be limited to normal distributions, because only mean and covariance appear, it is in fact true for all distributions for which mean and covariance are defined.

An important special case is that covariance is not defined for many fractal processes. Moreover, for the special case where X and Y are the same random process (so $Z = X^2$), then Equation (2.27) is a function of the variance (i.e. $\overline{Z} = \overline{X}^2 + \sigma_X^2$) and the variance is also not defined for a fractal X. Typically $\sigma_X^2 \propto T^\alpha$ where T is the period over which averaging is desired and α is a function of the fractal dimension so the variance approaches infinity as the length of data increases. Since sediment transport can be a function of Q^2 and discharge is asserted to be fractal over geomorphic time scales, this suggests that the average sediment transport will approach infinity with longer periods of record. We discuss the subtleties of this issue in Chapter 4.

2.4.2.2 Example 2: Linear Combination of Three Random Inputs

Example 1 showed a case where the result was exact. We will now consider a case where an exact answer can be obtained but where it is more common to use a simplification based on the normal distribution. Consider a random process Z that is the product of three random inputs U, X and Y as $Z = UXY$. Expressing this as perturbations

$$\overline{Z} + Z' = \left(\overline{U} + U'\right)\left(\overline{X} + X'\right)\left(\overline{Y} + Y'\right) \tag{2.28}$$

Expanding this and taking the expectation of the equation,

$$\begin{aligned}\overline{Z} &= \mathrm{E}\left[\overline{Z} + Z'\right] \\ &= \mathrm{E}\left[\left(\overline{U} + U'\right)\left(\overline{X} + X'\right)\left(\overline{Y} + Y'\right)\right] \\ &= \mathrm{E}\left[\overline{U}\,\overline{X}\,\overline{Y} + \overline{U}\,\overline{X}Y' + \overline{U}\,\overline{Y}X' + \overline{U}X'Y' + \overline{X}\,\overline{Y}U' \right.\\ &\qquad \left. + \overline{X}U'Y' + \overline{Y}U'X' + U'X'Y'\right] \\ &= \overline{U}\,\overline{X}\,\overline{Y} + \overline{U}\,\overline{X}\cdot\mathrm{E}[Y'] + \overline{U}\,\overline{Y}\cdot\mathrm{E}[X'] + \overline{U}\cdot\mathrm{E}[X'Y'] \\ &\quad + \overline{X}\,\overline{Y}{\cdot}\mathrm{E}[U'] + \overline{X}\cdot\mathrm{E}[U'Y'] + \overline{Y}\cdot\mathrm{E}[U'X'] + \mathrm{E}[U'X'Y']\end{aligned} \tag{2.29}$$

In contrast with Example 1 the substitution for the perturbations' mean and covariance

$$\overline{Z} = \overline{U}\,\overline{X}\,\overline{Y} + \overline{U}\sigma_{XY}^2 + \overline{X}\sigma_{UY}^2 + \overline{Y}\sigma_{UX}^2 + \mathrm{E}[U'X'Y'] \tag{2.30}$$

yields a new term that is a triple product of the perturbations. Its meaning is somewhat clearer if we look at the special case when U, X and Y are the same random variable, say, W (i.e. $Z = W^3$). In this case the equation becomes

$$\overline{Z} = \overline{W}^3 + 3\overline{W}\sigma_W^2 + \left(\sigma_W^2\right)^{3/2}\gamma \tag{2.31}$$

where γ is the skewness of W, suggesting that the third-order perturbation is related to deviations from normality.

We can also rearrange the original perturbation equation as

$$\overline{Z} = \overline{U}\,\overline{X}\,\overline{Y}\left(1 + r_{XY}^2 + r_{UY}^2 + r_{UX}^2 + \mathrm{E}\left[\frac{U'X'Y'}{\overline{U}\,\overline{X}\,\overline{Y}}\right]\right) \tag{2.32}$$

where r is the correlation coefficient. If the perturbations around the mean are small, then we can assert that

$E[U'X'Y'] \ll \overline{U}\,\overline{X}\,\overline{Y}$. Alternatively if the probability distributions of U, X and Y are all normal, then $E[U'X'Y'] = 0$. Either way Equation (2.32) can simplified as

$$\overline{Z} = \overline{U}\,\overline{X}\,\overline{Y} + \overline{U}\sigma_{XY}^2 + \overline{X}\sigma_{UY}^2 + \overline{Y}\sigma_{UX}^2 \tag{2.33}$$

In conclusion we note that this result is either

- Exact if the distributions of X, Y and Z are Gaussian or
- An approximation with the error of the approximation between a function of the degree of relationship between X, Y and Z in the third- and higher order moments (like skewness) and how large the random perturbations are around the mean.

This approach highlights one of the ways of arriving at an approximate solution by ignoring third-order and higher moments of the probability distribution.

2.4.2.3 *Example 3: Mean of a Single Nonlinear Input*

Finally we will consider a nonlinear transformation of a random input. We will consider a power law relationship, because many of the relationships in this book are of that form, but the generic approach is applicable to more complex functions:

$$Z = aX^b \tag{2.34}$$

where a and b are (nonrandom) parameters for the transformation. As a preliminary example we will consider a special case where the solution is particularly simple, $b = 2$. Substituting for the perturbations and solving for the mean of Z,

$$\begin{aligned}\overline{Z} &= E[\overline{Z} + Z'] \\ &= E\left[a(\overline{X} + X')^2\right] \\ &= a \cdot E\left[\overline{X}^2 + 2\overline{X}X' + X'^2\right] \\ &= a\left(\overline{X}^2 + \sigma_X^2\right) = a\overline{X}^2(1 + CV_X^2)\end{aligned} \tag{2.35}$$

where CV_X is the coefficient of variation of the random process X. Note that this is an exact solution that is independent of the form of the probability distribution of X, even though it uses only the mean and variance of X.

We now return to the original general problem in Equation (2.34). Substituting in the perturbations yields

$$\overline{Z} = a \cdot E\left[\left(\overline{X} + X'\right)^b\right] \tag{2.36}$$

where the problem we have solving this is that we cannot evaluate the expectation except in special cases of integer b, as we did for $b = 2$ above. To evaluate Equation (2.36) we make an approximation, which is to expand the power on b using Taylor Series around the mean $\overline{X}$:

$$\begin{aligned}\left(\overline{X} + X'\right)^b &= \overline{X}^b + bX'\overline{X}^{b-1} + \frac{b(b-1)}{2!}X'^2\overline{X}^{b-2} \\ &\quad + \frac{b(b-1)(b-2)}{3!}X'^3\overline{X}^{b-3} + \text{terms}\left(X'^4\right)\end{aligned} \tag{2.37}$$

where the 'terms (X'^4)' indicates terms consisting of perturbations to the fourth power and higher. This equation is then substituted into Equation (2.36) and evaluated using the standard results for mean, variance and skewness:

$$\begin{aligned}\overline{Z} &= a\overline{X}^b + \frac{ab(b-1)}{2!}\overline{X}^{b-2}\sigma_X^2 + \frac{b(b-1)(b-2)}{3!}\overline{X}^{b-3}\left(\sigma_X^2\right)^{3/2}\gamma \\ &\quad + E\left[\text{terms}\left(X'^4\right)\right]\end{aligned} \tag{2.38}$$

As many higher statistical moments (powers on the perturbation) as required can be included, but because of the division by the factorial the extra terms in the series rapidly decrease in size. A common assumption is to ignore all but the mean and variance (called a 'second moment approximation') so that we have

$$\overline{Z} \approx a\overline{X}^b + \frac{ab(b-1)}{2!}\overline{X}^{b-2}\sigma_X^2 = a\overline{X}^b\left(1 + \frac{b(b-1)}{2!}\frac{\sigma_X^2}{\overline{X}^2}\right) \tag{2.39}$$

Comparing this to the exact solution for $b = 2$ (Equation 2.35) there is a striking similarity in the form of the result. Moreover, the result of Equation (2.35) for $b = 2$ is exact because the skewness and higher-order terms all involve $(b - 2) = 0$. The second moment approximation of ignoring higher than second-order perturbation terms is quite powerful when used together with Taylor Series as it allows the development of explicit solutions to general stochastic problems, though as the coefficient of variation of X increases the error in the approximation also increases.

Finally we introduce the concept of an 'effective' parameter. We can rearrange Equation (2.39) as

$$\overline{Z} = a\overline{X}^b\left(1 + \frac{b(b-1)}{2!}\frac{\sigma_X^2}{\overline{X}^2}\right) = \hat{a}\overline{X}^b \tag{2.40}$$

where the new equation using parameter $\hat{a}$ is the same as Equation (2.34) if $\hat{a} = a\left(1 + \frac{b(b-1)}{2!}\frac{\sigma_X^2}{\overline{X}^2}\right)$. The relationship between a and $\hat{a}$ is simply a function of the nonlinearity of the relationship, b, and the variability of the process. If we are interested only in the mean effect on Z of the random input X, then this equation shows that if we use the original Equation (2.34) with $\overline{X}$ and parameter $\hat{a}$, then we would obtain the same $\overline{Z}$ as we would by using the random input X with parameter a and averaging the resulting random output Z. Accordingly $\hat{a}$ is called the 'effective parameter' and a the 'true parameter', where $\hat{a}$ captures all of the effect of randomness in X on the mean output $\overline{Z}$.

The concept of an effective parameter is an important and powerful concept. It implies that if we do the averaging correctly we can model the average effect of a random input by simply adjusting process parameters and use the statistics of the input, rather than having to do all the calculations involved in treating input randomness explicitly (e.g. Monte Carlo simulation). Moreover, in many cases the underlying functional form of the relationship is unchanged, though, unlike the simple example above, this is not always true. In many cases such unchanged functional forms have been implicitly assumed by process geomorphologists when using event-scale equations to average up processes to geomorphic time scales. Fortunately, they are often true.

Finally, if a model is calibrated to field data and the only available data are the averages $\overline{Z}$ and $\overline{X}$, then the calibrated parameter in Equation (2.34) will be the effective parameter $\hat{a}$ not the underlying true parameter a. This has implications for scaling up small-scale laboratory models, typically lacking field scale variability, to field scale. Equation (2.40) shows that the parameter a for the laboratory and field scales will be different. This difference results from the difference in variance of the input X at the laboratory and field scales. Equation (2.40) shows that even for modest nonlinearity and field scale variability (e.g. $b = 2$ and $CV = 1$) the difference between the true and effective parameters is a factor of 2.

2.4.2.4 *Example 4: Variance of a Single Nonlinear Input*

The previous example determined the mean of an input that is transformed by a simple nonlinear power law transformation. This example calculates the variance of that same series. Determining the variance of a time series involves additional steps. These steps are potentially quite messy even for simple relationships, so they will be demonstrated here with the simple example used above. More complex examples are analysed in the same way. Consider

$$Z = aX^2 \tag{2.41}$$

From the previous example we calculated the mean by substituting in the perturbations and evaluating the expectation

$$\overline{Z} = \mathrm{E}[\overline{Z} + Z'] = \mathrm{E}\left[a(\overline{X} + X')^2\right] = a\left(\overline{X}^2 + \sigma_X^2\right) \tag{2.42}$$

The variance of Z is $\sigma_X^2 = \mathrm{E}\left[(Z')^2\right]$, so the first step in evaluating the variance is to write an equation for the perturbation component of Z:

$$Z' = Z - \overline{Z} = aX^2 - a\left(\overline{X}^2 + \sigma_X^2\right) \tag{2.43}$$

We next substitute in the perturbation expression for X:

$$\begin{aligned} Z' &= a(\overline{X} + X')^2 - a\left(\overline{X}^2 + \sigma_X^2\right) \\ &= 2aX' + (X')^2 - a\sigma_X^2 \end{aligned} \tag{2.44}$$

We then square this equation and take the expectation:

$$\begin{aligned} \mathrm{E}\left[(Z')^2\right] &= \mathrm{E}\left[\left(2aX' + (X')^2 - a\sigma_X^2\right)^2\right] \\ &= 4a^2\sigma_X^2 - 2a\left(\sigma_X^2\right)^2 + 4a\cdot\mathrm{E}\left[(X')^3\right] + \mathrm{E}\left[(X')^4\right] \end{aligned} \tag{2.45}$$

If the skewness and kurtosis of X are zero, then the last two expectations are zero. If they are not zero, the second moment approximation is invoked, and they are set to zero. The variance of Z is then

$$\sigma_Z^2 = 4a^2\sigma_X^2 - 2a\left(\sigma_X^2\right)^2 \tag{2.46}$$

2.4.3 Conclusions

The examples above outline the most common techniques for analysis of stochastic processes using perturbation analysis. Many useful problems can be broken down into components where the examples can then be used to solve the problem, and certainly all applications in this book can be solved using these four case studies.

3 A Brief Hydrology and Geomorphology Primer

Water is central to many of the processes discussed in this book. Fluvial erosion on hillslopes is driven by overland flow during rainfall events. Soil weathering, vegetation and organic matter are driven by water in the soil profile. Accordingly a brief summary of the key aspects of hydrology that are important is warranted. Moreover a range of geomorphology concepts have permeated through the landform evolution literature, either because they triggered many of the original approaches or because they have been found useful when studying (either experimentally or mathematically) landform evolution. This chapter will highlight the most important hydrology and geomorphology concepts in soilscape and landform evolution. For more detail the reader is guided to one of the many introductory hydrology and geomorphology textbooks.

3.1 Hydroclimatology

3.1.1 Rainfall

The primary driver of hydrology is precipitation, and in particular rainfall. Snowfall is also important, but as a primary erosion agent it is the melting of the snow rather than the falling of the snow that is important (i.e. floods and soil saturation), and we will not discuss this melt behaviour. Rainfall on the other hand through both its long-term mean (yield hydrology, driving soil moisture) and its peaks (event hydrology, driving flood and erosion events) drives landscape response. In Section 3.2 we discuss the impact of (1) short-term, decadal scale, climate variability (e.g. El Nino, Indian Ocean Dipole [IOD], Interdecadal Pacific Oscillation [IPO], North Atlantic Oscillation [NAO]) and (2) long-term, millennial to epoch scale, climate variability (e.g. glacial-interglacial).

Two further factors are important because they drive the relative dominance of landscape and soilscape evolution processes in the field. The first factor is the relationship between the energetics of the rainfall (i.e. the kinetic energy of the raindrops when they hit the ground) and the amount of rainfall (which drives the amount of runoff and thus the amount of fluvial erosion). We will defer discussion of this issue until Chapter 4, where rainsplash erosion will be discussed.

The second factor is the spatial distribution of rainfall, and the interaction between rainfall location and (dynamic) topography, which we will discuss below. Rainfall and evaporation are related to elevation. Rainfall increases with elevation, but there is no consensus about the relationship. Potential evaporation also appears to increase with elevation, primarily because of increases in wind speed, though the number of studies is small.

Anders and Nesbitt (2015) used satellite remote sensing of rainfall to identify trends of rainfall with elevation and identified four climate regimes. For three of the climate regimes the rainfall increases up to a height threshold of about 1–2 km and then declines for heights above that threshold. For the tradewind regime there seems to be no trend with height, though this might simply be because the analysis stopped at an elevation of 1 km (the maximum elevation of the land in the two regions studied) and didn't reach a threshold height.

Hutchinson (1995, 1998) found a parabolic relationship between elevation and mean rainfall ($R^{1/2} = \alpha Z$ where R is rainfall and Z is elevation), through the coefficient α varied spatially, and he found aspect to be a second-order factor after elevation. Hutchinson (1995) analysed an area that is just south of Anders and Nesbitt's EAU region and with a similar exposure to the ocean, and about double the elevation range (i.e. maximum elevation about 2 km), so it is likely comparable with their EAU analysis, where Anders and Nesbitt found no trend with height. Hutchinson (1998) analysed a portion of the European Alps which has no equivalent in the Anders and Nesbitt analysis. In situations where the atmosphere is highly saturated with water, even small topographic variations (50–60 m elevation) can lead to a doubling of rainfall (Bergeron, 1961).

Kyriakidis et al. (2001) on the other hand found a relatively poor relationship with elevation (correlation coefficient $R = 0.22$) when 1 km spatial resolution elevation data were used. However, a better correlation was found when the elevation was averaged by a 13 km × 13 km window, suggesting that it was mesoscale topography that drives orography rather than the high-resolution topography.

Alpert (1986) and Roe (2005) outline a simple model for precipitation distribution driven by orography and based on an atmospheric lift model:

$$R = \alpha \bar{p} \bar{q} u \frac{\partial z}{\partial x} e^{-z_s/H_m} \quad (3.1)$$

where $\bar{p}$ is the depth averaged air density, $\bar{q}$ is the depth-averaged specific humidity (so $\bar{p}\bar{q}$ is the depth-averaged mass of water per cubic metre of air), u is the velocity of the wind at right angles to the mountain range spine and $\frac{\partial z}{\partial x}$ is the slope of the mountain range at right angles to the mountain range spine. The main assumption of the method is that a parcel of air approaching the range is raised at a velocity w with $w = u\frac{\partial z}{\partial x}$. The parameter α is the rainfall efficiency, which is the proportion of the condensation (i.e. raindrops formed in the cloud) that actually falls to the ground ($\alpha < 1$ and reported values are typically of the order of 0.2–0.5; e.g. Alpert, 1986; Yu and Cheng, 2013). The parameter H_m is a maximum precipitation height length scale that is of the order of 2–4 km.

Equation (3.1) is quite simple and one-dimensional, and can predict rainfall only on the windward side of the range. A limitation is that with increasing slope the rainfall rate increases without bound, and the equation does not allow for the possibility of depletion of water in the parcel of air that is being lifted by the mountain range. Specifically, $\bar{p}\bar{q}u$ is the mass rate of water entering the upwind boundary of the mountain range, while $\int_0^X R dx$ is the amount of rainfall falling between the upwind boundary of the range and a distance X into the range. At some distance X the rainfall integral (depending on how fast the exponential term in Equation (3.1) declines with increasing elevation) will equal the mass rate of moisture entering the upwind boundary of the range, so that no more atmospheric moisture can be condensed. This depletion of rainfall is commonly observed in the field with the maximum precipitation occurring on the windward side of the ridgeline (e.g. Alpert, 1986).

The use of the upslope lifting models like the Alpert model comes with a caveat. Yu and Cheng (2013) compared rainfall (observed using radar and rain gauges) in Taiwan with an upslope lift model and found a good match for the northern part of the mountain range and a poor match in the southern (for both the amount of rain and spatial distribution). The reason suggested for the poor performance in the south was that the mountain range was too narrow to allow time for the raindrops to form and fall in the location at which condensation occurred (i.e. where the lift was occurring). In this case the rainfall fell on the leeward side of the mountain range. This suggests that a length scale, L, might distinguish when the uplift model is appropriate, $L = ut$, where u is the wind velocity at right angles to the mountain range, and t is the time for raindrops to form and then fall to the ground. When L is large relative to the width of the range, then the uplift model is a poor approximation. Roe et al. (2002, 2003) added a tuneable advection-diffusion term to Equation (3.1) to adjust for the time to form raindrops, the advection of these raindrops by the prevailing wind and the spreading by gusting wind patterns. This allowed rainfall to spill over the mountain ridge into the rain shadow side of the range.

One constraint on the uplift model is that the stratosphere provides a lid on upward flow of air within the troposphere at an elevation of about 10,000–15,000 m. Some mountain ranges block a significant proportion of the lower troposphere (approximate range heights: Himalaya ~7,000 m, Andes ~5,000 m, Southern Alps NZ and Rockies ~3,000–4,000 m, Taiwan ~3,000 m) forcing wind to flow laterally along or around the barriers, reducing the amount of air being uplifted (Garreaud et al., 2009). This is not accounted for in the uplift model above.

It is also possible to take an altogether simpler approach to orographic rainfall and its effect of erosion rate: this is to simply assume a difference in erosion rate between the windward and leeward sides of the range. Willett (1999) assumed an erosion rate twice as high on the windward size than the leeward side (and independent of elevation or slope) as a first step to understanding the feedbacks between tectonics and erosion. However, like Equation (3.1), this assumes that there is a known, dominant direction from which rainfall comes and thus a known windward and leeward side of the range, and this would need to remain true over geologic time since he was interested in the effect of erosion on mountain range development.

This last point highlights one major constraint on rainfall-topography relationships for use in landform evolution studies. This problem is identifying what are the windward and leeward sides of topographic rises (hills, mountains and so on) since this distinction will follow from prevailing wind patterns, which themselves may vary with season and local topography (e.g. Espinoza et al., 2015). The recent availability of satellite rainfall products may lead to useful empirical relationships if prevailing wind directions are known (e.g. Nesbitt and Anders, 2009; Anders and Nesbitt, 2015), but the observed coupling between wind and topography suggests that ultimately a coupled landform-atmospheric model operating over geologic time may be required to fully resolve the relationship for any particular location.

3.1.2 Evaporation

Methods to estimate potential evapotranspiration (i.e. the maximum rate of evapotranspiration which occurs if the soil is saturated) and its coupling with the elevations of an evolving landform are less common than for rainfall. One of the most common models for potential evapotranspiration calculation is the Penman-Monteith model, which calculates the energy balance at the land surface. Kalma et al. (2008) provides a recent summary of how potential evaporation can be determined by using energy balance methods at a point (including Penman-Monteith), but concludes that uncertainties in methods for determining the environmental inputs into the energy equation (i.e. surface and atmospheric temperatures) limit methods for spatially extrapolating evaporation measurements. McVicar et al. (2007) determined the topographic influence on evaporation over the Loess Plateau in China. They spatially interpolated all of the input variables for Penman-Monteith (maximum and minimum temperature, near surface wind speed, atmospheric water content) using topography as one of the independent variables. The correlations of the input variables with topography varied from summer to winter by factors of two or more, and reductions due to the temperature gradient with increasing elevation were balanced by increases due to higher wind speed. Moreover, lapse rates for temperature changed from summer to winter. They did not find any consistent relationship between evapotranspiration and topography (McVicar, personal communication). A 30 km transect showed a factor of two variability in ET along the transect, a bias compared with field data of less than 5%, and visually the results showed a stronger correlation with aspect than with elevation.

3.1.3 Hydrology

When it rains, either the water infiltrates into the soil or it runs off as overland flow. Whatever cannot infiltrate either ponds on the soil surface or generates surface runoff downslope. This surface runoff drives erosion. It is an adage in the agricultural land management community that 'if you can control surface runoff then you can control erosion', so surface runoff is a central factor for landform evolution due to erosion.

We turn first to the infiltration process. The partitioning between infiltration and surface runoff is a function of the wetness of the soil and the unsaturated conductivity of the soil. If the initial soil wetness (called the antecedent wetness) at the start of a rainfall event is low, then the infiltration rate is high as the soil initially wets up. As the soil wets up the infiltration rate starts to drop, and over time the infiltration rate asymptotically approaches the saturated conductivity of the soil column. If the rainfall rate is higher than the infiltration rate, then the soil will become saturated. The saturated conductivity is a function of the soil properties, and the long-term infiltration rate increases for coarser soil grading (i.e. more sand and silt, and less clay), lower bulk density (so pores are more open), higher soil organic content (i.e. more open soil structure; see Chapter 10) and a higher density of macropores (i.e. old root channels, soil expansion/contraction cracks; Beven and Germann, 1982, 2013). Long-term multiyear drought has been shown to change runoff rates (Saft et al., 2015, 2016a), and while the exact mechanism is not well understood, it is suspected to be linked to vegetation changes because of strong relationships of surface runoff rate with spring rainfall and leaf area index (Tesemma et al., 2015; Saft et al., 2016b).

In the period after the rainfall event finishes, the soil water drains from the upper soil into the deeper soil layers until the upper soil moisture content reaches the field capacity. The field capacity is the soil moisture at which water can no longer drain downwards under gravity and the water is accessible only to plants. The field capacity is primarily a function of the grading of the soil and is 15–25% by volume for sandy soils and 45–55% for clay soils (NRCCA, 2016). The decrease of soil moisture immediately after rainfall ceases is a result of drainage and is a function of the saturated hydraulic conductivity of the soil and may occur quite quickly, in a few days. If there are roots from plants in the soil column, then the soil moisture will be further reduced by plants until the soil moisture content reaches the plant wilting point (a function of plant type, clay content and soil salinity; typically 5–10% for sandy soils and 15–20% for clay soils; NRCCA, 2016). This latter plant-driven soil moisture decrease is the water that is used by plants for transpiration, and the reduction typically takes days to weeks. Any water that penetrates the soil profile below the root zone becomes recharge to deeper groundwater.

An important historical distinction is made in the hydrology literature and divides hydrology into two application areas. The first is 'event' hydrology, which concentrates on the runoff in a single rainfall event to the exclusion of what happens between events, and the time resolution of modelling is typically of the order of hours or days depending on the response time of the catchment to rainfall (sometimes called the 'time of concentration'). The second is 'continuous' or 'yield' hydrology, which concentrates on simulation of continuous time series of runoff over (typically) years or decades with time resolution of days or longer. Historically, because of computer limitations, these have been pursued as separate fields, even though there have been significant, though not complete, overlaps in the techniques used. Event hydrology has a stronger emphasis on the hourly dynamics including infiltration dynamics and river routing. Yield hydrology has focussed more on the evaporation and transpiration dynamics that dominate in the inter-rainfall event periods at the daily and weekly resolution.

There are also two main types of hydrology process dominance in the field, and the distinction is a function of how dry the climate is (Figure 3.1).

3.1.3.1 Humid Zone Hydrology

This is generally the case when rainfall exceeds evapotranspiration, and results in 'saturation-excess' runoff generation (Figure 3.1a). In this case there is a net excess of water in the hillslope so a permanent water table is formed within the soil profile. Starting at the catchment divide as you proceed down the hillslope, this water excess increases and an increasing proportion of the bottom of the soil profile has a permanent groundwater table. The infiltration process described above occurs in the region above the water table, and any excess water recharges the water table. A key characteristic of this process is that toward the bottom of the hillslope in the valley bottom there is commonly a region of fully saturated soil. This saturated region is where most surface runoff is generated because rainfall cannot infiltrate into the soil. The water table moves up and down quite slowly (timescale of weeks) since rises are driven by rainfall infiltration and water table declines are driven by evaporation loss and recharge to the river. Thus the percentage of the catchment with the groundwater table at the surface changes quite slowly, so that percentage of catchment generating runoff does not change significantly during a single rainfall event. Within the fully saturated zone it is common for the soil chemistry to be anaerobic.

3.1.3.2 Arid Zone Hydrology

This is generally the case when rainfall is less than evapotranspiration and results in 'infiltration-excess' runoff generation (Figure 3.1b). In this case there is a net deficit of water within the profile, and there is rarely any permanent saturated zone within the profile (though saturation does occur during rainfall events). In extreme cases, other than the water that runs off during the rainfall event, all the soil water is returned to the atmosphere through evaporation or transpiration (commonly 80–90% of rainfall). Runoff tends to be generated in those areas where soil properties are such that infiltration rates are locally low (e.g. high clay content, low organic content) or vegetation is sparse (so that there are few old root channels for water to infiltrate through). It is possible for

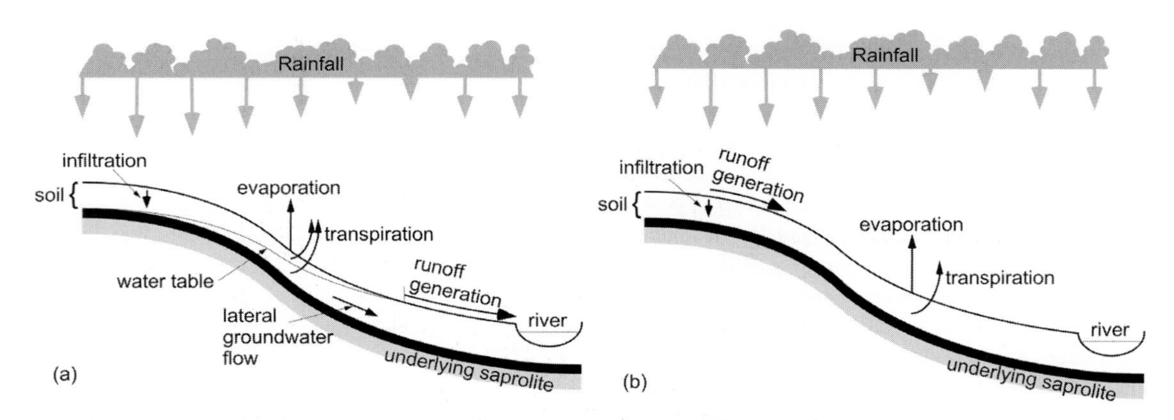

FIGURE 3.1: The two main hydrology process types: (a) humid catchment where there is a semipermanent water table within the soil profile and where any time it rains runoff occurs from those areas where the groundwater table has saturated the entire soil profile, (b) arid catchment where evaporation is greater than rainfall so a permanent water table cannot be created, and where runoff occurs during high-intensity rainfall events when the near-surface soil is saturated.

water to run off from localised low-infiltration zones into zones downstream with higher potential infiltration rates, and in this case the runoff can infiltrate in these high-infiltration rate zones. Since runoff is generated by local saturation at the surface of the soil profile, runoff generation can be very dynamic with infiltration capacity changing within a few minutes. The background soil moisture condition between rainfall events trends toward the wilting point of plants. Rivers and creeks are ephemeral, unless fed by deep groundwater sources. It is common for the soil profile to be aerobic, but with the soil atmosphere having high carbon dioxide and low oxygen levels in the root zone due to plant respiration (Chapter 14) and biodegradation of soil carbon (Chapter 10). The microbiology and plant growth dynamics are commonly limited by water availability.

3.1.3.3 Transitional Hydrology

The distinction of hydrology into 'humid zone' and 'arid zone' is a simplistic but useful shorthand for describing a catchment's dominant processes. In reality the distinction is not just about climate. If we consider Figure 3.1a, if the (1) soil was much deeper or (2) the hillslope much shorter, then the water table may not have reached the surface so that the saturated zone at the bottom of the hillslope may not have formed so that runoff would not be generated by the saturation-excess mechanism. Thus the drivers for saturation-excess runoff generation are not just climate (a permanent water table will not form without an excess of rainfall over evaporation) but also shallow soils and long hillslopes. Thus the drivers are a mixture of climate, soils and geomorphology.

Moreover, if we consider Figure 3.1b, infiltration-excess runoff will occur when the rainfall rate is high enough to saturate the surface of the soil. This process can also occur in Figure 3.1a in the parts of the hillslope upstream of the saturated zone shown in the figure generating runoff. While not shown in Figure 3.1b, it is possible that a water table exists that never reaches the surface.

Finally in a seasonal climate (where average rainfall and/or evaporation are seasonally variable) it is possible that during the seasons of excess rainfall that saturation excess may dominate, while during periods of excess evaporation there is a switch to infiltration excess. The classic example of this was the Tarrawarra catchment in southeastern Australia (Western et al., 1999) where during winter (low evaporation) there were persistent wet regions along the valley bottoms, while during summer (high evaporation) these patterns disappeared and relatively wetter regions were spread with no obvious pattern across all parts of the catchment.

3.1.4 Soil Profile Hydrology in the Partially Saturated Zone

The soil profile at the start of a rainfall event is normally dry, and the potential infiltration rate at the start of an event is quite high so that most water infiltrates. If the soil is wet before the event, then a greater proportion of rainfall runs off. A simple approximation for what happens with the infiltrated water within the profile is given by the Green-Ampt equation (Figure 3.2). This approximation assumes that at the start of a rainfall event the soil profile at the soil surface saturates, and a saturation front propagates down the profile as the infiltration at the surface continues. Behind the front the soil is saturated, while in front of it the soil moisture is whatever it was before the event. A simple implication of this is that the wetting front propagates down the profile at a velocity, v, determined by the infiltration rate, i, filling that part of the soil porosity that didn't contain water before the event, n_e, so $v = i/n_e$. The wetting front propagation stops when the infiltration at the surface stops (e.g. at the end of the rainfall event). For example, if the available porosity is 0.2 and the rainfall in the event is 10 mm, then the wetting front will propagate down the profile 10/0.2 mm = 50 mm. After the rainfall event, the soil water evaporates, is taken up by plants or drains down into lower parts of the soil profile below (e.g. in the example, below 50 mm). The sharp wetting front is maintained during infiltration by the

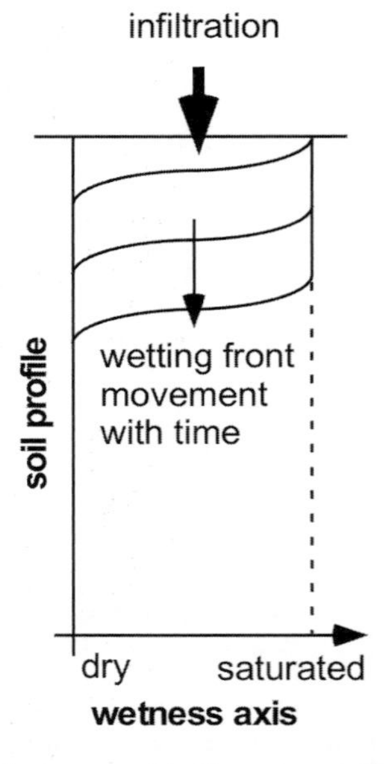

FIGURE 3.2: A schematic of the Green-Ampt infiltration equation, with the wetting front moving down the profile as the water infiltrates at the surface.

discharge increases. Equation (3.10) is true during a runoff event. In our evolution modelling, however, we are often interested in the cumulative effect of the range of storms, from small to big and from short to long, the so-called 'geomorphically effective' erosion rate. This requires averaging Equation (3.10) over time. The larger the value of m then the larger the amount of sediment transported by the larger discharges relative to smaller discharges (Section 2.4.2.3). As an end member we may ignore the lower discharges because they generate insignificant erosion relative to the high discharges. Many traditional agricultural erosion models (e.g. USLE) summarise this integration by using the peak 30 minute or 1 hour rainfall in a year (the so-called 1 in 1 year extreme rainfall) as a way of characterising rainfall erosivity for a site. Likewise we can do the same for discharge (and we will show how this can be used to simplify erosion modelling in landform evolution models in Chapter 4) so that the erosion potential of a site can be characterised by the mean annual peak discharge flood (the peak discharge of a 1 in 2.33 year extreme flood) so that the mean geomorphically effective sediment transport is then

$$\overline{Q_s} \sim Q_{2.33}^m \tag{3.11}$$

where $Q_{2.33}$ is the mean annual peak discharge. The significance of this equation is that the mean sediment transport is no longer a function of the mean discharge of the catchment (i.e. the total volume of water discharged divided by the time period) but of the mean peak extreme discharge, the peak discharge of the hydrograph (see examples in Solyum and Tucker, 2004, 2007). This mean peak extreme discharge is almost always nonlinear with area and is commonly modelled by

$$Q_{2.33} = \beta A^\delta \tag{3.12}$$

where δ is empirically observed to be in the range 0.5–0.8. It is important to note that $\delta \neq 1$ does not violate mass continuity. The total volume of water that leaves the catchment over the duration of the flood event complies with mass continuity (i.e. the total volume is proportional to area), it is just that the peak of the hydrograph does not increase linearly with area. There are a number of reasons for this nonlinearity with area.

One minor reason is that as the catchment area increases, the hydrograph tends to become longer in duration as well as higher. Since the volume of runoff from the event is the area under the hydrograph, this means that the peak cannot rise linearly with area. This effect is due to routing and temporary water storage on the hillslope and in the river network. This effect is normally modelled with either (1) a storage model (e.g. Muskingham-Cunge), (2) a kinematic wave model (where the flood wave speed is a function of the depth of flow) or (3) a diffusion wave model (where flood wave speed is a function of both the depth of flow and the longitudinal curvature of the flow depth profile down the channel). However, the contribution of this process to the nonlinearity of the discharge-area relationship is small, with $\delta > 0.9$ for kinematic wave routing, using typical surface roughnesses and channel/rill geometries (Huang and Willgoose, 1993; Willgoose and Kuczera, 1995).

The main contributor to the nonlinearity in Equation (3.12) is the spatial variability of rainfall. To explain this effect, consider a simple example of a thunderstorm (typically a few square kilometres in area). For a small catchment where the thunderstorm is larger than the entire catchment, the peak discharge can increase linearly with area (i.e. $\delta = 1$) because as we add extra area to the catchment this extra area is also covered by the thunderstorm. However, for a large catchment where the thunderstorm covers only a small proportion of the catchment area, an increase in the catchment area will not change the discharge because the extra area will not have any extra rainfall falling on it (i.e. $\delta = 0$). Clearly this example is rather simplistic, but it captures the idea that as the catchment area increases, the extra area added to the catchment falls outside the area of maximum rainfall so that, even if rain is falling, it is falling at a lower average rate than the average rate over the original, smaller, catchment. In event hydrology this behaviour, called a 'partial area storm', is modelled with an 'areal reduction factor' where the peak rainfall generating the storm for a given area is reduced as the area of the catchment is increased (e.g. Rodriguez-Iturbe and Mejia, 1974). The example above also suggests that the exponent on area may change with catchment area, being $\delta = 1$ for small catchments and declining with increasing area (Solyum and Tucker, 2004, 2007), which suggests that Equation (3.12) with a fixed exponent on area, while commonly used by hydrology practitioners, might be a poor approximation to the area dependence of $Q_{2.33}$. This is an area of ongoing research.

Finally, Equation (3.11) characterises the peak flow behaviour of the catchment and does not capture what happens for smaller and larger flows. Leopold et al. (1964) talks about the cumulative effect of the range of flood events, and introduces the concept of a 'dominant' discharge, the discharge that is most important for shaping the geomorphology. For river channels this is sometimes equated with the 'bankfull' discharge, which itself has been empirically observed to be about a 1 in 2 year discharge. While these are all empirical findings, the repeated appearance of a 1 in 2 year (or thereabouts) extreme discharge event suggests that frequent extreme runoffs are important in shaping the landscape.

3.2 Modelling Climate Temporal Variability

In many landscape and soilscape evolution problems it is sufficient, at least to first order, to assume that the climate does not change over time. However, there are applications where explicit modelling of climate variability is warranted. For instance, if we wish to characterise the effect of the different response times of the vegetation, soils and landforms and how their temporal changes will interact, this will depend on the rate of change of climate relative to the landscape response time. This section is a summary of methods for simulating climate variability.

In many cases we may know the climate history and wish to reproduce this exactly; this is deterministic variability. In other cases we don't know the exact details of the climate history, but we know statistics of the climate so that we can reproduce time series that have the correct statistics (e.g. mean, variance, autocorrelation) but not the exact year-to-year values. We typically model this latter case by stochastically generating a number of climate histories (each climate history is called a realisation of the climate) and then run the model with each of these histories and then assess the statistics of the soilscape and landform response; this is stochastic variability.

The accepted impacts of climate variability are on temperature and rainfall/runoff. However, there are significant, though poorly understood, impacts on vegetation and vegetation-fire feedbacks that will be touched upon in Chapter 14 (e.g. Robertson et al., 2016).

3.2.1 Deterministic Variability

In cases where it is required that an LEM be validated against behaviour of a monitored site (e.g. an experimental site), then the instrumental record (i.e. direct rainfall and evaporation measurements from instruments) should be used to reconstruct the runoff, erosion and soil moisture, and subsequent performance of the site (e.g. Bell and Willgoose, 1998). However, we know, particularly for landform evolution, that models are sensitive to initial conditions and forcings, so we should not necessarily expect that the model predictions would deterministically match the evolution observed in the field. This can be contrasted with statistical replication. For instance, matching the exact location of a gully with a given geometry is deterministic replication. In contrast, we may not be able replicate the gully location and/or geometry deterministically, but we may be able to replicate the probability of a gully of a given size occurs at that location, which is statistical replication.

More challenging is the case of validating models for field sites where the evolution is the function of pre-instrumental climate (i.e. instrumental records typically go back only 100 years, and sometimes less). In this case we must rely upon paleo-climate data. Well-known paleo-climate data are the 'Hockey Stick' curve for the last 1,000 years, and the ice core climate data going back about 800,000 years.

As an example we will consider the ice core climate record. To convert the ice core data into reconstructed climate data, concentrations of CO_2, isotopes of the gases, salt and dust trapped in the ice are related to climate. This is done with proxy relationships where known climate records (typically instrumental but sometimes paleo-climate data) are empirically related to atmospheric gas concentration, and this proxy relationship is used to reconstruct the climate from the ice core gas data. The accuracy of the reconstruction is limited by (1) the consistency of the gas record and dating between ice cores (e.g. Bazin et al., 2013), (2) the accuracy of the proxy relationship (e.g. Goodwin et al., 2004 shows correlation coefficients of 0.3–0.5) and (3) the temporal resolution of the ice core data. For instance, the Vostok record (Antarctica; Petit et al., 1999a, b) goes back to 414,000 years before present and has an average resolution of about 2,000 years, while the NGRIP record (Greenland; NGRIP, 2004a,b) goes back 123,000 years before present and has been gridded to a resolution of 50 years. Because of higher snow deposition rates, Greenland cores tend to have higher temporal resolution but shorter length of record than Antarctica (Alley, 2014). However, in both cases there is significant subgrid temporal variability (e.g. resolution finer than 50 years for NGRIP) not extractable from these records. We will return to this subgrid temporal variability in the next section.

Other climate histories use tree rings, coral and speleothem proxy data. The hockey stick curve is one example of these data (Mann et al., 1999). The principles are the same as for the ice core data: that is, some environmental indicator(s) that can be linked to climate, a proxy relationship and the longer term paleo-data record(s) from which climate is then reconstructed with some accuracy and temporal resolution. They tend to be location specific, so I leave it to the reader to investigate paleo-climate reconstructions for their site, or for methods to

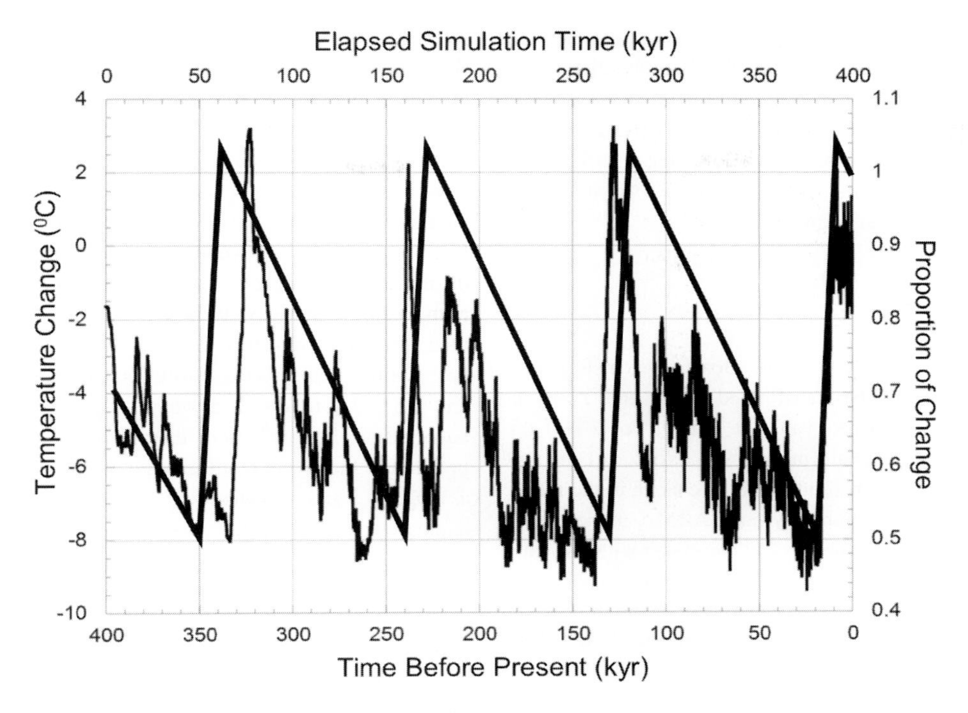

FIGURE 3.4: Ice core reconstructed temperature data for the last 420,000 years with current day at the extreme right-hand side of the graph, together with the Cohen saw-tooth approximation to that record (from Cohen, 2010).

geographically translate existing records to their site (e.g. Ho et al., 2014).

In landscape evolution studies it has been common to adopt simple sinusoidal or step-change climate change (e.g. Willgoose, 1994b; Moglen et al., 1998; Whipple and Tucker, 1999; Baldwin et al., 2003), either to examine the dynamics of numerical models or to develop analytical expressions for the dynamics, but it is less common to use reconstructed climates. Cohen (2010) and Cohen et al. (2013) used the Vostok record to examine the dynamics of soilscape evolution on a hillslope. Cohen (2010) also examined the impact of using a saw-toothed approximation to the Vostok record (Figure 3.4), and while short-term (<10,000 years) fluctuations were smoothed out the long-term trends were unchanged. This 10,000 year smoothing was consistent with simulations that used a simple step-change in the imposed climate, which indicated a circa 10,000 year response time for the soil profile.

3.2.2 Stochastic Variability

In many situations we do not have a deterministic climate record to use, but we may have climate statistics. Even in cases where we may have a deterministic record (as in the previous section), the low temporal resolution of the record may require infilling or disaggregation between data points to a higher temporal resolution required by the modelling. In these cases a range of stochastic climate simulators have been used in the hydrology community, and they are typically referred to by the generic terms 'climate simulator' or 'disaggregation model', or more specifically by the techniques used such as 'auto-regressive moving average (ARMA)', 'point process' or 'Markov chain'. This section cannot review all of the techniques, but will focus on the two most commonly used for rainfall simulation: ARMA and Markov chain models.

A discussion of disaggregation and ARMA models in hydrology can be found in Bras and Rodriguez-Iturbe (1985). A key feature of disaggregation models is that they take a low time resolution time series and break it up (disaggregate) into a higher resolution series, and the methods ensure that the high-resolution series replicates the low-resolution series: that is, if a yearly resolution series is disaggregated to a monthly series, then when the months are added up the yearly totals will exactly match the original yearly series. The techniques discussed below do not provide that assurance.

ARMA models are a simple technique for generating stochastic time series when the probability distribution function is Gaussian distributed and the time series is autocorrelated. In many cases the time series will not be Gaussian but can be transformed so that it is Gaussian after the transformation. A commonly used transformation is the Box-Cox transformation

$$X_t = \begin{cases} \dfrac{Y_t^{\lambda} - 1}{\lambda} & \lambda \neq 0 \\ \ln \lambda & \lambda = 0 \end{cases} \tag{3.13}$$

where the parameter λ is the Box-Cox transformation parameter, X_t is the transformed value and Y_t is the original untransformed value at time t. For $\lambda = 0$ this equation is a log transformation, while for $\lambda = 1$ the time series is untransformed other than a change in the mean.

A simple and commonly used ARMA model is one where the value at time $t+1$ is correlated with the value at time t: an autoregressive lag-1 (AR(1)) model. The simulation process is to take the value at time t and then simulate the value at time $t+1$ using the lag-1 correlation ρ. If X_t is the value at time t, then the value at time $t+1$ X_{t+1} is

$$X_{t+1} = (1-\rho)\mu_X + \rho X_t + \sqrt{\sigma_X{}^2(1-\rho^2)}\varepsilon_{t+1} \tag{3.14}$$

where ε_{t+1} is a Gaussian random variable with mean $= 0$ and *variance* $= 1$, and μ_X and $\sigma_X{}^2$ are the mean and variance of X. Higher order autoregressive models (where the value is correlated not just to the previous timestep but other previous times as well) and moving average (i.e. the MA part of ARMA) are possible. The reader is referred to Bras and Rodriguez-Iturbe (1985) for details.

It is also possible to model two or more variables simultaneously (e.g. rainfall and evaporation) where the variables are cross-correlated. In this case X_t and ε_t become vectors, and the equation involves matrix multiplication on $\underline{X}_t$ and $\underline{\varepsilon}_t$ where the matrices encapsulate both the cross-correlations and autocorrelations. This general multidimensional extension is the Yule-Walker equation.

ARMA models are quite simple to use and calibrate. MATLAB, Python and R all have packages to calibrate and simulate ARMA models. The main problem with ARMA models is the requirement for the variable(s) to be Gaussian. For instance, while mean annual rainfall is commonly close to Gaussian, daily rainfall is not. Daily rainfall has many days with zero rainfall, and it is not possible to transform these zeros in such a way that the time series can be made Gaussian. Daily resolution rainfall and runoff is a case for which Markov chain models are commonly used.

Markov chain rainfall models simulate a time series where a timestep (for simplicity of explanation let us assume a day but it can be any period) is randomly simulated to be either wet or dry. If the day is wet, then the amount of rainfall falling on that day is then sampled from a probability distribution (typically exponential or Gamma distributions provide a good fit to daily rainfall). If a wet day follows a previous wet day, then the rainfall on the second day may or may not be correlated to the rainfall on the first day. If the rainfall is Gamma distributed, then this model will have four parameters; the probabilities that a wet day follows a wet day ($p_{\text{wet-wet}}$) and a dry day follows a dry day ($p_{\text{dry-dry}}$), and the two parameters of the Gamma distribution. Note that the probability that a wet day follows a dry day is $p_{\text{dry-wet}} = 1 - p_{\text{dry-dry}}$ and likewise $p_{\text{wet-dry}} = 1 - p_{\text{wet-wet}}$. The simulation process is that at every timestep a uniformly distributed random number between 0 and 1 is generated, U_{t+1}. If the previous day was wet, then if $U_t < p_{\text{wet-wet}}$ the day will be wet, otherwise it will be dry. Likewise if the previous day was dry, then if $U_t < p_{\text{dry-dry}}$ the day will be dry, otherwise it will be wet. It is common to observe a seasonal variation through the year for the parameters (e.g. wet seasons typically show a larger value for $p_{\text{wet-wet}}$ and a smaller value for $p_{\text{dry-dry}}$, so wet periods last longer and are more frequent, and a Gamma distribution with a larger mean, so that wet periods are wetter). In this case the parameters can vary from season to season or from month to month (e.g. Chowdhury et al., 2017).

3.3 Geomorphology

This section will briefly outline some geomorphology techniques that are commonly used to describe geomorphology and LEMs. Some of these topics have developed in conjunction with LEM research in the last two decades, so to fully understand the background to these concepts requires knowledge from the chapters that follow. In this case the reader may wish to skim these sections now for the key concepts and outcomes but come back, as they are reading the remainder of the book, to fully appreciate some of the physics underpinning them. Most of the statistics below have strong physically-based causal explanations for homogeneous (rainfall, geology) catchments. When there is spatial variability, the behaviour departs from the explanations below, and this departure itself may be a good diagnostic of catchment process (e.g. Wu et al., 2006; Solyom and Tucker, 2007; Huang and Niemann, 2008; Roy et al., 2015).

3.3.1 Channel Network Classification Systems

Network classification systems are a means of conceptually organising the arrangement of the channel network. Some of the schemes involve physical length and areas, but most of them simply summarise the topology of the network. I will not provide a comprehensive summary here but only those that will be useful elsewhere in the book. A more comprehensive discussion can be found elsewhere (e.g. Abrahams, 1984).

A definition common to the classification systems below is that a 'link' is the channel segment between two adjacent tributaries. An 'exterior' link is a link between the upstream end of the channel where the hillslope transitions into a channel and the first tributary encountered travelling downstream, while an 'interior' link is any link between two tributaries.

The Strahler (1952) scheme classifies all exterior links as Strahler 'order' 1. Moving down a channel if this channel encounters a tributary of lower order, then the order of the next link downstream remains the same order. If the tributary is the same order, then the next link downstream is incremented by one. If the tributary is of higher order, then the next link downstream has the order of the higher order tributary. A Strahler 'stream' is then that river segment of connected links that going downstream all have the same order. A catchment or drainage network is said to have order k if the link at the catchment outlet has order k. Strahler defined the 'bifurcation ratio' as the average number of $(n-1)$ order streams that flow into a single n order stream. Shreve (1967) showed that an infinite topologically random network yielded a bifurcation ratio equal to 4. Tokunaga (1978) showed that the bifurcation ratio was scale dependent and that the bifurcation ratio changed with the order of the drainage network. He defined a new ratio that defined how many $(n-i)$ order streams drained into the single order n stream. Willgoose (1989) showed that as the network order increased, the Tokunaga definition asymptotically converged on the Strahler bifurcation ratio of 4.

Shreve (1967) introduced the concept of stream 'magnitude' and 'topologically random channel networks (TDRN)'. Every exterior link has magnitude 1, and the magnitude of an interior node is the sum of the magnitudes of the two immediately upstream links. If a network has n exterior links, then the magnitude of the network is $(2n-1)$, and if the area contributing to each link is the same, A_l, then the area of the catchment is $(2n-1)A_l$. Magnitude has sometimes been used as a proxy for catchment area. Shreve (1967) introduced a methodology to generate networks that were topologically distinct (i.e. the arrangement of the links was different) random networks (TDRNs) with a given magnitude/area. TDRNs have been the basis of much work on randomness in channel networks.

The main problem with these schemes is, no matter their enduring appeal in the geomorphology community, that they are coarse measures of network topology and have large random variability. Willgoose (1989) used a landform evolution model and showed that with random initial conditions the bifurcation ratio of the resulting network ranged between 2 and 6, this range overwhelming any signal resulting from changing process parameters. Using the Shreve TDRN model Kirchner (1993) reached a similar conclusion that randomness with order overwhelmed any deterministic signal.

3.3.2 Conceptual Channel Network Models

Prior to the current research using physically based landform evolution models a range of channel network models were used to simulate the blue lines on contour maps. They ranged from simple stochastic models, to models based on energy minimisation and entropy maximisation. Two models have stood the test of time and are summarised here.

The first model is the headward growth model (Howard, 1971). In this model a grid is defined and a value of 0 (hillslope) or 1 (channel) allocated to each node of the grid. Initially the entire grid is 0 except for one point on the outside edge of the grid that is 1, which is the starting point for the channel network. Headward growth of the channel network proceeds in a series of cycles. At each cycle one node that is adjacent to the existing channel network (i.e. that directly abuts to a node that is already 1) is randomly selected that becomes a new node in the channel network (i.e. goes from 0 to 1). In this way the channel network grows outward from the first initial starting node. Howard and subsequent authors have examined changing the rules (e.g. having more than one node become channel in each cycle), but the fundamental notion of a channel network growing headward from a starting node remains.

A similar approach, commonly used in the physics and cellular automata community, is based on diffusion-limited-aggregation (DLA) (Witten and Sander, 1981), where particles that are not connected to the network are moved around on the grid by diffusion until they hit the network at which time they become part of the network. Willgoose (1989) showed that for that part of the region that is not occupied by the network (i.e. where the particles are moving around by diffusion) this is a discrete approximation to

$$\frac{\partial z}{\partial t} = -D\left(\frac{\partial^2 z}{\partial x^2} + \frac{\partial^2 z}{\partial y^2}\right) \tag{3.15}$$

where z is the concentration of particles/unit area, and D is the diffusivity. Looking ahead to the discussion of soil creep (Chapter 9) this is the same equation used for a landform dominated by linear creep when z is defined as elevation.

The second model is the Optimal Channel Network (OCN) model. In this model a channel network is created by some method (typically by using a headward growth model, or by simply allocating every node a random flow direction while ensuring there are no closed loops), and then the network topology (i.e. how one node connects to another node) is modified until the total energy consumption of the network is everywhere equal and minimised. The energy consumed by the flow in each node is proportional to $Q^{0.5}$. The area draining through each node is derived at every point in the grid. The discharge at every node is determined from the area. The total energy consumed by the network is $E_p \propto \sum Q^{0.5}$ where the summation is over all the nodes in the grid. The network that minimises E_p is the OCN for that grid. The exponent of 0.5 is empirical but was derived by using empirical equations for how velocity and channel width change with area/discharge, and the slope-area relationship (Section 3.3.3). If any of these empirical equations change, then the exponent of 0.5 is changed, and the form of the resulting OCN will change. Ijjasz-Vasquez et al. (1993) compared the OCN to field catchments and found the differences between the field catchments and the OCN were small with the field catchments having an E_p about 5–10% higher than the OCN. Ibbitt et al. (1999) compared networks generated by the OCN and the SIBERIA landform evolution model, and found that SIBERIA and the OCN systematically had differences in E_p of about 10%, and that SIBERIA had an E_p closer to the field catchment than the OCN. When the OCN was used as input to SIBERIA, the channel network changed and E_p converged to the SIBERIA result. Ijjasz-Vasquez et al. (1993) speculated that their field catchments had not reached optimality, while the results of Ibbitt et al. (1999) suggested more fundamental differences.

3.3.3 Slope-Area Analysis

Probably the most significant advance in analysis techniques for landscapes to be developed in recent years is slope-area analysis. It has long been recognised that hillslope tops are concave down and that channels are concave up given the characteristic rolling hills and valleys, and that the rounded hilltops result from a transport process that increases with slope downstream (e.g. Gilbert, 1909). Empirical studies have also long shown that for rivers an approximately log-log linear relationship exists between distance downstream and slope (Hack, 1957; Tarboton et al., 1989). The availability of digital elevation models worldwide and GIS tools has made slope-area analyses easy to perform.

Kirkby (1971) showed mathematically how the relationship between the distance downslope and the gradient in the transport-limited erosion process determines the concavity of one-dimensional hillslopes (one-dimensional hillslopes have no flow divergence or convergence). Willgoose et al. (1991c) extended Kirkby's work by solving mathematically the catchment problem in two dimensions (of which Kirkby's solution is then a special case for one dimension) for dynamic equilibrium. Howard (1980) provides an equivalent two-dimensional solution for the case of detachment-limited erosion. The conclusion of these works is that the slope and drainage area of hillslopes and catchment can be plotted against each other for common transport- and detachment-limited erosion physics (see Chapter 4 for a discussion of erosion physics), and the slope and area are log-log linearly related (Figure 3.5). Willgoose (1994a, b) extended his 1991 solution for the cases of (1) zero continuing uplift and (2) cyclic uplift, and showed that the dynamic equilibrium relationship can be simply extended by including the average catchment elevation. Except for very low elevation catchments the effect of this average elevation correction is small, and less than the scatter typically encountered in the slope-area diagrams using field data. Finally, Whipple and Tucker (2002) compared the time-varying dynamics of elevations for transport- and detachment-limited erosion. They found that while the equilibrium trends were the same, the time-varying dynamics were different. Detachment-limited erosion created localised high-slope regions in river channels (with some resemblance to knickpoints) when adjusting between equilibria. For transport-limited erosion the transition was smooth and did not create knickpoints.

We will talk in detail about erosion processes in Chapter 4, but a simple calculation will highlight the process dependencies of the slope-area relationship for equilibrium landforms. If we assume a transport-limited sediment transport model where the sediment transport Q_s is

$$Q_s = K_1 Q^{m_1} S^{n_1} \tag{3.16}$$

and the discharge Q is related to area as

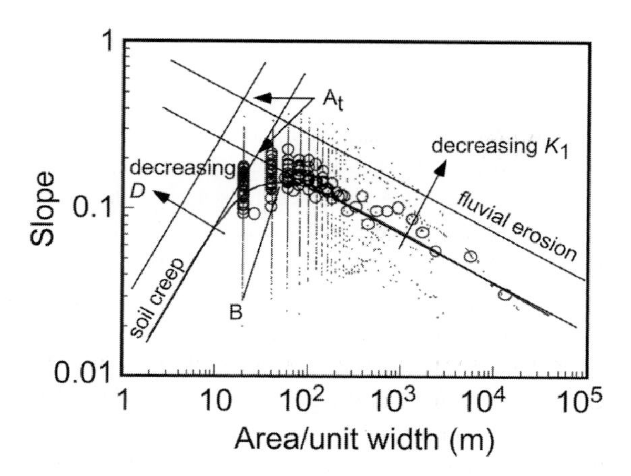

FIGURE 3.5: Slope area analysis with typical data derived from a ground survey also showing a best fit slope-area curve through these data for one (the straight lines; see text for explanation) or two processes (the curved line). The dots are the raw data from analysing a 20 m gridded DEM from the ground survey, while the circles are from averages of 20 pixels to better show the mean trend (after Willgoose, 1994b). The large scatter in the slope for any given area is not a reflection of low spatial resolution elevation data but seems to be fundamental because recent analyses using high spatial resolution LIDAR data exhibit similar scatter. The area at the intersection point of the fluvial and soil creep lines A_t is given by Equation (3.28).

$$Q = \beta_3 A^{m_3} \tag{3.17}$$

then it is possible to write an equation where the sediment transport out of the catchment is equal to the uplift U over the catchment (i.e. $Q_{s(\text{outlet})} = UA$). The transport-limited form of the slope-area relationship is then (Willgoose et al., 1991c)

$$K_1 {\beta_3}^{m_1} A^{m_1 m_3} S^{n_1} = UA \tag{3.18}$$

Rearranging this yields the classic form for the slope-area relationship

$$S = \left(\frac{U}{K_1 {\beta_3}^{m_1 m_3}}\right)^{1/n_1} A^{\alpha_1}$$

$$\alpha_1 = \frac{1 - m_1 m_3}{n_1} \tag{3.19}$$

This is a log-log linear relationship between area and hillslope gradient. The parameter α_1 is called the slope-area 'scaling exponent'. The term inside the parentheses reflects the balance between uplift and erosion rates and material erodibility, and determines the average slope of the catchment. Typical fluvial erosion parameters yield $\alpha_1 \approx -0.5$, which is approximately consistent with analyses of fluvial erosion–dominated river networks.

An alternative derivation can be performed for catchments where detachment-limited erosion occurs. For detachment limitation the erosion rate at every location of the catchment can be expressed as

$$E = K_2 Q^{m_2} S^{n_2} \tag{3.20}$$

When we substitute Equation (3.17) for discharge and assume that the erosion rate at every location in the catchment equals the uplift we obtain (Howard, 1980)

$$S = \left(\frac{U}{K_2 {\beta_3}^{m_2 m_3}}\right)^{1/n_2} A^{\alpha_2}$$

$$\alpha_2 = -\frac{m_2 m_3}{n_2} \tag{3.21}$$

The similarity between Equations (3.19) and (3.21) is striking, with the main difference being the change in the value of the slope-area scaling exponent. Again if we use typical parameters for detachment-limited erosion in rivers, we find that $\alpha_2 \approx -0.5$, which is consistent with observed exponents.

A number of authors have noted the similarity in the log-log linear form of Equations (3.19) and (3.21) and the typical slope-area scaling exponent, which means it is difficult to differentiate river process (i.e. transport- versus detachment-limited) solely from slope-area analyses (e.g. Whipple and Tucker, 2002). However, the physically based derivations above have provided a strong physical rationale for using slope-area analyses for analysing field catchments, and the application literature is now too big to summarise here.

When there is more than one process, there is an area threshold above and below which the different processes will be dominant (Willgoose et al., 1991c), leading to the possibility of there being different log-log linear relationships applying for different ranges of catchment area. We will see examples later in the section. Finally, it is possible to obtain slope-area relationships (albeit probably not log-log linear) even if the erosion processes don't have the nice power law relationship in Equations (3.16) and (3.20). All that is required is that the erosion processes depend only on area and slope.

If the erosion processes are not a function of slope and are only a function of area, then it is possible to determine a threshold uplift rate where for uplift rates higher than the threshold no equilibrium is possible. The elevations will increase without bound. Likewise, below this threshold elevations will decrease without bound. Fortunately for the case of high uplift, when this happens other processes such as landsliding (with its strong threshold slope dependence) and glacial erosion (with its strong threshold elevation dependence) (Mitchell and Humphries, 2015) become dominant, so mountains do not rise without limit.

One of the main problems with slope-area analysis is that there is always considerable scatter in the slopes of individual pixels for a specific area (typically one or more orders of magnitude; see Figure 3.5), and we still don't understand what the driver for this variability is. The variability can also make it statistically difficult to compare processes between field catchments, and between landform evolution models and the field. A number of authors have developed spatial averaging methods to address this (e.g. Cohen et al., 2008), but scatter about the mean slope-area trend when using field data remains a problem.

The slope-area relationship has been used in two significant landscape models that don't model landscape evolution. The first is the Optimal Channel Network (OCN) (see the previous section on Conceptual Network Models) where it is used to integrate out the elevation in the derivation of the spatial distribution of energies in the energy optimisation when the OCN generates its planar channel networks (Rodriguez-Iturbe et al., 1992b; Rigon et al., 1993). The second is the Quick Equilibrium Landforms (QUEL) model (Ibbitt et al., 1999; Willgoose, 2001), which uses a channel network as an initial condition and then reconstructs the elevations of the catchment using the slope-area relationship.

Finally, Tucker and Whipple (2002) reviewed field slope-area data (mostly for the United States) and used Tucker's GOLEM landform evolution model with estimated process parameters from the equilibrium slope-area relationships in Equations (3.19) and (3.21) to generate landforms with different process parameters. They then examined a range of geomorphic features of these landforms to assess whether these landforms were feasible in the field. In an earlier and more limited study Willgoose and Gyasi-Agyei (1995) showed similar results when they compared the differences in how landforms evolved when they chose process parameters that were consistent with either the USLE or CREAMS agricultural erosion models.

One problem in interpreting process functionality from the slope-area relationship is that it is not a one-to-one relationship. For instance, in Equations (3.19) and (3.21) an increase in uplift U can be balanced by an increase in erodibility K, leaving slopes changed. Moreover, as will be shown in Chapter 4 (where Equation (3.16) is derived from first principles) the parameter K includes a range of channel geometry parameters that are known to change with location (e.g. channel width, area, roughness), and that may confound any process estimation. Despite these problems researchers have used maps of the S/A^{α} to define spatial trends in uplift (Willgoose, 2001) and runoff/erosion (Roe et al., 2003) with varying success.

A recent methodological development of slope-area analysis is chi plots (not to be confused with the chi statistic from the statistics literature) (Perron and Royden, 2013). This technique addresses the scatter in the slope-area by calculating an integrated measure of slope-area from the outlet of the catchment to some point up a river profile, to yield average statistics for that reach of the river. The description of Perron and Royden unnecessarily limits the method to bedrock rivers at equilibrium. It is sufficient that the log-log linear slope-area relationship applies:

$$S = \frac{dz}{dx} = \gamma A^{\alpha} \tag{3.22}$$

where z is the elevation and x is the distance along the river positive in the upstream direction. Integrating this equation up from the catchment outlet yields

$$\int_{z_b}^{z} dz = \int_{x_b}^{x} \gamma A^{\alpha}\, dx \tag{3.23}$$

where (x, z) is the location and elevation at some point upstream of the catchment outlet (x_b, y_b). If the value of γ (the ratio of uplift to erodibility) is constant along the river, then γ can be moved outside the integral so that

$$z = z_b + \gamma \int_{x_b}^{x} A^{\alpha}\, dx \tag{3.24}$$

It is convenient to normalise the area with an area scale (e.g. the area of the catchment at the outlet) $\hat{A}$ so that

$$z = z_b + \gamma \hat{A}^{\alpha} \int_{x_b}^{x} \left(\frac{A}{\hat{A}}\right)^{\alpha} dx = z_b + \gamma \hat{A}^{\alpha} \chi \tag{3.25}$$

where χ is the chi integral for the location (x, y) on the river. The plot of z versus χ is the chi plot and is a straight line when Equation (3.22) applies because $\gamma \hat{A}^{\alpha}$ is a constant. This chi plot has the advantage of having much less scatter than the traditional slope-area plot (e.g. Figure 3.6) and still has an ability to map spatial variations in γ and α that may be indicative of spatial changes in process or uplift (Royden and Perron, 2013; Fox et al., 2014) though the quantitative interpretation of variations from linearity in Equation (3.25) due to spatial changes is not as direct as it is with traditional slope-area analysis. As previously noted, chi plotting is applicable to both transport- and detachment-limited landforms and requires only that the catchment comply with the log-log linear slope-area relationship.

Finally, we return to the physical explanation for the classical slope-area diagram for a short discussion of what

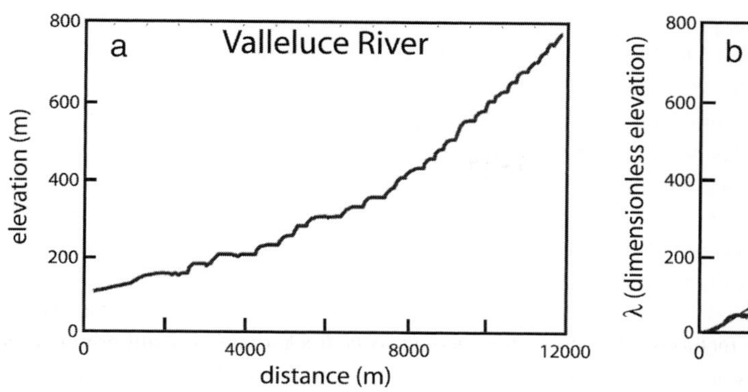

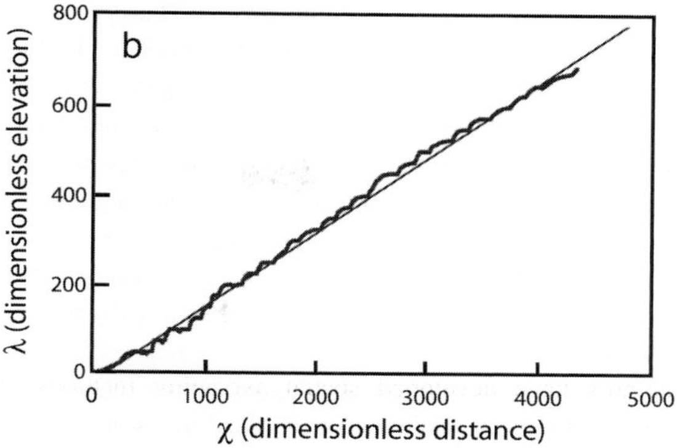

FIGURE 3.6: Chi plot (a) the river profile on traditional elevation and distance axes, (b) the same river profile transformed to chi coordinates (from Royden and Perron, 2013).

happens when multiple processes operate. Both the transport- and detachment-limited explanations (Equations (3.19) and (3.21), respectively) for fluvial erosion yield a negative value for α of about -0.5. If this equation is evaluated for decreasing area, the slope will increase, approaching infinity at the catchment divide. This is clearly unrealistic. Soil-mantled landscapes have rounded hilltops, while mountains have a maximum slope above which landslides occur. In both cases for high slopes and small contributing areas processes other than fluvial erosion become dominant. For soil-mantled landscapes, soil creep (Chapter 9) is important, so that the classical formulation for a transport-limited catchment (Willgoose et al., 1991c) is

$$Q_s = K_1 Q^{m_1} S^{n_1} + DS \qquad (3.26)$$

and following the derivation above for dynamic equilibrium

$$K_1 \beta_3{}^{m_1} A^{m_1 m_3} S^{n_1} + DS = UA \qquad (3.27).$$

This equation can only be solved analytically for slope for the values $n_1 = 1$ and 2, but can be solved numerically to fit data (Figure 3.5). If fluvial erosion is the dominant process, then the data follow the line labelled 'fluvial erosion', while if soil creep is dominant, then the data follow the line labelled 'soil creep'. When both of these processes are of similar magnitude, then the data follow the curved line and (1) for large area the catchment asymptotically approaches the fluvial erosion line because the area term in Equation (3.27) dominates, (2) for small area the catchment asymptotically approaches the soil creep line because the area is small so that soil creep dominates and (3) for intermediate areas there is a curved transition between these two end members. The exact shape of this transition depends on the ratio $m_1 m_3 : n_1$. This explanation is not restricted to the combination of transport-limitation and soil creep, but the exact form of the transition will be different for different combinations of processes. The key conclusion from this is that at some intermediate area there is a transition from concave down to concave up in the downslope profile of the landform (point B in Figure 3.5), and the area at which this occurs is a function of the relative magnitudes of the competing soil creep (commonly referred to as the 'diffusion process') and fluvial erosion (commonly referred to as the 'incision process'). This competition can be seen if Equation (3.27) is solved for the area at point A_t in Figure 3.5:

$$A_t = \left(\frac{D^{n_1}}{K_1 \beta_3{}^{m_1} U^{(n_1-1)/n_1}} \right)^{1/(n_1+m_1m_3-1)}$$

$$S_t = \left(\frac{U}{K_1 \beta_3{}^{m_1 m_3}} \right)^{1/n_1} A_t{}^{\alpha_1} \qquad (3.28).$$

For typical parameters the two exponents on the parentheses are positive:

$$n_1 \in [1,2]; m_1 \in [1,2]; m_3 \in [0.5.1];$$

$$(n_1-1)/n_1 \in [0, 0.5];$$

$$1/(n_1+m_1m_3-1) \in [0.3, 2]$$

$$\alpha_1 \in [-0.3, -0.6] \qquad (3.29)$$

so that as the ratio of the rate of soil creep D to the rate of erosion K_1 the area increases (the peak of the slope-area curve shifts right), and vice versa. When the runoff and erosion rates β_3 and K_1 increase, the area decreases. Thus it can be seen how the dominance of the processes changes the area at which downslope convexity changes sign and thus what percentage of the catchment is ridge/hilltop and what percentage is valley (see the cumulative area diagram in the next section to quantify this percentage). Note also that as the uplift rate increases (so the catchment becomes steeper) the area decreases. Willgoose et al. (1991c) provides a more general solution for A_t for when both processes are nonlinear functions of both area and slope.

One point that arises repeatedly in the pages that follow is what happens when (1) erosion rate decreases or (2) creep increases, with everything else unchanged. In these cases the transition area A_t in Equation (3.28) increases. If either creep rate or the erosion decrease, then the slope at the intersection point S_t will increase (Figure 3.5).

3.3.4 Cumulative Area Diagram

One of the key characteristics of catchments is that for the rounded hilltops near the catchment divide the catchment flowlines diverge. Moving downstream and entering the channel network, the flowlines converge. Much of the theoretical work for hillslopes in contrast assumes that flowlines downslope are parallel so that there is no flow divergence or convergence. A key challenge has been to characterise statistically the flow divergence and convergence properties of the flowlines on hillslopes.

The channel network in a catchment is generally convergent everywhere (exceptions include braided rivers and alluvial fans), creating the network of tributaries that the network classification methods in Section 3.3.1 characterise (historically by the mapping of blue lines on contour maps). With the advent of DEMs a finer resolution statistic was developed, the cumulative area diagram (CAD) (Figure 3.7). The CAD gives the percentage of the nodes in the DEM (y axis) that have an area draining through them greater than a given area (x axis). Normally (1) the areas on the horizontal axis are normalised by the maximum area in the DEM and (2) the analysis is done on individual catchments rather than DEMs as a whole. In the latter case ignoring catchment boundaries leads to many partial catchments, some of which will have area draining into them from off-DEM so the drainage areas cannot be calculated. Rodriguez-Iturbe et al. (1992a) found that there was a central log-log linear portion that had a slope of -0.5, and that this value has subsequently been confirmed by a number of authors to be an almost universal characteristic of catchments worldwide (e.g. de Vries et al., 1994; Moglen and Bras, 1995; Rodriguez-Iturbe and Rinaldo, 2001).

Perera and Willgoose (1998) showed that the linear central region can be derived as an asymptotic solution to the Tokunaga network classification scheme (Section 3.3.1), and that the slope of -0.5 follows when the network complies with the statistics of topologically random channel networks (Shreve, 1967). The importance of Perera and Willgoose's work is that it derives the relationship between drainage convergence properties and the slope of the CAD, and variations from -0.5 are indicative of systematic deviations in the convergence characteristics of the drainage network. For instance, the top left-hand corner of the CAD corresponds with the divergent hillslope, and the CAD shows a S-shaped region of varying slope with area that reflects the divergent flow pattern of

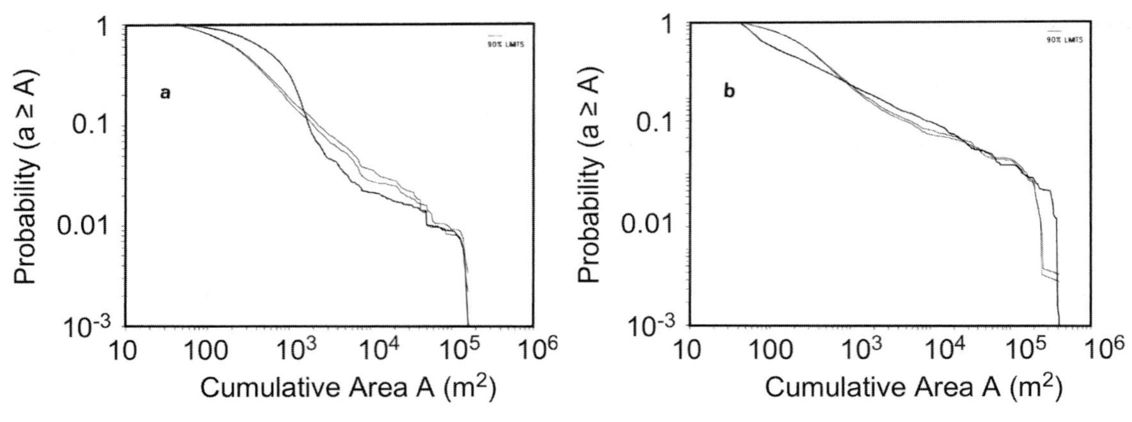

FIGURE 3.7: Cumulative area diagram (CAD) as derived from three different data sources for the same landscape as in Figures 3.5 and 3.9. The dark line is the CAD derived using a ground survey, while the finer lines are the 90% confidence limits from (a) the DEM was derived from interpolating a contour map, (b) the DEM was derived from digital stereo photogrammetry using stereo aerial photography (from Walker and Willgoose, 1999).

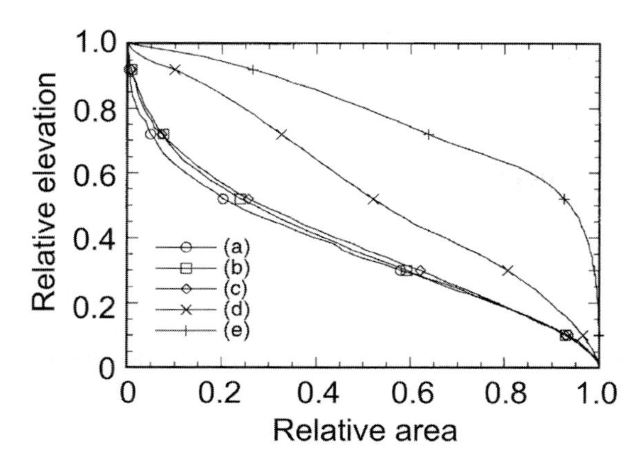

FIGURE 3.8: Hypsometric curve. The classical shape is indicated by curves (a)–(c). The different curves are for different ratios of diffusive erosion to fluvial erosion. Curve (a) is dominated by fluvial erosion, while curve (e) is dominated by diffusion, the ratio of diffusion to fluvial erosion increasing from (a) to (e). Key features are the knee at the bottom right due to internal hills and catchment divides in the lowlands of the catchment, the upward curvature in the top left for curves (a)–(c) due to fluvial erosion and the downward curvature everywhere in curve (e) due to the diffusion dominance (from Willgoose and Hancock, 1998).

the rounded tops of the hillslopes. The steep drop-off of the curve at the right-hand side is simply a reflection of boundary effects when pixel drainage areas become very close to the total area of the domain being analysed.

Moglen and Bras (1995) used the CAD to test the validity of a landform evolution model.

3.3.5 Hypsometric Curve

The hypsometric curve (Figure 3.8) characterises the proportion of a catchment that lies above a given elevation. It has long been used as an empirical measure of the distribution of landscape elevation (Strahler, 1952) but the linkage with catchment process was weak. It is possible to make some statements about the form of the hypsometric curve for catchments in dynamic equilibrium. The hypsometric integral is the area under the hypsometric curve.

In one dimension (i.e. a hillslope with parallel flow lines), the hillslope long profile is equivalent to the hypsometric curve, and the one-dimensional solutions of hillslope profile of Kirkby (1971) provide a direct link between the hillslope processes and hypsometric curve. A fluvial erosion-dominated hillslope with a concave up long profile will have a hypsometric curve that is concave up everywhere, while a creep-dominated hillslope with a concave down long profile will have a hypsometric curve that is concave down everywhere. For hillslopes where fluvial erosion and creep are both active, then the upper part of the hillslope is concave down and lower part is concave up (as discussed in Section 3.3.3), and the part of the hypsometric curve above the elevation of the inflection point in concavity of the long profile is concave down and the part below that elevation is concave up. One implication is that the exact shape of the hypsometric curve is a function of the length of the hillslope, with a shorter hillslope having a greater proportion of the hypsometric curve being concave down than a longer hillslope. This indicates that the shape of the hypsometric curve is scale dependent even in the absence of a change in process rates.

In two dimensions the shape of the catchment and the exact branching pattern of the network are important (Willgoose and Hancock, 1998). For long skinny catchments the hypsometric curve is similar to that for the one-dimensional hillslope discussed above. For more rounded in plan catchments where the channel network branches out in all directions, there is the appearance of a knee in the bottom right of the hypsometric curve for low elevations. This knee results from the small low internal hills that are formed in the central part of a catchment. While Willgoose and Hancock (1998) derived their results for transport-limited catchments, since their results depend only on the slope-area relationship, they are equally applicable for detachment-limited catchments.

The dependency of the hypsometric curve on catchment scale and the branching pattern of the channel network may suggest that it is not a good descriptor of geomorphology, but that is not the case. If the channel network is perfectly known (i.e. the flow directions at every node in the catchment are known), then the contributing area at every point can be determining simply by summing areas downstream from the source. Using the mean trend from the slope-area relationship we can deterministically reconstruct the elevations for the landscape working up from the elevation at the outlet, so that the hypsometric curve is a function of the channel network and the slope-area relationship. Moreover, we have found that the scatter in the hypsometric curve from catchment to catchment is small because it is an integrated measure of the catchment elevations. It is useful as an indirect test of the slope-area relationship. This reconstruction of the catchment elevations up from the outlet using the drainage pattern and the average slope-area curve is the basis of the QUEL model (Ibbitt et al., 1999; Willgoose, 2001) discussed in Section 3.3.3.

Finally we note some statistical limitations on the rescaling that is necessary to plot the hypsometric curve and the subsequent calculation of the hypsometric integral, which have implications for its reliability as a testing statistic. The use of relief for the vertical rescaling, while the standard method, is problematic because the highest elevation of the landscape is the highest peak, which may not be strongly related to the average elevation of the remainder of the landscape (see Section 2.3.3).

3.3.6 The Width Function

In its classical form the width function is a plot of the number of channels versus distance along the channel upstream from the catchment outlet. Historically it has been drawn by plotting contours of distance upstream from the outlet and counting the number of blue lines that crossed each contour, then plotting distance from the outlet versus the number of blue lines (Surkan, 1969). The main attraction of the width function is that it can be transformed into the unit hydrograph for the catchment. If the velocity of the flood wave within the catchment is constant (Section 3.1.7), then the distance from the outlet can be transformed into the travel time to the outlet. If the hillslope lengths are the same everywhere, then the number of channels is area/unit length of channel which when unit runoff is applied to that area is the volume of water contributed. Thus the width function characterises the runoff routing behaviour of the catchment. Kirkby (1976) used the TDRN random network model (Section 3.3.1) to generate catchment width functions and evaluated the effect of catchment runoff rate and time to peak.

Since the arrival of digital terrain analysis the width function has been generalised so that the width is the number of drainage paths, rather than the number of channels or blue lines. In this latter form the width function provides a high-resolution view of the drainage pattern of the catchment (both the channels and the hillslopes). However, in this form it is quite sensitive to random noise effects (e.g. errors in digital elevation models), so typically it requires some smoothing to identify the overall trend (Figure 3.9) (Walker and Willgoose, 1999). Rinaldo et al. (1995) found that the shape of the catchment and the shape of the width function are strongly linked and thus the unit hydrograph of the catchment is strongly linked to catchment shape.

Three statistics measured from the width function are commonly used. The maximum width is called the catchment width, the maximum length is the catchment length and the ratio of catchment width to length is the catchment roundness. All three are dependent on the extremes of the width function and so have significant sampling error and have proven to be relatively poor measures of catchment topological shape (see Section 2.3.3)

3.4 Conclusions

This section has provided a very brief overview of hydrology and geomorphology concepts that will arise later in this book. By necessity the discussion has been mathematically superficial because of the brevity of the treatment, and the presentation has concentrated on providing an intuition of how the processes work.

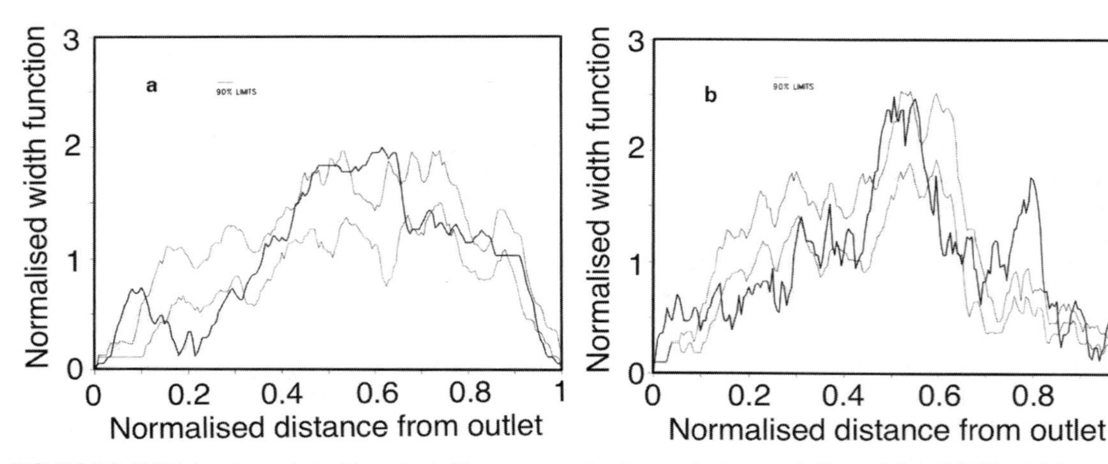

FIGURE 3.9: Width function as derived from three different sources for the same landscape as in Figures 3.5 and 3.7. The dark line is the best estimate of the width function derived using a ground survey, while the finer lines are the 90% confidence of width function from (a) the DEM was derived from interpolating a contour map, (b) the DEM was derived from digital stereo photogrammetry using stereo aerial photography (from Walker and Willgoose, 1999).

If we focus first on the hydroclimatology, the key common feature of all processes discussed here is that they all feed into the long-term behaviour of the soil or landscape. The discussion of extreme events is important because they are involved in the triggering and threshold behaviour of pedogenic or geomorphic events that have long-term consequences (e.g. chemical weathering, landslides). Mostly, however, the issues discussed are the same ones that hydrologists study as part of long-term catchment yield studies used in water supply assessment, and these issues likewise determine the rates of geomorphic processes once they are triggered. These processes are the runoff generation during rainfall events and the inter-event drying that then determines how wet the landscape is at the start of the next rainfall event. One of the most challenging aspects of these hydrology calculations, common to many of the calculations in this book, is that it is rarely feasible to directly model all the processes at the level of detail at which we would like to model them because (1) the computational demands are excessive and (2) the inputs for these detailed models are climate series that in general we don't know deterministically. In the latter case we would have to generate several parallel stochastic time series (e.g. rainfall and evaporation) that are generally autocorrelated in time and space, and cross-correlated between each other. In many cases hydrologists know that these correlations are important, but the correct modelling of them is problematic (e.g. correct modelling of spatially distributed rainfall time series with their combination of time-varying spatial and temporal correlations is still a work in progress), so there may be low confidence in the validity of hydrology time series modelled by this direct modelling approach, quite apart from the mathematical and computational complexity.

Accordingly, many existing landform evolution and soil evolution models use simplified conceptualisations of hydrological behaviour, commonly called 'effective parameterisations'. In the view of the author these effective parameterisations are likely to be more reliable for prediction (or at least more transparent about their reliability) because of their observed empirical basis. Combining these effective parameterisations with the physical intuition provided in the sections above should allow us to test the sensitivity of these effective parameterisations to changes in the environmental and climate drivers. The challenge then is finding the right balance of direct versus effective modelling. And, of course, this doesn't preclude further research aimed at elucidating the physical and stochastic drivers of these effective parameterisations.

The geomorphology topics briefly reviewed here are not inputs to LEMs but their outputs. They provide a framework within which to assess the overall geomorphology of LEMs. There are, of course, many other properties of landscapes (e.g. ones that assess site-specific characteristics such as regions of high risk of erosion when examining waste containment structures) that have been, or are being, used to assess both geomorphology and LEMs, but the ones discussed above form a set commonly used by researchers.

One thing to note about these geomorphology properties is that they are all statistical properties of the landform rather than deterministic descriptors. The evolving landscapes created by LEMs are in many cases sensitive to the exact details of unknowable inputs (e.g. the initial shape of the landform, climate history). If we explicitly model these unknowable inputs with random inputs (replicating their statistical characteristics), we find that the LEM-generated landscapes can be deterministically different (e.g. hills and valleys are in different locations) but that the statistical properties of the hills and valleys are the same. Geographers have a concept called 'uniqueness of place' to describe this (see most introductory physical geography texts, or page 5 in Beven, 2000 and Petersen et al., 2012 for a deeper discussion). If you imagine standing at two different lookouts in the Yorkshire Dales, then it is possible just by looking to say that both views are of the Yorkshire Dales, even though the exact details of the views are different. Thus the statistics of the views are the same (and somehow uniquely identify the view as 'Yorkshire Dales', and not, e.g. 'Colorado Rockies') while the deterministic details are different.

The follow-on question is 'How well does the statistic distinguish the "Yorkshire Dales" from the "Colorado Rockies"?' This is a question of the signal-to-noise in the statistic. The noise is how much of the deterministic detail exhibits itself in the statistical value, while the signal is how much of the generic Dales-iness or Rockies-iness characteristics are exhibited. The classic technique to reduce noise is to take many samples (i.e. lots of sites) and average the results. A good statistic is one that requires only a few samples to distinguish the sites. But it is also a function of how different the landscapes are. It would be a poor statistic that couldn't distinguish the Dales from the Rockies with a small number of sites, but a more challenging task would be to statistically distinguish the Yorkshire Dales from the North York Moors, adjacent regions that are distinguishable by eye but similar in geology and climate. Even if the reader is not familiar with the beauty of North Yorkshire,

I'm sure you can construct similar examples from your own locale.

When we discuss model testing these issues of signal-to-noise and what makes a good geomorphology statistic will become very important.

3.5 Further Reading

Many introductory hydrology textbooks cover parts of this chapter, though they do not emphasize the aspects of yield hydrology we are interested in. A good overall hydrology textbook which is full of sage practical advice is Beven (2012), though it concentrates on event modelling rather than the continuous yield modelling we are interested in. It also has a stronger emphasis on the hydrology of humid climates than arid climates. The review of Roe (2005) provides an excellent treatment of orographic rainfall. Ruddiman (2013) provides a good overview of paleoclimate and its reconstruction. Unfortunately there is no up-to-date reference that can be recommended for the modern geomorphology techniques discussed here. Abrahams (1984) provides a good review of the older work, but it lacks the recent insights into the effectiveness of these older statistics that has come from work in digital elevation model analysis and landform evolution modelling.

soils cemented by secondary minerals) and entrainment mechanisms.

It is convenient to define a third type of sediment transport that does not fall neatly under either of the definitions above:

3. Source-limited: This is where the rate of erosion over some period is limited by the ability of other processes (typically weathering) to produce transportable material for the flow to entrain. For example, in a highly seasonal monsoonal environment it is possible for the transportable material to be created during the dry season (e.g. rocks being transformed into sandy debris from weathering), and this new transportable material is a store of sediment ready to be transported at the start of the next wet season. Once this material is stripped away in the early part of the wet season (typically by transport-limited erosion), the store is depleted. At this stage the surface will be covered by a coarse armour that cannot be transported, so erosion stops or slows down. Under some circumstances the flow may become detachment-limited, but more typically either (1) the armour is too coarse to be transported even if its entrained in the flow (see the Shield's stress threshold in the following sections) or (2) the entrained material is so coarse that the transport capacity of the flow drops to a very low value given the coarse particle size grading that is entrained into the flow.

Finally, it should be noted that mass is transported out of a catchment by three main mechanisms. Their estimated global rates of delivery to the ocean (Syvitski et al., 2003) are (1) as suspended load within the flowing water column (18 Gt/year), (2) as bedload where the bed of the river itself is moving downstream (2 Gt/year) and (3) as dissolved load where it is dissolved in the water (5 Gt/year). A further 3.1 Gt/year is delivered to the oceans from ice, wind and coastal erosion. Given the relative magnitudes of the processes it is unsurprising that most research has concentrated on suspended sediment transport. The Syvitski et al. load estimates are based on sediment gauging stations, and so include anthropogenic effects and may not be typical of the long-term geologic history. Reliable long-term data on total sediment load (both suspended and bed load) is difficult to quantify, and catchment-scale data are scarce in the literature (Hancock et al., 2017a).

The sections below fill in the details of how the three transport-, detachment- and source-limited mechanisms work and the mathematics used to model them. The treatment below assumes that you have read Chapter 3.

4.2.1 Transport-Limited Erosion

4.2.1.1 Equations

The fundamental processes that determine the transport capacity of a flow have been studied for many years. Many specific forms of the sediment transport capacity have been developed, so we will focus here on the common fundamental features. The sediment transported by the flow increases when (1) the shear stress applied to the soil by the water at the soil-water interface increases, (2) the diameter of the transported sediment decreases and (3) the specific gravity of the sediment particles decreases. Moreover, there is a shear stress threshold, normally called the Shield's stress threshold, below which sediment cannot be entrained into the flow. The equation normally used to model transport-limited sediment transport in LEMs is a form that does not explicitly include shear stress

$$q_{sc} = KQ^{m_1}S^{n_1} \tag{4.10}$$

where K is the erodibility, m_1 and n_1 are parameters and the subscript c on q_{sc} indicates that this is the sediment transport capacity to distinguish it from the actual sediment transport q_s which in later sections in this chapter will not necessarily be equal to sediment transport capacity. The form of this equation has the advantage of having a direct and simple dependency on the discharge Q (and thus drainage area) and slope S, but hides many important aspects of the physics of sediment transport. Accordingly we will show how this equation can be obtained from a more fundamental representation of fluvial sediment transport.

The nondimensional sediment transport $\widehat{q_{sc}}$ is related to the nondimensional shear stress $\hat{\tau}$ by (Vanoni, 1975)

$$\widehat{q_{sc}} = K\hat{\tau}^p \tag{4.11}$$

where K and p are parameters, and the hat indicates the nondimensional quantities

$$\widehat{q_{sc}} = \frac{q_{sc}}{\rho_w g s F \sqrt{g(s-1){d_s}^3}}$$

$$\hat{\tau} = \frac{\tau_0}{\rho_w g(s-1)d_s} \tag{4.12}$$

where ρ_w is the density of water, g acceleration due to gravity, s the specific gravity of the sediment, q_s the sediment transport in mass/time/unit width of flow, τ_0 the bottom shear stress at the soil-water interface and d_s the representative diameter of the sediment particles (e.g. median or mean diameter). The parameter F is a value weakly dependent on the diameter of the sediment and is about 2/3. For a given sediment density and size,

Equations (4.11) and (4.12) can be simplified (noting that $\tau_0 = \rho_w gRS$) to

$$\begin{aligned} q_{sc} &= K'(RS)^p \\ K' &= \left(\frac{K}{(s-1)d_s}\right)^p \rho_w gsF\sqrt{g(s-1){d_s}^3} \end{aligned} \tag{4.13}$$

where R is the hydraulic radius of the flow, S is the slope of the flow and K' is a multiplicative constant dependent on sediment grading and density. Henderson (1966) indicates that $p \approx 3$ so sediment transport capacity increases with decreasing diameter and specific gravity of the sediment grains. To replace the dependency of Equation (4.13) on R with a dependency on Q we use the Manning equation for flow

$$Q = \frac{R^{5/3} S^{1/2} P}{n} \tag{4.14}$$

where P is the wetted perimeter of the flow, and n is the Manning roughness coefficient for the soil surface. Rearranging this equation we substitute for R in Equation (4.13):

$$\begin{aligned} q_{sc} &= K'' Q^{3p/5} S^{7p/10} \\ K'' &= \frac{K' n^{3p/5}}{P^{3p/5}} \end{aligned} \tag{4.15}$$

For $p = 3$ (the value for Einstein-Brown sediment transport) the equation becomes

$$q_{sc} = K'' Q^{1.8} S^{2.1} \tag{4.16}$$

If we expand the constant K'' and make the terms for specific gravity, grain diameter and soil surface roughness explicit, then

$$q_{sc} \propto \frac{1}{{d_s}^{1.5}} \frac{s}{(s-1)^{2.5}} n^{1.8} Q^{1.8} S^{2.1} \tag{4.17}$$

Equation (4.17) shows that the sediment flux increases if the sediment diameter decreases, the specific gravity decreases and the roughness increases. The trend with roughness appears at first to be anomalous, but it arises because as the roughness increases, the velocity decreases, and for a given discharge this means the depth of flow increases, leading to an increase in the bottom shear stress τ_0. These results are specific for the Einstein-Brown sediment transport equation, but the general trends with sediment characteristics and bed roughness are the same for many sediment transport equations. Finally, Equation (4.15) shows a dependency on the wetted perimeter of the flow P. Normally P increases with discharge, and regime equations (see Section 4.5.1) have shown a dependency of P on discharge of the form

$$P = K_w Q^\alpha \tag{4.18}$$

where K_w and α are coefficients. This results in the exponent on the discharge term in Equations (4.15) and (4.17) being reduced to $3p/5(1-\alpha) = 1.8(1-\alpha)$. Published regime equations based on field data suggest α is in the range 0.4–0.5 (Graf, 1984). For instance, for $\alpha = 0.5$ the exponent on discharge in Equation (4.17) is $0.50p = 0.9$.

One of the problems with the empirical equations for wetted perimeter like in Equation (4.18) is that they typically have a dependence solely on discharge, yet there should also be a dependence on slope. If discharge is fixed as the slope decreases, the depth of water in the channel will increase, so the wetted perimeter will also increase. An equation for P that includes slope can be derived using the Manning equation (following Julien, 2014). If the wetted perimeter of the channel is given by the equation (and hydraulic radius R follows because $R = A_{cs}/P$),

$$\begin{aligned} P &= \beta {A_{cs}}^\delta \\ R &= \beta^{\frac{-1}{\delta}} P^{\frac{1-\delta}{\delta}} \end{aligned} \tag{4.19}$$

where A_{cs} is the cross-sectional area. Substituting into the Manning equation (Equation (4.14)) yields

$$P = \left[\beta^{\left(\frac{5}{5-2\delta}\right)} n^{\left(\frac{3\delta}{5-2\delta}\right)}\right] Q^{\left(\frac{3\delta}{5-2\delta}\right)} S^{-\left(\frac{3\delta}{2(5-2\delta)}\right)} \tag{4.20}$$

If the geometry of the channel is such that $\delta = 0.5$ (e.g. triangular channel cross section), then this reduces to

$$P = \left[\beta^{1.25} n^{0.375}\right] Q^{0.375} S^{-0.188} \tag{4.21}$$

Julien (2014) presented an equation very similar to this (and tested it against field measurements for 835 sites) and which also included an empirical factor allowing for the d_{50} of the sediment (the grain diameter 50% of the mass is less than the median grain diameter by mass) in the channel. Equation (4.21) can be simplified if the log-log linear slope-catchment area relationship of Equations (3.14) and (3.16) is used with a typical observed scaling exponent on area (i.e. $S \sim A^{-0.5} \sim Q^{-0.5}$ where A is the catchment area). This allows us to eliminate the slope dependence and calculate an exponent on Q of 0.47 for Equation (4.18), which is within the range of empirical values in equations that ignore slope. However, this latter result will be true only when the catchment complies with the log-log linear slope-catchment area relationship, otherwise Equation (4.20) applies.

Most of the theory here was been explicitly developed for channels. However, it is also broadly true for hillslopes. Hillslopes are rarely perfectly flat, so flow concentrates into rivulets and the hydraulics of each of those rivulets is the same as the larger channels. The wetted perimeter increases as the discharge and the depth of flow increases until for extreme flows all the microtopography on the hillslope is submerged. The main difference between hillslopes and rivers is the importance and permanence of microtopography (e.g. leaf litter debris dams, animal hoof prints) on hillslopes, and the lack of a mobile bed on the hillslope so there is no bedload. The hydraulics and erosion characteristics of hillslopes have been extensively studied (Abrahams and Parsons, 1990, 1991a,b, 1994; Parsons et al., 1990, 1992, 1997; Willgoose and Kuczera, 1995; Willgoose and Sharmeen, 2006), and much of the fundamental physics of transport-limited fluvial erosion on hillslopes is the same as rivers.

However, there is an import caveat for hillslopes. All of the equations above were derived from experiments with sediments that were all one diameter. When the soil being eroded has a range of sizes, there is the possibility of selective entrainment of the finer fractions, and the creation of a coarser armour that resists erosion on the surface. Section 4.2.4 discusses this process in detail, but it is important to note here that long-term erosion can sort the hillslope surface such that steeper slopes are covered by coarser material than flatter slopes, so that relative to an unsorted hillslope (e.g. Equation (4.16)) for a sorted slope the dependence on slope is reduced. Willgoose and Riley (1998a) observed this for degraded mine waste slopes where the equation they fitted to data was

$$q_{sc} = 3.55Q^{1.42}S^{0.66} \tag{4.22}$$

where it can be seen that the slope dependency is significantly less than Equation (4.16). Subsequent field work to assess the surface grading on the slopes and the use of other erosion models (Evans and Loch, 1996) confirmed this behaviour. Further study by Sharmeen and Willgoose (2007) with their ARMOUR soil armouring model showed that if the physics of Equation (4.16) was allowed to evolve over time after 25 years (the age of the Willgoose and Riley experimental sites), the discharge and slope exponents were about those fitted in Equation (4.22), and that at 100 years the equations stabilised to Equation (4.10) where m_1 and n_1 were about 1–1.2 and 0.5–0.7, respectively, for two different contrasting spoils. This implies that the use of the unsorted sediment parameters in Equation (4.16) for hillslope erosion on old (more than a few years) hillslopes is inappropriate, and the user either needs to either (1) measure surface armour diameter versus area and slope in the field and include this in the model, (2) use generalised results for diameter, area and slope (see Chapter 7; Cohen et al., 2009, 2010; Welivitiya, 2017) as part of the erosion model or (3) use an equation with parameters as in Equation (4.22) while being mindful that Equation (4.22) was derived for only two sites. The observed short response time (25 years) of a few years of changes in m_1 and n_1 to the equilibrium between armour and erosion suggests a strong dynamical response to short-term climate variability, which has also been observed in our coupled soilscape-landscape evolution model SSSPAM (Welivitiya, 2017).

Before we begin to use these equations, it should be noted that there are many equations for sediment transport. They vary in their rates and their dependency on discharge, slope and sediment characteristics. However, they are all fundamentally based on underlying principles similar to those outlined here. There is considerable scatter in the predictions of these equations. Without calibration to field or experimental data it is rare for any of these sediment transport equations to be able to predict sediment transport rates to better than a factor of 2 or 3 (e.g. Hancock et al., 2015a). Thus the sediment transport models used in professional practice all have a strong empirical basis, underpinned by extensive field data collection aimed at reducing this predictive uncertainty.

4.2.1.2 *Time and Space Variation*

The equations above give the sediment transport capacity at a given point in space and time. During a rainfall event the discharge will be varying on a second/minute/hourly time scale. We typically cannot resolve this temporal resolution since we run LEMs for years to millennia. However, the discussion in the previous section suggests that accurate modelling requires (1) modelling the evolution of landforms over thousands of years at minute resolution and (2) the need to be able to estimate the water and sediment discharges at minute resolution may involve rainfall prediction at minute resolution. Even if we can compute the erosion loss at minute resolution, the generation of a minute resolution rainfall series that has the correct correlation structure in space to drive a runoff generation model is beyond the state-of-the-art. We return to this difficult challenge later in this chapter after we have discussed the other types of erosion mechanisms, because it is a challenge common to all erosion mechanisms.

The next section deals with question of what happens as the sediment transport varies in space (i.e. downstream along a drainage path).

4.2.1.3 *Entrainment- and Deposition-Limited Systems*

The discussion above suggests that when the flow has a transport capacity that is more than the amount of sediment in the flow, then the flow erodes to instantaneously make up the deficit. Likewise when the transport capacity is less than the amount of sediment in the flow, sediment is deposited. However, the flow cannot instantaneously entrain or deposit all of the sediment required, so there is a limitation on the rate at which the transport capacity can be satisfied when the transport capacity is changing downslope.

Deposition limitation is where the rate of deposition is limited by the fall rate of particles. For example, let's assume that the sediment is equally distributed through the depth of the flow with concentration c (in units of mass/volume of water, e.g. mg/l). We will also initially assume that all the particles have the same fall velocity in water w. Then the maximum possible rate of deposition per unit bed area is $D = wc$, where D is the deposition rate in mass/unit time/unit area. If we assume that during the time that the sediment is being deposited that no sediment is also being entrained, then this means that the distance the flow travels to deposit a mass of sediment ΔD is

$$L = \frac{v\Delta D}{wc} \tag{4.23}$$

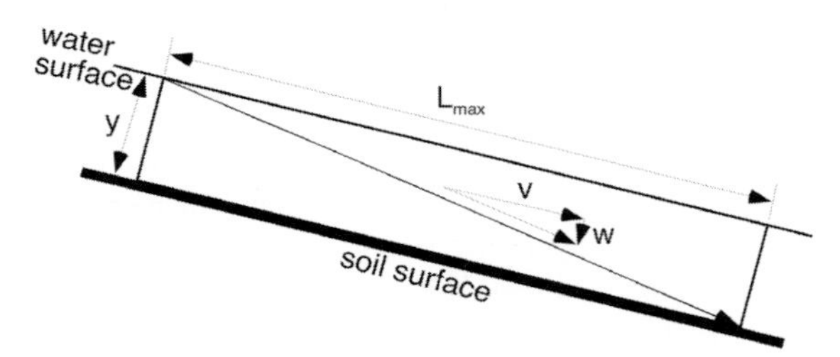

FIGURE 4.3: Deposition of particles showing how a particle that starts at the water surface settles linearly with time and distance downslope. L_{max} is the longest distance that a particle can travel, and defines the maximum settling distance for a particle with settling velocity w in a flow with velocity v.

and the distance that the flow travels to drop all sediment (e.g. in a purely depositional environment like an alluvial fan) is $L_{\max} = {}^{vy}/_{w}$ (Figure 4.3). In a numerical model, if L and $L_{\max} = {}^{vy}/_{w}$ are less than the grid resolution, then from the viewpoint of the modelling, the deposition is all inside the model pixel (i.e. it is a subgrid process) and deposition is not deposition-limited. If L is longer than the grid resolution, then this process may need to be modelled explicitly (e.g. a large alluvial fan with a coarse spatial discretisation).

A model implementation issue is illustrated in Figure 4.4. Assume that the sediment concentration profile at point A is well mixed vertically. If all sediment settles with settling velocity w, then at the downstream end of the pixel A-B in Figure 4.4a the top of the flow down to depth $w\Delta t$ will contain no sediment. Accordingly the concentration profile entering pixel B-D is no longer well mixed but is clear at the top, and the deposition rate at the bottom of the flow is unchanged from pixel A-B since the concentration at the bottom of the flow does not change. However, if there is vertical mixing by turbulence (Figure 4.4b), then the concentration profile at C will be the same through the profile, but the concentration will have decreased by the factor $\left((D_f - w\Delta t)/D_f\right)$ where D_f is the depth of the flow, and as a result the deposition rate in pixel C-D is reduced from that occurring in node A-C. Typically models don't track the concentration profile and implicitly assume that the profile is always well mixed at all times (i.e. Figure 4.4b). For Figure 4.4a the deposition downstream will be a constant deposition rate until we reach point E, and then deposition stops because all sediment has been removed from the flow. For Figure 4.4b the deposition downstream will initially be the same rate as Figure 4.4a but will exponentially decline with distance downstream.

There are a number of complications of this simple conceptualisation. The concentration profile is not

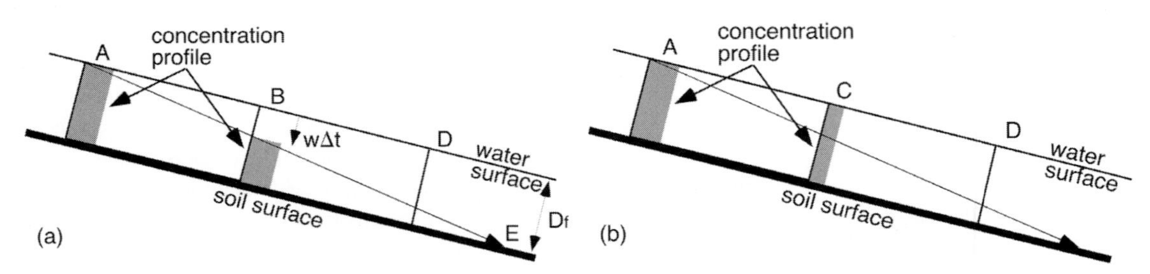

FIGURE 4.4: Concentration profile of sediment down the profile in a purely depositional environment. (a) At point A the profile is fully mixed from top to bottom. At point B after time Δt the top of the profile down to $w\Delta t$ is clear of sediment since the whole profile has settled by $w\Delta t$. (b) However, if point B in (a) is the edge of one node and the entry to another, it is common to consider the profile well mixed at entry to the next downstream node so that the profile at C is what the model actually assumes.

normally the same through the profile, but rather the concentration is higher at the bottom of the flow, and is commonly assumed to be exponentially increasing with depth (Vanoni, 1975). In that case the concentration in Equation (4.23) is the concentration of sediment in the flow at the soil-water interface, and the rate of deposition will initially be fast and decrease with time as the concentration at the interface decreases with time. Finally the fall velocity of the sediment particle in the river is not the same as in a still water settling column in the laboratory because of the effect of turbulence (Dietrich, 1982; Boilatt and Graf, 1982; Willgoose, 1997).

The application to a mixed sediment load is conceptually straightforward with larger diameter or denser particles with higher settling velocity settling out faster than smaller less dense particles. Thus what happens with a flow that is carrying a mixed sediment load is that the deposition process not only reduces the load in the flow, but it also preferentially reduces the proportion of coarse dense particles in the mixture because they settle out faster.

We now turn to the altogether more difficult issue of entrainment limitation. As with deposition, entrainment of sediment does not happen instantaneously. In contrast to deposition there is no generally applicable physically based means to calculate the rate of entrainment. However, at transport capacity sediment is depositing from the flow as discussed above, and in Equation (4.23) and Figure 4.5, and this loss of sediment from the flow is exactly balanced by new sediment entrained into the flow. If we assume that the entrainment process is solely driven by the shear stress of the flow, then the rate of entrainment for a given discharge q, which has a corresponding sediment discharge q_s and sediment concentration c in the flow ($c = q_s/q$), is equal to the rate of deposition of sediment when the sediment load is at equilibrium, and this rate is independent of the concentration of sediment in the flow at that current time (which in general is not the equilibrium concentration) (Figure 4.5). Accordingly

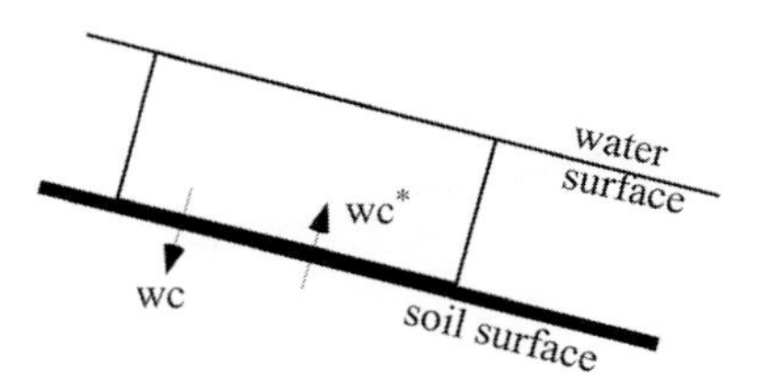

FIGURE 4.5: Schematic showing the (im)balance between the rate of deposition from the bottom of the profile, wc, and the rate of entrainment of sediment, wc^*.

after a change in the sediment transport capacity (with a consequent change in c^*)

$$\begin{aligned} \text{entrainment} &= wc^* \\ \text{deposition} &= wc \\ y\frac{dc}{dt} &= wc^* - wc = w(c^* - c) \end{aligned} \tag{4.24}$$

where y is the depth of the flow, c^* is the new equilibrium concentration and c is the concentration in the flow at any time after the change in transport capacity. Solving this differential equation yields the solution for how concentration (and thus the sediment load) in the flow changes after the change in transport capacity

$$c = c^* + (c_0 - c^*)e^{-wt/y} \tag{4.25}$$

where c_0 is the concentration before the change in transport capacity. This shows that the concentration asymptotically approaches the new equilibrium concentration. Our interest is to determine the net erosion or deposition (in units of mass/unit time/unit plan area). If entrainment is defined positive and deposition negative, then

$$\text{net entrainment} = w(c^* - c) = w(c^* - c_0)e^{-wt/y} \tag{4.26}$$

which is the rate of erosion or deposition downstream of the change in sediment transport capacity. The time scale of the change is $t = y/w$, which can be used to calculate the distance downstream, L, the adjustment propagates, $L = vt = vy/w$, where v is the velocity of the flow. If the velocity of flow is high, the depth of the flow is high, or the settling velocity is small, then the distance of adjustment is large.

Equation (4.24) can be rearranged to show how the rate of detachment changes with time (noting that $q_s = cq$)

$$\begin{aligned} \text{rate of detachment} &= y\frac{dq_s}{dt} = w(q_{sc} - q_s) \\ &= wq_{sc}\left(1 - \frac{q_s}{q_{sc}}\right) \end{aligned} \tag{4.27}$$

This form of the equation emphasizes that the rate of detachment is a function of the difference between the actual sediment transport and the sediment transport capacity.

Some example time scales are (1) for a 0.1 mm fine sand particle (settling velocity is about 0.01 m/s) in a 1 m deep flow (e.g. a channel) the time scale is just over 1 day and (2) for a 10 mm flow (e.g. a hillslope) the corresponding time scale is about 20 minutes. In both examples the time and the distance that will be travelled during that time by the flow is large. Finer sediments will have even longer time scales. It is thus reasonable to assert, given how variable sediment transport capacity is downslope, that it will be rare for the sediment concentrations in the flow to achieve the sediment transport capacity, though, of course, the details will depend on the sediment, downstream long profile and flow hydraulics.

Equation (4.27) might appear at first glance to be in conflict with Equation (4.26), but Equation (4.26) has been derived for the case where there is no entrainment and the deposition rate is given by the initial rate of deposition, and so has been derived for a simpler scenario.

It is possible to construct both the transport-limited and the detachment-limited sediment transport equations from this conceptualisation. In fact, this approach was how the Einstein (1950) transport-limited sediment transport equation was originally formulated (e.g. Henderson, 1966). Davy and Lague (2009) explored the implications of this approach in the field. They found that the main advantage of this approach is that the transition from transport- to detachment-limited transport was more gradual and less abrupt, which may also explain why researchers have failed to find an abrupt change in the field.

4.2.2 Detachment-Limited Erosion and Bedrock Incision

Detachment-limited erosion occurs when the constraint on sediment transport is the ability to detach particles from the surface being eroded. In contrast with transport-limited erosion, once detachment occurs, there is no constraint on the ability of the flow to transport sediment downstream. These models are sometimes called incision models because they only downcut and cannot deposit sediment. Howard and Kerby (1983) and Howard (1994) defined a detachment-limited erosion model for rivers flowing over rock or regolith

$$\frac{\partial z}{\partial t} = -\frac{\partial q_s}{\partial x} = \begin{cases} -K_t(\tau_0 - \tau_c) & \text{for } \tau_0 > \tau_c \\ 0 & \text{for } \tau_0 \leq \tau_c \end{cases} \tag{4.28}$$

where it is the incision rate rather than the sediment transport rate that is defined explicitly, K_t is the incision rate parameter and τ_c is the critical shear stress threshold below which sediment cannot be detached from the bed and incision does not occur. This equation is defined for a river, and so has a downstream direction x but no cross-stream component (Chapter 2). This can be converted into a relationship dependent on discharge and slope using the same methodology used for the transport-limited erosion equations so that

$$\frac{\partial z}{\partial t} = -K_t(K_z Q^{m_2} S^{n_2} - \tau_c) \tag{4.29}$$

where $K_z = \left(\frac{\rho_w g n^{3/5}}{K_w^{3/5}}\right)$, $m_2 = 0.6(1-\alpha)$, $n_2 = 0.7$ and K_w and α are defined in Equation (4.18). Tomkin et al. (2003) suggested that $m_2 = n_2$ and are in the range 2/3–5/3. This equation of incision as a function of discharge (or area) and slope has become known as the stream power equation (e.g. Sklar and Dietrich, 1998), and should not be confused with the unit stream power equation for sediment transport capacity of Yang (1973).

Now we can note that the detachment-limited equation is not incompatible with the transport-limited equation since, as is clear in Equation (4.29), both mechanisms must comply with the mass balance equation. A simple form of the combination of transport- and detachment-limited transport can then be written as

$$\frac{\partial z}{\partial t} = -\frac{\partial q_s}{\partial x} = \begin{cases} -\frac{\partial q_{sc}}{\partial x} & \text{when } D_t > \frac{\partial q_{sc}}{\partial x} \\ -D_t & \text{when } D_t \leq \frac{\partial q_{sc}}{\partial x} \end{cases} \tag{4.30}$$

where

$$D_t = \begin{cases} K_t(\tau_0 - \tau_c) & \text{for } \tau_0 > \tau_c \\ 0 & \text{for } \tau_0 \leq \tau_c \end{cases} \tag{4.31}$$

This is not a comprehensive form for the equations because it assumes in the transport capacity term in Equation (4.30) that the upstream flow is also at transport capacity but regardless it captures the first-order idea that detachment from the surface may limit the sediment that can potentially be transported by the flow. This limitation is in addition to any entrainment limitation discussed in the transport-limited erosion section, which is specific to the potential effects of downstream changes in transport capacity.

Also note that while Equation (4.30) has been written using the detachment-limitation formulation of Howard, the equation is easily generalised to any type of detachment-limitation D_t by substituting a different equation for D_t (e.g. rainsplash).

A slightly different model of detachment limitation used in the WEPP agricultural erosion model (and has thus been optimised for hillslope erosion rather than channel erosion) combines the transport capacity of the flow with the detachment capacity of the shear stress of the flow (Nearing et al., 1989)

$$\begin{aligned} D_f &= -D_c\left(1 - {}^{q_s}/_{q_{sc}}\right) \\ D_c &= K_r(\tau_0 - \tau_c) \end{aligned} \tag{4.32}$$

where D_f is the detachment rate for the flow (equivalent to $-D_t$ in Equation (4.30)), K_r is the soil erodibility of the flow, and noting that the sign convention here is that D_f is negative where the surface is eroding for consistency with Equation (4.30) (but of opposite sign to that used in Nearing et al., 1989) and D_c is the detachment capacity for a clear flow with no sediment. The major difference between Equation (4.30) and (4.32) is that in Equation (4.32) the amount of detachment is a function of the difference between the actual sediment transport rate and capacity, while Equation (4.30) has a detachment rate that is a constant rate independent of the amount of sediment in the flow already. The formulation in Equation (4.32) is conceptually similar to the entrainment limitation for transport-limited erosion in Equation (4.25). Though the rate constants may be different, in both cases the rate of detachment is modified by the difference between actual sediment transport and the sediment transport capacity.

The main weakness of the detachment-limited erosion equation dependent on shear stress as formulated by Howard for bedrock rivers is that it was developed empirically from fieldwork at a small number of sites, and it has lacked a fundamental process-based cause and effect. A large number of studies have tested the model in the field, and Lague (2014) provides a comprehensive summary of these works. A major research effort has gone into developing process-based models based on impact dynamics (the effect of impacts between particles in the flow and particles on the bed at the water-soil interface) to give Howard's shear stress model, in addition to its conceptual appeal, a stronger physical basis (e.g. Whipple et al., 2000).

The two mechanisms responsible for incision are (Lamb et al., 2008, 2016) (1) rock particles, which are plucked from the bedrock surface by the flowing water, and (2) sediment in the flow, which impacts the bedrock surface and detaches fragments from the surface (the 'ballistics model'). It is generally accepted that mechanism (1), on its own, cannot generate the rates of incision that are observed in the field. Thus most recent research has been on the second mechanism, the ballistics model.

There is a long history of research into erosion of surfaces resulting from ballistics outside the geomorphology community (e.g. the erosion of pipelines carrying particulate load). In the context of bedrock river erosion, Sklar and Dietrich (2004) summarise the fundamentals of the ballistics model. The erosion by ballistics consists of two components: (1) the volume eroded per impact of a sediment particle with the bed, V_s, and (2) the number of impacts per unit time, the impact rate, I_r. Sklar and Dietrich derive the ballistics transport equation assuming all sediment is one size and density.

The volume eroded per impact is a function of the kinetic energy of the sediment particle and how much of that kinetic energy is transferred to the bed surface, and the ability of the bedrock to absorb energy before fracturing under tensile stress so that

$$V_s = \frac{M_p {w_z}^2 E}{k_v {\sigma_T}^2} \tag{4.33}$$

where M_p is the mass of the sediment particles impacting the surface, w_z is the vertical component of the particle velocity, E is the Young's modulus for the bedrock matrix, σ_T is the tensile strength of the bedrock matrix and k_v is an empirical parameter that relates the rate of particle generation relative to the energy per unit volume stored by the bedrock at the moment of tensile failure.

The impact rate is

$$I_r = \frac{q_s}{M_p L_s} \tag{4.34}$$

where L_s is the average hop length for the particle, the average distance between impacts with the stream bed by a sediment particle.

These equations yield the rate of incision as a result of ballistic impact

$$\frac{\partial z}{\partial t} = V_s I_r = \left(\frac{{w_s}^2 E}{k_v {\sigma_T}^2 L_s} \right) q_s = K_b q_s \tag{4.35}$$

where K_b is a rate constant for the ballistics impact process and is a function of sediment characteristics (w_s), bedrock characteristics (k_v, E, σ_T) and sediment-flow interactions (L_s).

Returning to Howard's shear incision equation one of the first improvements to Howard's model was the 'tools' model (Sklar and Dietrich, 2004, 2012). The idea of this mechanism is that even for detachment-limited transport, sediment is being carried in the flow (but the rate is less than the sediment transport capacity) and is being exchanged between suspension/saltation and the bed. The cover of the bed by sediment is a function of the sediment being transported: low at low transport rates and high at high transport rates, peaking at 1.0 when the transport capacity limit is reached. At low sediment loads the number of particles (the 'tools') impacting the bedrock surface is low even though the amount of bedrock exposed is high and incision rates are low. At high sediment loads, near the transport capacity limit, the bed is mostly covered by sediment, so while there are plentiful particles/tools, they rarely impact exposed bedrock on the channel bottom because it is covered by sediment most of the time. The maximum incision rate occurs at an intermediate load when there are plentiful sediment particles to impact the bedrock and the bed is not always covered by sediment.

The cover effect is modelled by a linear function of the actual sediment transport divided by the sediment transport capacity

$$F_e = \left(1 - {}^{q_s}/_{q_{sc}}\right)^{\eta} \tag{4.36}$$

where F_e is the fraction of exposed bedrock, and Sklar and Dietrich assumed $\eta = 1$. Sklar and Dietrich assumed that incision occurs only when the bedrock is exposed, so F_e is a multiplicative factor on the incision rule (e.g. Equation (4.35)). Chatanantavet and Parker (2009) compared Equation (4.36) against river abrasion data and found the best fit was for $\eta = 1$.

Combining Equations (4.35) and (4.36) yields the incision rate in the presence of the cover effect

$$\frac{\partial z}{\partial t} = K_b q_s \left(1 - {}^{q_s}/_{q_{sc}}\right) \tag{4.37}$$

when $\eta = 1$. This shows that when $q_s = 0$ and $q_s = q_{sc}$ the incision rate is zero and the incision rate is maximum when $q_s = 0.5q_{sc}$. Note that this equation is fundamentally different from Howard's Equation (4.28) because of the dependence on the sediment flux, which involves an integration of the incision/erosion rates for the entire upstream catchment, while Howard's equation depends only on the flow rate from upstream. Building on Sklar and Dietrich's work there have been other models for the cover effect (e.g. Lamb et al., 2008; Chatanantavet and Parker, 2009, 2011), experiments to better characterise the empirical parameters for bedrock erosion (Sklar and Dietrich, 2001), experiments to characterise particle fragmentation characteristics leading to different functional forms for Equation (4.35) and data on the change in cover and relative magnitude of tools versus cover with sediment supply (Nelson et al., 2009; Turowski and Rickenmann, 2009). All maintain the underlying relationship between sediment load and transport capacity, with a midrange high incision rate.

An unresolved problem with both the shear stress detachment and the tools-cover models is that they do not appear to be able to reproduce the log-log linear slope-area relationship of catchments (Section 3.3.3). Tomkin et al. (2003) evaluated several variants of the shear stress model against field data using slope-area profile data and found that none of the models fitted the observations. They assumed that discharge was proportional to area. They attribute the differences to incorrect representation of the discharge relationship with area, and their results were consistent with a power law relationship with power less than 1 (e.g. Equation (3.10)).

Whipple and Tucker (2002) and Sklar and Dietrich (2008) explored the slope-area properties of rivers subjected to their model. They found that the scaling exponent α in the slope-area relationship decreased with increasing area so that the log-log slope-area relationship was concave upward rather than linear. Using a sensitivity study, they showed that this change in α with area could potentially be offset by orographic changes in rainfall, downstream fining and changes in the bedload grading. However, it would appear that a number of compensating processes need to be 'just right' to generate the log-log linear slope-area relationship. For this to be broadly true across many catchments suggests that there is a self-organisation principle driving this balance. This begs the question of what the principle is and why the log-log linear slope-area relationship is central to this perfect balance (e.g. Rodriguez-Iturbe and Rinaldo, 2001).

Another challenge in using the 'tools' model is that the impact energy is a function of the mass of the particles, so there is a need to estimate the grading of the material being transported. In general this material will be material from the hillslopes, since if erosion is the same everywhere, then almost 100% of the material in the channel must have come from the hillslopes. Thus it appears that the only way to parameterise the in-channel grading is to couple the channel model with the hillslopes, and for the hillslope model to be able to predict the grading of the material it delivers to the channel (e.g. Michaelides and Singer, 2014). This will likely involve coupling with a soils model much like will be discussed in Chapter 7, where the grading is a function of the in-profile soil weathering processes breaking down the saprolite into smaller fragments with time. Perhaps some of the grading-slope-area relationships derived by Cohen et al. (2009, 2010, 2013) and Welivitiya et al. (2016) may be useful effective parameterisations of this process. Sklar and Marshall (2016) have proposed a model based on empirical data for soil grading from landslide deposits. These works suggest that as incision rate increases (and erosion rate on the hillslope consequently increases), then the grading of the material delivered to the channel will coarsen. Welivitiya (2017) found that as the hillslope erosion rate increased, the grading of the hillslope surface, and thus the eroded sediment, also increased because the resultant thinner soils were subjected to less weathering. Brocard et al. (2016) found that the fine nature of the eroded hillslope deposits in Puerto Rico (due to the wet humid climate enhancing soil weathering) resulted in a low rate of downcutting in a bedrock channel because of the lack of abrasive power of the fine sediments. To date

achieve this balance, and the balance is a function solely of the soilscape evolution models to be described later in this book.

Finally, conceptual models for armouring are also possible. A simple model employed within the SIBERIA LEM uses a reduction in erodibility that is a function of cumulative erosion since the start of the simulation (cumulative erosion being a proxy for the amount of fines stripped from the surface) to estimate the erosion rate reduction. Hancock et al. (2017c) demonstrated that this approach works well for a mine trial plot site.

4.2.4.2 *Particle Density and Enrichment*

One aspect of sediment transport that is unique to hillslope erosion is that many of the particles that are transported are soil aggregates. As discussed in Section 10.6.1 soil organic matter (mostly exudates from plant roots and fungal hyphae), clay, fine roots and fungal hyphae combine to form larger particles called water-stable aggregates. These aggregates are also less dense (~1,200–1,300 kg/m^3) than disaggregated (also referred to as 'dispersed') soil particles (~2,500–2,700 kg/m^3). Aggregates are enriched in soil organic carbon so that erosion of these is an important component of carbon loss from soils. Examining Equation (4.13) it can be seen that depending on the relative impact of increased particle size (reducing particle transport rate) and decreased density (increasing particle transport rate) these aggregates may be more or less transportable than the dispersed soil particles. Using Equation (4.13) the transport rate for soil aggregates relative to dispersed soil is

$$\frac{q_{sA}}{q_{sD}} = \left(\frac{d_{sA}}{d_{sD}}\right)^{1.5-p} \frac{s_A}{s_D} \left(\frac{s_A - 1}{s_D - 1}\right)^{0.5-p} \tag{4.42}$$

where the notation is the same as in Equation (4.13) except that the subscripts A and D indicate the aggregates and dispersed particles, respectively. Figure 4.6 plots Equation (4.42) for $p = 3$ to show the sensitivity of the sediment transport rate to the diameter and density of the aggregates, and it shows how sensitive the transport capacity of the aggregates is to the density of the aggregates. The size range used for the diameter axis uses aggregates of 2.5 mm diameter, and the dispersed soil ranges from silt to sand. Figure 4.6 shows that for most of the range of values of diameter and density, the water stable aggregates are more transportable than the dispersed particles so that the contents of the aggregates (and thus soil organic matter) will be more concentrated in the flow than the soil surface. The ratio of the concentration (mass of the constituent per unit mass of sediment) in the flow relative to the concentration in the surface is called the 'enrichment factor'.

4.2.5 The Transition from Detachment-Limited to Transport-Limited

Howard (1998) discusses what happens in the field when rivers transition from transport-limited to detachment-limited and vice versa, and how transport and detachment

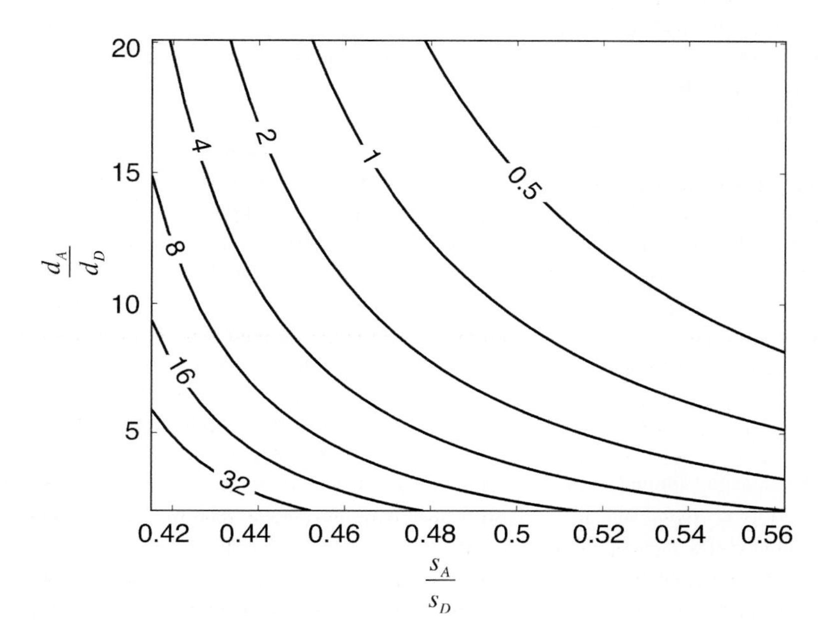

FIGURE 4.6: Contours of relative mobility showing the relative entrainment of sediment for varying diameter (vertical axis) and varying specific gravity (horizontal axis). The top right corner is heavy large particles, and the bottom left corner is light small particles.

limitation can coexist in the same stream. Patches of sediment cover the streambed, and between the patches the bare bedrock is exposed. The bare rock will be incised by the flow while the sediment patches protect the bedrock (and presumably result from either localised deposition or waves of sediment bedforms moving down the stream). The similarity with the discussion in Section 4.2.2 regarding the tools and cover model for bedrock rivers should be clear, though in this section the sediment patches simply cover the underlying bedrock and do not contribute to bedrock incision.

The key issue then is determining what percentage of the bed, p, is covered by sediment. If the patches of sediment occur randomly in space, then a time-space substitution is possible, and we can then say that $(1-p)$ is the proportion of the time that the bed is detachment-limited, while the remainder of the time the sediment transport is transport capacity-limited so that the lowering rate of the stream bed then becomes (from Equations (4.10) and (4.29))

$$\frac{\partial z}{\partial t} = -(1-p)K_t(K_z Q^{m_2} S^{n_2} - \tau_c) \tag{4.43}$$

and the actual sediment transport averaged over time is

$$q_{\mathrm{sa}} = \iint\limits_{A_c} (1-p)K_t(K_z Q^{m_2} S^{n_2} - \tau_c)dA + \iint\limits_{A-A_c} (1-p)E dA + pKQ^{m_1} S^{n_1} \tag{4.44}$$

where the first integral is the lowering/erosion over the area of channel during the period of bedrock erosion, the second integral is the erosion over the hillslopes in the period of bedrock erosion (E is the hillslope erosion rate) and the third term is the sediment transport capacity during the period between bedrock erosion events. Note that $(1-p)$ is inside the integral as, in general, p will vary spatially. At first sight it would appear that p is free and cannot be determined, but this is not the case. First, note that by definition

$$\iint\limits_{A_c} (1-p)K_t(K_z Q^{m_2} S^{n_2} - \tau_c)dA + \iint\limits_{A-A_c} (1-p)E dA < KQ^{m_1} S^{n_1} \tag{4.45}$$

otherwise the catchment will be transport-limited. Both the detachment-limited and the transport-limited (the left-hand and right-hand side of Equation (4.45), respectively) terms do not change in response to changes in the upstream geomorphology. Consider now what happens if the amount eroded on the hillslopes upstream increases (and assuming none of this hillslope material is deposited upstream). The amount of sediment exiting the catchment (i.e. Equation (4.44)) must also increase. Since the transport terms are fixed, the only way this can occur is if p increases (since if p increases the sediment load increases, as a result of Equation (4.45)) and an increasing proportion of the time is being spent in the transport-limited condition.

The pure detachment-limited equation (Equation (4.29)) applies no matter what the upstream transport is, provided only that the sediment load is less than the transport capacity. Accordingly it must be true when there is no hillslope erosion and the only incision is occurring in the stream, and this is the case for $p = 0$. As hillslope erosion increases then p increases until $p = 1$. If the hillslope erosion continues to increase, then deposition must occur upstream because the stream sediment transport capacity is insufficient to transport all the sediment out of the catchment. Furthermore by the time-space substitution, this argument also applies for the proportion of the bed covered by sediment with it increasing from $p = 0$ (for no hillslope erosion) up to $p = 1$. The relationship between the hillslope erosion rate and p is linear. We thus conclude that the value of p follows directly from the ratio of hillslope erosion rate relative to the channel transport capacity. This linear variation is consistent with experiments that have specifically examined the variation of channel coverage with bedforms versus changing sediment inputs into the flume (e.g. Nelson et al., 2009).

One intriguing consequence of this relationship is that if the hillslopes steepen and erosion subsequently increases, then p will increase and the average rate of incision in the channel will decrease (Equation (4.43)), tending to flatten the hillslopes. In many cases it is active channel incision that destabilises or steepens the hillslope leading to high hillslope erosion, so a reduction in the incision rate will result in a reduced hillslope erosion rate. This negative feedback means that over time the channel incision rate and the hillslope erosion rate will reach an equilibrium, where the rate of downcutting of the river equals the rate of erosion from the hillslopes (Kirkby and Willgoose, 2005). However, Gasparini et al. (2007) used a LEM to show that the transient behaviour before equilibrium is reached and where equilibrium is disturbed by changes in external forcings (e.g. climate, tectonics) of the landforms where both detachment- and transport-limited are operating can be quite complex. The complexity arises because the incision by detachment-limitation upstream drives what parts of the downstream reaches are transport-limited.

The derivations above do not include the tools-cover model extension to the detachment-limited model, but an extension to this case can be done using the same approach.

4.2.6 Weathering-Limited Erosion

The third type of sediment transport limitation is where the source of transportable sediment provides a limit on how much sediment can be transported. The classic case of this is where weathering breaks down material at a rate that is less than the sediment transport capacity. In this case the surface of the soil is covered by a coarse armour that is too coarse to be transported, but weathering fragmentation creates smaller particles that are fine enough to be transported. In Chapter 7 a range of potential fragmentation mechanisms will be discussed.

Sharmeen and Willgoose (2006) discussed a case where this occurs. They studied a mine site consisting of a rapidly weathering waste rock from a mine. In the monsoonal environment they found that during the dry season the rock fragmented from physical weathering due to magnesium sulphate precipitation in micro-cracks in rock fragments, and then during the wet season the fine fragmentation products were transported during the early part of the wet season. The erosion didn't stop completely when the fine particles were depleted, but the decline in transport was quite rapid. They coupled a fragmentation model, as will be discussed in Chapter 7, with an erosion and armouring model based on the hiding model of Parker and Klingeman (1982). By varying the rate of weathering they showed a transition between transport limitation and weathering limitation (Figure 4.7). When the fines generated by weathering were dramatically smaller in size than the armour (e.g. spalling-like behaviour), then the transition between limitation modes for changes in the weathering/erosion rates is abrupt (type I weathering limitation in Figure 4.7), but where the weathering products are similar to the size of the original particles (e.g. when particles break in half), then there is a more gradual transition (type II weathering in Figure 4.7). Type II weathering limitation occurs because the weathering results only in a gradual change in transportability of the material so that the change in the transport rate is more driven by the gradual change in the grading of the surface as a result of weathering. In contrast, for type I weathering limitation there is an immediate source of very highly transportable material which is either available or not, leading to a threshold behaviour.

Type II weathering limitation can also lead to source limitation that is a function of chemical weathering (Riebe et al., 2003, 2004). In this case the chemical weathering through the profile (that generates fine material when the particles are exhumed at the surface) is in balance with the erosion. If erosion for some reason increases, then the soil will thin, soil particles are exposed to less time weathering within the soil profile and the grading of the material on the surface will become coarser stabilising the erosion rate. Likewise if erosion drops, then the soils will thicken, time spent weathering in the profile will increase and the soil surface will become finer so that the soil erodibility will become higher, stabilising the erosion rate from further reductions. Thus there is a balance between the source of the potentially transportable material (i.e. weathering) and the erosion and armouring process. Riebe concluded that field data supported this hypothesis (Figure 4.8). In Chapter 7 we will discuss how this process can be modelled using a soil depth and a physical weathering fragmentation model and how simulation results with these models also support this mechanism (Cohen et al., 2010; Welivitiya et al., 2016).

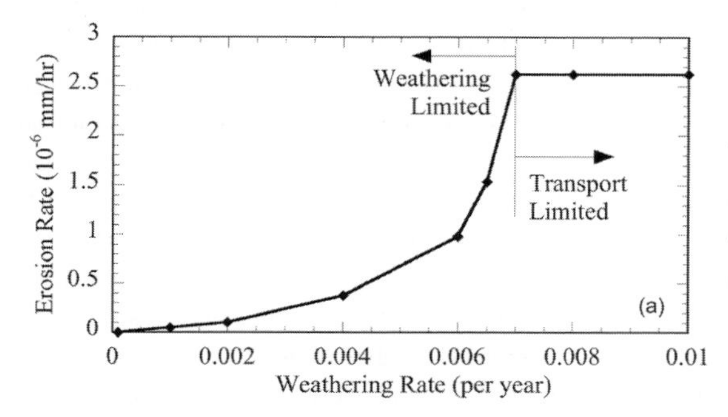

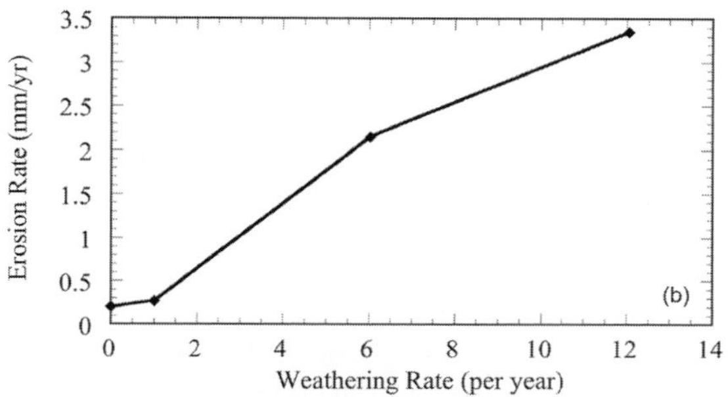

FIGURE 4.7: Physical weathering-limited erosion showing (a) type I (spalling) weathering limitation, and (b) type II (body fracture) weathering limitation (from Sharmeen and Willgoose, 2006).

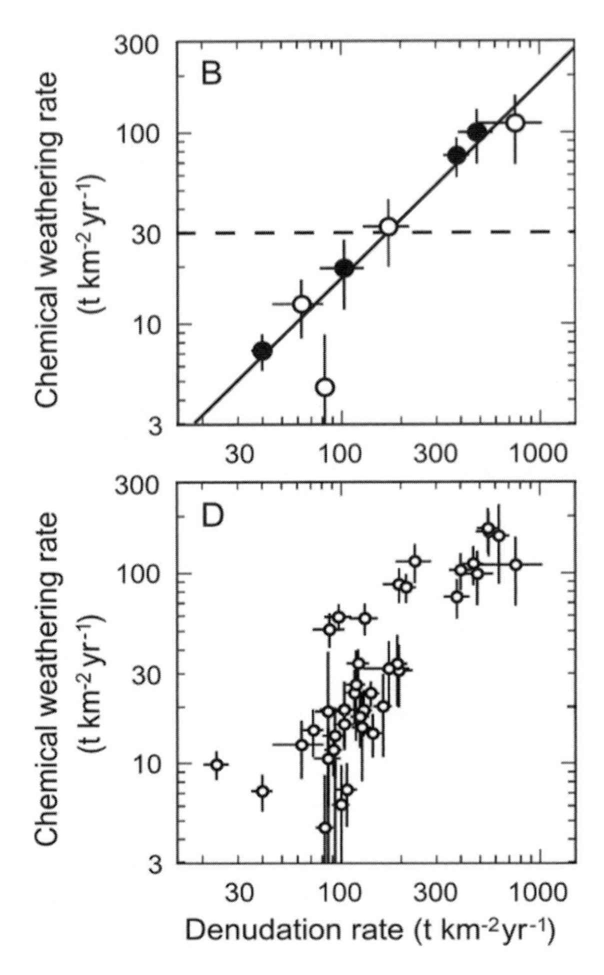

FIGURE 4.8: Chemical weathering-limited erosion showing how the erosion rate (denudation rate in the figure) is limited by the chemical weathering (from Riebe et al., 2004).

4.2.7 Aeolian Erosion and Deposition

In this book we will not discuss aeolian processes in detail. However, many predominantly fluvially dominated sites have aeolian inputs that cannot be completely ignored. These sites do not need to have fields of sand dunes, but it may simply be that a portion of the mineralogy of the field site may have been generated or sourced in dry past climates. Even under current-day climate it is believed that dust from Saharan Africa provides essential nutrients for the soils of the Amazon (Gläser et al., 2015).

It has been long recognised that the principles of erosion for atmospheric dust are much the same as for sediment in rivers. Much of the original research on sediment transport was done with atmospheric experiments (Bagnold, 1936). The main difference is that the viscosity and density of the atmosphere are different so that the coefficients on transport rates scale differently (Sørensen, 2004).

As far as deposition is concerned, the same issues with process rates arise as for fluvial sediment transport. The impact of aeolian deposition on soils has been shown to vary. For small deposition rates relative to fluvial erosion rates there may be little impact, but when rates of deposition of fine sediment are comparable with rates of fluvial erosion, it is possible for the armours that develop on the surface due to fluvial erosion to be drowned and for the erosion rates to proceed at rates much in excess of the rates that would apply in the absence of the deposition of the fine material (Cohen et al., 2015, 2016).

4.3 Temporal Averaging of Erosion

The transport-limited and detachment-limited erosion mechanisms discussed previously apply for the instantaneous rate of sediment transport during a runoff event, and the dependence on instantaneous discharge and/or bottom shear stress in the equations characterises this temporal dependence. Accordingly the sediment load varies during the runoff event in response to the changing discharge. The discharge varies in space and time both within and between rainfall events as outlined in Chapter 3. In landform evolution we are interested in the net effect of these processes over long periods (years to centuries and longer), and in many cases the details of the intra-event processes is only a necessary step toward determining the 'geomorphologically effective' equation, which is a temporal integral across the distribution of all the runoff events.

If we are simply running a computer model, and not needing any deeper analytical insight, then a simple, though rather brutal, solution would be to use a hydrology model and a rainfall time series to generate a runoff time series at each node in the landscape discretisation and then use this time-varying hydrology as input to the erosion models in this chapter. This time series of erosion at every pixel can then be used to drive landform evolution (Coulthard and Skinner, 2016). This approach avoids the need to explicitly develop a geomorphologically effective equation because the cumulative erosion is simply the result of the time-varying simulation, provided the time resolution of the simulation is sufficient to capture the dynamics of the runoff and erosion processes. This is the approach that has been adopted by many modern agricultural erosion models (e.g. WEPP, SWAT), but is fundamentally dependent on having either (1) high-quality, high

sensitivity to Manning n. If one of these equations defines the threshold of transition to channel, then the only remaining question is to define the discharge used in the equation, that is, mean discharge, dominant discharge, one-in-two-year discharge or bankfull discharge (Boardman et al., 2003). This discharge definition is an unresolved issue, but will impact on both β_5 and m_5 (see Section 3.1.9).

Montgomery and Dietrich (1988, 1989) plotted field channel data against area and slope, and found results that are functionally consistent with Equation (4.59) if the common assumption of discharge linear with area is made. While functionally consistent, the parameters were very different from those above as they found $m_5/n_5 \approx 0.5$ for their Tennessee Valley site. Montgomery and Foufoula-Georgiou (1993) present a landform slope-area analysis (see Section 3.3.3) for the same Tennessee Valley site, and this analysis shows, within the accuracy of the DEM data they used, that the channel head area and slope data plots parallel to the slope-area relationship for the topography, so their channel head relationship is likely a reflection of the erosion processes driving the slope-area relationship for the catchment rather than any fundamental relationship for channel head dynamics (e.g. Kirkby et al., 2003; Figure 8 in Poesen et al., 2003). Most subsequent studies have found it is difficult to identify a channel head relationship, but if any relationship can be identified, it is typically in terms of contributing area and slope, and of the form of Equation (4.59) (e.g. Torri and Poesen, 2014; Dewitte et al., 2015). Moreover, these studies have typically failed to show that the relationship is different from the landform slope-area relationship, so it is possible their relationships simply reflect points from the landform slope-area relationship. Almost all studies indicate a strong relationship with area (e.g. Placzkowska et al., 2015) and use of a 'support area' alone with no slope dependency to define the hillslope-channel transition is common in LEMs. Prosser and Abernethy (1996) found a good correlation between channel heads and the shear stress definition if they predicted runoff using saturation-excess runoff generation (see Section 3.1.3.1) rather than infiltration-excess runoff so that discharge was not proportional to area.

Willgoose et al. (1990) suggested a solution to the apparent contradiction between the stability and threshold approaches to defining the channel head. The Smith and Bretherton stability criterion defines the regions where channels can potentially form, while the threshold criterion determines where they can actually occur within that potential region. The threshold criterion remains poorly resolved.

4.4.3 Temporal Dynamics

It is common to observe that the gully/channel network expands rapidly (i.e. drainage density increases) after land disturbance and during this extension most sediment comes from gully excavation rather than sheet erosion (Prosser et al., 1995; Wasson et al., 1998; Poesen et al., 2003). This extension after disturbance is consistent with both gully head criteria. For the threshold criteria the land disturbance may result in either a higher discharge or a lower threshold. For the stability approach, if the discharge increases, then the region dominated by incisive processes expands with a larger region that can potentially channelize.

The discussion above assumes that the channel heads are always in equilibrium with the landform and its hydrology and erosion. If channels heads are in disequilibrium and are advancing upslope as a result of a change in the environment, it is generally observed in the field (Radoane et al., 1995; Vandekerckhove et al., 2001; Vandekerckhove et al., 2003; Rengers and Tucker, 2014) that the strongest covariate for gully advance rate is the contributing area to the gully head and that the form of the relationship is

$$E_r = KA^{\alpha} \tag{4.60}$$

where E_r is the extension rate or velocity of the gully head upslope, α is an area exponent and K is a rate constant dependent on runoff rate (Vanmaercke et al., 2016). The value of α is observed to be in the range 0.38–0.79 (Radoane et al., 1995; Vandekerckhove et al., 2003; Rengers and Tucker, 2014; Vanmaercke et al., 2016).

4.4.4 Mass Balance Considerations

Ultimately the balance of erosion dynamics (typically the transport-limited sediment transport physics described in Section 4.2.1) at the headcut must drive (1) the velocity of the hillslope-channel transition upslope for extending networks and (2) rates of infilling for shrinking or retreating networks. To date there is no consensus about the details of gully extension dynamics (Poesen et al., 2003; Valentin et al., 2005). For instance, for a gully to extend and maintain a sharp gully head, two things must happen: (1) there must be a step change in sediment transport at the gully head to maintain the distinct head and (2) the sediment transport processes in the gully downstream of the gully head must be more efficient than upstream so that the sediment being generated by the excavation of the extending gully head can be removed. Typical explanations for the step change involve breaching of the vegetation cover (e.g. Prosser et al., 1995), but this doesn't explain other cases, such as gully formation in bare agricultural fields. Finally, when the gully stops advancing, there must be a change in the transport capacity in the gully downstream (e.g. gully widening downstream of the gully head), otherwise the gully will erode the gully downstream of the gully head to compensate for the reduction of sediment delivery by the gully head that is no longer extending.

Finally, there is one aspect of the hillslope-channel transition that is of specific interest to LEM users: the mass balance impacts of the transition from hillslope to gully/channel. In the LEM it is not simply a matter of identifying where channels occur and then changing the dominant process modelled in the model depending on whether that location is hillslope or channel. The channel also has width and depth dimensions. Typically the width of the channel is less than the grid resolution of the model so is not explicitly resolved. As a consequence the change in mass in the transition between hillslope to channel is not resolved (i.e. mass = width × depth × density; Figure 4.9). Generally a model without some explicit allowance for this effect will overestimate the amount of sediment lost in the transition from hillslope to channel. In the agricultural erosion community it is generally believed that the mass of sediment eroded by rill and gully erosion is of a comparable magnitude to sheet erosion (e.g. Vandaele and Poesen, 1995; Martinez-Casasnovas, 2003; Poesen et al., 2003), so this mass balance error may be significant for cases with active gully development.

4.5 Channel-Specific Issues

This section addresses issues of channels in alluvial floodplains. It is sufficient at this stage to think of these channels

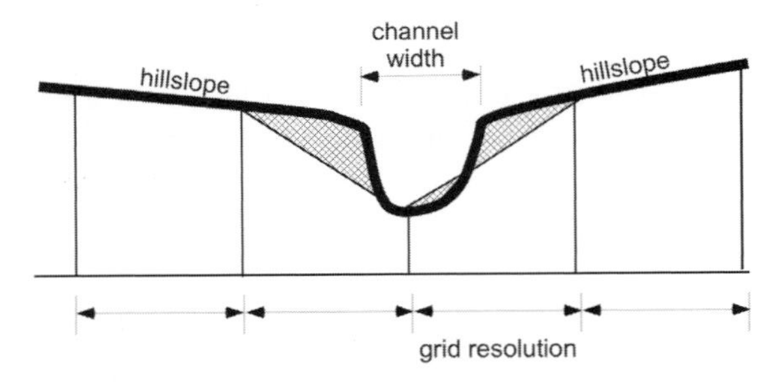

FIGURE 4.9: Channel and hillslope cross section at right angles to the flow. The heavy line shows the actual profile while the light lines shows how it is interpolated in a digital elevation model with resolution coarser than the width of the channel. The hatched cross-sectional area is material that the LEM will erode that is not actually eroded in the field.

as being transport-limited, though we will see when we look at the long-term evolution of floodplains that this is unlikely to be true at all times. As we have seen in previous sections, whether for transport- or detachment-limited conditions, a description of the channel geometry is required. At a minimum, the channel width is needed.

4.5.1 Channel Geometry

Two approaches can be taken to describe channel geometry. They are (1) use empirical equations for channel geometry (e.g. regime equations) that are typically parameterised by discharge and channel slope or (2) use physically based dynamical models to explicitly model the evolution of channel geometry and couple it to the evolution of the broader landform and contributing hillslopes. The first approach has the advantage of simplicity, but regime-like equations are implicitly assumed to be equilibrium properties of the channel, so it is not possible to model dynamic interactions between channel geometry and landform evolution using them.

An example where a dynamic description of channel geometry might be useful is the common observation of changes in channel width with reductions in sediment load, and those reductions in sediment load could result from changes in the erosion rate (and thus landform evolution) on the hillslopes. As the previous derivations have shown, a change in channel width will result in a change in sediment transport capacity, a subsequent change in channel evolution, which will feedback into hillslope evolution. An obvious question is whether it is necessary to model the relatively 'fast' evolution of the channel geometry if we are interested only in the relatively 'slower' evolution of the landform. The downside of using the second, dynamical approach is that (a) the processes controlling geometry dynamics are not well understood quantitatively and (b) physically based dynamical models will need to be run at the time scale of the evolution of the channel cross section, which will typically result in very small model timesteps and long compute times.

We will not discuss regime equations in detail here (Parker et al., 2007 revisits both the data and the physics underlying them) other than to note that they characterise the bankfull dimensions of the channel as a function of the bankfull discharge Q and are of the form

$$\begin{aligned} B &= K_b Q^b \\ H &= K_h Q^f \end{aligned} \tag{4.61}$$

where B is the bankfull width, H is the bankfull depth, K_b and K_h are proportionality constants and the exponents b

and f are typically about 0.5 and 0.4, respectively (Millar, 2005 in a review quotes ranges 0.45–0.55 and 0.33–0.40, respectively). It has been noted by many authors over many years (e.g. Henderson, 1966) that these equations are not dimensionally consistent (e.g. both K_b and K_h have dimensions that change with b and f). Also the exponents b and f are not independent since $Q = BHv$ where v is bankfull velocity.

Under some circumstances these equations can be dimensionally consistent. As shown in Section 3.1.7 for catchments at dynamic equilibrium (so slope approximately declines with the square root of area), velocity is weakly dependent on discharge when travelling downstream in the channel ($v \approx K_v Q^{0.1}$; Equation 3.4). Substituting this into the equation for discharge and summing the exponents from the regime equations

$$b + f + 0.1 = 1 \tag{4.62}$$

we see that the typical values for b and f are approximately dimensionally consistent. The need to assume the slope-area curve (which determines discharge dependence of the velocity equation) at dynamic equilibrium to achieve dimensional consistency highlights the dependence of the regime equations on channel slope, an output of LEMs. When regime equations are used in river engineering practice, two further equations, one for sediment grain size and one for channel slope, are normally specified to address this potential dimensional inconsistency. Both slope and grain size are an output from LEMs rather than specified by empirical equations. This may lead to dimensional inconsistencies in channel geometry if the regime equations in Equation (4.61) are used in LEMs (where slopes are an output rather than specified), a potential issue that is typically ignored by LEM developers.

We will now discuss the dynamical channel geometry models in detail. Later in this chapter we will construct a model for river meandering and floodplain evolution, and this meandering model is built on the principles used for the channel geometry models. While the mathematics of channel geometry evolution can be quite challenging, the underlying principle is relatively simple. The underlying principle is that channels evolve until the shear stress applied to the boundary of the channel (i.e. both the channel bottom and the sides) is constant. This means that the mobility of the sediment forming the channel boundary is equal and that sediment transport occurs at the same rate on all parts (i.e. both the bed and the banks) of the channel boundary.

A simple example will demonstrate this equal shear stress principle. Consider the cross section in Figure 4.10

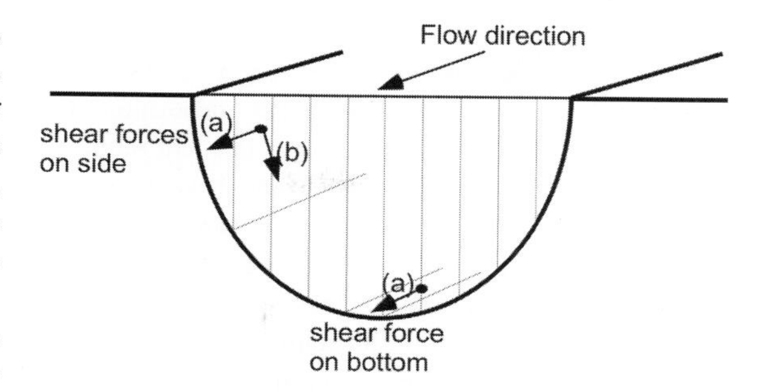

FIGURE 4.10: Schematic of shear stresses being applied by (a) the flow to the particles on the bed and (b) gravity down the side of the channel.

where the cross section is divided into vertical slices as shown. If we consider each slice independently of the others, then the shear stress applied by the flow on the channel boundary is

$$\tau = \rho g y S \Big/ \left| \frac{dy}{dx} \right| \tag{4.63}$$

where the absolute value on the derivative is to remove the change in sign that occurs at the channel centreline. At the edges of the channel where the flow is shallower, this longitudinal shear stress is less. However, there is also a shear stress being applied to the particles on the boundary by gravity driving particles to move down the channel boundary at right angles to the flow toward the bottom of the channel. The total shear stress on the boundary is then the vector sum of (1) the downstream shear stress applied by the flow and (2) the gravity force applied at right angles to the flow. At equilibrium the channel cross section geometry is such that the magnitude of the total shear stress is the same everywhere. This approach is the basis of a number of simple models for optimal channel geometry (e.g. Lane, 1955). However, this simple example ignores a number of important processes. For example, each slice has a different average velocity, so there are shear stresses (as a result of the velocity gradient across the slice boundary) applied at the vertical interface between each slice. However, to first order it captures the accepted drivers of channel geometry. This simple model also ignores the stabilising effect of vegetation and increased shear resistance of vegetated banks (Pollen-Bankhead and Simon, 2010; Camporeale et al., 2013).

This approach can be extended to transient conditions:

- If there are regions on the channel cross section with localised high shear stresses (high relative to the rest of the channel cross section), then these regions will erode until the shear stress is reduced back to that being applied to the rest of the cross section. Likewise for localised low shear stresses the erosion will be reduced relative to the rest of the cross section (deposition may even occur) until the cross section is in equilibrium.

- If the entire cross section has a higher shear stress than the equilibrium shear stress, then the channel geometry will erode until the shear stress is reduced to the equilibrium value. Generally this involves widening of the channel. In the derivation for transport-limited erosion (Equation (4.15)) the wetted perimeter (approximately equal to channel width) appears in the denominator of the sediment transport rate constant so that for constant discharge and longitudinal slope the sediment transport rate decreases (and as a consequence so does the erosion rate) as the channel widens. Thus the channel geometry response to excess shear stress and so excess sediment transport capacity is for the channel to widen, flow to become shallower and the average cross-sectional velocity to decrease.
- Likewise if the shear stress is lower than the equilibrium shear stress, the cross section will narrow and deepen until the depth of flow is such that the shear stress applied to the channel bed is equal to the equilibrium value.

The derivations above are normally discussed in the context of alluvial channels and transport-limited sediment transport. However, nothing in the derivation invalidates its use for detachment-limited conditions in bedrock channels. The shear stress threshold could as easily be the threshold for plucking of particles from the rock (since the rate of plucking in bed rock rivers is normally parameterised by use of shear stress, with or without a shear stress threshold below which plucking does not occur; Whipple et al., 2000; Lamb et al., 2015). One interesting aside for bedrock rivers is that using flume studies, Johnson and Whipple (2007) found that the channel cross section evolved from initially being detachment-limited to a condition where the roughness of the channel was sufficiently high that the transport capacity dropped and the channel became transport-limited. When they increased the sediment load to the channel, sediment deposition occurred in the scour holes of the river, smoothing the roughness. We speculate that this feedback between detachment limitation increasing bedrock roughness and transport limitation smoothing the roughness may lead to bedrock channel cross sections, equilibrating at the boundary between detachment and transport limitation. This suggests a coupling between these, local, cross-sectional dynamics with upstream sediment processes (which determine whether enough sediment is being delivered from upstream to make a cross section transport-limited).

4.5.2 Floodplains and Meandering

We now build on the channel geometry models to build a physically based model for river meandering. Consider a minor lateral (i.e. at right angles to the direction of flow) perturbation of flow such that the highest velocity is no longer exactly in the centre line but is slight offset to one side. In this case the shear stresses will now be slightly higher on the side of the channel with the higher velocity. This results in higher erosion on that side of the channel so that the channel cross section moves laterally in the same direction as the perturbation in the velocity, a positive feedback. These perturbations are the trigger for the beginnings of meandering in what was initially a straight channel. This is the basis of all the physically based models of channel meandering within a floodplain. The full mathematics for this process is quite challenging, and to date no general solution has been developed. The differences between the approaches adopted in the literature are in the approximations in the equations used for the modelling, and the field and experimental data to justify those approximations. Because of this mathematical complexity, this section concentrates on the underlying concepts, and the reader is referred to the key literature for the detailed derivations and field data. The two main approaches are (1) the HIPS equation (named for the developers; Hasegawa, 1977; Ikeda et al., 1981; Parker et al., 1982) and (2) topographic steering (Dietrich and Whiting, 1989).

Parker et al. (2011) summarises the HIPS approach. Figure 4.11 shows the plan of a river bend. The key point is that the average velocity on the outside of the channel centreline is larger than the average velocity on the inside so that

$$\Delta\overline{u} = \frac{1}{2}(\overline{u}_o - \overline{u}_i) \tag{4.64}$$

where the subscripts o and i refer to the outside and inside of the meander bend, respectively, and the rate of migration of the meander is then

$$\zeta = K\Delta\overline{u} \tag{4.65}$$

where ζ is the rate of migration of the meander bend and K is an 'appropriately chosen dimensionless coefficient' of the order of 10^{-8}–10^{-7} (Parker et al., 2011). Parker notes that while the mathematics leading to Equation (4.65) yields the physical dependency of K, experience has found that the value needs to be determined by calibration in the field. That still leaves the issue of determining the velocities in Equation (4.64). A variety of means are available of determining velocities, but all require cross-sectional geometry. This cross-sectional geometry can either be assumed or derived with a coupled erosion model (Parker, personal communication). Once a geometry is determined, the velocities can be calculated by solution of the shallow water equations, or, given the empirical basis of Equation (4.65) which requires calibration in the field (Parker et al., 2011), something more approximate such as the application of the Manning equation on either side of the centre line. The key concept is that the velocity of the outside of the cross section is higher than on the inside. The HIPS model is based on a linearisation of the flow equations. Güeralp and Marston (2012) review a number of extensions that use higher order approximations and that claim to better model details of the shape of the meander bends.

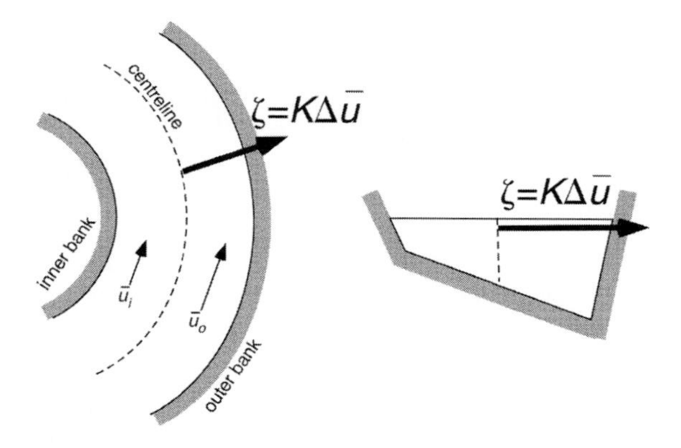

FIGURE 4.11: Schematic in plan of the forces and velocities in a meander bend: (a) plan, (b) cross section (after Parker et al., 2011).

The topographic steering model depends on the difference in the depth of the flow from the inside and outside of the bend (Figure 4.12). As the cross section changes from being symmetrical at the approach to a meander bend to being asymmetrical within the bend (with the deepest depth on the outside of the bend), the river flow (and the thalwag) is steered toward the outer bank. This steering causes the flow to hit the outside of the bend at an angle so that erosion occurs on the outer bank. It is this steering of the flow toward the outer bank that triggers erosion. Parker et al. (2011) notes that the main LEM implementations of topographic steering (Lancaster and Bras, 2002; Tucker et al., 2001a), while based on different physics from HIPS, are functionally equivalent to the HIPS equations as summarised in Equation (4.64).

The main functional difference between HIPs and topographic steering is that HIPS can generate meandering from a symmetric channel cross section, while topographic steering requires the asymmetry to start the meandering. The debate about which of the HIPs and topographic steering models better simulates meander

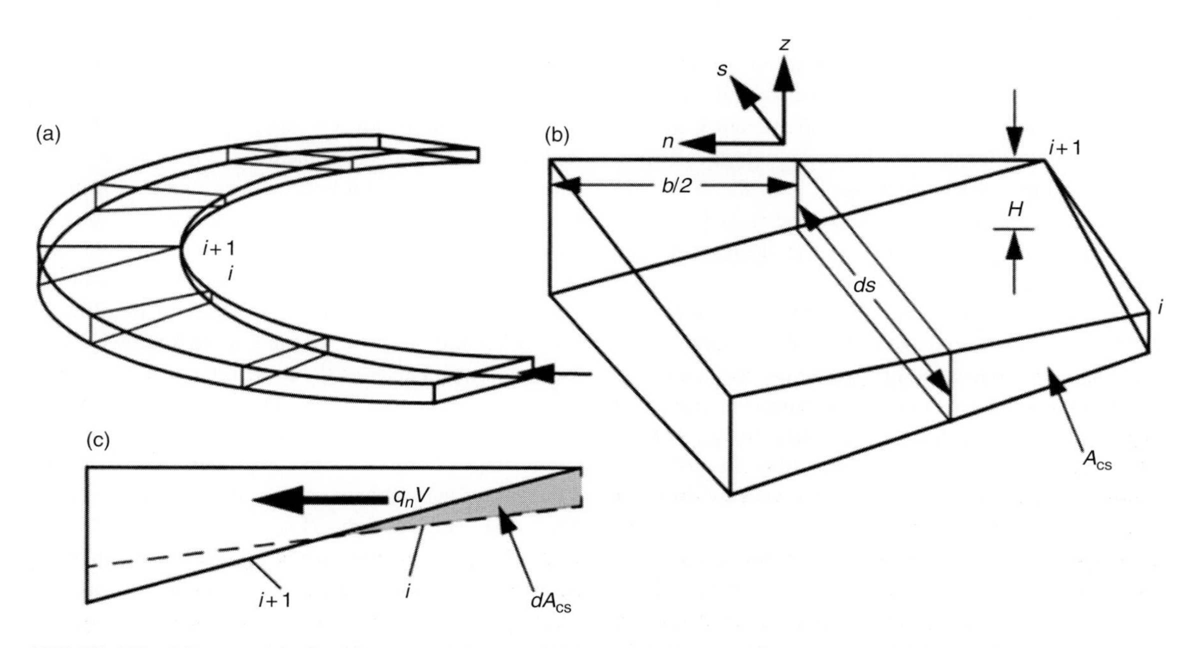

FIGURE 4.12: Schematic of the flows directions in the topographic steering mechanism (from Lancaster and Bras, 2002).

dynamics is now more or less settled in favour of HIPS so that river velocity differences rather than depth differences are believed to be the dominant driver of meander development (Parker et al., 2011).

Howard (1992) suggested an equation for meander lateral movement based on a combination of HIPS velocity and topographic steering cross-sectional change drivers that was

$$\zeta = K\left(a\Delta\overline{u} + b\Delta\overline{h}\right) \tag{4.66}$$

where $\Delta\overline{h}$ is the difference in the flow depth between the outer and inner half of the cross section, and a and b are parameters. The HIPs component is the first term inside the parentheses, while the topographic steering component is the second term. Subsequent analyses of field and laboratory data (for a discussion of this see Howard, 1996 and Parker et al., 2011) has shown the topographic steering term is dominated by the HIPs velocity term (i.e. $b = 0$) leading to Equation (4.65).

Howard and Knutson (1984) presented a meandering model that captures many aspects of the physics discussed above. Even though it predates much of the more recent theoretical work it still provides insight into how to construct a meandering model. There are a number of competing processes. The channel is modelled as a series of links downstream that can move in response to bank erosion. The rate of migration is a function of the radius of curvature of the meander at that location and the upstream geometry of flow. A model based solely on radius of curvature yielded unrealistic meanders. An implicit assumption is that the channel geometry and meandering evolve much faster than the floodplain so that the meandering can be considered to occur on a fixed elevation floodplain. As the sinuosity of the stream increases, the streamwise length of the river increases while the valley slope remains unchanged. Thus the slope, and the erosive potential of the river decreases, and the flow depth increases in proportion to the decrease in the streamwise slope.

Howard (1996) and Lancaster and Bras (2002) note that for meanders to migrate down-valley as observed in the field, then the location of maximum ζ must be offset downstream of the location of maximum river curvature. If there is no offset, then the meanders migrate only at right angles to the direction down-valley. In HIPS this offset downstream occurs because the velocity does not respond instantaneously to the change in channel cross section, and the distance from the minimum radius downstream to the peak velocity is a function of the rate of acceleration of the flow and distance travelled downstream during that acceleration (and is itself a function of the average velocity of the cross section). Lancaster and Bras (2002) and Tucker et al. (2001a), while they used the topographic steering mechanism, also noted that the peak erosion needs to be downstream of the peak curvature of the meander bend, and they offset the location of the peak migration rate downstream by a term that was a function of the average velocity of flow.

Van De Wiel et al. (2007) and Coulthard et al. (2007) describe a meandering model in their CASEAR LEM which is a function of the radius of curvature of the flow direction and where the maximum rate of migration is at the location of the maximum curvature. These authors have not done any studies of meander migration or shape with their model. Given that Howard and Knutson (1984) concluded that a model based solely on radius of curvature and ignoring the upstream geometry (and the lags introduced) did not produce downstream migrating meanders, this would suggest that the Coulthard model would be unable to simulate down-valley migration of meanders.

Güeralp and Marston (2012) noted that, in general, model-generated meanders are too regular compared with observed meanders, and that there is some evidence that at least part of this field irregularity is a function of variability of soil and vegetation on the floodplain, both of which are ignored in the models above. One of the implicit assumptions of both HIPS and topographic steering is that the dominant mechanism for meander migration is the erosion rate of the bank. If the erodibility of the bank is spatially variable, then meander migration will be constrained. This variability could be due to vegetation (e.g. Pollen-Bankhead and Somin, 2010), floodplain soil variability from previous meander migration or floodplain topography. For floodplain topography, if the channel bank is high, then it is possible that the bank erosion will undercut the bank and a large amount of sediment from the collapsed bank will slump into the river and protect the bank from further migration until the slump is eroded away. Thus the migration rate is a function not just of the bank erosion rate but also of the ability of the flow to remove the full height of the eroding bank (Howard and Knutson, 1984; Parker et al., 2011; Bufe et al., 2016).

Finally, in almost all cases the modelling has assumed that the width of the channel does not change with meandering. Changing width complicates the theory, but a number of authors have attempted to extend the HIPS theory for variable width (Luchi et al., 2010; Zolezzi et al., 2012). The main impact of varying width is to make meander bends wider and shorter than is the case for constant width channels. Parker et al. (2011) present a framework where the outer bank is eroding and the inner bank is depositing independently and where both are explicitly modelled so that width variations can arise naturally out of the formulation. With the exception of Parker et al. (2011) the models discussed above assume that the deposition on the inner bank keeps pace with the erosion on the outer bend because they implicitly assume that the channel width doesn't change. An advantage of the Parker formulation is that it is then quite natural to allow the outer bank to be more or less resistant due to vegetation or slumping without implicitly assuming an exactly compensating change in the deposition for the point bar on the inner bank.

One of the problems with the modelling of meandering is the difficulty of testing the models against data, and thus determining what is the best modelling approach. For instance, Parker et al. (2011) noted that none of the models work well without significant site-specific calibration, and in many cases the criterion for success is simply replication of the rates and direction of the meander movement (e.g. from aerial photos over 10's of years), which may not be sufficient to model long-term behaviour. Teles et al. (1998) propose a test using the stratigraphic properties of the floodplain deposits. While their work was very preliminary, there seems some potential in this approach if large enough areas of floodplain can be analysed. To collect data for the large areas required ground penetrating radar or recent developments in the use of seismic surveying for shallow depths might be suitable.

4.5.3 Alluvial Floodplains

The modelling of meandering rivers in LEMs discussed in the previous section deals almost entirely with the river itself and does not explicitly model the evolution of the floodplain within which the meandering river is situated (e.g. Howard, 1992). The meandering river simply moves from side to side within the preexisting floodplain, and the elevation of the floodplain outside the channel does not change. That leaves two unresolved modelling tasks: (1) aggradation or degradation of the floodplain and (2) widening of the floodplain.

Floodplain widening can be modelled by adding a multiplicative term in Equation (4.65) to account for differences in the bank erosion resistance of the rock outcrops surrounding the floodplain relative to the floodplain material (Howard, 1996).

The overbank deposition and scouring on the floodplain could be calculated by doing a two-dimensional coupled floodplain hydraulics and sediment transport model, but this has two disadvantages for a LEM: (1) it requires event by event modelling of runoff, and probably good intra-event hydrology to capture the portion of the hydrograph that generates overbank flow and the hysteretic sediment transport of the rising and falling limbs of the hydrograph, and (2) even for a single flood event, the computations for these types of models are high, so modelling multiple events over 100's to 1,000's of years is likely to be infeasible. Lauer and Parker (2004) present a box model that explicitly captures the event dynamics but without the complex floodplain hydraulics. Typically, however, the few LEMs that explicitly model floodplain dynamics use empirical models for floodplain deposition and are based on general observations of floodplain deposition patterns. Existing models are based on models proposed by Howard (1992, 1996). Howard's model captures two empirical observations:

- The rate of overland deposition diminishes with floodplain age and floodplain elevation, and this is because the higher elevation locations on floodplains are flooded less frequently.
- Floodplain deposition rates diminish with distance from the channel in the floodplain.

To capture these observations Howard (1996) proposed an equation for deposition

$$\phi = \left[v + \mu e^{-d/\lambda}\right] e^{\left[-\gamma\left(z_{fp} - z_{bed}\right)^{w}\right]} \tag{4.67}$$

where ϕ is the overbank sedimentation rate, v and μ are rate parameters for sediment deposition for fine and coarse sediment respectively, λ is a length scale controlling the rate of decline of coarse sediment deposition away from the river, d is the distance to the closest part of the channel, $z_{fp} - z_{bed}$ is the difference in height between the channel bed and the floodplain and γ and w are parameters controlling the rate of deposition with floodplain depth. The first term (in the square brackets) controls how the sedimentation changes with distance from the river. The second term (the exponentiation) controls how the sedimentation changes with increasing height of the floodplain above the river bed, and implicitly reflects the probability distribution of the river discharges, the bankfull conveyance capacity of the river and thus how often the river floods the floodplain. The main difference of Equation (4.67) from that proposed earlier by Howard (1992) is that the earlier work had a linear function for the distance decay, rather than an exponential term.

Tucker et al. (2001a) modified the Howard (1996) model by removing the distinction between fine and coarse sediment and simplifying the function for floodplain elevation relative to river elevation. Their relationship was

$$\phi = \mu e^{-d/\lambda}\left(z_{\text{fp}} - z_{\text{wl}}\right) \tag{4.68}$$

where the notation is the same as in Equation (4.67) and z_{wl} is the elevation of the water level in the river. The water level in the river is determined by

$$z_{\text{wl}} = z_{\text{bed}} + k_h Q_b^{\delta_b}\left(Q/Q_b\right)^{\delta_s} \tag{4.69}$$

where the parameters k_h, δ_b and δ_s can be determined from regime equations as discussed in Section 4.5.1. In contrast with Howard's equation, Equation (4.68) requires knowledge of the probability distribution of discharges in the river to determine the distribution of Q and thus z_{wl}.

Finally, we observe that the elevation of the floodplain bottom (i.e. the floodplain deposit-saprolite interface) is not normally modelled. We might reasonably assume that the processes that convert saprolite to soil in Chapter 6 are inactive because the floodplain deposits are too deep for them to be active. The floodplain bottom can then only decline only as a result of incision of the saprolite interface by the flowing water in the river. The implication of this incision control is that floodplain lowering is a function of the rate at which the bottom of the meandering river is able to incise the floodplain bottom as it meanders back and forth across the floodplain (called 'bevelling'; Bufe et al., 2016). While for much of the time alluvial rivers have a sand or gravel bottom (so the bedrock is protected from incision), during floods the channel cross section is excavated exposing the underlying bedrock and the exposed bedrock base is incised. Thus bedrock lowering of the floodplain bottom is driven by the small proportion of time that the bedrock is exposed by the cross-sectional changes during flood events, and the small proportion of the area of the floodplain covered by the incising channel. These regions of bedrock incision will control the rate of lowering of the floodplain. If these locations are distributed down the river valley, then they will also control the evolution of the valley long profile. These cross-sectional dynamics are the domain of mobile bed channel models, where the channel cross section dynamically responds to the time-varying flow (Julien and Wargadalam, 1995; Julien, 2014). To date no LEM models this process, so no current LEM can model floodplain decline.

4.5.4 Alluvial Fans, Deltas and Downstream Fining

Alluvial fans are primarily depositional structures, with entrenched streams that avulse across the surface of the fan as the fan builds up. Most existing landform evolution models that model transport-limited sediment transport (i.e. processes that allow deposition) generate structures whose elevations look like alluvial fans (e.g. Willgoose and Riley, 1998; Coulthard et al., 2002). Recent work (Welivitiya, 2017) has developed a coupled soilscape and landform evolution model that implements (1) mixed sediment size erosion and deposition and (2) tracks the full particle size distribution in the flow. This model generated realistic landforms (cross- and long-section concavities) and realistic patterns of particle size distributions that qualitatively match observations including downstream fining, and lenses of coarse and fine materials within the fan itself (Figure 16.5). The internal structure of their fan appears qualitatively consistent with previous modelling work of submarine fans that used an empirical model for sediment size fractions (Koltermann and Gorelick, 1992). Gasparini et al. (1999, 2004) also observed downstream fining within an erosional drainage network. Welivitiya (2017) also qualitatively observed other features of the simulated alluvial fan such as longitudinal radial filaments of coarse material. These match features observed in the field (Gómez-Villar and García-Ruiz, 2000; Clarke, 2015).

4.6 Some Synthesis of Fluvial Erosion

Before discussing the details of numerical implementations and the issues they address, it is worth summarising how the many topics discussed in this chapter fit together. It will then be clear how many of the concerns expressed by Chen et al. (2014), while true in some cases for the early versions of LEMS, have been, or are in the process of being, addressed in recent years. The reader might also like to look at the discussion about erosion in LEMs in Tucker and Hancock (2010), which complements the discussion below.

When modelling transport-, detachment- and weathering-limited transport mechanisms the early implementations of the LEMS typically only allowed one of these mechanisms to be active at any specific time and at any specific node, and if more than one process was modelled, there was a switch between them based on a threshold of some kind. For instance, early versions of SIBERIA (Willgoose et al., 1991a) had a switch that changed the erosion model applied on hillslope and channels based on whether a node was a channel of hillslope, or whether a node was transport-limited or detachment-limited. Most modern models have abandoned that approach because it can lead to mass balance problems when the dominant process switches. Nowadays they track the amount of sediment being carried in the flow and use a formal detachment process to model entrainment, and compare the potential load at every node against the transport capacity (which may itself change based on the grading of the material being carried) to determine the amount of erosion or deposition. This explicit entrainment mechanism is consistent with modern agricultural erosion models.

Some models track the grading of the sediment being carried, but the tracking can introduce severe nonlinearities that significantly slow numerical solvers. For instance, SIBERIA V8 (i.e. after 2001; Willgoose, 2005b) optionally tracks the mean diameter (version 7 and earlier did not track sediment grading), CHILD (Tucker et al., 2001a) tracks the relative proportion of two particle size fractions (sand and gravel), while ARMOUR (Willgoose and Sharmeen, 2006), mARM (Cohen et al., 2009, 2010), SSSPAM (Welivitiya et al., 2016) and CAESAR (Coulthard et al., 2002) track the full particle size distribution. All the models mentioned here allow the fluvial sediment transport capacity to respond to changes in the grading of the sediment in the water since, as we saw in Section 4.2, there are relationships that express how the sediment transport rate changes with sediment d_{50}. Even if grading is tracked, there are complexities because the grading of the material available for erosion (the source for the material tracked downstream) then needs to be known, necessitating that either (1) the soil surface grading is specified a priori or (2) a soil submodel is incorporated that estimates the surface grading over time and that responds to erosion and deposition (e.g. the SEM models that will be discussed in Chapters 5–11)

To average over time, two main approaches have been adopted. Most models determine a relationship between the geomorphologically effective discharge/transport and catchment area, and use this relationship to determine the geomorphologically effective discharge or geomorphologically effective sediment transport rate using the analysis for area draining through each node. Some explicitly model the intra-event dynamics using a coupled rainfall-runoff model. The level of sophistication of this relationship varies between models (1) *purely conceptual* (e.g. DELIM, Howard, 1994), (2) *calibrated* to output from a distributed rainfall-runoff model (e.g. SIBERIA, Willgoose and Riley, 1998) and (3) *fully coupled* that are coupled to a distributed rainfall-runoff model (e.g. CAESAR, Coulthard et al., 2007). Only the last kind can fully model the intra-event flood dynamics (e.g. flood wave routing down the channel), but it comes at a significant computational cost: runs that take 15 minutes in SIBERIA (discharge-area relationship calibrated to a distributed rainfall-runoff model) can take 1–2 weeks with CAESAR (fully coupled with the LISFLOOD rainfall-runoff model) with little difference in the net erosion estimates (Hancock et al., 2010). But as we saw in the discussion of meandering processes, the behaviour of the floodplain deposition is dependent upon correctly modelling the probability distribution of the runoff events that created flow depths in the channel that spilled over onto the floodplain, so to model floodplain dynamics, temporal variability needs to be explicitly modelled.

It is not just floodplains that require a probability distribution for discharge. Other possible applications include post-failure behaviour after breaching of flow or erosion control structures (e.g. contour banks, farm dams, armoured constructed channels, drop structures, moonscaping on mine sites). In many of these cases it is extreme floods, ones that occur only every few years or even less frequently, that are critical. Typically once failure occurs, the erosion protection they provide is removed (and in some cases, such as contour bank and moonscaping failure, erosion may be worse than without the contour bank being built in the first place), and the long-term erosion response of the landform may change markedly post-failure (Willgoose and Gyasi-Agyei, 1995).

For the conceptual and calibrated models (e.g. SIBERIA, CHILD) this random discharge is achieved by randomly varying in time the runoff rate in the discharge-area relationship. It is important to understand that this is different from fully coupled discharge modelling (using a time-varying rainfall input) because the former do not explicitly model the runoff-routing process (even though for the calibrated models it will be 'cooked into' the calibration parameters, specifically those relating discharge to area). Some authors (e.g. Chen et al., 2014) have misunderstood the use of geomorphologically effective discharge and confuse it with the average discharge (i.e. the annual discharge divided by 365 days), but Section 4.3 shows how the concept of geomorphologically effective discharge is related to extreme flow events rather than the mean flow.

4.7 Numerical Issues

Many of the concepts discussed here, such as spatial and temporal discretisation and numerical solvers, have significant implications for the validity of results from LEMs. In some cases incorrect or poor choices may invalidate the results of the modelling entirely, even if the underlying science is completely adequate. The issues discussed briefly below are, in the author's view, so important, and so widely misunderstood, that a complete chapter (Chapter 15) is devoted to them. Issues specific to erosion modelling, such as numerical stiffness, are discussed below, and in Chapter 15 they will be placed into the context of a larger coupled landscape evolution model.

4.7.1 Spatial Discretisation, Drainage Analysis Algorithms

The first of the numerical challenges is to determine the catchment, or discharge at every point of the domain, so the erosion equations can be applied. This is normally done with a drainage direction search algorithm on a spatial discretisation.

Spatial discretisation has been done in two ways:

- A rectangular grid with uniform spacing in both directions. There may be an implicit assumption that the grid is oriented in the E-W and N-S directions. Many codes have used square grids because of their simplicity. However, many field digital elevation data sets are delivered on a degree, minute, second (DMS) grid (e.g. the Shuttle Radar Topography Mission [SRTM] data). which means that a square grid in DMS units will be rectangular in units of metres depending on the latitude of the site, so there is benefit in using rectangular grids. For instance, at 40°S or N a 1 second square grid (the highest resolution SRTM data) is 30 m in the N-S direction and 23 m in the E-W direction.
- Irregularly spaced data that are discretised with triangles into triangulated irregular networks (TIN). While more flexible than grids, significantly more information needs to be stored to do with how the triangles are connected (Tucker et al., 2001b).

Once the grid discretisation is determined, the next decision is how to determine the area draining through each node. The default, though not necessarily the best, approach is to determine at every node the steepest downslope direction to an adjacent node and rout all flow from the current node to this adjacent node. Once the directions are known at every node, it is a simple matter of simply summing up the areas contributing to the current node. For gridded DEMs this is referred to as the D8 algorithm because it deterministically routs all flow to one of the eight adjacent nodes. There is no equivalent name for D8 on TINs, but this approach of routing all flow in the steepest downslope direction is used in the CHILD LEM, where all the flow is routed along the boundary between the triangles that the current node is a vertex of.

The main problem with D8 (O'Callaghan and Mark, 1984) (and presumably the equivalent on TINs) is to do with its response to changing elevations. Inherent in the method is a threshold where the drainage direction does not change immediately with incremental changes in adjacent elevations but will change suddenly when the adjacent node with the steepest downslope direction changes. So for small changes in elevations the drainage direction (and thus areas draining through nodes downstream) is less responsive, and then when the threshold is exceeded there is a sudden change in area. Since excavation of a valley has as a key component the convergence of flow into the valley, this means that because D8 is initially unresponsive to flow convergence, valley formation is slower than it should be. Sediment transport capacity is proportional to $Q^m \sim A^m$ where m is 1.5–2, so that when, for example, two nodes of equal area converge into a single node downstream, the sediment transport capacity at the downstream node is twice (for $m = 2$) the sediment draining into that node, and the downstream node has to erode the deficit, excavating a valley in the process.

An alternative drainage direction algorithm is D∞ (Tarboton, 1997), which weights the drainage between the adjacent node with the steepest downslope direction and the node next to it that has the next steepest direction (Figure 4.13). D∞ yields a more accurate definition of the drainage direction and areas, but for LEMs it has a more important property. The weighting of the proportion of area/discharge/sediment flux between the two nodes downstream immediately responds to changes in elevations, so even small changes in elevations result in changes in the areas downstream. Willgoose (2005a) compared D8 and D∞ on a constructed landform and found that over 1,000 years the D∞ simulations evolved

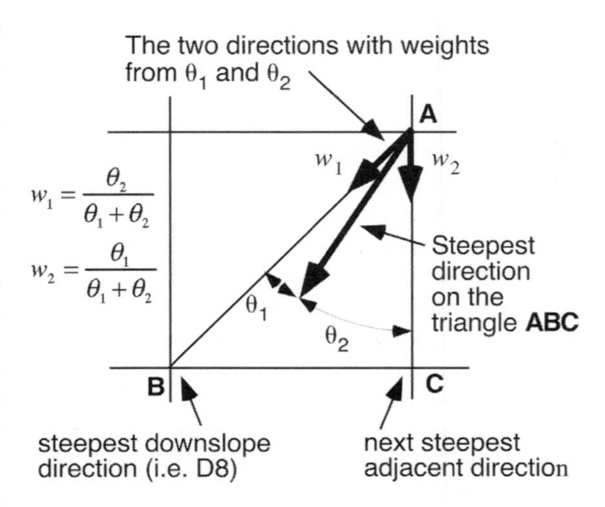

FIGURE 4.13: The weighting scheme for D∞.

about 15% faster than the D8 simulation. There were a large number of valleys generated on the landform, and their excavation dominated the sediment mass balance over the 1,000 years of the simulation. The only apparent explanation was the threshold in the D8 that suppresses flow convergence and valley formation. A side benefit of D∞ was that the valleys curved more smoothly in plan rather than looking blocky and gridded.

It is likely that the TIN equivalent of D8 and D∞ will also show this latter blocky behaviour, though the random location of the grid points likely obscures the 'gridded' flow patterns. No such comparison (i.e. between TIN versions of D8 and D∞) has been done with TINs (Tucker, personal communication), but it seems reasonable to suspect that a similar threshold convergence mechanism may occur for TINs as well.

The area at every node needs to be recalculated at every timestep, and a LEM simulation may involve a million or more timesteps so efficient calculation of area is required. During the original construction of SIBERIA (i.e. Willgoose, 1989) a large number of algorithms for calculating area were tested and the fastest one for D8 was

1. Determine the D8 directions
2. Search for all sources of drainage lines in the grid. These are the nodes that don't have any nodes draining into them.
3. Starting from each source in turn, step down the drainage lines summing the area of all nodes draining into each node (i.e. including any tributaries) as you follow the drainage directions downstream until you come a node where one or more of the tributaries have not yet had areas calculated for them. At this point stop and repeat step 3 for the next source in the list.
4. When you do step 3 for the last source, you are guaranteed to have calculated the area for every node.

Our experience is that this algorithm is not just fast, but that it scales well to large domains (we have used it for up to 3,000,000 node domains). Braun and Sambridge (1997) independently arrived at a very similar algorithm for their TIN-based DEM and called it CASCADE.

For the calculation of areas with D∞ there seems to be no alternative to using the recursive algorithm of Tarboton (1997). Informal benchmarking using SIBERIA indicates that the area calculation with D∞ is about four times slower than with the D8 algorithm above. More critical is that it is not possible to parallelise the recursive solution on modern multicore processors, whereas the D8 algorithm can be. Calculating areas using D∞ is the only component of SIBERIA that has not been able to be parallelised in some form.

4.7.2 Timestepping, Landform Time Scales and Numerical Stiffness

The erosion models we have discussed here have some characteristics that make them expensive to solve in time, and difficult to achieve an accurate mass conservation. The dependence of sediment transport on area (through the discharge dependence in Equations (4.10) and (4.29)) means that the rate of change of elevation (which is related to the response time) for large areas near the catchment outlet is fast and for small areas on the tops of hillslopes is slow. This range of response times means that small timesteps are required to solve the equation for elevation change near the catchment outlet while large timesteps can be used for hilltops, while the amount of time to run a landform to equilibrium is determined by the slowest part of the landform, the hilltops. Numerical problems with large differences in response times are grouped under the heading of 'stiff problems'. The problem can be quantified by examining a simple problem with transport-limited erosion and tectonic uplift in one dimension (Equations (2.1), (3.12) and (4.10)):

$$\begin{aligned}\frac{\partial z}{\partial t} &= U - \frac{\partial q_s}{\partial x} = U - \frac{\partial}{\partial x}(KQ^m S^n) = U - \frac{\partial}{\partial x}\left(K\beta^m A^{m\delta} S^n\right) \\ &= U - K\beta^m \frac{\partial}{\partial x}\left(A^{m\delta} S^n\right)\end{aligned} \tag{4.70}$$

If we nondimensionalise this equation using

$$\begin{aligned} z &= \bar{z}z' \\ A &= \bar{A}A' \\ t &= \bar{t}t' \end{aligned} \tag{4.71}$$

where $\bar{z}$ is a vertical length scale characterising the elevation of the problem (e.g. mean elevation, relief), $\bar{A}$ is a horizontal length scale characterising the area of the problem (e.g. catchment area) and $\bar{t}$ is some time scale of the problem (e.g. rate of change of elevation), then we can rewrite Equation (4.70) as

$$\frac{\partial z'}{\partial t'} = \frac{\bar{t}}{\bar{z}} U - \left[K\beta^m \bar{A}^{(m\delta)} \bar{z}^{(n-1)} \bar{t}\right] \frac{\partial}{\partial x}\left(A'^{m\delta} S^n\right) \tag{4.72}$$

where the term inside the square brackets parameterises the scaling of the rate of change of elevation for different area and elevation from erosion. For typical values of $m \approx 2$ and $m\delta \approx 2$ then as the area of the catchment increases (and with slope constant) the rate of change of elevation on the left-hand side increases with the square of area. This is shown graphically in Figure 4.14a where a node with large area changes elevation relatively faster than one with a smaller area. The implication is that if an Euler timestepping algorithm is used for large timesteps,

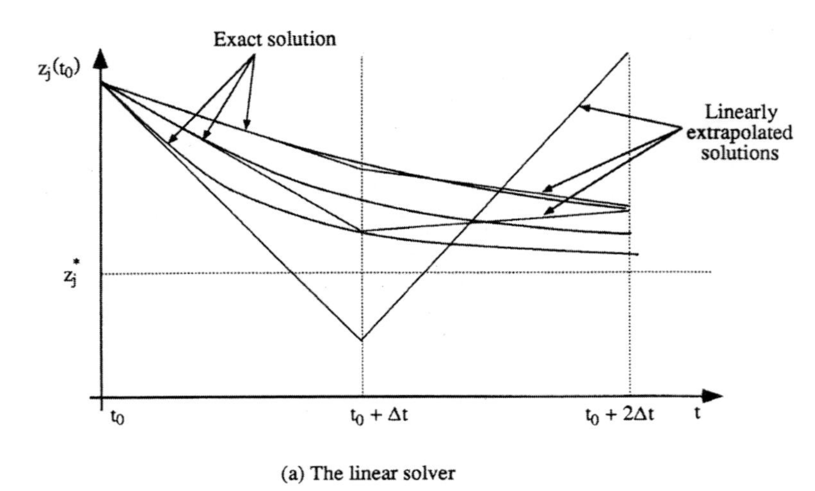

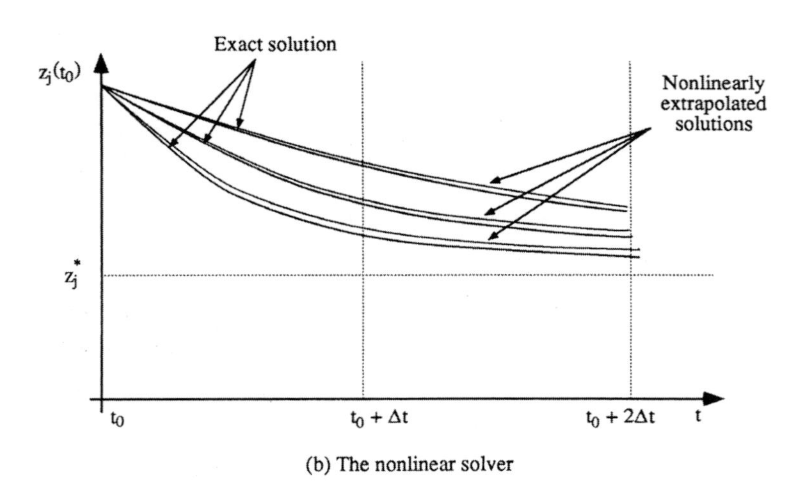

FIGURE 4.14: The schematic of the approximate analytic solution for the time evolution of elevations for fluvial erosion (from Willgoose, 1989).

oscillations will be created in the solution because the projection forward in time will overshoot the correct answer for nodes with large areas.

The significance of this overshoot is for the analysis of the drainage pattern at the next timestep. All current LEMS use drainage direction algorithms (D8, D∞, as discussed in the previous section), and the overshoot creates a localised pit in the elevations, which breaks the drainage line so that flow and sediment transport is interrupted. This means that areas draining through the nodes downstream of the pit drop dramatically, so that the sediment transport balance in these downstream nodes is changed dramatically. An inevitable consequence is to trigger elevation instabilities downstream, and catastrophic failure of the elevation solver follows very quickly. Some users have suggested pit-filling these pits, but this only wallpapers over the underlying problem and introduces mass conservation problems (since pit-filling fills up the pits with sediment that is not accounted for in the mass conservation equations).

The common, brute force, solution to these overshoots and oscillations is to use smaller timesteps, but since the power on area is about 2, this means that even a modest-sized problem (e.g. 100 $\times$ 100 nodes, with a single catchment outlet so the maximum catchment area is 10^4 nodes) will have a range of time scales (and thus rates of change in elevations) of 10^8, and this range increases for bigger problems. This scaling problem is not as severe in the case of stream power detachment-limited Equation (4.29) because the power on the area in this case is about 1–1.5, so in the example the time scale range is only 10^4–10^6.

Willgoose (1989) proposed a solution using an approximation to the analytical solution of elevation with time (Figure 4.14b). He noted that for a one-dimensional problem, if we assumed that the elevations immediately upstream and downstream were fixed, then with time the elevations will converge to an equilibrium solution where the elevation is not changing. For a one-dimensional problem this equilibrium solution can be derived analytically by equating the sediment inflow with the sediment outflow so that

$$Q^m(z_{i-1} - z^*{}_i)^n = (Q + \Delta Q)^m (z^*{}_i - z_{i+1})^n \tag{4.73}$$

where the asterisk superscript indicates the equilibrium value and ΔQ is the extra increment of discharge contributed by node i, and

$$z^*{}_i = \left(\frac{Q^{m/n} z_{i-1} + (Q + \Delta Q)^{m/n} z_{i+1}}{Q^{m/n} + (Q + \Delta Q)^{m/n}} \right) \tag{4.74}.$$

The approximate solution that is then used has a slope at the start of the timestep that is given by Equation (4.70) and that converges to Equation (4.74) over time. In two dimensions the approximate analytic solution requires the use of Taylor series to solve the equivalent equation to Equation (4.73) where there are multiple inflows of sediment from upstream nodes, but the principle remains. Experience in using this solution technique in SIBERIA, and incremental improvements since its original derivation in Willgoose (1989), indicates that the criterion that limits the size of timesteps is no longer stability and potential oscillations, but acceptable mass balance (Section 15.3.2). The mass balance errors arise because the equilibrium solution in Equation (4.74) assumes that (1) the elevations of the upstream and downstream nodes are not changing with time and (2) the drainage directions do not change during the timestep. This is typical of Euler numerical solvers, but with increasing timestep size the approximate analytical solution is an increasingly poor approximation. However, the mass balance error does not scale so severely as the Euler solver with the domain size making very large grids and higher spatial resolutions more feasible.

An alternative solution to the stiffness problem is to use implicit solvers. Implicit solvers are more stable allowing larger timesteps. Fagherazzi et al. (2002) and Perron (2011) outline implicit solvers for the transport-limited sediment transport equations.

Fagherazzi et al. (2002) used Taylor series to linearise Equation (4.70) in slope S and then solved for the future elevation in the classical way that implicit solvers are formulated including the need to invert an $N \times N$ matrix where N is the number of nodes. The method is unconditionally stable (i.e. there will be no oscillations), but like the SIBERIA method it assumes that the drainage network does not change during the timestep, ultimately limiting the maximum timestep that can be used just as it does for the SIBERIA method. Comparing the implicit method and a simple Euler solution, they found a factor of 100 times speed-up with less than a 5% mass balance error.

In Chapter 9 we will talk about diffusive transport processes (processes that don't have an area dependence, only slope), but it is convenient to jump ahead a bit at this stage. Both Willgoose's and Fagherazzi's methods are applicable to diffusive transport as well as the fluvial erosion, but without the area dependency the stiffness problems disappear, and many of the complexities and approximations of these methods become unnecessary. Perron (2011) proposed an implicit solver for nonlinear diffusion (Equation 9.20). Perron used a Taylor series expansion of the nonlinear diffusion equation and then formulated an implicit solver based on this linearization. The nonlinear diffusion problem does not have an area dependence, so there is no need for drainage directions to be defined to determine area, and accordingly Perron's solution does not assume a fixed drainage network during the timestep. Perron does solve an example involving fluvial erosion, but at each node he assumes that either diffusion or fluvial erosion is active but not both (the switching criticised by Chen et al. (2014) discussed in Section 4.6) and then solves the fluvial erosion equations with a simple Euler explicit solver using operator splitting (see Section 2.2.2). Thus Perron's solution is not designed to address the stiffness problems of the fluvial erosion process.

4.7.3 Catchment Divide Migration

LEMs approximate the location of the catchment divide. In D8 for each node the area of the node is assumed to flow is a single downslope direction, and for D∞ two downslope directions. The drainage divide is then along the boundary of the cells halfway between nodes. In reality the drainage divide will not be exactly halfway between the nodes. Stream capture occurs when a node switches its drainage directions from one subcatchment to another. For D8 all the area of a node flows into one or the other subcatchment. More realistically as the divide moves there will be gradual change over time in the apportioning of the nodal area from one subcatchment to another.

Goren et al. (2014) proposed a method for modelling the gradual change in the location of the drainage divide. It modelled the evolution of the rivers and then applied an analytical solution for the long profile of the hillslope between the rivers, giving a precise location for the divide based on the hillslope profiles. As the river elevations and locations changed, so did the hillslope profile and the location of the catchment divide. Goren and his coworkers tested it against a traditional LEM. They found that it modelled the catchment divide better at coarse resolution (8 km), and that the results of the two models were similar at higher resolutions (1 km, with a hillslope length in both cases of 500 m). The method was not tested at the resolutions normally used in LEMS that resolve the hillslopes (10–100 m, or 0.01–0.2 times the hillslope length).

4.7.4 Tracking Sediment Grading

It should be clear that to simulate the sediment selectivity of armouring and deposition it is necessary to track the entire grading distribution of the sediment. For erosion, without the full grading distribution it is not possible to model the entrainment of the finest fractions and the subsequent reduction in mobility of remaining sediment on the armouring surface to erosion. For deposition to model the faster deposition of coarse fractions (and the spatial patterns of deposited sediment on alluvial fans), and the subsequent relative increase in the mobility of the sediments remaining in the flow, the deposition of the coarsest fractions need to be tracked.

However, tracking the full grading introduces nonlinearities in the sediment transport equation that, for a full physically based treatment, can be computationally overwhelming. For instance, the author's ARMOUR model (Willgoose and Sharmeen, 2006) tracks the entire grading distribution and simulates selective entrainment and deposition. ARMOUR runs many orders of magnitude slower than SIBERIA, which doesn't track the full grading distribution. In Chapter 7 we will discuss one approximate solution to this problem that allows us to track the full grading distribution. SIBERIA uses a different, more limited, approach.

The author's SIBERIA model implements a simple, physically based, approximate algorithm to track the changes in the transport capacity as a result of changes in the average grading of the sediment. SIBERIA tracks the mean of the grading so that when two flow paths (let's say (1) and (2)) merge, then the grading of the sediment after the merger (3) is

$$d_3 = \frac{q_1 c_1 d_1 + q_1 c_1 d_1}{q_1 c_1 + q_1 c_1} \tag{4.75}$$

where d is the average diameter of the sediment, q is the discharge and c is the sediment concentration. The transport capacity of the flow in SIBERIA changes with the mean diameter of the sediment, as given in Equation (4.12). For the eroding surface the average grading of the surface sediment is a model input, and it is assumed that the surface grading is not changed by selective entrainment. Likewise for deposition, selective deposition of coarse particles does not occur, so both the grading in the flow and the deposited sediment is unchanged by the deposition process.

Without the ability to model the selective entrainment of fines, this approach cannot armour the eroding surface. This results in erosion depths (both sheet and gully erosion) that are too high. To address this, SIBERIA implements a conceptual armouring model. The erodibility is modelled to decrease with increasing cumulative depth of erosion. The rationale for this is that erosion initially removes the finest particle size fraction from the surface layer. When 100% of this fraction is removed, then the next coarser fraction is removed in turn. As this process proceeds, the eroding surface is gradually coarsened (the mean diameter is increasing), and this coarsening is determined by the cumulative amount of erosion. This relationship between erodibility and cumulative erosion is easily determined from the full grading distribution of the surface material, and the rate of 'armouring' is dependent only on one parameter, the depth of the surface armour layer assumed. A deeper armour has a greater mass of fine material, and so requires more cumulative erosion to reach a given mean diameter armour than a thinner layer. A thinner armour means the surface armours faster.

4.8 Further Reading

The best up-to-date discussion of rainsplash mechanics and field experimental data is Dunne et al. (2010). Southard (2006) provides a beautifully intuitive discussion of sediment transport fundamentals. For those who are interested in aeolian transport processes, a good place to start is the special issue on 'Aeolian Research: processes, instrumentation, landforms and palaeoenvironments', *Geomorphology*, 59 (1–4), 2004. For agricultural erosion, one consistent, coherent reference does not exist, but the WEPP model documentation is probably the most comprehensive up-to-date reference (online, but see Laflen et al., 1991 for an overview).

5 Soils: Constructing a Soilscape Evolution Model – Basic Concepts

5.1 Overview

From the previous chapter on erosion it can be seen (and will also be seen in later chapters) that information about soils is required to fully parameterise many processes. Without some soil dynamics model that interacts with the landforms, the user of a landform evolution model needs to specify the soil properties.

But there are many other reasons above and beyond landform evolution to be interested in soils and their dynamics. Soils cover 80% of the earth's terrestrial surface (the remaining 20% is predominantly covered by ice, permafrost and alps) (FAO, 2009), and by understanding the temporal dynamics and spatial coupling of soils we can develop a quantitative understanding of (1) the spatial distribution of soils, (2) the links between soil properties, the underlying geology and climate and (3) their resilience to environmental change and human impacts.

A compelling example of this need for deeper understanding of soil dynamics is recent work examining the uncertainties in the models of the global carbon cycle. Our current best estimate of sequestration of anthropogenic carbon emissions is that 55% of the extra carbon dioxide emitted into the atmosphere that is sequestered, is sequestered on the land (Stocker et al., 2013) as enhanced soil carbon, microorganisms and plant biomass. This terrestrial rate of sequestration per square kilometre is more than three times that of the ocean. Yet our understanding of the rates of sequestration by these terrestrial processes is poor, and we do not have a clear understanding of their sensitivities to climate change. It is conceivable that this 55% terrestrial sequestration rate may change because of the impacts of climate variability and climate change, with consequent knock-on effects for atmosphere carbon dioxide levels and climate change. For example, Friedlingstein et al. (2006) examined 11 biogeochemical models used to model the global carbon cycle and models for global climate projections out to 2100. The projected carbon dioxide concentrations in the atmosphere ranged from a very slight increase of about 50 ppm above current day levels to a more than tripling to more than 1,000 ppm. They concluded that the main reason for this variability was differences in each model's projection of soil moisture and knock-on impacts (particularly in semiarid climates) on vegetation growth and soil microbiology. Likewise, Peng et al. (2014) found that changes in soil moisture, after changes in mean temperature, were the most significant factor in determining global soil carbon storage. Results from soil moisture research generally conclude that the parameterisation of the soil functional properties (e.g. hydraulic conductivity) is the single biggest uncertainty after climate in estimating soil moisture (e.g. Chen et al., 2015). The conclusion is that improving our predictions of atmospheric carbon at the end of the century is crucially dependent on improvements in the representation of soils in coupled global biogeochemical and climate models. This recognition is one of the motivations for the digital soil mapping (DSM) community to develop global soil maps focussing on functional soil properties (e.g. the GlobalSoilMap initiative). To date DSM has been based on interpolation of field data, and a process-based explanation for the spatial distribution of soils would be a useful tool to complement existing DSM statistical approaches. Moreover, DSM assumes that the soil properties do not change, and a dynamic soilscape model may provide deeper insight into the impacts of climate change of the soil properties themselves.

One of the challenges of modelling soil evolution is that soil has many facets that are important (e.g. soil physical properties such as particle size distribution, soil functional properties such hydraulic conductivity and soil chemical properties such as pH). For a comprehensive review of soil processes and the challenges in modelling them see Vereecken et al. (2016). Vereecken is a consolidated overview as perceived by the soil science community, and other fields could add even more aspects that need to be modelled (e.g. mine waste rehabilitation requires pedogenesis models, which has only passing mention in Vereecken). In many cases we need to know both their distribution in space (e.g. downslope and across hillslopes) and with depth down the profile, as well as how they may change with time. The following soil process chapters will discuss these properties in order of increasing complexity. The order in which they will be discussed is

- Soil depth
- Soil physics properties such as particle size distribution, clay content, porosity and bulk density
- Soil physics functional properties such as hydraulic conductivity, porosity, shear strength, rheology and water-holding capacity
- Soil chemistry and its impacts on weathering processes
- Soil carbon and soil organic matter and
- Soil transport downslope.

The initial chapters will discuss modelling of residual soil development, where there is no soil movement across or downslope. Initially we focus on processes in the soil profile that operate vertically. We will then move on to processes that move soil downslope – erosion at the soil surface and movement of the soil mass downslope by creep or flow.

Finally, there is the question of what is soil. Everything above solid bedrock is 'regolith', and if there is biological material as part of the regolith, it is 'soil' (Ollier and Pain 1996; Scott and Pain, 2009). Regolith is also sometimes referred to as saprolite, and underneath saprolite is lightly weathered fractured rock called saprock. The models described in the following chapters for soil depth and mineral matter do not explicitly model the biological component and so, strictly speaking, should be called regolith models rather than soils models, but this distinction will not be stressed.

Anderson and Anderson (2010) provide another definition. Saprolite consists of the rock in its original unweathered location, shape and organisation. Soil is that same material but no longer spatially organised as it was in the underlying saprolite. Using an analogy of children's building blocks, the saprolite has all the blocks nicely packed away with the blocks tightly packed together and organised with no space or gaps in between (i.e. porosity is low), while soil is as if those blocks were randomly tossed into the box so that there are random gaps and spaces between the blocks (i.e. porosity is high). The implication of this is that there is a step increase in the porosity (and without chemical alteration) from saprolite to soil. If the chemical dissolution loss of the rock is small (e.g. Anderson et al., 2002), this bulk density decrease is accompanied by a volume increase when saprolite is transformed into soil.

The typical mathematical approach for modelling environmental systems is based on mass balances. However, there has also been a history of research into energy and entropy methods, but they are not as enlightening about soil processes as mass balance methods. Minasny et al. (2008) performed an approximate energy balance for soil formation and estimated that the energy required for mineral chemical weathering was 0.01% of the total energy for soil formation, which makes it very hard to measure in the field, and very hard to formulate defensible mathematical equations using energy methods.

5.2 The Interaction between the Soil Profile and Weathering

Weathering of rock into its residual soil profile is a complicated process involving physical (e.g. fracturing by imposed stresses), chemical (e.g. transformation of minerals into clays) and biological processes (e.g. modifying the soil atmosphere by respiration). Any modelling of weathering is bound to be a simplification of these complex and interacting processes. The chapters that follow will look at these three components in turn.

We also simplify the discussion by treating physical and chemical weathering separately though there are no doubt strong interactions. The justification for this separation is that physical weathering is the first process to operate on particles because large particles have a low specific area (i.e. surface area per unit mass of rock), so chemical processes are relatively slow. As the rock breaks down from the action of physical processes, the specific area rises and at some time (depending on the relative intensities of physical and chemical processes) the specific area will be high enough that chemical processes will dominate the physical processes. Thus initially physical weathering dominates (though chemical weathering is still occurring at relatively low rates), and then at some stage chemical weathering dominates (though physical weathering is still occurring at a relatively low rate).

To further simplify the discussion that follows, the rate of weathering has been disaggregated from the products of weathering. This allows us to simplify the task of modelling weathering, and this separation follows the approaches and emphases of the literature. We focus on soil grading because many pedotransfer functions of soil functional properties that are used in environmental models (e.g. hydraulic conductivity, soil water-holding capacity) are based, at least in part, on the particle size distribution of the soil. This is not to downgrade the importance of other soil properties (e.g. organic matter, cation exchange capacity), only that it provides a starting point for modelling.

With respect to the evolution of the particle size distribution for physical weathering, there is a small amount of quantitative data on the rate of weathering but very little on the products of weathering. However, the recent work of Wells et al. (2005, 2006, 2007, 2008) has clarified, for one rock type, the grading characteristics of the weathering products. Again, for chemical weathering there is considerable data on the rates of weathering (though there is debate about the consistency of the data from laboratory and field experiments) but only qualitative data on the grading of the products of the weathering processes. Both bodies of work provide pathways for weathering but little information on the grading characteristics of the resulting particulate matter and its chemical reactivity.

Finally I will deviate slightly from traditional terminology for classifying weathering processes.

The chapter on 'physical weathering' is about how the geologic material in saprolite is physically transformed into finer material, primarily by fragmentation processes. In this physical transformation process we model only the transformation of the physical properties of the soil, for example, particle size distribution, porosity and bulk density. The mineralogy is unchanged. It is important to note that the processes causing this fragmentation may be physical (e.g. freeze/thaw), chemical (e.g. volume change as a result of hydration of minerals) or biological (e.g. fungal hyphae physically wedging cracks open).

The chapter on 'chemical weathering' is about the chemical transformation of the rock fragments in the soil into clays and other secondary minerals. While this is fundamentally a chemical process, the processes involved may have both chemical (e.g. carbonic acid infiltration as a result of rainfall equilibrating with carbon dioxide in the atmosphere) and biological components (e.g. carbon dioxide and anoxic soil atmosphere generation within the soil as a result of biodegradation). There may also be physical components if the chemical process triggers fragmentation (e.g. many hydration products increase in volume), thus generating more (chemically) untransformed surfaces available for further chemical transformation.

In the soils literature the use of 'physical' or 'chemical' descriptors is normally used to distinguish the processes driving the weathering. Here we are focussed on the results of the processes (and the modelling techniques to describe these transformations), and it is more convenient to distinguish them by how we model the dynamics of the physical and chemical products of these processes.

Because this book is about techniques for process modelling, many soil processes which we know operate (e.g. soil microbiology) but have not been described quantitatively (or in some cases even qualitatively described) have been dealt with here in a cursory fashion. Accordingly, the reader may find it useful to read the soilscape chapters that follow in conjunction with a more holistic description of soil processes such as Paton et al. (1995), or Schaetzl and Thompson (2015) to gain a complete view of soil evolution.

6 Soils: Soil Depth

6.1 Introduction

Soil depth is a fundamental property of soils. Sensitivity tests have shown that uncertainty in our knowledge of soil depth is a significant contributor to uncertainty in estimating carbon and water fluxes between the terrestrial environment and the atmosphere (Knorr and Lakshmi, 2001; Peterman et al., 2014), and thus the soil moisture and the biosequestration potential of the terrestrial environment.

Yet even something as apparently simple as soil depth can be difficult to define objectively, as the definitions discussed in the previous chapter for regolith, soils and saprolite indicate. The models described below define soil as all materials above the unweathered bedrock or saprolite, so soil depth is that depth to the bedrock or saprolite interface. In the field such a distinct boundary may or may not exist. For instance, the regolith may contain significant corestones surrounded by soil, and the corestones may become bigger and more frequent with depth until there is a transition to solid bedrock (e.g. Fletcher et al., 2006; Graham et al., 2010). Measurements of changes in porosity and bulk density down through the soil profile in the field show mixed results even when allowing for the impacts of agricultural landuse. Some profiles show no significant change down the soil profile suggesting that all change occurs at the base of the profile (e.g. Angers et al., 1997; Corti et al., 2001), while others show a trend of increasing bulk density with depth suggesting that changes may occur gradually throughout the profile (e.g. Unger and Jones, 1998; Graham et al., 2010). Ouimet (2008) noted a sharp boundary in bulk density within his profiles and attributed this to tree throw physically mixing (i.e. bioturbation) the soil. Finally, Richter and Markewitz (1995) report a gradual transition in the soil biogeochemistry with depth and argue that physical properties underestimate the depth of the biochemically active zone. Despite these conflicting data the literature discussed in this chapter typically assumes that a distinct interface between saprolite and soil exists.

6.2 Soil Depth and Bedrock Conversion to Soil

Ahnert (1976) first described a model for the evolution of soil depth:

$$\frac{dD}{dt} = \frac{\rho_s}{\rho_b}(P - E) \tag{6.1}$$

where D is the depth of the soil, P is the rate of conversion of bedrock to soil (depth/time), E is the erosion rate (depth/time) and ρ_s and ρ_b are the bulk densities of the soil averaged over depth and bedrock density at the soil-bedrock interface, respectively. Some early equations ignore changes in porosity and bulk density that occur in the conversion from bedrock to soil. Equation (6.1) includes this bulk density change. Ahnert did not consider bulk density changes resulting from soil production but they are included above for consistency with Chapter 2. The general form of Equation (6.1) (either with or without the density change) is the basis of all the soil depth modelling that has followed on from Ahnert.

This equation ignores the effect of dissolution because, as we see when we discuss bulk density evolution (Section 7.6), there is the possibility of complex interactions between dissolution and bulk density changes. However, in a model of chemical dissolution in soil Mudd and Furbish (2004) and Yoo and Mudd (2008) assumed that bioturbation would mix the soil from top to bottom and ensure that bulk density remained the same throughout the profile (Brimhall et al., 1992). Brimhall et al. (1992) found a complex pattern of dilation (e.g. increase in porosity) and collapse (decrease in porosity) based on depth, age and mineralogy, and attributed many of the changes to bioturbation and related processes, which will be discussed in Chapter 7.

The function P is (interchangeably) called the bedrock conversion function, bedrock weathering function, bedrock erosion function or soil production function. In recent years the soil production function (SPF) has been the most commonly used description, and this is the terminology used in this book. If, as is commonly accepted, the SPF declines with the depth of the soil-bedrock interface below the surface, then an equilibrium depth can be attained. A particularly simple solution arises when the 'exponential' SPF is used:

$$P = P_0 e^{-\lambda z} \tag{6.2}$$

where P_0 is the soil production rate when there is zero soil coverage, λ is the rate at which the conversion rate decreases with increasing soil depth and z is the depth below the soil surface. For equilibrium when the left-hand side of Equation (6.1) equals zero,

$$P_0 e^{-\lambda D} = E \tag{6.3}$$

where D is the depth below the surface of the bedrock interface (i.e. the soil depth), or rearranging

$$D = \lambda \ln \left(\frac{P_0}{E} \right) \tag{6.4}$$

This equation shows that if the SPF and erosion rates are the same everywhere, then the soil depth will also be the same everywhere. Also if the SPF rate increases or the erosion rate decreases, then the soil depth will increase. Since it is rare for soils to be constant depth everywhere, one of the conversion or erosion rates must vary in space if Equation (6.4) is to be true. Equation (6.4) also says that everything else being equal, then higher erosion rates should lead to shallower soils, which has been observed in the field (Cox, 1980; Dietrich et al., 1995; Heimsath et al., 1999).

The conceptual appeal of the exponential form of the SPF is that the conversion rate decreases with the depth of the bedrock interface below the soil surface. If soil production is driven by temperature (e.g. driving cycles of internal thermal stresses in rock particles) and soil wetness variations (e.g. driving cycles of salt crystallisation), both of which become less extreme with depth in the soil profile, this function is qualitatively consistent with observed behaviour. Recent research using cosmogenic nuclides for the dating of soil profiles has provided strong empirical evidence that, to first order, the exponential decline with depth in Equation (6.2) is well founded (Heimsath et al., 1997, 1999). There do, however, appear to be significant variations in both P_0 and λ from site to site (Anderson et al., 2002). Recent field research has clarified some causal factors for variations from site to site (Navarre-Sitchler et al., 2013) with

- the underlying geology and soil moisture being highlighted as dominant and
- less clear-cut conclusions on the role of temperature, with some authors showing a strong impact (e.g. Eppes and Griffing, 2010) while others show a relatively minor impact for sites below the snowline (e.g. Rasmussen et al., 2010).

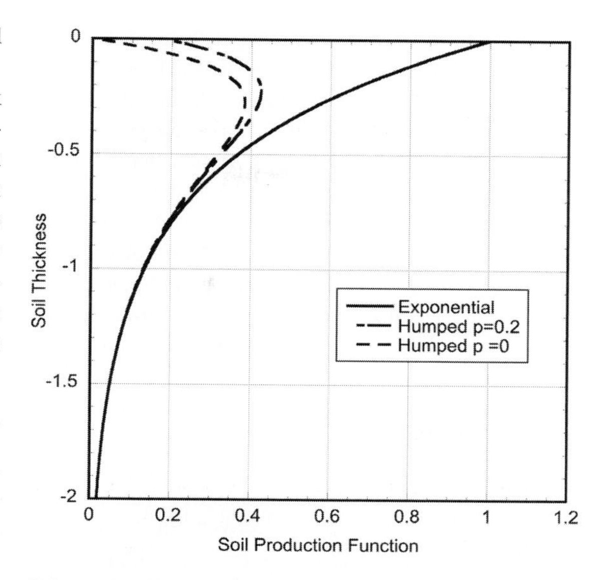

FIGURE 6.1: Example exponential and humped soil production functions (SPFs). Two humped functions are shown: (a) $p = 0$ is zero at the soil surface, (b) $p = 0.2$ is nonzero at the soil surface; all other parameters are the same. The depth decay parameter is the same for all three functions (i.e. $\lambda = \delta_2$), so, in this example, the exponential and humped functions converge to the same value for deeper soils.

There are other postulated SPFs. One criticism of the exponential function is that it has the highest weathering rate when there is no soil. If weathering is also driven by the time that the rock is wet (e.g. by chemical reactions), then the rate should be highest just below the surface when there is sufficient soil to store some of the infiltrated water (Carson and Kirkby, 1972). To model this, the "humped" function was developed (Figure 6.1). A variety of mathematical forms have been presented (e.g. Ahnert, 1977; Anderson, 2002), all of which are empirical. While Heimsath et al. (2009) and Stockmann et al. (2014) provided experimental evidence which supports the exponential function, their data are also consistent with humped behaviour because there is significant scatter in their data for shallow soil depths.

The key differences between the mathematical formulations of the humped SPF are whether (1) the function goes to zero or not, for zero soil depth (i.e. whether a bare rock surface can weather at all), (2) it asymptotes to zero for large depths (i.e. weathering declines to zero at depth like the exponential function) and (3) the function is continuously differentiable over its range (differentiability can be useful for numerical solvers). Many of the published equations do not meet the third point because they provide different equations for above and below the depth of maximum weathering rate. One equation that satisfies all three requirements is a modification of that used by Minasny and McBratney (2006) and Cohen et al. (2010):

$$P = P_0\left(1 - e^{-\delta_1 z - P_a}\right)e^{-\delta_2 z} \tag{6.5}$$

where δ_1, δ_2 and P_a are all parameters with positive values, where $\delta_1 > \delta_2$, and P_a controls the weathering rate for zero soil depth. The depth D^* at which maximum weathering rate occurs is

$$D^* = \frac{1}{\delta_1}\left[\ln\left(\frac{\delta_1 + \delta_2}{\delta_2}\right) - P_a\right] \tag{6.6}$$

Furbish and Fagherazzi (2001) present another equation, but it has the disadvantage that it cannot control the rate of decline with depth (i.e. the δ_2 in Equation (6.5)) independently of the weathering rate at the surface (i.e. P_a in Equation (6.5)), so we will not discuss it. Finally some studies suggest that soil production rates are independent of depth (Wilkinson et al., 2003), though they are in the minority and the reasons for their observed behaviour is not known.

All these equations are empirical and based on field experiment data including cosmogenic nuclide dating (CN), optically stimulated luminescence (OSL) dating and geochemical modelling of rivers and soil profiles (Minasny et al., 2015). One consideration with the dating methods is they rely upon assumptions about rates of exposure to the atmosphere. Bioturbation potentially mixes the soil, bringing deeper material to the surface and burying surface material. If the soil is fully mixed from top to bottom, then the cosmogenic dating methods will yield the average age of the soil, whereas OSL will yield the time since the particle(s) were last at the surface (Dunai, 2010).

One of the limitations of the SPF above is that they lack any explicit feedbacks with soil moisture and temperature. The work of Freer et al. (2002) provides empirical evidence of the importance of spatial feedbacks that are not included in the above formulations. They mapped the bedrock topography and soil depth at the Panola site and showed that the spatial pattern of the bedrock topography and soil depth appeared to be unconnected to the surface topography. The bedrock surface had patterns of drainage with valleys and holes in the bedrock surface suggestive of concentrated water flow over the bedrock surface (James et al., 2010). This disconnect between the surface topography and the bedrock topography, but where the patterns were clearly not random nor linked to geology, suggested feedbacks in the soil thickness dynamics that cannot be modelled by the simple models above.

Saco et al. (2006) investigated whether the bedrock patterns observed at Panola are a result of soil water where the lateral soil water drainage patterns are determined by the bedrock topography, not the soil surface topography. They extended the exponential SPF by making it a function of soil moisture so that the SPF became

$$P = f(M)P_0 e^{-\lambda z} \tag{6.7}$$

where M is the soil moisture as determined by using the wetness index

$$M = K\frac{A}{S} \tag{6.8}$$

but where the contributing area A and slope S used in the wetness index formulation were derived from analysing the bedrock topography rather than, as is traditionally done, the soil surface topography. Depending on the functional relationship between soil moisture and weathering used (i.e. the functional form of $f(M)$, e.g. using absolute depth of soil water versus percentage saturation of the soil profile) spatial patterns in the bedrock topography and soil depth qualitatively similar to that observed by Freer could be generated (Figure 16.1).

There are ecological feedbacks that I will touch on briefly here but will return to later in the book. In arid zones where the vegetation spatial distribution is banded, it has been observed (Ludwig et al., 2005) that infiltration was higher (and therefore so was soil moisture) within the bands of vegetation (enhanced organic matter changed the soil structure and increased the infiltration capacity) than between the vegetation bands. Underneath the bands of vegetation the soil depth was greater than in the interband areas where there was lower soil moisture and organic matter. The model formulations of Saco et al. (2006, 2007) are consistent with this behaviour.

Gabet and Mudd (2010) have proposed a tree throw mechanism that results in a distinct soil-saprolite boundary, a distinct change in bulk density at this boundary and which naturally leads to a humped SPF. In essence tree roots rip out rock from the saprolite during

rainfall. From this runoff time series they used a variety of published detachment and selective entrainment mechanisms to entrain the size fractions into the flow while recharging the surface layer from the layer below to maintain a mass balance in the surface layer. In this way they allowed the surface grading to evolve as a result of runoff events.

They identified that the Parker and Klingeman (1982) model best fit the data, and Sharmeen and Willgoose (2006, 2007) explored the long-term implications of this model for erosion rates for this site. Fitting the empirical fluvial transport-limited equation (Equation (4.10))

$$Q_s = KQ^m S^n \tag{7.1}$$

they found that not only did the erodibility K decline with time, as expected, but the parameters m, n and $\frac{m-1}{n}$ also changed significantly with time. The parameter m declined from 1.8 to a minimum of 1.0, then stabilised at around 1.2, while n declined from 2.1 to a minimum of -0.5, then stabilised at around 0.5, both over 100 years. The parameter trends matched field data from 20-year-old sites (Evans and Loch, 1996; Willgoose and Riley, 1998a). The changes in m and n resulted from longer and steeper slopes developing a coarser armour (so becoming relatively less erodible), and meant that fluvial erosion model parameters derived for unsorted sediments in flume studies (as is normally done; e.g. the initial $m = 1.8$ and $n = 2.1$ above resulted from using the Einstein-Brown sediment transport model on unsorted sediments) may not be appropriate for equilibrium soils on hillslopes (Section 4.2.4).

Their results also indicated that it would take about 200 years to stabilise to these equilibrium values, at which stage about 30 mm of cumulative erosion had occurred. We will return to the question of the rate at which these soils equilibrate later in the book, but we will note here that 30 mm erosion is quite small relative to elevations of landforms, providing evidence that the surface erosion properties of the soil would equilibrate long before the erosion created any significant landform evolution. The downside of the Sharmeen approach was the computations at each time step were very intensive, and the time resolution required was high (seconds during runoff events) so that it was not feasible to simulate more than a few short hillslopes for a few hundred years.

Cohen et al. (2009) using a new, more efficient, approach, mARM1D (to be described in detail in the next section), was able to replicate the results of Sharmeen when only armouring occurred on the surface. In an extension he included a weathering model that broke down the armour particles that was calibrated to laboratory weathering experiments (Wells et al., 2005, 2006, 2007, 2008) and found that the equilibrium time for the surface was longer than for the no weathering case, on the order of 500 years.

7.3 The Evolution of the Full Soil Profile

We now turn to models of the grading properties of the full soil profile, rather than only the surface.

There are a range of analytical particle size distribution functions that have been used to fit experimental particle size distributions, but it is difficult to distinguish them on causal grounds so typically authors have used the functions that fit their data best. For instance, Sanchidrian et al. (2014) compared 17 particle size distribution functions against 1,234 data sets and found different functions fit different parts of the distribution best, but none did best overall. Accordingly our focus has been on modelling the particle size distribution explicitly using physical principles, and that is the basis of the treatment that follows.

Legros and Pedro (1985) modelled the evolution of the grading of a soil column by physical and chemical weathering. They modelled the soil column as one lumped whole and compared the pedogenesis trajectories with field data. They modelled the soil as being broken up into 1,000 particle size fractions from 2 μm to 2,000 μm diameter. They ignored size fractions greater than 2 mm, and simulated processes that transformed particles in one size class into particles in another smaller size class. These processes included the following:

1. *Fragmentation*: A particle of a particular size is fragmented into a number of smaller particles of the same total mass as the original particle and
2. *Dissolution*: Which dissolved material from the surface of a particle, and when the particle was small enough it transitioned to the next smaller size class. The dissolved material was lost to the system.

Legros and Pedro did not detail their fragmentation and dissolution processes (e.g. size and number of particles, rate of processes), but their model exhibited a change of soil texture over time that they postulated was an analogue for field sites they presented.

Subsequent work in the field has either explicitly followed the methodology of Cohen et al. (2009, 2010) (i.e. mARM and mARM3D) or can be recast into Cohen's framework. The remainder of this section draws from Cohen's mARM3D model unless otherwise stated. We will start with a qualitative description of how the model is constructed, and only then will we fill in the mathematical details of how the model is solved.

A soil profile is broken into a series of layers (Figure 7.1), with a thin layer on the surface that directly interacts with water flowing over the surface, and a semi-infinite layer of fractured rock underlying the profile. Each layer has a particle size distribution describing the mass proportion of each size within the soil grading. Pedogenic processes are modelled within each layer, and vertical interactions between the layers are modelled so that material can be moved between layers. To model a catchment the catchment is discretised into a spatial grid of nodes so that the catchment consists of a number of profiles, one profile for each node. Each layer is fully mixed vertically. Using the surface elevations of each node, the surface water drainage pattern is modelled and surface water flows from node to node. This surface water erodes and/or deposits sediment at each node based on the geometry of the surface flow network, the sediment being transported within the flowing water and the local transport capacity at that node. This used the erosion models discussed in Chapter 4. mARM3D ignores groundwater flows between the nodes (though, in principle, there is nothing to stop this being modelled), and there was no interaction between the soil layers at one node with the soil layers at another node. The main limits on the number of soil layers and the number of nodes spatially are computer storage and compute times.

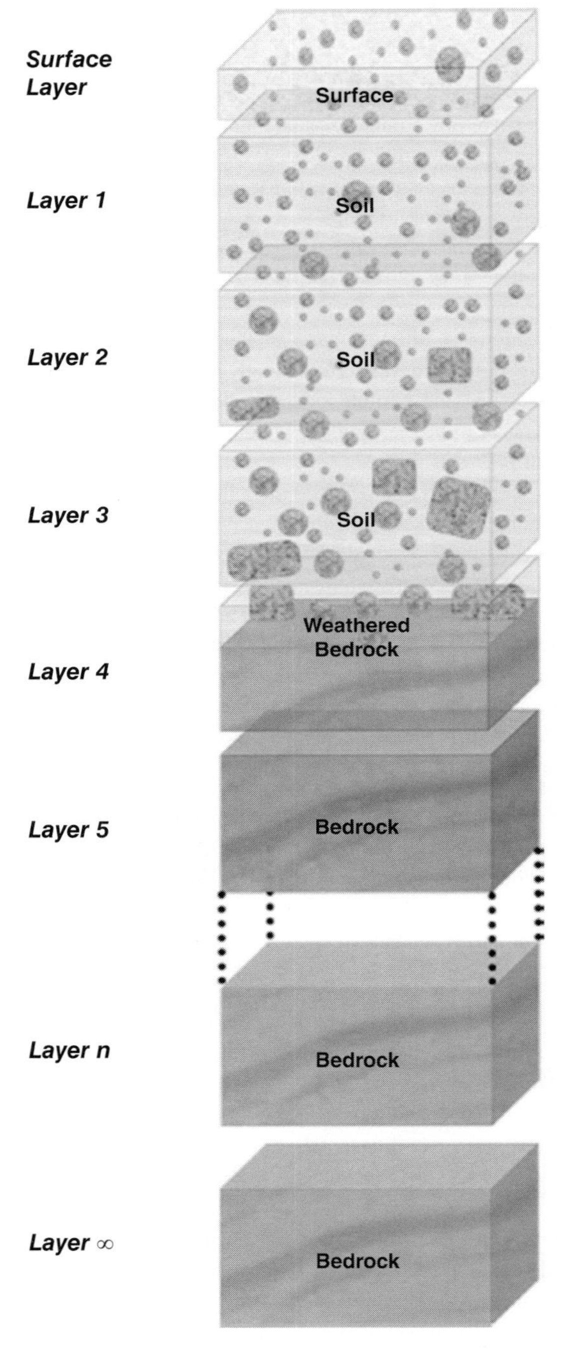

FIGURE 7.1: Schematic of the discretisation of the mARM and SSSPAM soil profile pedogenesis models (after Cohen et al., 2010).

7.3.1 Dynamics of a Single Soil Layer

We will first describe how mARM3D simulates each layer, and then we will discuss how the layers interact vertically. We start with the thin surface layer. This layer is the layer that interacts with the water flowing over the surface. In the most general case if the sediment transport capacity of the flow is more than the amount of sediment being carried by the flow, then the flow will erode material from the surface, while if the transport capacity is less than the sediment in the flow, then it will deposit sediment. If erosion occurs, then there is preferential entrainment of the finest fractions from the surface layer into the flow. To maintain the mass of the surface layer when there is erosion, material is transferred from the layer directly below to exactly balance the material being removed by erosion. If deposition occurs, then there is preferential deposition of the coarsest fractions (the coarsest particles settle out fastest) from the flow into the surface layer. To maintain the mass balance in the surface layer during deposition, material from the surface layer is transferred into the layer directly below to balance that material being deposited.

For the layers below the surface layer the mass balance is maintained at every time step. Thus if material is transferred from the layer to be put into another higher layer, then material is entrained from the layer below to balance the mass removed. Likewise if material is transferred into the layer from the layer above, then material from that layer is pushed down into the layer below. In the simplest case, if material is removed from the top layer by erosion, then material from the next layer below is transferred up to balance the mass lost to erosion. The next layer below that then has material transferred upwards to balance the material lost from the layer above, and this processes cascades down the profile to the bottom of the soil profile so that the soil-bedrock interface moves closer to the surface and the soil thins.

Within each layer weathering can occur and that weathering changes the grading distribution of that layer. Each layer weathers independently of the other layers. Mass conservative (i.e. physical weathering) and non-mass conservative (i.e. dissolution) weathering can be modelled, but to date full chemical weathering where dissolution is followed by precipitation of secondary minerals has not been modelled, because mARM does not have a coupled biogeochemical model to simulate in-profile geochemistry (Chapter 8). The weathering rate is typically depth dependent and can also be a function of particle grading and time. The model does not have a coupled groundwater model and so cannot model the effects of the interaction between soil moisture, soil grading and weathering rates, though a known, specified soil moisture spatial pattern (and thus a specified spatial weathering pattern) can be input.

The bottom layer of the soil profile has the grading of the underlying rock, rock being defined as material that is 100% the coarsest size fraction in the particle size distribution. That layer becomes soil as the weathering function breaks down the material into smaller size fractions. Any layers below cannot be soil until a specified proportion of the rock fraction in the layer above is broken down. Thus the bedrock-soil interface arises naturally as a result of the weathering process and is not explicitly modelled by the soil production function from Chapter 6.

Having outlined the conceptual approach used in mARM3D the mathematical details follow. The approach draws heavily from the state-space literature, and some of the terminology of this literature will be used where appropriate.

The soil grading at any given time and in any given layer is represented by a vector called the state vector $\underline{g}$:

$$\underline{g} = \begin{bmatrix} g_1 \\ g_2 \\ \vdots \\ g_{m-1} \\ g_m \end{bmatrix} \tag{7.2}$$

and the entries in the state vector, g_i, are the mass of sediment in each grading size range i where there are m size fractions for the grading. Cohen et al. (2010) used the proportion by mass of the layer in each size range as the state, but subsequent experience has found that using the actual mass in the layer in each size range makes it easier to apply mass conservation principles directly to the construction of the transition matrices (see below). The transition from the grading at any given time to the grading at the next time step is described by a matrix equation. It describes both how the grading changes with time and how the grading at one time, t_1, is related to the grading at some time in the future t_2:

$$\underline{g}(t_2) = \mathbf{R}\underline{g}(t_1) \tag{7.3}$$

where in the state-space literature the matrix $\mathbf{R}$ is called the transition matrix and is typically a function of the size of the timestep $(t_2 - t_1)$. Note that any set of coupled differential equations can also be expressed with Equation (7.3). The advantage of the matrix formulation is that we can explicitly formulate all the processes that change the grading with the same methodology.

To represent the soil profile, we write one of these equations for each layer so that we can write Equation (7.3) for each layer down the profile, where $\mathbf{R}_j$ is different in each layer reflecting how weathering processes change down the profile. For each layer j we then have

$$\underline{g}_j(t_2) = \mathbf{R}_j\underline{g}_j(t_1) \tag{7.4}$$

The discussion in the remainder of this section will consider only what happens in a single layer, so we will drop the j subscript for the moment.

We simulate each physical process as a multiplicative change to the state. If we have a single process, let's call it A, with a corresponding transition matrix $\mathbf{R}$, then the evolution of the grading vector over one timestep is (using subscripts for the timestep)

$$\underline{g}_{t+1} = \mathbf{R}\underline{g}_t \tag{7.5}$$

which is simply Equation (7.3) where one timestep is $(t_2 - t_1)$. To demonstrate how the method works for more than one process, consider the case where there are two independent physical processes, called A and B with corresponding transition matrices in Equation (7.5) of $\mathbf{R}$ and $\mathbf{S}$; then the combination of these processes on the soil grading is

$$\underline{g}_{t+1} = \mathbf{S}\left(\mathbf{R}\underline{g}_t\right) = \mathbf{SR}\underline{g}_t \tag{7.6}$$

where the order of the matrix multiplication implies that process A acts on the grading first and B operates second on the result of process A. This can be generalized to any number of processes, and is the key to the simplicity and generality of the approach. This splitting of processes into independent operations is 'operator splitting' (Section 2.2.2), and this splitting considerably simplifies the task of constructing the matrices for the combined effect of simultaneous physical processes.

Equations (7.3) to (7.6) follow the traditional presentation of transition matrices. In the discussion that follows it is more convenient to work with the marginal transition matrix than the actual transition matrix. In this case

$$\mathbf{R} = \mathbf{I} + \mathbf{A} \tag{7.7}$$

where $\mathbf{I}$ is the identity matrix (where $\underline{g} = \mathbf{I}\underline{g}$) and $\mathbf{A}$ is the marginal transition matrix for process A. The change in grading over a timestep is proportional to matrix $\mathbf{A}$. The advantage of formulating the matrices in this form is that the rate of the process represented by $\mathbf{A}$ (e.g. the weathering rate) can be changed simply by multiplying the marginal matrix by a scaling factor. For example, doubling the process rate is achieved by applying Equation (7.5) twice (the same as doing two timesteps), and if the timestep is small, then this is the same as multiplying the marginal transition matrix by 2 so that

$$\begin{aligned} &\text{process } A(\textit{nominal rate}) = \mathbf{R} = \mathbf{I} + \mathbf{A} \\ &\text{process } A(\textit{twice nominal rate}) = \mathbf{R}^2 = (\mathbf{I} + \mathbf{A})^2 \cong \mathbf{I} + 2\mathbf{A} \end{aligned} \tag{7.8}$$

where the $\mathbf{A}^2$ term is dropped since it is small for a small timestep. In the discussion that follows, the marginal transition matrix is used unless otherwise noted. In marginal transitional matrix form, Equation (7.6) is

$$\underline{g}_{t+1} = \mathbf{SR}\underline{g}_t = (\mathbf{I} + \mathbf{B})(\mathbf{I} + \mathbf{A})\underline{g}_t \tag{7.9}$$

where $\mathbf{A}$ and $\mathbf{B}$ are the marginal transition matrices for processes A and B, respectively, and correspond to the transition matrices $\mathbf{R}$ and $\mathbf{S}$ in Equation (7.6). Representing weathering and other processes within a layer is then a matter of constructing the marginal transition matrices and repeatedly applying the equations above.

Before we discuss how soil layers interact, a simple example will be useful to explain how the details of this process work. Consider an example where there are only three grading size ranges (for convenience let's call them $i = 1, 2, 3$, 'small', 'medium' and 'large' grain sizes), and we will construct a marginal transition matrix for a weathering process that breaks large particles into medium particles, medium particles into small particles, and leaves the small particles unchanged. If at any one timestep 2% of the mass of the large particles are weathered into medium sized particles and 1% of the medium into fine, then the marginal transitional matrix is

$$\mathbf{A} = \begin{bmatrix} 0 & 0.01 & 0 \\ 0 & -0.01 & 0.02 \\ 0 & 0 & -0.02 \end{bmatrix} \tag{7.10}$$

where the diagonal element says how much the mass changes in that grading range and the off-diagonal terms say how much of a different size range is added to it. For instance, for the medium (i.e. $i = 2$) size range

$$g_{2,t+1} = (1 - 0.01)g_{2,t} + 0.02g_{3,t} \tag{7.11}$$

where $\mathbf{A}_{2,2} = -0.01$ indicates that 1% of the mass in the medium size range is removed from it and $\mathbf{A}_{3,2} = 0.02$ indicates that 2% of the mass in the large size range is added to the medium size range.

Note that for mass conservation each column of the marginal transition matrix must add to zero. If mass is lost (e.g. by dissolution of particles), then the column(s) will sum to less than zero. Also note that none of the diagonal terms can be less than -1 because otherwise the equation is transforming more mass in the layer than actually exists in that layer. In this latter case a smaller timestep is required so that all the elements of the marginal transition matrix are smaller.

Finally it should be noted that, within the physical constraints above, there is considerable flexibility for the contents of the matrix, and therefore what processes can be simulated. The entries below the diagonal will normally all be zero because otherwise the process being represented by the matrix will be making larger particles from smaller particles (albeit if you are modelling soil aggregation or particle cementation this may be entirely reasonable). In the example in Equation (7.10) particles changed only to particles in the next size class down, which may not be the case, for instance, for particles that fragment into many smaller particles, or for spalling where there is a single large particle and many smaller particles resulting from the weathering. We will discuss these cases later.

While not presented this way the work of Salvador-Blanes et al. (2007) can be cast into this matrix formulation. Salvador-Blanes modified the approach of Legros and Pedro (1985) to include the modelling of the breakdown of particles less than 2 mm in diameter and used Legros's method whereby a particle was transformed into a particle in the next size class down (he defined that as 2 μm smaller). Both Legros and Salvador-Blanes used 1,000 size fractions for particles less than 2 mm, and this is easily transformed into Equation (7.10) (the $\mathbf{A}$ matrix will be $1{,}000 \times 1{,}000$) with only the diagonal terms and that entry directly above the diagonal term being nonzero, and the values being the rates at which one size fraction is transformed to the next smaller size fraction per timestep.

The recasting of Salvador-Blanes's approach highlights an important but implied aspect of the matrix methodology. The matrix $\mathbf{A}$ is not describing how an individual particle is breaking down, but the aggregate result of the breakdown of the many particles in a size fraction. It is not possible to move a single particle to the next size class smaller and maintain mass conservation without generating some other smaller particles. However, the matrix method describes the aggregate of a given mass of particles, and it is possible to have many particles transform to the next size class lower without generating an array of fine particles.

7.3.2 Interactions between Soil Layers

We now describe how one layer within the soil profile interacts with another layer within the soil profile. Let us consider two layers j and k. We can construct a matrix equation that describes how layer j changes layer k in one timestep:

$$\underline{g}_{k,t+1} = \mathbf{L}_{k,j}\underline{g}_{j,t} \tag{7.12}$$

This equation says how much mass in layer j in each of the grading size fractions is added to or subtracted from layer k in one timestep. Equation (7.12) models only how material is moved between layers and not weathering. We have used the matrix notation $\mathbf{L}$ here to distinguish this movement between layers from the weathering process matrices (that transform the particle size distribution within a single layer). For mass conservation in layer k, if mass is added to layer k from layer j, then an equal amount of mass must be removed from layer k and moved to another layer. The combination of the weathering and movement can be expressed in matrix form if we construct a grading vector that merges all the grading vectors for each of the individual layers, and the marginal transition matrix is constructed from the layer transition matrices and interlayer movement transition matrices so that

$$\begin{bmatrix} \underline{g}_1 \\ \underline{g}_2 \\ \vdots \\ \underline{g}_n \\ \underline{g}_\infty \end{bmatrix}_{t+1} = \begin{bmatrix} \underline{g}_1 \\ \underline{g}_2 \\ \vdots \\ \underline{g}_n \\ \underline{g}_\infty \end{bmatrix}_t + \begin{bmatrix} \mathbf{A}_{1,1} & \mathbf{L}_{1,2} & \cdots & \mathbf{L}_{1,n} & \mathbf{L}_{1,\infty} \\ \mathbf{L}_{2,1} & \mathbf{A}_{2,2} & \cdots & \mathbf{L}_{2,n} & \mathbf{L}_{2,\infty} \\ \vdots & \vdots & \ddots & \vdots & \vdots \\ \mathbf{L}_{n,1} & \mathbf{L}_{n,2} & \cdots & \mathbf{A}_{n,n} & \mathbf{L}_{n,\infty} \\ [0] & [0] & \cdots & [0] & \mathbf{A}_{\infty,\infty} \end{bmatrix} \begin{bmatrix} \underline{g}_1 \\ \underline{g}_2 \\ \vdots \\ \underline{g}_n \\ \underline{g}_\infty \end{bmatrix}_t \tag{7.13}$$

where the state vector is the gradings for all the layers from the surface armouring layer (subscript 1 indicates the surface armouring layer), through the profile layers and including the semi-infinite underlying layer (subscript ∞). The state vector is thus a vector of length $m(n+1)$ and the matrix is of dimension $(m(n+1)) \times (m(n+1))$. The notation $[0]$ indicates a matrix that is $m \times m$ and where all matrix entries are zeros. Equation (7.13) shows that all layers can interact with all other layers both above and below, including the semi-infinite underlying layer. Looking at the bottom row of the matrix, the semi-infinite layer can change through time (as a result of the matrix $\mathbf{A}_{\infty,\infty}$), but it cannot be influenced by any of the overlying layers (since all the entries are $[0]$). Equation (7.13) can be written in a more compact form

$$\underline{\underline{g}}_{t+1} = (\mathbf{I} + \mathbf{B})\underline{\underline{g}}_t \tag{7.14}$$

where the double underbar notation distinguishes the grading vector for an individual layer (one underbar) from the vector for the gradings for all the layers (two underbars). The latter is sometimes called a supervector in the modelling literature (because it is a vector of vectors).

The construction of matrix $\mathbf{B}$ appears at face value to be rather daunting simply because of its size. However, Equation (7.13) is the most general statement of the problem, and in many cases simplifications are possible. For example,

- *If interactions between layers occur only between adjacent layers:* In this case for layer j, the only matrices that are nonzero are the matrices $\mathbf{A}_{j,j}$ for the weathering within that layer, $\mathbf{L}_{j,j-1}$ and $\mathbf{L}_{j,j+1}$ which describe how the layer above and below, respectively change layer j, and $\mathbf{L}_{j-1,j}$ and $\mathbf{L}_{j+1,j}$ which describe how layer j changes the layer above and below, respectively:

$$\begin{bmatrix} \mathbf{A}_{1,1} & \mathbf{L}_{1,2} & [0] & \cdots & [0] & [0] & [0] \\ \mathbf{L}_{2,1} & \mathbf{A}_{2,2} & \mathbf{L}_{2,3} & \cdots & [0] & [0] & [0] \\ [0] & \mathbf{L}_{3,2} & \mathbf{A}_{3,3} & \cdots & [0] & [0] & [0] \\ \vdots & \vdots & \vdots & \ddots & \vdots & \vdots & \vdots \\ [0] & [0] & [0] & \cdots & \mathbf{A}_{n-1,n-1} & \mathbf{L}_{n-1,n} & [0] \\ [0] & [0] & [0] & \cdots & \mathbf{L}_{n,n-1} & \mathbf{A}_{n,n} & \mathbf{L}_{n,\infty} \\ [0] & [0] & [0] & \cdots & [0] & [0] & \mathbf{A}_{\infty,\infty} \end{bmatrix} \tag{7.15}$$

- *If in addition to interactions only between adjacent layers, there is no change in the grading over time within a layer:* This might happen when there is only mixing of the soil (e.g. bioturbation) and no breakdown of the mineral matter. The matrix in Equation (7.15) simplifies even further to

$$\begin{bmatrix} [0] & \mathbf{L}_{1,2} & [0] & \cdots & [0] & [0] & [0] \\ \mathbf{L}_{2,1} & [0] & \mathbf{L}_{2,3} & \cdots & [0] & [0] & [0] \\ [0] & \mathbf{L}_{3,2} & [0] & \cdots & [0] & [0] & [0] \\ \vdots & \vdots & \vdots & \ddots & \vdots & \vdots & \vdots \\ [0] & [0] & [0] & \cdots & [0] & \mathbf{L}_{n-1,n} & [0] \\ [0] & [0] & [0] & \cdots & \mathbf{L}_{n,n-1} & [0] & \mathbf{L}_{n,\infty} \\ [0] & [0] & [0] & \cdots & [0] & [0] & [0] \end{bmatrix} \quad (7.16)$$

- *Adjusting the layers in response to erosion at the surface:* When material is removed from the surface layer, then material to balance that lost to erosion must be removed from the layer below to make sure the mass in the layer does not change. This cascades down through all layers in the profile:

$$\begin{bmatrix} -\frac{E}{d_1}\mathbf{A}_{1,1} & \frac{E}{d_1}\mathbf{I} & [0] & \cdots & [0] & [0] & [0] \\ [0] & -\frac{E}{d_2}\mathbf{I} & \frac{E}{d_2}\mathbf{I} & \cdots & [0] & [0] & [0] \\ [0] & [0] & -\frac{E}{d_3}\mathbf{I} & \cdots & [0] & [0] & [0] \\ \vdots & \vdots & \vdots & \ddots & \vdots & \vdots & \vdots \\ [0] & [0] & [0] & \cdots & -\frac{E}{d_{n-1}}\mathbf{I} & \frac{E}{d_{n-1}}\mathbf{I} & [0] \\ [0] & [0] & [0] & \cdots & [0] & -\frac{E}{d_n}\mathbf{I} & \frac{E}{d_n}\mathbf{I} \\ [0] & [0] & [0] & \cdots & [0] & [0] & [0] \end{bmatrix} \quad (7.17)$$

where E is the erosion in one timestep in depth units, d_i is the depth of layer i (which converts sediment mass due to erosion E into a proportion of the layer mass) and $\mathbf{I}$ is a $k \times k$ identity matrix. The matrix $\mathbf{A}$ is the armouring transition matrix for the surface layer and determines the size selectivity of the sediment entrainment due to erosion. As an aside, this is the first time that bulk density of the soil appears in the matrix methodology. Bulk density is the conversion factor between depth of soil eroded and mass of soil eroded. If soil erosion is expressed in mass rather than depth units, then the bulk density conversion is not required.

7.3.3 Constructing the A Matrix for Weathering

We will now discuss how the matrix $\mathbf{A}$ is formulated to model weathering. Unless otherwise stated, the discussion in this section is for one layer, and we drop the layer subscript j for clarity of discussion. We will first discuss mass conservative weathering and then generalise the discussion to non-mass conservative weathering afterwards.

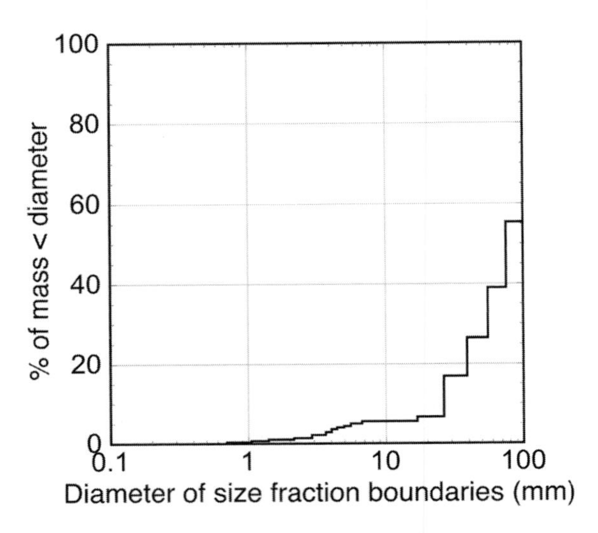

FIGURE 7.2: How soil grading is conceptualised in mARM and SSSPAM as uniformly distributed within size classes. Data are the Ranger Mine grading used in Willgoose and Sharmeen (2006) and Cohen et al. (2009, 2010).

The basis of the conceptualisation of weathering is that particles are spread uniformly within each size range within which they fall (Figure 7.2). Thus some particles will be at the lower boundary of the size, some in the middle and some at the top of the size range. We conceptualise the fragmentation process as a parent particle breaking into a number of daughter particles (Figure 7.3). Figure 7.3 shows that depending on the fragmentation mechanism there may be a range of different types of daughter particles created. If we assume that the density distribution within a single particle size fraction is constant, then mass conservation implies volume conservation. In this case the total volume of the daughter particles is the same as the volume of the parent particle.

We now extend our discussion from what happens to a single particle to what happens to all the particles within a size class range. Figure 7.4 shows what will happen to the larger size range when the particles break into two particles of different diameter, if they all follow the rules for breaking of a single particle (i.e. all particles break with exactly the same geometry). In the discussions that follow we assume that all particles are spherical. Initially we will consider the case where a parent particle breaks into two daughter particles. Figure 7.4 shows how the geometry of a single particle breaking allows us to map the parent particle size grade to the smaller dimensions of the two daughter particles' size grades. Likewise we can map the largest size from the parent size gradings to the largest size of the daughter size grading. From the geometry of breaking of the single particle we know what the proportions of the mass of daughter particles should be in each of the two daughter particle gradings ranges. Figure 7.4 shows that if we take one parent size fraction, then the daughter particle's size distribution will not typically neatly fit into a single size fraction, but will need to be interpolated across a number of size fractions. These proportions will be a function of the diameters of the daughter particles and of the size grading fractions adopted by the user in the modelling. Finally, if the distribution between the lower and upper size of the parent particles is uniform, then the distribution of the daughter particles is also uniform between the lower and upper size grading.

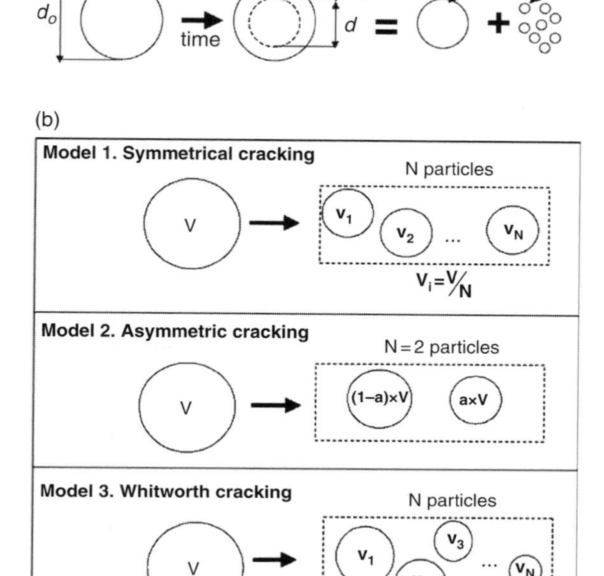

FIGURE 7.3: Schematic of fracturing models for rock particles (a) breaking of small particles from a rind and (b) body fragmentation (from Sharmeen and Willgoose, 2006).

Using these assumptions it is relatively straightforward, though tedious, to construct a matrix to simulate weathering. The calculations in Figure 7.4 are done for each grading range and the results summed.

However, a conceptual simplification is possible. There is a combination of grading size fractions and weathering processes that leads to particularly simple results, and which allow us to derive some analytical results that provide useful insight into the time variation of grading under the action of weathering. We will use this to demonstrate how the weathering matrix works and derive some simple weathering results. We will then show that this simple model can be used to derive more complex models.

For this example we will define the fractions in the grading size distribution such that the lower diameter limit of each size fraction is $\left(0.5^{1/3}\right)$ times the upper diameter limit for that same size fraction. If we take the maximum size limit as 2 mm, then this yields the limits for the size fractions as in Table 7.1. There is no special significance to the upper and lower limits of 2 mm and 0.125 mm in the table; they are simply used to show what this grading looks like. It is useful to examine this size grading because this grading fractionation is different from that normally

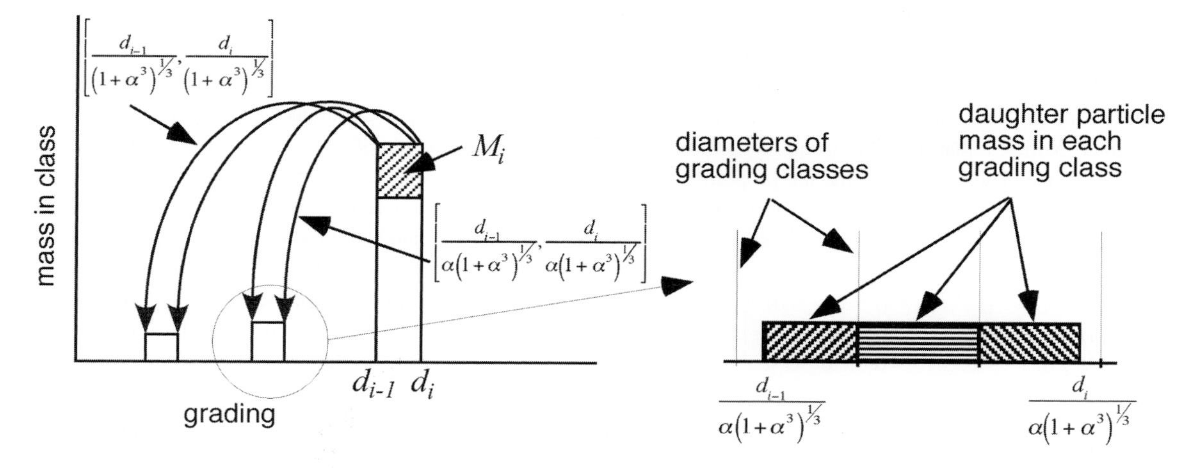

FIGURE 7.4: Conceptualisation of how size fractions in the grading are transformed in each time step of mARM (from Cohen et al., 2009).

TABLE 7.1: Size grading classes for different weathering mechanisms for two equal particles and three equal particles fracturing

Size grading class	Upper-lower size limit (mm)	
	2 daughters	3 daughters
12	2.0–1.587	2.0–1.387
11	1.587–1.260	1.387–0.962
10	1.260–1.000	0.962–0.666
9	1.000–0.794	0.666–0.462
8	0.794–0.630	0.462–0.321
7	0.630–0.500	0.321–0.222
6	0.500–0.397	0.222–0.154
5	0.397–0.315	0.154–0.107
4	0.315–0.250	0.107–0.074
3	0.250–0.198	0.074–0.051
2	0.198–0.158	0.051–0.037
1	0.158–0.125	0.037–0.025

used for soils analysis (e.g. phi grading) and thus will be unfamiliar to readers. The significance of the grading in this example is that if a spherical particle splits into two equal volume spherical daughter particles, then for conservation of volume all the particles in one size grading will fall exactly into the size grading of the next size fraction smaller. For example, for the grading class 12 in Table 7.1 a particle with the largest diameter of 2 mm will split into two particles of diameter 1.587 mm, and a particle with the smallest diameter of 1.587 mm will split into two particles of diameter 1.260 mm. These upper and lower size limits define the size range for the next smaller Grading Class 11. This is true for all Grading Class ranges in Table 7.1.

Accordingly, to assemble the weathering matrix with this grading when a parent particle fractures into two equally sized daughter particles is straightforward because the daughter particles only ever fall into the size fraction below, and this size fraction below can receive only daughter particles that fracture from the size fraction directly above. Thus if we say that the proportion of the particles that fracture for one timestep is α, then the weathering matrix is

$$\mathbf{A} = \begin{bmatrix} 0 & \alpha & 0 & \cdots & 0 & 0 & 0 \\ 0 & -\alpha & \alpha & \cdots & 0 & 0 & 0 \\ 0 & 0 & -\alpha & \cdots & 0 & 0 & 0 \\ \vdots & \vdots & \vdots & \ddots & \vdots & \vdots & \vdots \\ 0 & 0 & 0 & \cdots & -\alpha & \alpha & 0 \\ 0 & 0 & 0 & \cdots & 0 & -\alpha & \alpha \\ 0 & 0 & 0 & \cdots & 0 & 0 & -\alpha \end{bmatrix} \tag{7.18}$$

The diagonal and off-diagonal α terms mean that α of the mass in the larger size grading is all added to the next smaller size grading. The diagonal element for the smallest fraction is zero because we assume that particles this size cannot weather any smaller. Equation (7.18) is mass conservative, so all the columns of the matrix add to 0.

It is possible to calculate the time variation of the mean diameter of the particle size distribution purely from the $\mathbf{I}+\mathbf{A}$ matrix using Equation (7.18). With the exception of the smallest size grading, the mass g_i that is in any given size fraction i will change after one timestep so that $(1-\alpha)$ remains in size fraction i while α will now be in the next size fraction smaller $i-1$, which in the case of Equation (7.18) is $\left(0.5^{1/3}\right)$ smaller than the size fraction above, and from the way the size fractions are defined in Table 7.1 that is true for all size fractions i. For the smallest size fraction there is no change. If we consider the case where all the mass is initially concentrated in the largest size fractions and there is an insignificant mass in the smallest fraction (this can be achieved by defining the smallest size fraction to be very small), then in one timestep (time changing from t to $t+1$) the mean diameter of the soil changes from

$$\begin{aligned} d_{\text{mean},t+1} &= \left((1-\alpha)+\alpha 0.5^{1/3}\right) d_{\text{mean},t} \\ &= (1-0.2063\alpha) d_{\text{mean},t} \end{aligned} \tag{7.19}$$

which allows us to derive an equation for the evolution of the mean diameter of the soil T timesteps into the future:

$$d_{\text{mean},t=T} = (1-0.2063\alpha)^T d_{\text{mean},t=0} \tag{7.20}$$

which is a semi-log linear relationship between diameter and time, where the slope on a semi-log plot is -0.2063α. This can also be expressed as an exponential so that

$$d_{\text{mean},t=T} = d_{\text{mean},t=0} e^{\ln(1-0.2063\alpha)T} \tag{7.21}$$

Figure 7.5 shows this curve for a number of values of α and with a starting mean diameter of 3.5 mm.

The exact details of Equations (7.19) to (7.21) are a function of the specific assumptions in Equations (7.18), but it is straightforward to extend this analysis to any distribution of daughter products, and the only part of Equation (7.19) that will change is the number inside the parentheses; the relationship itself will still be log-log linear. This is true provided only that (1) the definition of the size fractions is defined as in Table 7.1 for two daughter particles, (2) the fracture model is independent of the diameter of the particle so that the daughter particles are always the same size relative to the parent particle and that this fracture model is independent of

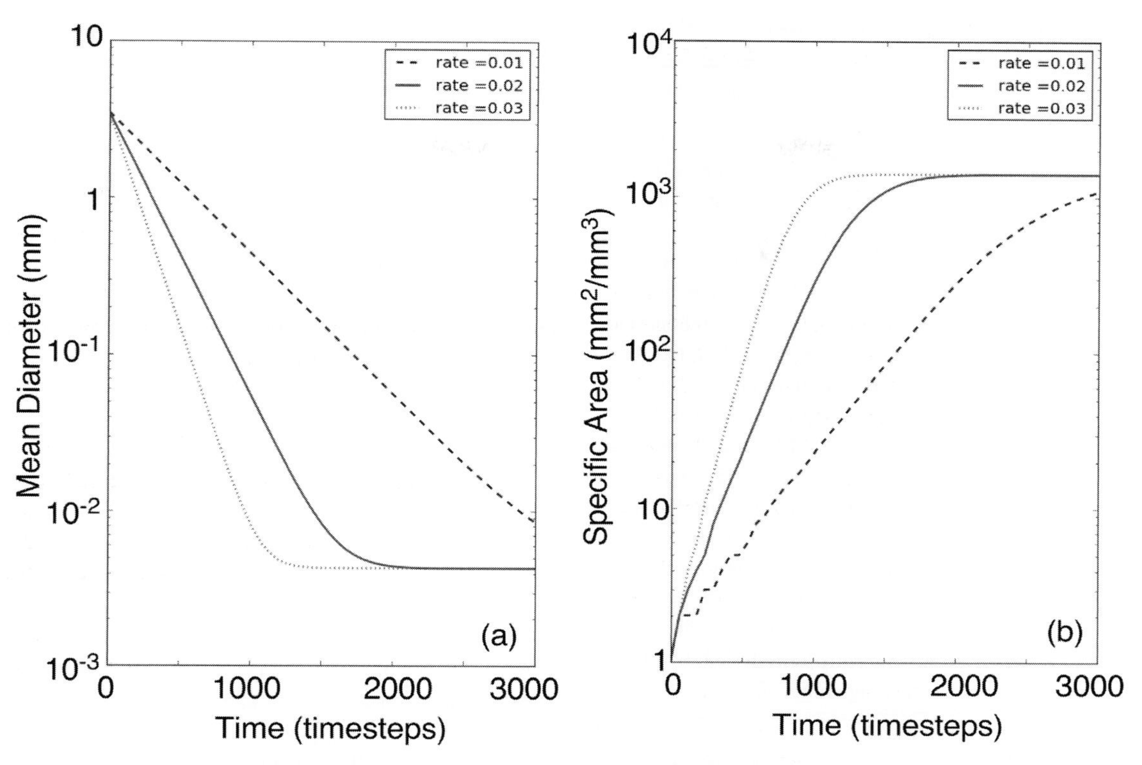

FIGURE 7.5: Numerical solution of the time evolution of the mean of the particle size distribution for a range of weathering rates. The parameter 'rate' is $rate = 0.2063\alpha$ in Equation (7.19). The levelling off on the right-hand side of each graph is because at this stage all soil is in the finest size fraction in the modelled soil grading.

the diameter of the parent particle and (3) the smallest size fraction is small enough that only a minor amount of the mass occurs within the smallest fraction (otherwise the curves will start to level out as the particles age as seen in Figure 7.5 on the right-hand side).

The assumption that particles split into two equally sized particles leading to Equation (7.21) is rather restrictive, but this result can be generalised as follows. Consider first the case where N equally sized particles are generated at each time step instead of the two in the example above. This means that the diameter of the daughter particles is now $N^{-1/3}$ the diameter of the parent particle. In the same way that we developed a grading range definition for two daughter particles fracturing in Table 7.1, we can do this in general for all values of N. In Table 7.1 fracturing for three equal daughters is shown. Note that the total size range covered by the 12 size fractions in Table 7.1 is different, with the three daughters column going down to 0.025 mm, while the two daughters range goes down to only 0.125 mm. Thus if you wish to cover a specific total size range (e.g. for a given soil), the two daughters size range requires more size fractions. If the proportion of a given size fraction that breaks down each timestep is α as before, then Equation (7.18) is still applicable except now the definition of the size grading ranges is different and the matrix will be of a different dimension. The equivalent three daughters result to Equation (7.19) is

$$\begin{aligned} d_{\text{mean},t+1} &= \left((1-\alpha) + 0.333^{1/3}\alpha\right) d_{\text{mean},t} \\ &= (1 - 0.3066\alpha) d_{\text{mean},t} \end{aligned} \tag{7.22}$$

and Equation (7.20) becomes

$$d_{\text{mean},t=T} = (1 - 0.3066\alpha)^T \, d_{\text{mean},t=0} \tag{7.23}$$

In general for N daughter particles

$$d_{\text{mean},t=T} = \left(1 - \left(1 - N^{-1/3}\right)\alpha\right)^T d_{\text{mean},t=0} \tag{7.24}$$

so it is clear that if particles break into a larger number of smaller particles at each timestep, then the soil will become finer at a faster rate, and Equation (7.24) defines what that speedup will be.

In the discussion that follows we further generalise the fracture geometries considered. To simplify the discussion

TABLE 7.2: Fragmentation notation examples

Fragmentation notation	Physical example for daughter particle geometry
p2-0–100	2 particles of 50% volume of parent. The split in two geometry of Wells et al. (2008).
p2-0–50-50	1 particle of 50% volume, and 2 particles of 25% volume of parent.
p2-0–50-0–50	1 particle of 50% volume, 0 particles of 25% volume and 4 particles of 12.5% volume of parent. Spalling-like behaviour.
p2-0-0-0–100	0 particles of 50% and 25% volume, and 8 particles of 12.5% volume of the parent. Used by Finke (2012) in the SOILGEN model.
p2-0–50-25-12.5-12.5	1 particle of 50%, 1 particle of 25%, 1 particle of 12.5% and 2 particles of 6.25% volume. A scaling fracturing similar to Whitworth cracking (Figure 7.3).

of geometry we introduce a shorthand for fragmentation geometry, which indicates how, on average, the volume of particles will be distributed after a fragmentation event assuming the size grading definitions in Table 7.1. The general form is pX-AAA-BBB-CCC-DDD where X is the number of daughter particles the grading fractions have been defined for (i.e. X = 2 for the two particle grading in Table 7.1, X = 3 for the three particle grading), AAA is the percentage of volume that remains in the parent size grading fraction after fragmentation, BBB is the proportion in the next size grading smaller, CCC in the next size down again and so on. In this notation where all of the particles split into two with two equally sized particles, then the shorthand is p2-0–100. Some geometry examples are listed in Table 7.2. Some simple examples of fracture geometries and their fragmentation notation are also shown. While all the examples in Table 7.1 have no mass left in the parent grading after fragmentation (they all have a leading zero), at the end of this section an important, but more complex, fragmentation geometry that leads to a nonzero percentage in the parent grading will be discussed.

Instead of two equally sized daughter particles, let us now break the particle into three daughters where one daughter is half the volume of the original and the other two daughters are a quarter of the volume of the original (i.e. p2-0–50-50). Mass conservation still applies, but what now happens is that half the weathered volume goes into the next class down from the parent, while half the weathered volume goes into the class 2 classes below the parent so the **A** matrix is

$$\mathbf{A} = \begin{bmatrix} 0 & \alpha & \alpha/2 & \cdots & 0 & 0 & 0 \\ 0 & -\alpha & \alpha/2 & \cdots & 0 & 0 & 0 \\ 0 & 0 & -\alpha & \cdots & 0 & 0 & 0 \\ \vdots & \vdots & \vdots & \ddots & \vdots & \vdots & \vdots \\ 0 & 0 & 0 & \cdots & -\alpha & \alpha/2 & \alpha/2 \\ 0 & 0 & 0 & \cdots & 0 & -\alpha & \alpha/2 \\ 0 & 0 & 0 & \cdots & 0 & 0 & -\alpha \end{bmatrix} \quad (7.25)$$

and Equation (7.19) becomes

$$\begin{aligned} d_{\text{mean},t+1} &= \left((1-\alpha) + \alpha/2\, 0.5^{1/3} + \alpha/2\, 0.25^{1/3}\right) d_{\text{mean},t} \\ &= (1 - 0.2882\alpha) d_{\text{mean},t} \end{aligned} \quad (7.26)$$

As expected this case weathers faster than the two equal daughters case. In a similar fashion you could consider that this three daughters case can be extended where one of the two smaller particles itself breaks into two, so the daughter particles are $1 \times (1/2)$ volume, $1 \times (1/4)$ volume and $2 \times (1/8)$ volume and so on.

Similarly the three equal daughters particle fracturing could be extended by having one of the daughters break into three equal volume smaller particles, so that the daughters are two particles with 1/3 the parent volume (i.e. 2/3 of the parent volume) and three particles with 1/9 the parent volume (i.e. 1/3 of the parent volume), yielding p3-67.7-33.3. In this case Equation (7.25) becomes

$$\mathbf{A} = \begin{bmatrix} 0 & \alpha & \alpha/3 & \cdots & 0 & 0 & 0 \\ 0 & -\alpha & 2\alpha/3 & \cdots & 0 & 0 & 0 \\ 0 & 0 & -\alpha & \cdots & 0 & 0 & 0 \\ \vdots & \vdots & \vdots & \ddots & \vdots & \vdots & \vdots \\ 0 & 0 & 0 & \cdots & -\alpha & 2\alpha/3 & \alpha/3 \\ 0 & 0 & 0 & \cdots & 0 & -\alpha & 2\alpha/3 \\ 0 & 0 & 0 & \cdots & 0 & 0 & -\alpha \end{bmatrix} \quad (7.27)$$

and Equation (7.26) becomes

$$\begin{aligned} d_{\text{mean},t+1} &= \left((1-\alpha) + 2\alpha/3\, 0.333^{1/3} + \alpha/3\, 0.333^{2/3}\right) d_{\text{mean},t} \\ &= (1 - 0.3775\alpha) d_{\text{mean},t} \end{aligned} \quad (7.28)$$

If two of the three daughters break into three (i.e. p3-33.3–66.7), then

$$\begin{aligned} d_{\text{mean},t+1} &= \left((1-\alpha) + \alpha/3\, 0.333^{1/3} + 2\alpha/3\, 0.333^{2/3}\right) d_{\text{mean},t} \\ &= (1 - 0.4484\alpha) d_{\text{mean},t} \end{aligned} \quad (7.29)$$

and as for the two daughters cases, both of these cases weather faster than the case where all the daughters are of equal sizes, and when more fine particles are generated (i.e. Equation (7.29) versus (7.28), or Equation (7.28) versus (7.26)) then the soil weathers faster even when the rate of breakdown per timestep is unchanged.

It should be clear from these examples that the combinations of sizes of daughter particles possible are quite extensive. The only constraint is a geometry constraint driven by the grading size definition adopted from Table 7.1. If the two daughter grading range in Table 7.1 is adopted, then particles must break into particles where the volumes of the daughters are related to the parent particle volume by integer powers of 1/2. Likewise for the three daughter grading range in Table 7.1, then all the daughters must have volumes that are integer powers of 1/3. This integer power constraint is so that all the daughters fall entirely and only into one of the grading size ranges and do not span two grading size ranges.

This section has shown how to construct the weathering **A** matrix using a conceptualisation of the fracturing of individual parent particles into a range of daughter particles. Thus everything in this section has been strongly physically based on fracturing mechanisms that are, in principle, observable in the laboratory or the field. This presentation was intentional because it highlights the link between fracturing of individual particles and the rather less tangible mathematics of the **A** matrix.

We will deviate from this philosophy for a moment to generalise Equation (7.18), which will then allow us to provide a general analytical solution to the change in grading over time. When constructing the **A** matrix, to ensure mass conservation we need to ensure only that (1) the summation of each column should be zero, (2) the diagonal term is between -1 and 0 and (3) the off-diagonal terms are greater than or equal to 0. Thus if we can construct a generic column for matrix **A**, that is the same for all grading fractions

$$\text{column } i \text{ in } \mathbf{A} = \begin{bmatrix} a_{i,1} \\ \vdots \\ a_{i,i-2} \\ a_{i,i-1} \\ a_{i,i} \\ 0 \\ 0 \\ \vdots \\ 0 \end{bmatrix} \tag{7.30}$$

we can then write Equation (7.19) as (using the two daughter size grading)

$$d_{\text{mean},t+1} = \Big((1 - a_{i,i}) + a_{i,i-1} 0.5^{1/3} + a_{i,i-2} 0.5^{2/3} + \ldots + a_{i,1} 0.5^{(i-1)/3} \Big) d_{\text{mean},t} \tag{7.31}$$

and if the values of $a_{i,j}$ are small when j is small (i.e. large parent particles do not generate many very small daughter particles), then we can write

$$d_{\text{mean},t+1} \cong \bar{a} d_{\text{mean},t} \tag{7.32}$$

where $\bar{a}$ is a constant reflecting the structure of **A** so that we can write a general equation for how the mean diameter of the soil will change over time

$$d_{\text{mean},t=T} = (\bar{a})^T d_{\text{mean},t=0} \tag{7.33}$$

which shows that the mean of the soil grading is an exponential function with time of the initial grading irrespective of the structure of **A**. Note that the restrictions on the form of **A** are relatively modest (just that weathering cannot generate a large mass proportion of small particles), so this is quite a powerful and general conclusion. Note also that the use of the two daughter size fraction definition in Table 7.1 is a convenience rather than a necessity, as we will see in the next section. The fragmentation notation for this case is p2-(1-$a_{i,i}$)-$a_{i,i-1}$- ⋯ -$a_{i,1}$.

The discussion above presumes that all particles fracture with exactly the same geometry. This is not essential. One simple extension is to allow particles to break with two possible geometries at different rates. The **A** matrix for the combined process is then

$$\mathbf{A} = w_1 \mathbf{A}_1 + w_2 \mathbf{A}_2 \tag{7.34}$$

where the subscripts are the weathering rate w and weathering matrix **A** for each of the two processes. This can be generalised to as many processes as required.

Figure 7.6 extends the discussion by examining a number of fracture geometries, and shows how (1) the different fracture geometries yield different rates of evolution of the mean grading and (2) all the geometries evolve as a semi-log of time as in Equation (7.33). The meaning of the coding used for describing fracture geometry is explained in the figure caption, but is based on the fracture geometry where particles are 1/2, 1/4, 1/8 and so on of the volume of the parent particle.

One of the fracture geometries in Figure 7.6 is philosophically different from the ones described above. In all the fracture geometries above the geometry of the daughter particles is deterministically fixed as a proportion of the volume of the parent particle. The aggregated behaviour of all the particles is then just the sum of all the particles in that grading range. However, consider the case where a single particle breaks into two particles but where the fracture location in the particle is randomly distributed within the particle (e.g. a schist where the particle cleaves along the layering but where the cleavage plane is randomly located within the particle). The probability

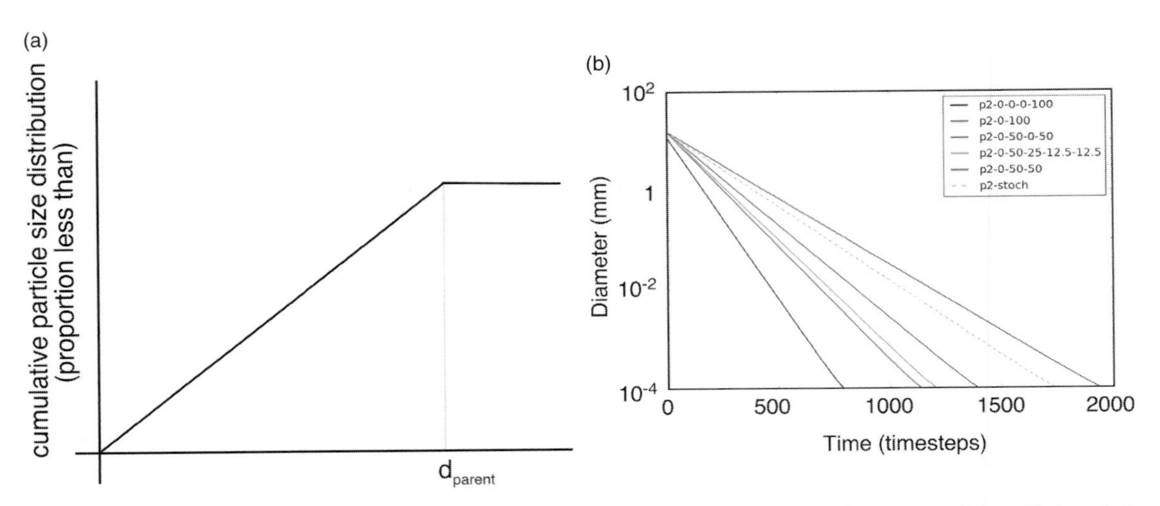

FIGURE 7.6: (a) The average particle size distribution of the daughter particles resulting from random fragmentation (P2-stoch), d_{parent} is the diameter of the parent particle, (b) Numerical solution of the time evolution of the mean of the particle size distribution for the same weathering rate (as used in Figure 7.5) and a range of fragmentation models. The weathering model p2-0–50-50 is the same as 'rate = 0.03' in Figure 7.5.

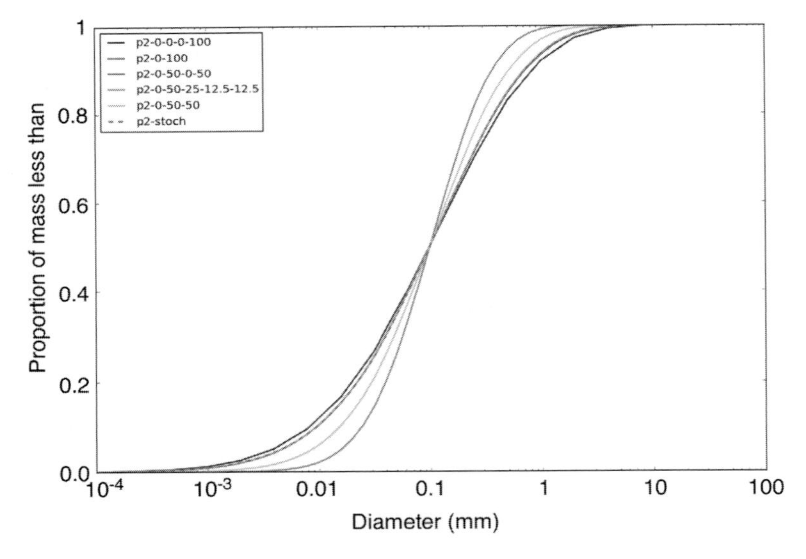

FIGURE 7.7: These are the same simulations as in Figure 7.6 (i.e. same weathering rate, different fragmentation models) but showing the particle size distribution when the d_{50} of the all fragmentation models is equal to 0.1 mm. Note that Figure 7.6 shows that the age of each of the distributions will be different since the different fragmentation models generate a different rate of change in the d_{50} for the same weathering rate.

distribution of the volume of daughter particles derived from particles in a specified particle range is shown in Figure 7.6a. It is convenient here to define the x axis as the volume of particles rather than their diameter because the probability distribution of the daughter particles is then particularly simple. If a large number of particles fracture, then the probability distribution function gives the mean of the volume distribution of the daughter particles, from which **A** can be easily derived. In fragmentation notation it is p2-30.7-34.6-17.3-17.4, if we lump the volume of all particles more than three grading fractions smaller into the third grading fraction (i.e. the 17.4). Other than the proportion remaining in the parent grading range, the relativities between the daughter size fractions are the same as the scaling model p2-0–50-25-12.5-12.6.

Figure 7.6 shows that those fractions that generate a greater proportion of volume in smaller size fractions fine faster, which is consistent with Equation (7.24). Figure 7.7 shows the particle size distribution resulting from using a number of different fragmentation models. Initially all

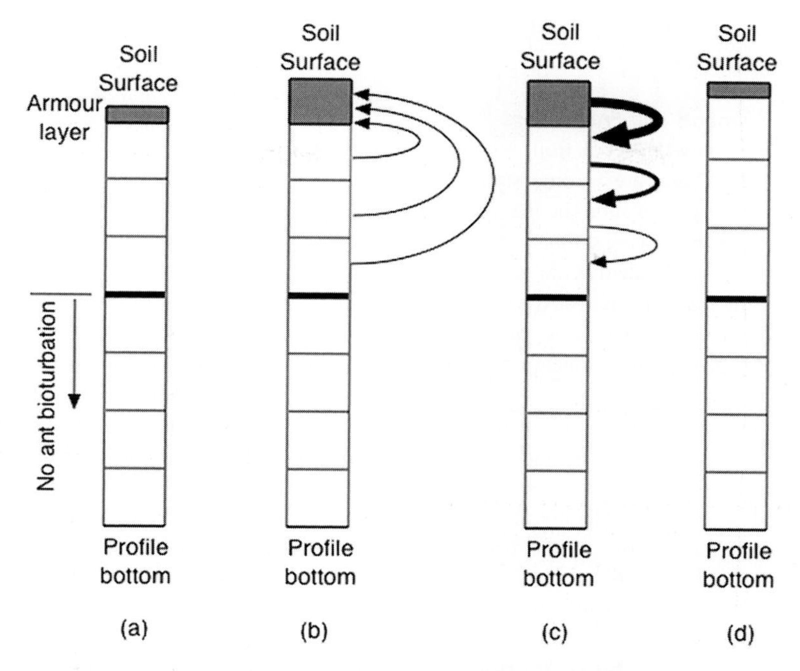

FIGURE 7.9: Schematic of how layers interact in ant and termite bioturbation. While the calculations are done using mass, the figure shows the layer thicknesses: (a) the soil profile before ant bioturbation, including the thin surface armour layer, (b) the movement of material by ants, from the top three layers into the surface armour layer (those layers above the thick line are subject to ant transport, those below not; the grey region is the surface armour layer; note that the only layer that changes thickness is the surface armour layer, so the bulk density in the lower layers is reduced), (c) the redistribution of excess sediment from the surface armour layer (the thickness of the lines and arrows indicate quantities of soil being moved between layers) and (d) the reinstatement of the armour layer, which is now a mix of the original armour layer and the ant transport material. The soil surface has risen because of the reduction of the bulk density of the layers from which ants removed materials.

below and the fourth term is the material moved by the ants/termites. The zeros in two of the equations are placeholders so that the third and fourth terms are the same processes in all four equations. Constructing the transition matrix from these equations involves filling the matrix according to Equation (7.51):

$$\mathbf{B} = \begin{bmatrix} -\left(\dfrac{\sum_{i=2}^{4}\Delta m_i}{m_1+\sum_{i=2}^{4}\Delta m_i}\right)\mathbf{I} & B\varphi_2\mathbf{A} & B\varphi_3\mathbf{A} & B\varphi_4\mathbf{A} & [0] & \cdots \\ \left(\dfrac{\sum_{i=2}^{4}\Delta m_i}{m_1+\sum_{i=2}^{4}\Delta m_i}\right)\mathbf{I} & -\left(\dfrac{\sum_{i=3}^{4}\Delta m_i}{m_2+\sum_{i=3}^{4}\Delta m_i}\mathbf{I}+B\varphi_2\mathbf{A}\right) & [0] & [0] & [0] & \cdots \\ [0] & \left(\dfrac{\sum_{i=3}^{4}\Delta m_i}{m_2+\sum_{i=3}^{4}\Delta m_i}\right)\mathbf{I} & -\left(\dfrac{\Delta m_4}{m_3+\Delta m_4}\mathbf{I}+B\varphi_3\mathbf{A}\right) & [0] & [0] & \cdots \\ [0] & [0] & \left(\dfrac{\Delta m_4}{m_3+\Delta m_4}\right)\mathbf{I} & -B\varphi_4\mathbf{A} & [0] & \cdots \\ [0] & [0] & [0] & [0] & [0] & \cdots \\ \vdots & \vdots & \vdots & \vdots & \vdots & \ddots \end{bmatrix} \tag{7.52}$$

where, for brevity, only the first five rows and columns of the transition matrix are shown.

7.3.5.4 Gophers, Wombats and Prairie Dogs

Burrowing animals also move material from within the profile to the surface. There are strong similarities between the ant/termite representation in the previous section and that for larger burrowing animals. The main differences are (1) the size and extent of the tunnels, (2) the size selectivity of the material excavated and (3) that there tends to be more lateral (either across or up/down slope) movement of material because tunnels are predominantly horizontal.

In North America pocket gophers are an important mechanism for laterally moving soil (Gabet, 2000) impacting on vegetation (Huntly and Inouye, 1988). Their burrows extend up to 100 m laterally though they do not dig very deep (<1 m). It thus seems reasonable to assume that they would also mix the soil vertically (Hole, 1981), though there is less quantitative literature on this possible impact (Gabet et al., 2003).

In a review of bioturbation in North American soils Zaitlin and Hayashi (2012) concluded that prairie dogs had a significant vertical mixing effect down to about 2 m, but little impact on hillslope movement because they prefer to live in flat areas.

For other environments we have less data. Field and Anderson (2003) provide comparative figures for volumes of soil bioturbed per hectare for an Australian catchment. Tree throw was the most significant mechanism, but bioturbation by wombats was the second most important and of a comparable magnitude to termites.

In addition to tunnel digging there was also the impact of surface foraging where the surface soil is mixed by the digging for roots and leaves. Mitchell (1985) (in Paton et al., 1995) found that the soil volume moved during foraging (down to depths of 15 cm) by wombats was 20–50 times the volume moved per year by tunnel digging. The depth of disturbance by wombats by tunnelling (down to 2.4 m) was significantly higher than tree throw (down to 0.42 m), suggesting a greater impact from wombats than tree throw on vertical mixing of soils.

In the absence of more definitive data it is suggested that the vertical mixing of the profile from burrowing animals should be handled with the same approach as for ants and termites (Equation (7.52)). The one difference is that there is likely to be less grain size selectivity for the burrowing animals because of the animals' large size. To model the surface foraging where the surface materials are mixed together, the approach used in tree throw (Equation (7.43)) seems appropriate.

7.4 Sediment Deposition

Sediment deposition was discussed in Chapter 4 in the context of the sediment in the flow and how it impacts on transport capacity. In this section we discuss what happens when that sediment is deposited. As discussed previously there are two aspects to sediment to be deposited: (1) the amount to be deposited at that point and (2) the grading that will be deposited at that point. The grading changes due to selective deposition are discussed in Chapter 4. Here we simply take the grading vector that gives us the mass of the deposited material, $\underline{g}_d$, in each of the size fractions (as derived in Chapter 4), and the total mass of deposited sediment is $m_d = \sum_i g_{d,i}$ where the subscript d indicates that the grading vector is the deposited material.

This problem bears strong mathematical parallels with the bioturbation by ants, where material was deposited on the surface by the ants. The difference is that there is no balancing loss of mass from the layers underneath. The deposition stage of the process is for the surface layer

$$\underline{g}_{\text{after}} = \underline{g}_{\text{before}} + \underline{g}_d \tag{7.53}$$

This sediment then needs to be distributed down the profile. Because this sediment is an addition to the profile, the soil profile will become deeper:

$$\underline{\underline{g}}_{t+1} = \begin{bmatrix} -\frac{m_d}{m_1}\mathbf{I} & [0] & [0] & [0] & [0] & \cdots \\ \frac{m_d}{m_1}\mathbf{I} & -\frac{m_d}{m_2}\mathbf{I} & [0] & [0] & [0] & \cdots \\ [0] & \frac{m_d}{m_2}\mathbf{I} & -\frac{m_d}{m_3}\mathbf{I} & [0] & [0] & \cdots \\ [0] & [0] & \frac{m_d}{m_3}\mathbf{I} & -\frac{m_d}{m_4}\mathbf{I} & [0] & \cdots \\ [0] & [0] & [0] & \frac{m_d}{m_4}\mathbf{I} & -\frac{m_d}{m_5}\mathbf{I} & \cdots \\ \vdots & \vdots & \vdots & \vdots & \vdots & \ddots \end{bmatrix} \underline{\underline{g}}_t + \begin{bmatrix} \mathbf{I} \\ [0] \\ [0] \\ [0] \\ [0] \\ \vdots \end{bmatrix} \underline{g}_d \tag{7.54}$$

where the definitions of m_i are the same as that used for Equation (7.48). The diagonal and off-diagonal terms repeat all the way to the bottom corner (i.e. bottom layer) of the matrix because the deposition pushes down the material in all layers by an amount equal to the mass of deposition. The second term in the equation in $\underline{g}_d$ simply ensures that all the deposited material goes into the top layer and nowhere else during that timestep. The similarity with the erosion equation in Equation (7.17) should be clear.

The preceding discussion makes no assumptions about the process driver of deposition. Classically it is sediment being deposited out of fluvial transport. It could also be from aeolian transport (Cohen et al., 2015). In this case (aeolian) deposition and fluvial (erosion) can be occurring simultaneously. Cohen showed that if the aeolian material is very fine it can enhance erosion and destroy the armour that might have formed in the absence of aeolian deposition. No matter what the grading of the aeolian material is, it changes the characteristics of the armour layer and potentially changes the fluvial erosion characteristics of the surface.

7.5 Differentiating between Mineral Components

The discussion above has been about differentiating the behaviour of the different fractions within the soil solely on the basis of size. However, the state-space matrix approach can be easily extended to include differentiation based on some other property of the soil. The example we will use here is mineral fractions (e.g. quartz, k-feldspar, biotite etc.), but it will become clear that the techniques outlined below are suitable for labelling any characteristic of the soil (e.g. particle density), or for explicitly modelling nonmineral fractions (e.g. particulate organic matter). Finally the terminology of referring to the $\underline{g}$ state vector as the grading vector, while convenient so far, will become a bit misleading because the state vector will now contain other, additional, information about the soil, so we will adopt terminology used in the state-space literature and hereafter refer to the grading vector $\underline{g}$ as the soil state vector.

In the previous sections the soil state vector $\underline{g}$ contains the mass of material in each size range without regard to what the characteristics of that material are other than its size. Consider the case where we have three different minerals in the soil that we wish to distinguish (e.g. perhaps they weather at different rates and we would like to track the effect of these different weathering rates on the soil grading as it evolves). Let's call these minerals Q, K and B to distinguish them. We can then extend our definition of $\underline{g}$ so that it now gives the mass in each size range for each of the three minerals for each layer in the soil profile. The vector will then be

$$\underline{g} = \begin{bmatrix} g_{1,Q} \\ g_{2,Q} \\ \vdots \\ g_{m-1,Q} \\ g_{m,Q} \\ \cdots\cdots \\ g_{1,K} \\ g_{2,K} \\ \vdots \\ g_{m-1,K} \\ g_{m,K} \\ \cdots\cdots \\ g_{1,B} \\ g_{2,B} \\ \vdots \\ g_{m-1,B} \\ g_{m,B} \end{bmatrix} = \begin{bmatrix} \underline{g_Q} \\ \cdots \\ \underline{g_K} \\ \cdots \\ \underline{g_B} \end{bmatrix} \tag{7.55}$$

so that using m size fractions the vector will be $3m$ in dimension. The horizontal dotted lines in the vector of Equation (7.55) are added simply to highlight the organisation of the vector with the grading of the first mineral first, the second mineral second and the third mineral last. Thus the full $\underline{g}$ vector is split into three parts with the first third being the $\underline{g_Q}$ grading vector for the Q mineral, $\underline{g_K}$ the grading vector for the K mineral and $\underline{g_B}$ the grading vector for the B mineral. This organisation first by mineral and then by grading will be convenient shortly. The **A** matrix will also be correspondingly bigger and will be of dimension $3m \times 3m$. While the vector and matrix are larger, the formulation of the **A** matrix is no more difficult than it was for the nonmineral case. For physical weathering there is no transfer of mass between the mineral fractions, so each mineral fraction is mass conservative in its own right, and then for each mineral fraction we can write the weathering transition matrices

$$\begin{aligned} \underline{g_Q}_{t+1} &= (\mathbf{I} + \mathbf{A}_Q)\underline{g_Q}_t \\ \underline{g_K}_{t+1} &= (\mathbf{I} + \mathbf{A}_K)\underline{g_K}_t \\ \underline{g_B}_{t+1} &= (\mathbf{I} + \mathbf{A}_B)\underline{g_B}_t \end{aligned} \tag{7.56}$$

where the **A** matrices are the matrices that would apply for that mineral if that mineral was 100% of the mass content of the soil layer. We can then write the matrix equation for the full soil state vector for a single soil layer

$$\underline{g}_{t+1} = \begin{bmatrix} (\mathbf{I} + \mathbf{A}_Q) & [0] & [0] \\ [0] & (\mathbf{I} + \mathbf{A}_K) & [0] \\ [0] & [0] & (\mathbf{I} + \mathbf{A}_B) \end{bmatrix} \underline{g}_t \tag{7.57}$$

The zero off-diagonal matrices indicate that none of the mineral fractions transform into a different mineral fraction during weathering. For physical weathering that is the expected behaviour, but for chemical weathering this may not be the case. If, for instance, one of the mineral fractions is a chemical weathering product of another fraction, then there will be off-diagonal matrices that parameterise the rate and grading characteristics of that weathering process. We can show how this would work in principle by extending our example in Equation (7.57). Imagine that mineral component Q is a weathering product of mineral K, then it is possible to write a matrix equation relating the effect of weather from K to Q on the grading of both Q (the destination of the chemical weathering) and K (the source for the chemical weathering) so that

$$\underline{g}_{t+1} = \begin{bmatrix} (\mathbf{I}+\mathbf{A}_Q) & (\mathbf{I}+\mathbf{A}_{QK}) & [0] \\ (\mathbf{I}+\mathbf{A}_{KQ}) & (\mathbf{I}+\mathbf{A}_K) & [0] \\ [0] & [0] & (\mathbf{I}+\mathbf{A}_B) \end{bmatrix} \underline{g}_t \tag{7.58}$$

where the transition matrix $\mathbf{A}_{QK}$ calculates how much of mineral K is transformed into mineral Q, what the grading of the source material K was and how it is transformed into the grading of the destination mineral Q, while $\mathbf{A}_{KQ}$ calculates the inverse of how K is transformed into chemical weathering product Q. In general this chemical weathering transformation will not be mass conservative with respect to the soil components because the chemical reactions will typically involve constituents and reactions that have not been tracked in the matrix equations such as hydration (i.e. with water), oxidation (i.e. with oxygen) and carbonation (i.e. with carbon dioxide). All these reactions will change the mass of the soil mineral constituents and can be accounted for with $\mathbf{A}_Q$, $\mathbf{A}_K$ and $\mathbf{A}_B$. We will focus on chemical weathering in detail in the next chapter (Chapter 8).

In a model for the evolution of organic matter within soils Kirkby (1977) proposed a similar method using matrices to differentiate the organic fractions (carbohydrates, amino acids and lignins) and mathematically described how they transformed from one fraction to another. He also proposed a specific structure for the transition matrices for the different types of soil humus, organic sourcing from leaf fall, and examined what the long-term equilibrium organic content of the soil was.

Finally, we note that this section has focussed on the weathering process and differences in weathering rates, but this approach can also be taken with erosion and armouring processes. For instance, it is common for agricultural erosion models to distinguish sediment classes based on their relative transportability (e.g. CREAMS, WEPP; Knisel, 1980; Laflen et al., 1991). This transportability is a function of the diameter of the particles and their specific gravity, so these erosion models are potentially capable of distinguishing erosion rates for high specific gravity particles (e.g. high in iron) from midrange specific gravity (e.g. silicates), and low specific gravity particles (e.g. organic matter and soil aggregates).

7.6 The Evolution of Porosity

The layers in the matrix model described above are defined based on the mass per unit area in the layer. To convert these masses per unit area to a depth of soil we need to use the bulk density of the soil in that layer. If the density of particles doesn't change, then this is equal to the change in the porosity. The main drivers of changes in the density of particles are chemical weathering and the amount of soil carbon. In general the porosity of the soil will change over time simply from the change in the grading and the particle-packing arrangement that is possible with that grading. But porosity will also change with macropore development and changing mineral constituency. If either of these latter properties is being tracked, then these effects can be modelled. For instance, the bioturbation by ants discussed above involves the removal of fine materials from a layer leaving behind voids between the large particles, and will accordingly increase the porosity of the layer from which material is being removed. Note that by defining the soil state vector by mass, this porosity change does not appear in the matrix equations, and appears only at the end when a conversion from mass of soil to depth of soil is required.

For a granular media with two size particles (e.g. a binary media) a relationship due to Fraser (1935) and Clarke (1979) based on the percentage of the two sizes is commonly used. This equation has been used (e.g. Morin, 2005) and studied by a number of subsequent researchers (Zhang et al., 2009). For a granular media made up of a range of particle sizes, a number of empirical extensions of this model (see a comparison by Tranter et al., 2007) can be used to estimate the porosity, ϕ, including (Koltermann and Gorelick, 1995)

$$\phi = \begin{cases} \phi_c - cy_1\left(1-\phi_f\right) + (1-y_1)\phi_f & c < \phi_c \\ \phi_c(1-y_2) + c\phi_f & c \geq \phi_c \end{cases} \tag{7.59}$$

where

$$\begin{aligned} y_1 &= c\left(\frac{y_{\min}-1}{\phi_c}\right) + 1 \\ y_2 &= (c-1)\left(\frac{1-y_{\min}}{1-\phi_c}\right) + 1 \\ y_{\min} &= 1 + \phi_f - \frac{\phi_{\min}}{\phi_c} \\ \phi_{\min} &= \phi_c(1-y_{\min}) + \phi_c\phi_f \end{aligned} \tag{7.60}$$

which is applicable for gradings where there are two distinctly different size fractions (e.g. sand/clay, gravel/clay)

assume that there is only one dissolved product and one solid product.

The physics of the model below follows that in the CrunchFlow reactive transport model (Steefel, 2009). CrunchFlow is a multicomponent (i.e. more than one mineral) model. This multicomponent capability complicates the model mathematics considerably, so to highlight the key concepts the presentation below is simplified to model only one component. CrunchFlow is missing one important process, which is the exhumation of fresh saprolite from below, and the model here has been extended to include this process. If the soil depth is constant with time and erosion is occurring, then a consequence is that an equal amount of saprolite exhumation must occur to balance erosion.

We can write the chemical equation for the weathering reaction as

$$aA + S = bP + cL \tag{8.5}$$

where A is the acid loaded water, S is the unweathered rock (the substrate), P is the solid product and L is the dissolved product, which we will refer to as leachate. Equation (8.5) is a mass balance equation, and the parameters a, b, c are needed to balance the equation and quantify how much mass of component is created per unit mass of substrate. The parameters a, b, c will be different for different mineral weathering reactions, so we will not be more specific at the moment, and this will allow the exploration of the equation behaviour for the full range of parameter values. For congruent weathering $b = 0$.

The governing equation we will solve for the chemical weathering reaction will be one-dimensional in the direction of gravity, with the positive z direction vertically upwards:

$$\begin{aligned}
&\frac{\partial A}{\partial t} + v\frac{\partial A}{\partial z} - D\frac{\partial^2 A}{\partial z^2} = -k_2 R \\
&\frac{\partial S}{\partial t} - \mathrm{E}\frac{\partial S}{\partial z} = -R \\
&\frac{\partial L}{\partial t} + v\frac{\partial L}{\partial z} - D\frac{\partial^2 L}{\partial z^2} = k_3 R \\
&P = k_4(S_0 - S) \\
&R = k_1 A^{\alpha}(k_5(S - S_{\min}))
\end{aligned} \tag{8.6}$$

where t is time and z is distance from the bottom of the soil profile (i.e. positive upwards). The velocity v is the planar average velocity per unit area, called the specific discharge in the groundwater literature.

In the acidic water, A, equation (Equation (8.6a)), the first term of the left-hand side is the rate of change of the concentration, the second term is the advection term that gives the rate at which the acid is moving down the profile, and the third term is a diffusion/dispersivity term that defines the rate at which the acid spreads within the profile. The right-hand side, R, is the reaction term discussed below. Equation (8.6a) is a classic advection-diffusion/dispersion-reaction transport equation with advection occurring down the profile.

In the substrate, S, equation (Equation (8.6b)), the first term on the left-hand side is the rate of change of the substrate, and the second term is the rate of exhumation of the saprolite due to erosion. If the soil depth is constant with time. then the rate of erosion, E, is equal to the rate of exhumation of the saprolite into the soil profile. This second term thus adds fresh substrate mass to the base of the soil profile, exactly equalling the mass of material lost by erosion at the surface. The term on the right-hand side of this equation is the rate of transformation of the substrate. This is a classic advection-reaction equation where advection is occurring up the profile. The only difference between the right sides of Equations (8.6a) and (8.6b) is the constant of proportionality k_2, which reflects the differences in the mass balance in Equation (8.5b) so that

$$k_2 = \frac{1}{a} \tag{8.7}$$

The dissolved product, L, equation is not necessary at this stage, but it is convenient to present it now; it will become important later. If the dissolved product-loaded water leaves the profile unaltered, then this will be the leachate from the profile. The velocity and diffusivity terms are assumed to be the same as for the acid solution A since they are both transported by the same solution (and we assume they have the same sorptivity potential so their advection velocities are the same). The rate of reaction on the right-hand side is the same except for the mass balance in Equation (8.5) so that

$$k_3 = \frac{1}{c} \tag{8.8}$$

Note that in Equation (8.6) the value of L does not influence the values of A and S. Later we will extend these equations to cases where this is not the case. As for the acid equation, this is a classic advection-diffusion/dispersion-reaction equation with advection occurring down the profile.

The solid weathering product, P, is simply defined by the amount of the original rock that is weathered, where the amount of original rock before weathering is S_0

resulting from the rock properties at the saprolite-soil boundary, and the mass balance in Equation (8.5) so that

$$k_4 = \frac{1}{b} \tag{8.9}$$

This equation assumes that the solid weathering products move upwards through the soil profile (by exhumation) at the same rate as the unweathered rock, S. Thus this equation does not model cases where dissolved products may be precipitated away from the site of the chemical reaction. In the figures in this chapter the values of P are not plotted because of the simplicity of the relation between S and P, where trends in P can be derived by inspection from the plots of S.

Finally the reaction term, R, is a combination of the acid concentration, A, and the reactive surface area of substrate, $k_5(S - S_{min})$. Solid-fluid reaction rates are determined by the reactivity between the two components, k_1, the acid concentration in the fluid, A, and the SSA. We will assume that the SSA is the same everywhere, so that the surface area dependency can be given by $\text{SSA} = k_3(S - S_{min})$. We will generalise this surface area assumption later. The value S_{min} is a threshold that serves several purposes. This represents a proportion of the substrate that cannot be weathered so that as the substrate approaches the value S_{min} the reaction shuts down. In terms of physical processes this may represent substrate that is not accessible to weathering. For instance, in arid areas it is common to find minerals coated in an iron oxide, so they will not be accessible to environmental processes (this process is called occlusion), and the surface area associated with those minerals is not available for reaction. In the discussion below we will assume $\alpha = 1$, though some authors have suggested other values.

Solving this set of equations with time leads to an equilibrium balance between the reactive loss of S, the erosion of S from the surface and the addition of S from exhumation. An example of this evolution is shown in Figure 8.2, which uses the nominal parameter set in Table 8.5. The general trends are the following:

- The concentration of the acidic water, A, decreases from 1 to near 0 at the base of the soil profile. This shows that as the acid reacts with the rock it diminishes in strength.
- The substrate rock, S, becomes more weathered as you approach the surface. In this example about 60% of the rock is untransformed, while about 40% has been

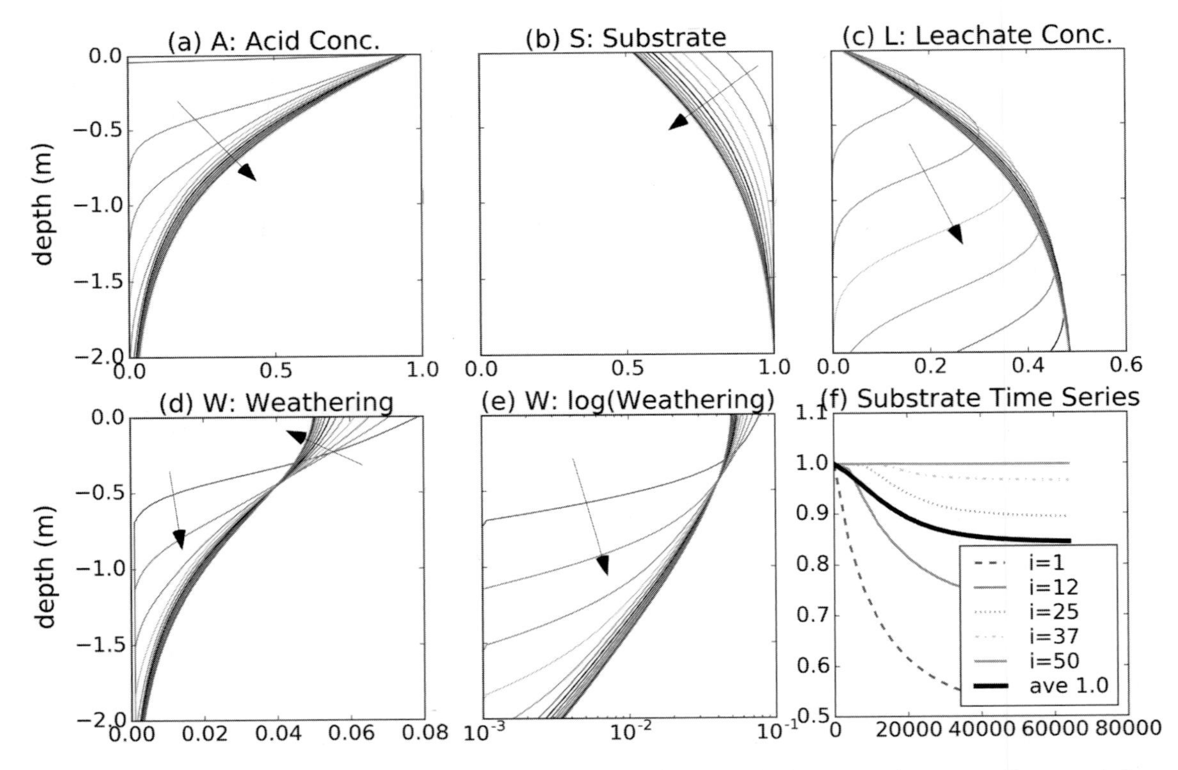

FIGURE 8.2: Example output of the profile evolution from the chemical weathering model with the nominal parameters. The arrows indicate the direction of evolution with time. For a detailed explanation of the panels see Figure 8.4.

TABLE 8.5 Parameter sets for the sensitivity study

Parameter (Equation (8.6))	Nominal parameter set	Full depletion parameter set
k_1	0.1	0.5
k_2	2.0	2.0
k_3	1.0	1.0
k_4	1.0	1.0
k_5	1.0	1.0
V	0.1	0.1
D	1.0e-6	1.0e-6
E	0.1	0.1
α	1	1
S_{min}	0	0
$L_{equilib}$ in Equation (8.13) (Leachate Control)	1,000 (0.4)	1,000 (0.4)

transformed. The curve of unweathered substrate with soil depth is sometimes called a 'depletion curve'.

- The temporal trend at any given depth is that the rock starts off as being unweathered, and the weathering approaches an equilibrium profile. Likewise at early times the acid is consumed more rapidly (because the substrate and thus the reaction rates are high) so there is less acid in the water at the base of the profile. The details of the transients are a function of the initial conditions and so are only indicative of likely behaviour in the field.
- The nominal parameter set in this simulation has a very low diffusion/dispersivity, D, so mass transport within the profile by water is primarily by the advection of the solution down the profile.

The weathering profile in Figure 8.2 is calculated as the rate of change of the substrate:

$$W(z) = R \tag{8.10}$$

Figure 8.2 is one example of the model behaviour, but it is clear that the log weathering function is not linear with depth, so in this case weathering is neither exponential nor humped with depth.

8.3.2 Limits on Profile Behaviour

Before the sensitivity studies are presented in later sections, it is worth discussing some forms of limitation on profile behaviour that are assumed or have been demonstrated in the field. This will allow the interpretation of the sensitivity studies in terms of terminology used in the chemical weathering literature.

The first concept is 'transport-controlled' or 'transport-limited' profiles. Equation (8.6) can reach an equilibrium state where the concentrations on the right-hand side of the equation have reached a maximum value (e.g. for dissolution imagine a saturated solution where no more solid can be dissolved in the solution, $L = L_{equilib}$). As a thought experiment, imagine water is stationary within the profile. Once this equilibrium condition had occurred. then no more weathering can occur because the weathering reaction is limited by the concentration of the weathering products in the solution. The only way that the weathering reaction can continue is if some of the solute (i.e. the leachate) is flushed out of the bottom of the soil profile by the infiltration, in which case the reaction proceeds enough to replace the leachate lost from the profile. A mass balance can be done on the entire profile by noting that the mass lost from the profile at the bottom is

$$M = vL_{equilib} \tag{8.11}$$

where M is the mass of leachate leaving the profile, v is the velocity in Equation (8.6) and $L_{equilib}$ is the equilibrium concentration of the leachate. For steady state conditions this mass leaving the profile must be replaced by the amount generated by the weathering reaction within the profile, so that the rate of weathering of the profile is limited by the rate at which leachate is transported out of the profile, thus the name 'transport-controlled'. Whether a profile is transport-controlled is dependent on the relativity between the residence time of the water (related to the velocity through the profile) and the time it takes for the weathering reaction to reach equilibrium. If the residence time is long or the rate of equilibration of the reaction fast, then transport control will occur (e.g. White et al., 2008). In a later section we will extend Equation (8.6) to examine transport control. Finally note that this definition of transport-controlled or transport-limited is not related in any form to transport-limited sediment transport.

Another condition is the more obvious 'rate-controlled', 'kinetic-controlled' or 'weathering-limited' profile (Brantley and Lebedeva, 2011), where the amount of weathering is controlled by the reaction rate of the weathering and not by the rate of flow through the profile. Typically experiments are rate-controlled because they have high ratios of water to rock in the reactor vessels so that dissolved product concentrations are low and unlikely to reach the equilibrium concentration. Equation (8.6) is implicitly rate-controlled because of the formulation chosen for the reaction term R. We will generalise Equation (8.6) later.

A final condition is where the rate of weathering of the profile is limited by the rate of exhumation/erosion of the saprolite. Riebe et al. (2004) called this 'supply-limited'. This condition can occur when the weathering reaction equilibrates quickly and the only way for the weathering reaction to proceed is for new material to be made available for weathering.

As discussed in the previous section, weathering rates decline with time. It is thus possible for a profile to be transport-limited at early time (when reaction rates are fast and leachate concentrations high) and transition to a weathering-limited profile as the reaction rate declines (Brantley and Lebedeva, 2011).

Finally there is 'isovolumetric weathering'. This is the case when the weathering reaction occurs without any change in volume. For example, in cases where there are large quantities of leachate product and very little solid, it is possible that the soil may collapse because of the voids generated in the particles and the volume will decrease. Any change is volume is parameterised by strain ε_i for mineral i in the soil

$$\varepsilon_i = \frac{V_{s,i} - V_{p,i}}{V_{p,i}} \tag{8.12}$$

where $V_{s,i}$ is the volume of mineral i in the soil and $V_{p,i}$ is the volume of mineral i in the saprolite (Brimhall and Dietrich, 1987; Brantley and Lebedeva, 2011). Note that strain is defined with respect to a mineral fraction, not the soil as a whole. This definition of strain should not be confused with that used for stress-strain calculations in mechanics, despite the superficial similarity.

8.3.3 Parameter Sensitivity Studies

In the following examples we present only the final equilibrium results and not the transient behaviour. The intent of the following examples is to explore the range of behaviours that the systems of equations in Equation (8.6) can exhibit. Changing the values for most of the model parameters can dramatically change the qualitative behaviour in Figure 8.2.

For instance, if we increase the ability of the acid to completely consume the substrate we find dramatically different transient behaviour. Figure 8.3 has the same parameters as Figure 8.2 (i.e. the nominal parameter set in Table 8.5) with the exception that the rate of reaction, k_1, is increased by a factor of 5, and the amount of initial substrate has been reduced from 1.0 to 0.25, so it is easier for the acid to completely deplete all the substrate (this parameter set will be hereafter called the 'full-depletion' parameter set; see Table 8.5). The S-shaped substrate curves are typical of field data collected by White et al. (2008) (Figure 8.18), and we find a humped weathering function, where the location of the peak of the hump moves down the profile as the substrate is consumed. In this case the hump is a transient feature, but we will show that it is also possible to generate this hump, albeit with some difficulty, at steady state. The hump will occur when the parameters are such that both the substrate and the acid are completely consumed within the profile.

Figures 8.4 to 8.7 show the sensitivity of the results for a change of $\pm 50\%$ in the parameters k_1, k_2 and k_3. In Figures 8.4 and 8.6 the qualitative trends in A, S and L do not markedly change: (1) A still decreases nonlinearly from the surface to the base of the profile, (2) S still increases nonlinearly from the surface to the base though there is a marked reduction in the amount of S at the surface as k_2 increases (i.e. how much substrate is used per unit reaction) and (3) dissolved product P simply reflects the changes in A because the relationship between A and P is unchanged. As expected Figure 8.7 shows a change only in the dissolved product P and not in A or S, and this is a consequence of Equation (8.6), where the value of the dissolved product does not influence A and S.

The effect of increasing the weathering rate k_1 is to sharpen the weathering front. This is particularly obvious in Figure 8.5 using the full depletion parameter set (Table 8.5), but Figure 8.4, which is for the nominal parameter set, is also consistent with this sharpening.

The appearance of a humped like weathering function in Figure 8.6 is an important result and highlights one important aspect of the behaviour of Equation (8.6). In Figure 8.6 even a modest increase in k_2 of 50% leads to a humplike weathering function with the maximum weathering rate at a depth of about 0.5 m. This is a result of subtle interactions within the reaction term on the right-hand side of Equation (8.6). The key components of this interaction are (1) fresh acidic fluid enters at the surface and the acid potential is consumed as it infiltrates down the profile and (2) fresh unweathered rock is exhumed at the profile base and is depleted by the acid reaction within the profile. The reaction rate (i.e. weathering rate) is a function of the multiplication of the acid concentration, A, and unweathered rock, S. The weathering is low if either of them is low and is only relatively high if both of them are high simultaneously. In Figure 8.6 A is low at the profile base and S is low at the surface, leading to low weathering rates at the surface and profile base, and the combination of A and S mid-profile

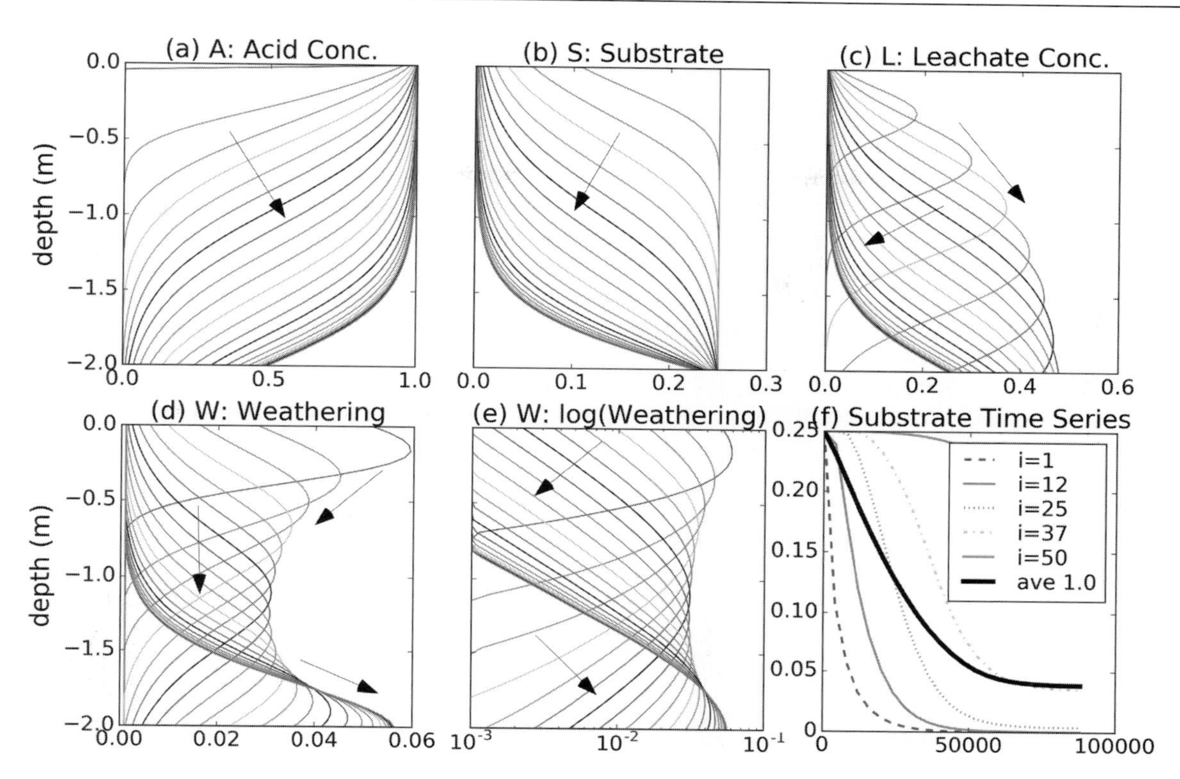

FIGURE 8.3: Example output of the profile evolution from the chemical weathering model with the full depletion parameters. The arrows indicate the direction of evolution over time. Unlike the full depletion simulations later in the chapter the initial substrate concentration was $S_0 = 0.25$, rather than the value $S_0 = 0.5$ used in the remainder of the chapter for depletion parameter simulations. This smaller value $S_0 = 0.25$ was used because it allows us to show in panel (d) how the weathering function changes over time, from one where weathering is highest at the soil surface, to one where there is a humped function with a peak in mid-profile, and finally to the equilibrium where the weathering rate is greatest at the bottom of the profile. For a detailed explanation of the panels see Figure 8.4.

is sufficient to lead to the weathering rate being highest there. Moreover, the main effect of modifying k_2 is to substantially reduce the substrate near the surface (since it controls the rate of substrate consumption), so the reason for the appearance of a humped weathering function is the low proportion of substrate at the soil surface because of the relatively higher rate of substrate weathering within the profile. We will repeatedly see this behaviour arising out of the multiplicative combination of A and S.

Figure 8.8 shows the effect of changing the water infiltration velocity. As the velocity increases the acid has less time to react with the near surface rock and is able to penetrate deeper into the profile. This means that the acid is more concentrated when it reaches the deeper fresher rock. The effect of this can be seen in the substrate where the amount of weathering at depth is higher at higher velocities, and that at the higher velocities the weathering function in the deeper part of the profile is higher, while near the surface it is relatively unchanged. It may seem rather unexpected that if the velocity is higher, then the weathering rate is higher. However, with a fixed concentration of reactant at the upper boundary condition (i.e. a fixed pH), a higher velocity means that more acid is being flushed through the profile, so that the cumulative acid load through the profile increases with the velocity. A 50% increase in velocity results in a 50% increase in acid load. If the acid concentration is a result of rainfall equilibration with atmospheric CO_2 then this is correct behaviour. This relationship between velocity and rate of weathering is consistent with column tests by van Grinsven and van Riemsdijk (1992). However, as a thought experiment, it is enlightening to separate the velocity from load effects. Figure 8.9 isolates the effect of increased velocity from increased load by reducing the concentration of A as velocity increases. For instance, the 50% increased velocity has a soil surface boundary condition of $A = 0.6667$ the nominal A concentration, so that the total load is the same. This figure shows that at the surface there are differences for the different velocities, but that

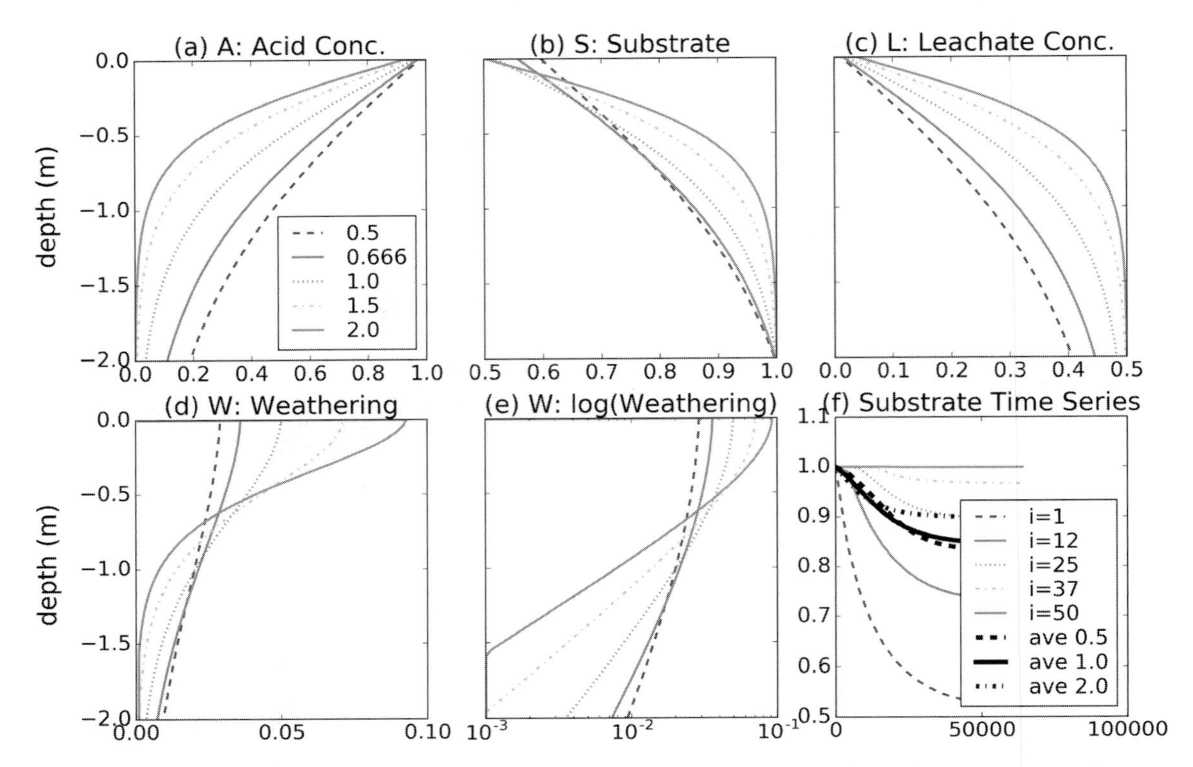

FIGURE 8.4: Sensitivity analysis for the weathering model with nominal parameters for changes in the weathering rate parameter k_1. The rapid approach to leachate concentration of 0.5 for the higher reaction rates in panel (c) is because with the parameters $k_2 = 2.0$ and $k_3 = 1.0$, the maximum amount of leachate that can be generated is $L = 0.5$ when all the acid is used (i.e. in panel (a) at depths greater than about 1 m). This is also the reason that the substrate concentration is approximately unchanged for depths greater 1 m.

Explanation of the panels: These are the final equilibrium profiles. For the first five panels the legend gives the multiplier that is used to modify the parameter being changed in the sensitivity study, so 0.5, for instance, means that the nominal parameter is multiplied by 0.5 for that run. (top left) The concentration of the acid in the infiltration water; (top middle) the concentration of the substrate (typically saprolite, but it could be a mineral or cation), and this curve is commonly referred to as a depletion profile; (top right) the concentration of the leachate (dissolved weathering product); (bottom left) the rate of weathering reaction; (bottom middle) the rate of the weathering reaction plotted on a semi-log horizontal scale (exponential weathering functions will plot as a straight line on this plot); (bottom right) a time series of substrate at various locations in the profile. The thin lines are for five layers equidistant from the top of the profile ($i = 1$ in the legend) to the base of the profile ($i = 50$ in the legend). The thick lines are the substrate concentration average over the whole profile for three parameter multipliers corresponding to the legend in the other five panels.

by the time the water has reached the base of the profile there is almost no difference in acid concentration. There is a small difference in the substrate at the soil surface. However, the weathering functions are very different, with the slowest velocity having the highest weathering rate. This is because everything else being equal, the acid has more time to react with the rock and so is able to transform a greater proportion of the rock, and that is reflected in the slightly lower substrate at the soil surface for the lower velocity.

Figure 8.10 shows the effect of varying the exhumation rate (and, as the soil depth is assumed constant with time, varying the erosion rate). The amount of substrate that makes its way to the surface is greater for a greater exhumation rate. This is consistent with the age of the rock being less, and being exposed to weathering for less time, for higher exhumation rates. This is also shown in the weathering function, which has lower weathering rates for higher exhumation. Again we see a humped weathering function appear for the higher rates, and this is because we have a larger substrate higher in the profile, and this coincides with the higher acid concentrations. In contrast to the higher velocity case above though, here the higher weathering rates in the hump are largely driven by the higher substrate, because there is only a small change in the acid concentrations.

Figure 8.11 shows the effect of increasing the dispersivity by six orders of magnitude. Other than a slight

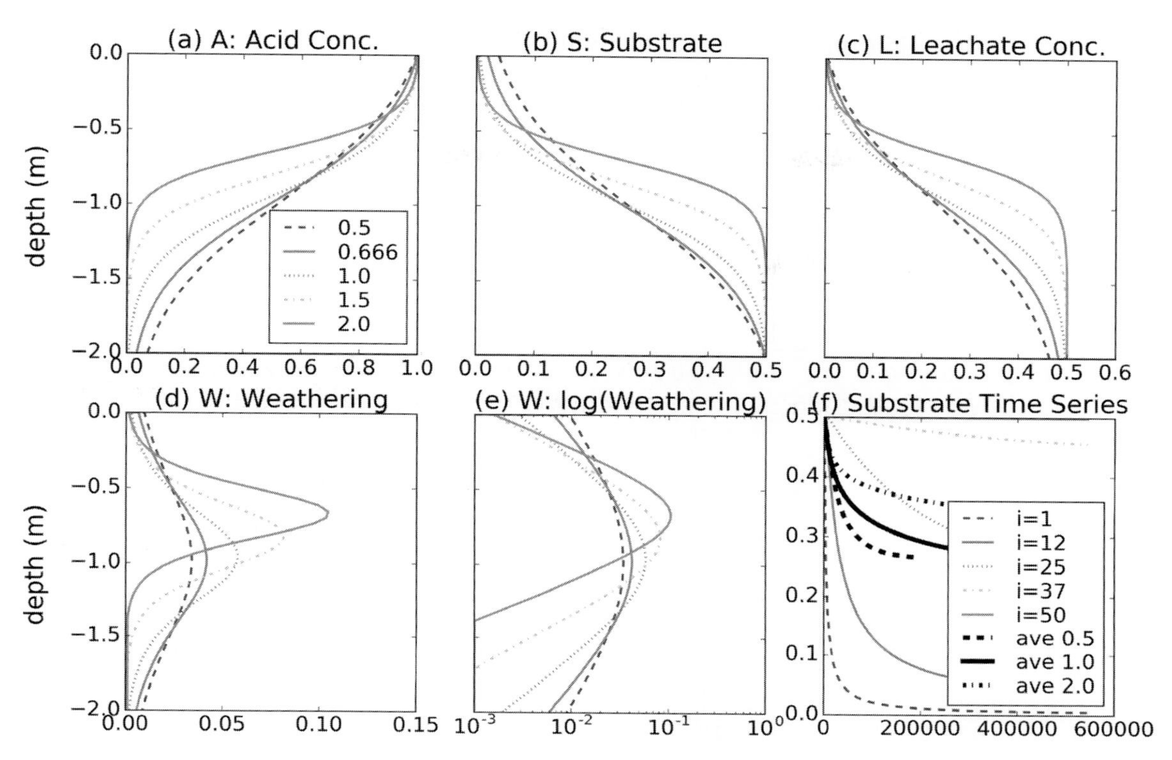

FIGURE 8.5: Sensitivity analysis for the weathering model with the full depletion parameters for changes in the weathering rate parameter k_1. This can be compared with Figure 8.4. The main effect of an increase in the weathering rate is that the weathering front is sharper and more distinct. The weathering peak also occurs slightly higher in the profile. For a detailed explanation of the panels see Figure 8.4.

smoothing that makes the profiles slightly more linear, there is no significant difference for different dispersivities.

We conclude this part of the sensitivity study of Equation (8.6) by noting that in none of the cases shown here do we see evidence of a weathering function with an exponential decline with depth. However, we do see, for high leaching velocities and uplift rates, that a humped weathering function can occur.

8.3.3.1 Leachate Concentration Rate Control and Transport-Limitation

The main shortcoming of Equation (8.6) is that the reaction rates are unlimited. In reality if rock fragments are placed into an acid solution and the acid solution concentration is maintained constant throughout the experiment, then the weathering reaction will eventually equilibrate at an equilibrium leachate concentration. Initially the reaction rate will proceed at its maximum rate (as defined in Equation (8.6)) but the reaction rate will slow asymptotically as the equilibrium is approached. Following Lasaga (1984), Steefel (2009), and Maher (2010) the reaction term can be reformulated as

$$R = k_1 A(k_5(S - S_{\min}))\left(1 - \left({}^{L}\!/_{L_{\text{equilib}}}\right)^{M}\right)^{n} \tag{8.13}$$

where L_{equilib} is the equilibrium concentration of the leachate in the fluid and ${}^{L}\!/_{L_{\text{equilib}}}$ is called the saturation index Ω (White et al., 2008). The parameters M and n are fitted to experimental data (Steefel, 2009) but we will assume, for simplicity, they are both 1, as Steefel indicates this is the case for kaolinite. This leachate concentration-dependent term now couples the leachate production with the rate of transformation of S and depletion of A. As L approaches the equilibrium concentration, the reaction term converges to zero. Note that for low concentrations of leachate the leachate-limitation term is approximately 1.0 and the reaction rates are not limited, which is what was assumed in the sensitivity studies in the previous section. A more sophisticated formulation, based on surface chemistry fundamentals (Lasaga, 1984), highlights that there is also a temperature dependence in this equation (both rate and equilibrium concentration, e.g. silica equilibrium concentration increases by a factor of two from 10° to 20°C), which we will ignore.

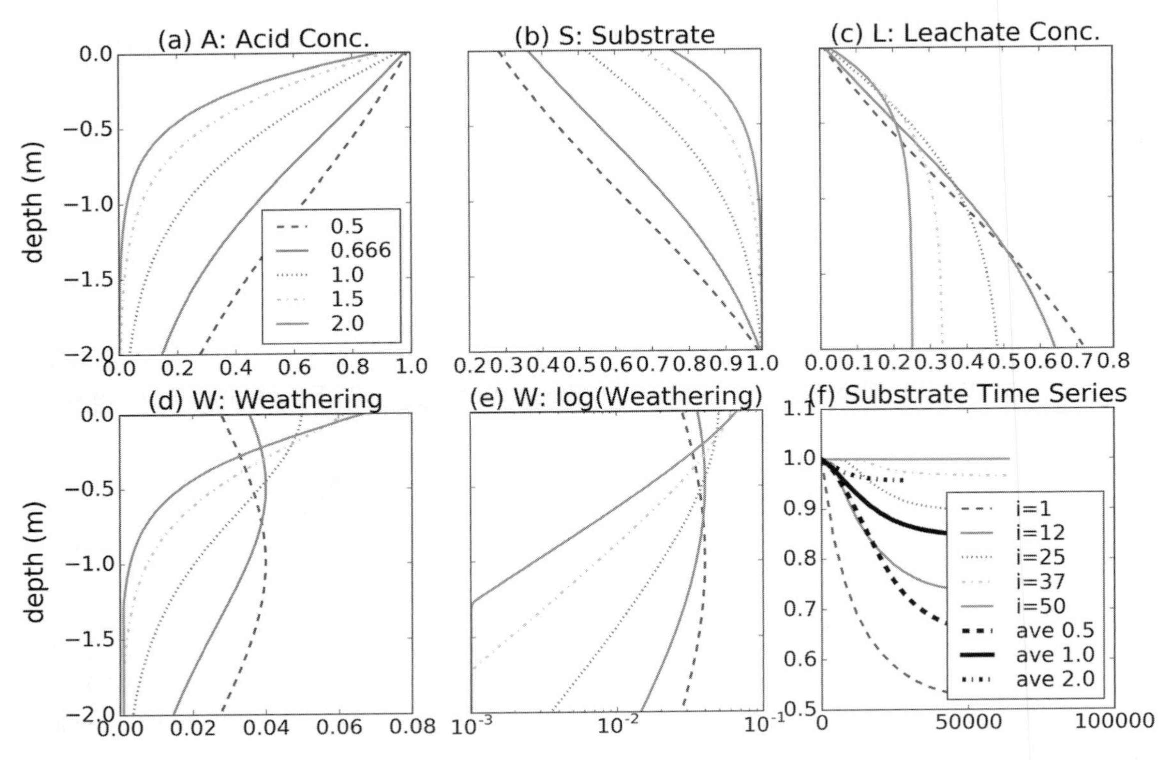

FIGURE 8.6: Sensitivity analysis for the weathering model with nominal parameters for changes in the acid consumption rate parameter k_2. For a detailed explanation of the panels see Figure 8.4.

In the sensitivity studies that follow, $L_{\text{equilib}} = 0.4$. This value was chosen solely to demonstrate the effect of the leachate-limitation term. The value of 0.4 is below the peak values of leachate concentration in the previous sensitivity studies, so comparisons to the simulations above will lead to visible effects that can then be interpreted. Figures 8.12 and 8.13 show the sensitivity analysis for k_1 and v, and are the leachate-limited equivalents to Figure 8.4 and 8.8. Leachate limitation has a marked difference in the weathering behaviour.

In Figure 8.12 the most obvious difference is that the rate of consumption of acid and transformation of substrate has been substantially reduced, and as expected the concentration of leachate levels off near the equilibrium value of 0.4. While the maximum weathering (at the surface) is not very different from the unlimited case, the weathering rate drops more rapidly with depth than for the unlimited case. The log(weathering) function is almost linear, and so is close to exponential with depth. Finally the change in shape of the acid curve is quite marked. This is related to the shape of the leachate curve where leachate doesn't increase with depth because it is near equilibrium, so that the natural decrease in acid with depth can accelerate because it is not being offset by the feedback that increases leachate. At the bottom of the profile the geochemistry is transport limited because the limit on weathering is the ability to transport the leachate out of the bottom of the soil profile. Finally the reduced scatter between the curves for different values of k_1 indicates that leachate limitation reduces the sensitivity of the model to changes in k_1, which is surprising because k_1 determines the rate of the reaction. This suggests that the model quickly transitions down the profile from reaction rate limited to leachate concentration limited (and transport limitation), which is consistent with Maher et al. (2009).

Figure 8.13 shows the sensitivity analysis for velocity with leachate limitation. Many of the conclusions from the k_1 sensitivity test above (i.e. Figure 8.12) carry over to this case. However, this figure shows that all signs of the humped weathering function (Figure 8.7) have disappeared. The reduction in the sensitivity of acid concentration with changing velocity is similar to that for k_1 but the reduction in sensitivity in substrate is not as marked. As for k_1 sensitivity plots in the bottom

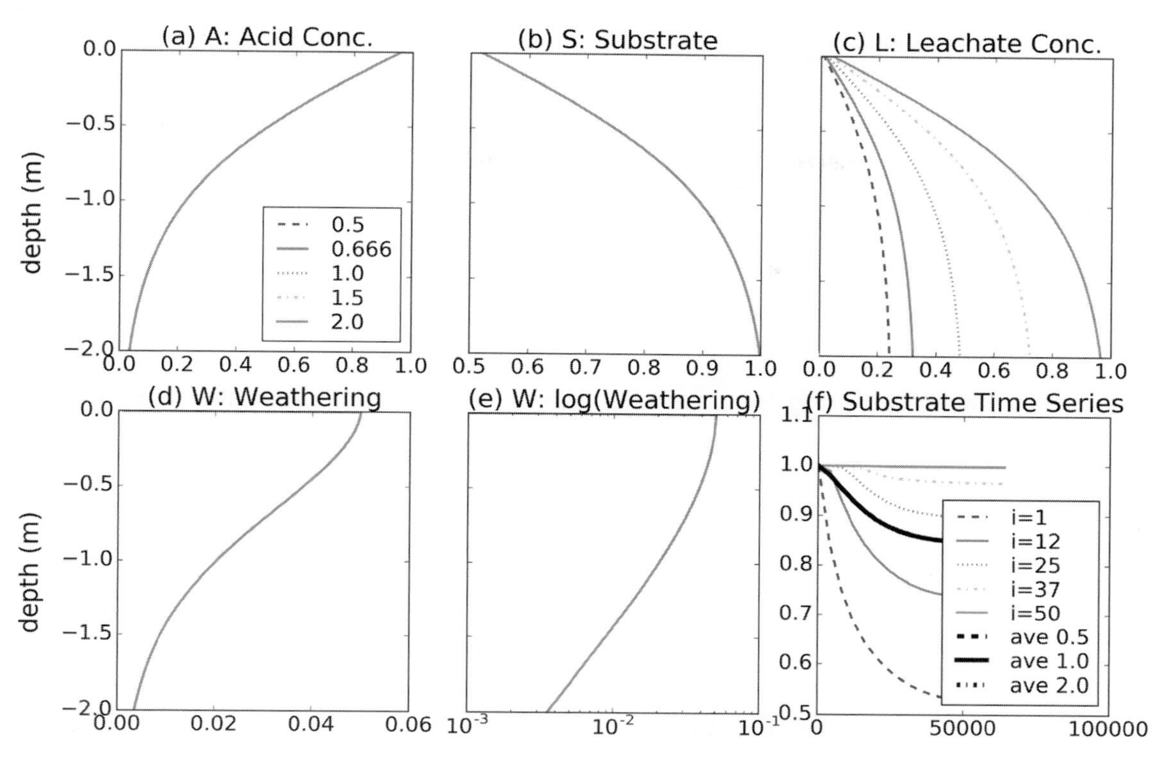

FIGURE 8.7: Sensitivity analysis for the weathering model with nominal parameters for changes in the leachate generation rate parameter k_3. For a detailed explanation of the panels see Figure 8.4. For A, S and W all values of k_3 plot on the same line.

half of the soil profile the leachate concentration is limited by the equilibrium value and geochemistry is transport limited.

8.3.3.2 Relative Concentrations of Acid and Substrate

The simulations above were performed with initial concentrations of acid and substrate both equal to one, and where the relative reaction rate between them was parameterised by $k_2 = 0.5$. These parameters were chosen because they consumed most, but not all, of the acid and substrate. The equations used are the original Equation (8.6) without leachate limitation.

Figures 8.14 and 8.15 show the same velocity sensitivity results in Figure 8.8 but with half the substrate and a higher reaction rate ($k_1 = 0.5$ rather than $k_1 = 0.1$). The difference between Figures 8.14 and 8.15 is that Figure 8.15 zooms in on the parameter values for velocity that yield a hump in the weathering function. Likewise Figures 8.16 and 8.17 show the same erosion sensitivity results as Figure 8.10 but with the same substrate and reaction rate changes. Because of the higher reaction rate, the acid is depleted by the time it reaches the bottom of the profile, and the slower the velocity the higher up in the profile this depletion occurs. Moreover because of the lower substrate, all of the substrate is transformed by the time it reaches the surface. Thus at the top of the profile the substrate is zero, and at the bottom of the profile the acid is depleted. Because of the multiplicative term in the reaction equation this means that the reaction rate (and thus the weathering rate) is near zero at both the top and bottom of the profile, with a maximum within the profile. Thus the weathering function in this case is a humped function with the maximum rate within the profile. For the mid-range velocity this humped peak occurs almost in the centre of the profile. For a higher velocity the acid doesn't deplete as quickly down the profile so the peak is toward the bottom of the profile, while for the lower velocity the depletion occurs higher in the profile and the hump occurs higher in the profile. Similar behaviour, not shown here, also occurs for different exhumation rates. Note that it is the combination of (1) depleted substrate at the surface and (2) depleted acid at the profile base that creates the hump in the weathering profile.

The S-shaped substrate curves are consistent with data from a series of chronosequences on terraces in White et al. (2008). White fitted a multicomponent

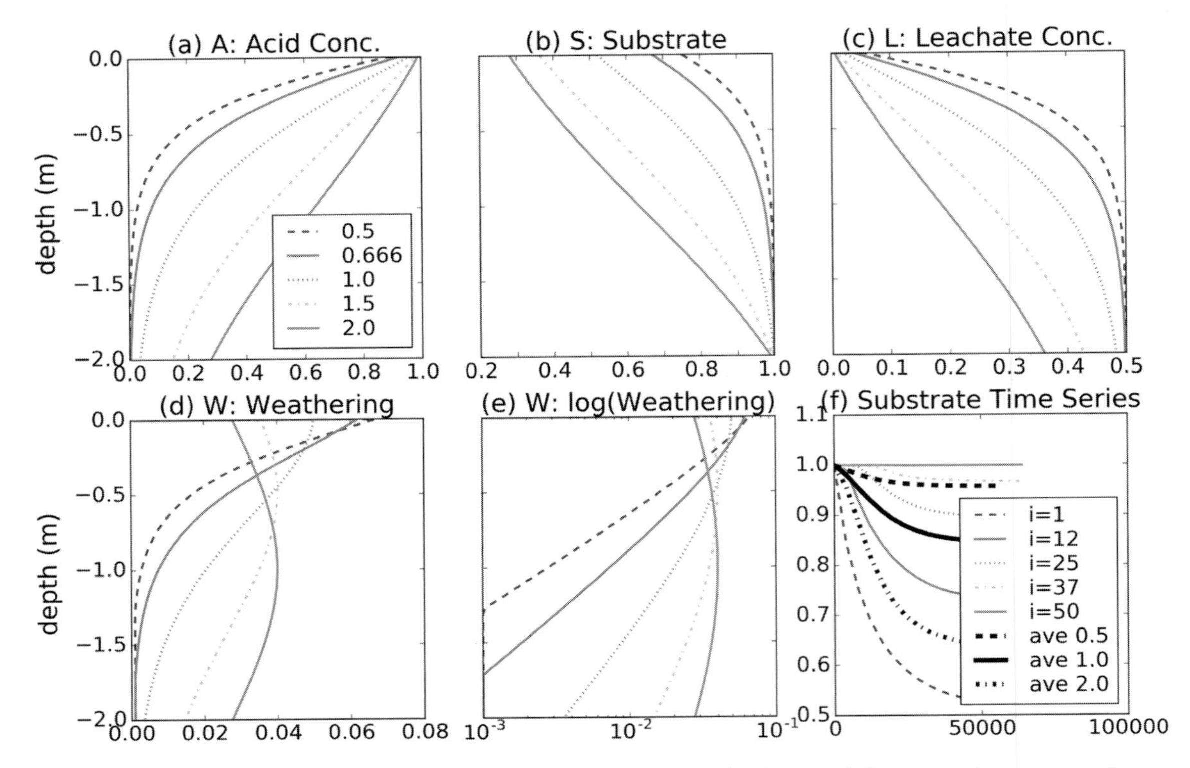

FIGURE 8.8: Sensitivity analysis for the weathering model with nominal parameters for changes in infiltration velocity parameter v. For a detailed explanation of the panels see Figure 8.4.

model very similar to that presented here to the data (Figure 8.18).

8.3.3.3 *Variation of Soil Depth*

The depth variation of weathering depends on the balance between erosion/exhumation and acid infiltration. The exact form of the curves will be dependent on the depth of the soil and the boundary conditions at the top and bottom of the soil profile. Figure 8.19 shows the variation in the weathering when the depth of the soil is varied from 1.0 m to 4.0 m. In this case the main difference is in the substrate curves. For the shallower soils the proportion of substrate weathered at the surface is less than for the deeper soils, for all depths within the soil profile, because the shallower soil has been exposed to weathering for less time than the deeper profiles. The weathering plots also show that the main differences are in the top 0.5 m of the soil, where the acid is freshest, because there is more unweathered substrate to weather for the shallower soils. Below 2 m all the acid is depleted, and it makes no difference how much substrate there is to weather because there is no acid for the weathering reaction.

Figure 8.20 shows the variation with soil depth of the full depletion parameters as in Figure 8.14. As the soil deepens, the peak of the hump occurs at a deeper depth. The weathering rate at the surface is reduced for the deeper soils because more of the substrate has been weathered. While the depth of the soil ranges from 1.0 m to 4 m, the depth at which the peak of the hump occurs varies only from 0.5 m to 0.8 m, suggesting that the peak is largely tied to the soil surface rather than the base of the soil profile. This suggests that the major contributor to the peak behaviour is the acid concentration, not the supply of substrate.

8.3.3.4 *Variability of Specific Surface Area*

The simulations above assume that the specific area of the substrate is the same throughout the profile. If physical weathering is the primary mechanism breaking down the rock fragments, then the rock fragments will become finer as they are exposed for longer to weathering, so that the finest particles will be in the layer below the soil surface armour layer in the soil (Chapter 7). How the specific area changes through the profile will depend on the physical weathering function (e.g. exponential or humped), the

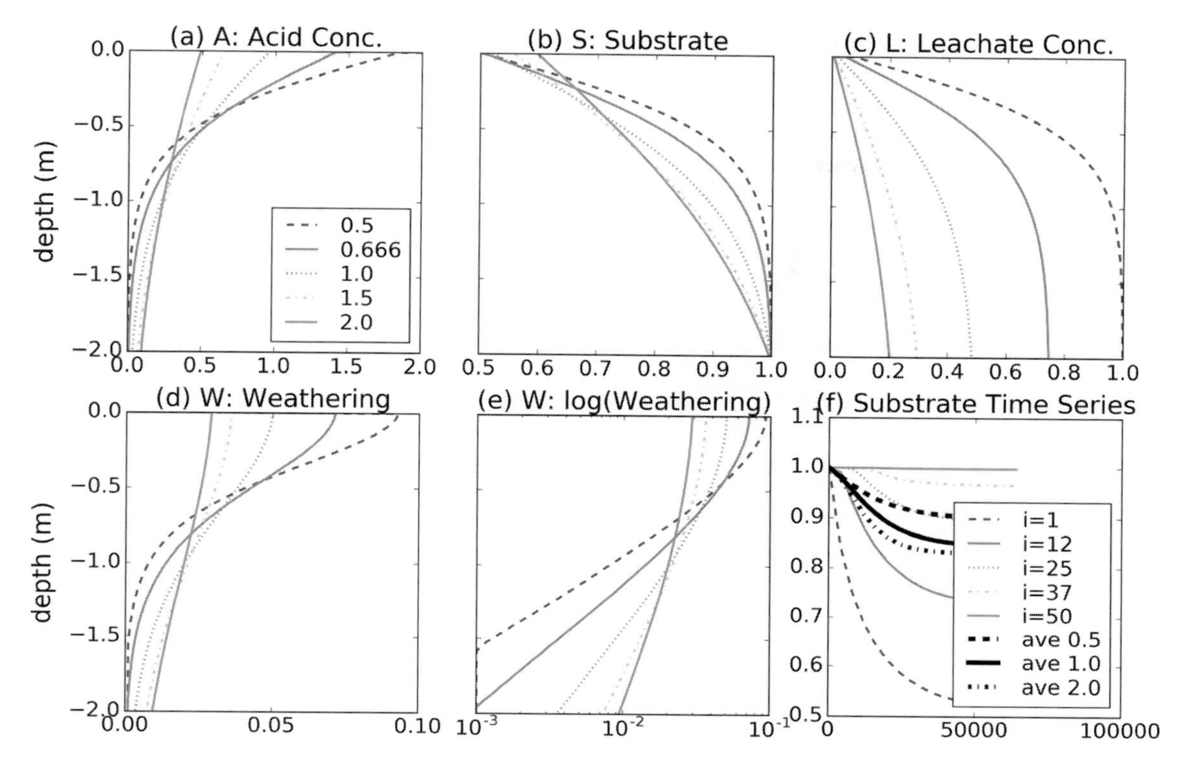

FIGURE 8.9: Sensitivity analysis for the weathering model with nominal parameters for changes in infiltration velocity parameter v and fixed cumulative acid load. Note that the fixed acid load condition means that the concentration of acid at the surface changes with the inverse of the velocity. Also note that the result for multiplier = 1.0 is the same as the curve for multiplier = 1.0 in Figure 8.8. For a detailed explanation of the panels see Figure 8.4.

parameters of the physical weathering function and the erosion rate (since this determines the rate of exhumation). With these variables in mind we will carry out a simple sensitivity study where specific area is allowed to vary with depth down the profile with a reaction term that is

$$R = k_1 A\left(k_5(S - S_{\min})\left(\frac{0.5\text{depth}}{z}\right)^{\varepsilon}\right)\left(1 - \left({}^{L}/_{L_{\text{equilib}}}\right)^{M}\right)^{n} \tag{8.14}$$

where ε is a parameter we vary to observe the impact on the model results. A large value means that shallow depths will have finer soil, higher specific area and high reaction rate. Figure 8.21 shows the results for different values of ε. The results for $\varepsilon = 0$ have a specific area independent of depth, and are the same as the nominal parameter simulations in Figure 8.2. For all positive values the specific area is increasing as you get closer to the surface, and that behaviour is more pronounced for larger values of ε. As expected there is more depletion of both acid and substrate near the surface, and this is balanced by less transformation of substrate within the profile at deeper depth. Also the weathering function shows the weathering becoming more and more concentrated in the near surface (and at a higher rate within this concentrated zone), where specific area is highest. If the physical fragmentation of particles is driven by chemical weathering, then there will be a positive feedback between chemical weathering rate and the fineness of the particles.

8.3.3.5 *Within Soil Acid Generation*

It's common to observe in agricultural soils that the carbon dioxide content of the pore space within the soil is near 100%, compared with the content in the atmosphere of 0.4%. This carbon dioxide is generated by the breakdown of organic matter in the soil by microorganisms in the soil (Lucas, 2001) and plant respiration (Chapter 14). If this is the case, then the acid in the soil water may be dominated by the in-profile generated carbon dioxide rather than the carbon dioxide dissolved in the rainwater that infiltrates (Little et al., 2005). If the rate of consumption of the carbon dioxide is much less

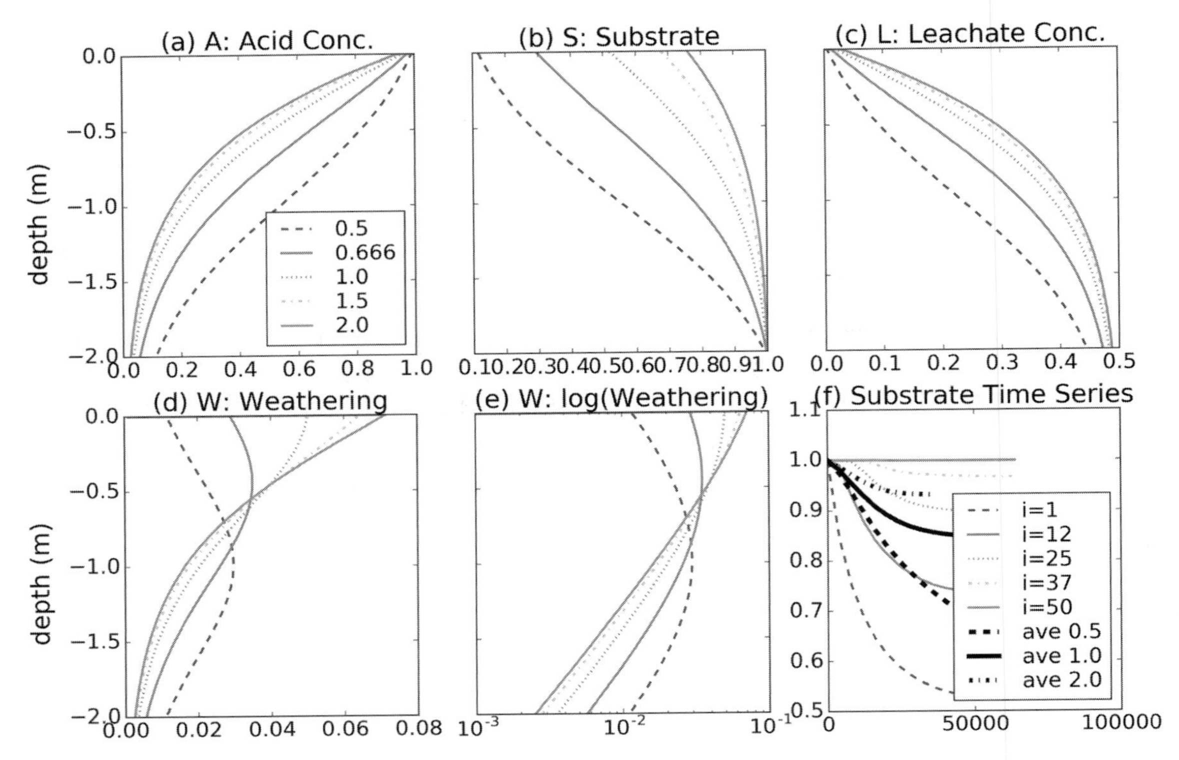

FIGURE 8.10: Sensitivity analysis for the weathering model with nominal parameters for changes in erosion rate parameter E. For a detailed explanation of the panels see Figure 8.4.

than the rate of generation by organic matter breakdown, then Equation (8.6) can be simplified to

$$
\begin{aligned}
A &= A_{\text{equilib}} \\
\frac{\partial S}{\partial t} - e\frac{\partial S}{\partial z} &= -R \\
\frac{\partial L}{\partial t} + v\frac{\partial L}{\partial z} - D\frac{\partial^2 L}{\partial z^2} &= k_3 R \qquad (8.15) \\
R &= k_1 A^{\alpha}(k_5(S - S_{\min})) \\
P &= k_4(S_0 - S)
\end{aligned}
$$

where A_{equilib} is the acid concentration when equilibrated with the soil atmosphere. If, as an example, we set $A_{\text{equilib}} = 1.$ then Figures 8.22 and 8.23 shows the results, with three different exhumation rates. Figure 8.23 shows the result with leachate control and Figure 8.22 without it. What is marked about this figure is a reversed exponential weathering function with the weathering rate being greatest at the soil-saprolite interface and declining exponentially toward the surface. The reason for this is with a constant A the reaction with substrate with time can be simplified to

$$\frac{\partial S}{\partial t} = -(k_1 k_5 A)S = -KS \qquad (8.16)$$

where all the terms inside the parentheses are constant. The solution to this equation is $S = S_0 e^{-Kt}$ where S_0 is the initial concentration in the substrate. If we note that the exhumation rate is equivalent to distance up the profile from the base, we can substitute for time ($z = Et$ so z is the distance above the soil-saprolite interface) and the weathering function is

$$W = \frac{\partial S}{\partial z} = \frac{\partial}{\partial z}\left(S_0 e^{Kz/E}\right) = \frac{S_0 K}{E} e^{Kz/E} \qquad (8.17)$$

This shows that the reverse exponential weathering profile is a direct consequence of the fixed acid concentration and constant exhumation rate. This reversed exponential was used as a weathering function by Welivitiya (2017) in his soilscape model.

More sophisticated approaches to the prediction of the soil atmosphere are based on mass balances of organic matter decay, root respiration and exchange fluxes between the soil atmosphere and the above-ground atmosphere (e.g. Pumpanen et al., 2003) and will not be discussed here.

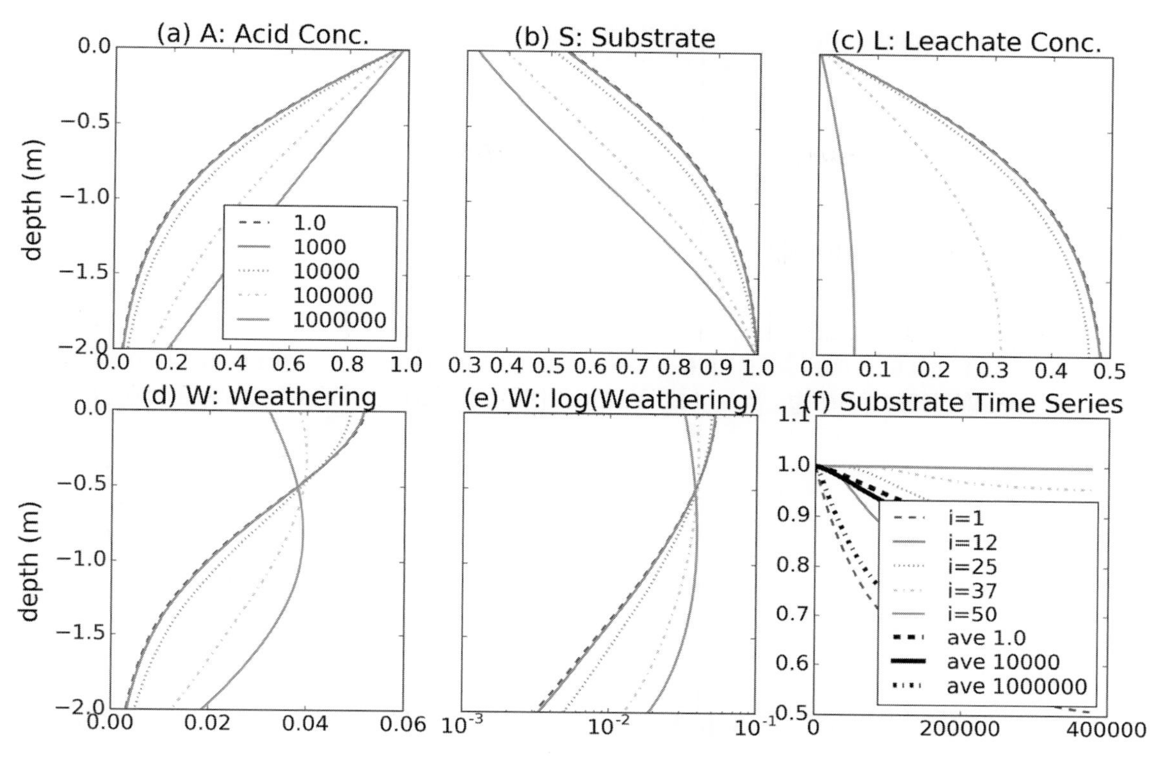

FIGURE 8.11: Sensitivity analysis for the weathering model with nominal parameters for changes in dispersivity D. For a detailed explanation of the panels see Figure 8.4.

The discussion above focussed on in-profile generation of acid as a result of carbon dioxide. Another mechanism is oxidation of sulphide minerals to generate sulphuric acid, which reacts with minerals to generate sulphate minerals and leachates. This is the mechanism involved in acid mine drainage (AMD), but it also occurs in natural environments (e.g. Chigira and Oyama, 1999; Oyama and Chigara, 1999). The weathering products and leachate and rates of weathering are different from carbon dioxide, but the generic governing equations in Equation (8.15) are still applicable. Chigara and Oyama note that there are two zones in the soil profile. The first zone is the region of active weathering where the strength of rock is reduced by weathering. Above this zone is an oxidised zone where the iron in the pyrite is precipitated, and precipitated iron cements the soil particles so that the soil strength is increased. The presence of two zones is because the AMD process results from the interaction of two chemistries. The first process, dominant in the upper oxidised zone, is a result of the availability of oxygen but depletion of minerals that can react with the produced acid. The second process is a result of the acid generation (which itself is a result of oxygen availability) and dominates in the weathering zone.

8.3.4 Bioturbation

Bioturbation was discussed in Section 7.3.5 in the context of physical weathering and evolution of the particle size distribution. The model discussed in that section was a simple Fickian diffusion equation for the mixing of the layers. It is likewise possible to implement a Fickian diffusion model for the bioturbation of the substrate on the chemical weathering model. We can modify the substrate equation in Equation (8.6) as

$$\frac{\partial S}{\partial t} - e\frac{\partial S}{\partial z} - D_s\frac{\partial^2 S}{\partial z^2} = -R \tag{8.18}$$

where D_s is the diffusivity for the mixing of the soil due to bioturbation. The weathering product equation is similarly changed. The equations for acid and leachate remain unchanged. As for the physical weathering case, the substrate diffusivity can vary with depth and time depending on the bioturbation mechanism (Section 7.3.5), and may

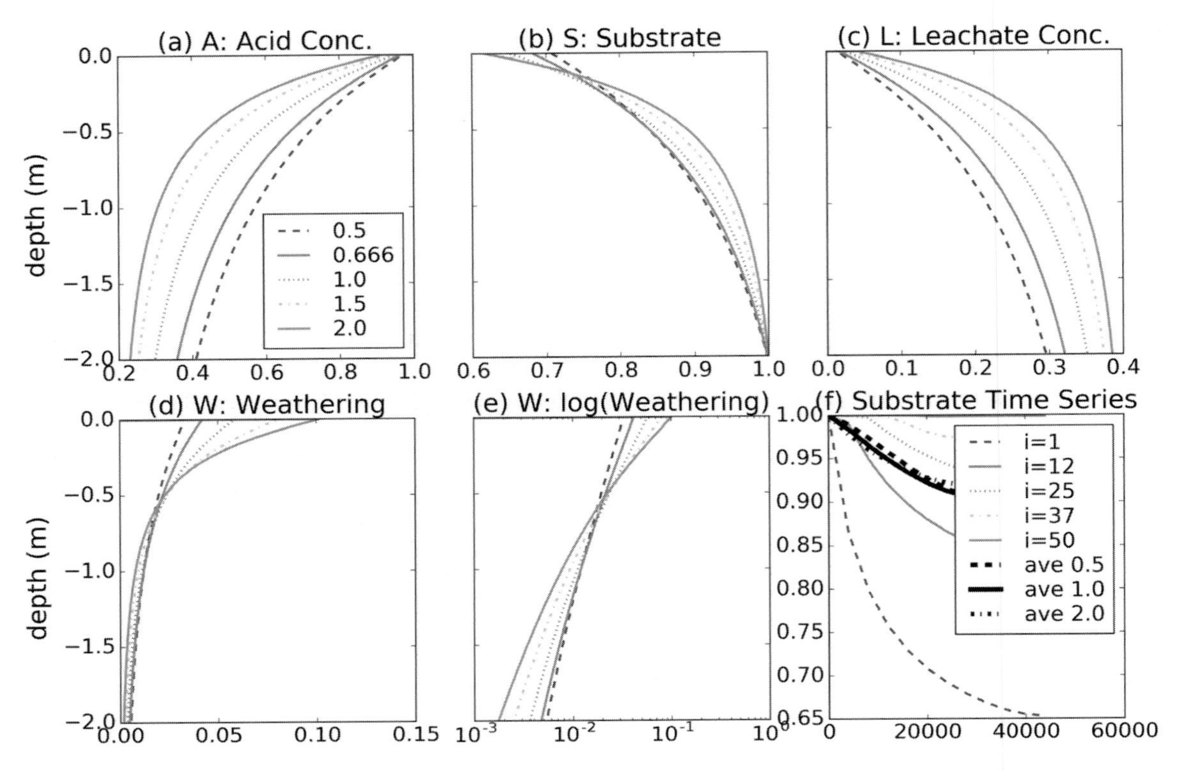

FIGURE 8.12: Sensitivity analysis for the leachate-limited weathering model with nominal parameters, for changes of the reaction rate parameter k_l. This figure is identical to Figure 8.4 other than the leachate limitation. For a detailed explanation of the panels see Figure 8.4.

interact with physical weathering processes and/or the grading of the soil.

Figure 8.24 shows a sensitivity run with the full depletion parameters. The nominal value is $D_s = 0.005$. It can be seen that as the bioturbation increases the weathering function becomes less humped and more exponential (the convergence to linearity in the log (weathering) panel), because the bioturbation mixes unweathered soil through the profile, so the humping, which results from depletion of saprolite near the surface (note the larger values for saprolite near the surface for the high diffusivities), does not occur. This reduction in saprolite depletion near the surface means that acid consumption rates increase in the surface layers so the acid concentration also declines more rapidly near the soil surface.

The figure shows that everything else being equal, the proportion of unweathered saprolite in the soil profile increases with increasing bioturbation, and the region of most active weathering is closer to the surface. For the nominal parameter set similar behaviour is observed (Figure 8.25). The commonality of the results for the two parameter sets suggests that these trends with increasing bioturbation may be robust against changes in the rates of the other in-profile processes.

8.3.5 The Hydrology Conceptualisation

The simple conceptualisation of hydrology in the model in the previous sections assumes that water is flowing through the soil at a constant rate. This is a poor approximation to the way infiltration actually occurs, with high flows through the soil during rainfall events and long periods of no flow between rainfall events. The conceptualisation also ignores transpiration, which will extract water from the soil in the root zone. And as noted in Chapter 3, in arid zones it is common that infiltration from small events does not penetrate all the way down the profile, and it is only large rainfall events that wet the entire profile from top to bottom.

To address both of these time- and profile-varying processes would require sophisticated unsaturated flow modelling within the soil layer with a full Richards

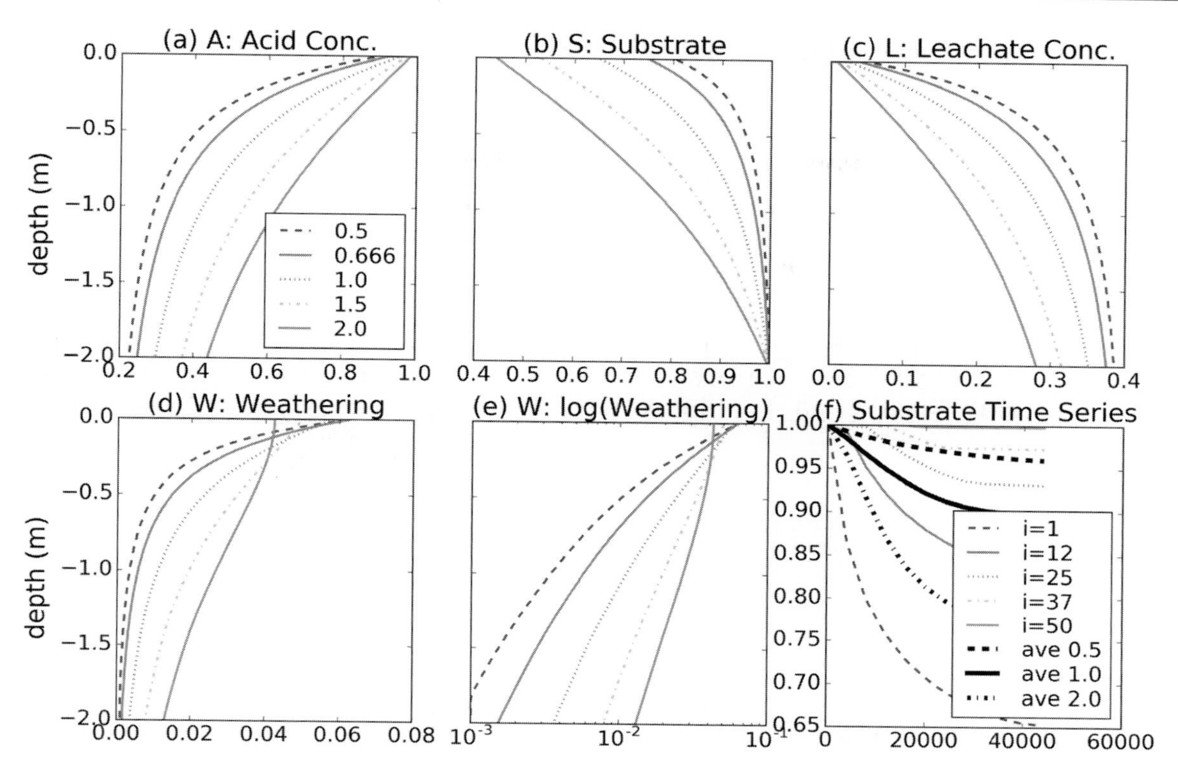

FIGURE 8.13: Sensitivity analysis for the leachate-limited weathering model with nominal parameters for changes in the infiltration velocity parameter, v. This figure is identical to Figure 8.8 other than the leachate limitation. For a detailed explanation of the panels see Figure 8.4.

equation solver and plant transpiration dynamics (e.g. using the HYDRUS model). Such a model is beyond the scope of this book, but some simple approximations can be made that give insight into what a full modelling exercise might display.

We will consider first the case of event-based infiltration. The velocity term in Equation (8.6) defines how much time the water is in contact with the substrate, and therefore how much weathering can occur. For instance, if we consider a small volume of soil of height Δz, then with velocity v the time the water is in that section of the profile is $\Delta t = {}^{\Delta z}/_{v}$. In the existing time-invariant model this is the residence time in the Δz volume of soil.

Consider now event-based infiltration. During the rainfall event water will fill the soil voids with water down to a depth determined by the depth of rainfall and porosity of the soil (Section 3.1.4). Using the Green-Ampt approximation to infiltration, if the infiltration from the rainfall event is R, and the porosity of the soil is n then this depth will be $z = {}^{R}/_{n}$. Immediately after the rainfall stops, most of that water drains deeper into the profile, but a small amount of water remains attached as a thin film to the soil particles' surfaces, called the field capacity (typically about 10–30%), which cannot drain out under gravity because it is held onto the soil particles by surface tension effects. This water will continue to react with the substrate until it (1) depletes the acid or the substrate, (2) reaches leachate limitation, (3) is evaporated or transpired or (4) is replaced by new water from a subsequent infiltration event (the water from the event must of course penetrate to that depth). In this case the residence time of water, T_r, is the mean time between flushing events at that depth in the soil profile. This mean residence time will be short at the top of the profile (because most rainfall events will saturate the surface) and longer at the base of the profile (because only large rainfall events will infiltrate this deeply). To first order the average effect of this behaviour would be the same as if the water were slowly flowing through the profile but where the velocity reduces with depth (it becomes slower with depth as the flushing events get less frequent) so that

$$v = {}^{\Delta z}/_{T_r} \tag{8.19}$$

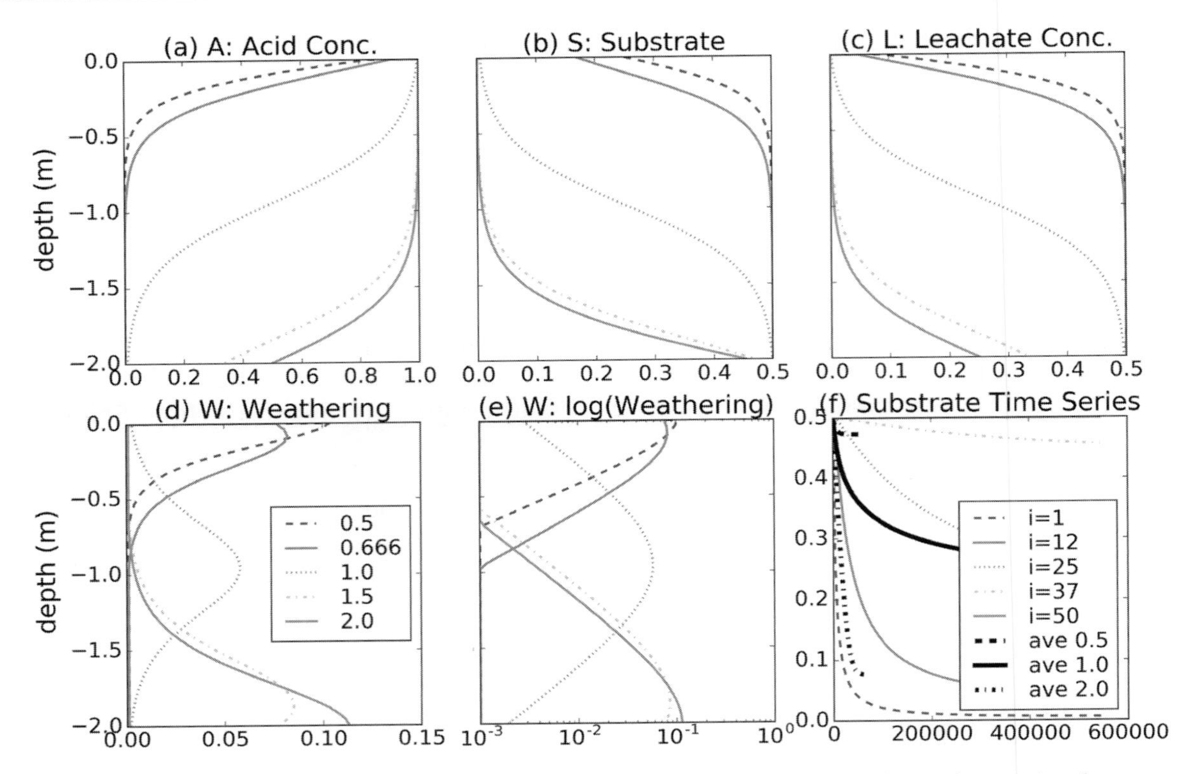

FIGURE 8.14: Velocity sensitivity analysis for the weathering model with the full depletion parameters showing the generation of a humped weathering profile for changes in the infiltration velocity parameter, v. This figure can be compared with that for the nominal parameters in Figure 8.8. For a detailed explanation of the panels see Figure 8.4.

This, of course, averages out the variability between infiltration events, which may be important if the fluid becomes leachate concentration limited between infiltration events. Note also that the time of residence will be related to climate, with wetter climates having lower residence times and higher velocities.

We turn now to how transpiration extracts water from within the soil profile. Plants cannot extract all the water in the profile, being unable to extract water below the wilting point of the plant. This wilting point soil moisture varies between plants, soil salinity and soil properties. Typically it is lower than the field capacity, and of the order of 1–20%. The effect of plants is to remove water (and some of the dissolved weathering products) and reduce the amount of water infiltrating below the root zone. Thus the effect of the root zone is to reduce the number of infiltration events that penetrate beyond the root zone. Thus the mean residence time of the water below the root zone will be longer than it would be in the absence of transpiration. A crude first approximation would be to reduce the velocity of the water below the root zone by the ratio of the volume of transpiration (which is equal to the water extracted from the profile) to the volume of infiltration per unit time so that below the root zone

$$v_{\text{below}} = v_{\text{above}} \left(\frac{I - T}{I} \right) \tag{8.20}$$

where I is the infiltration (mm/day) and T is the transpiration (mm/day), and the subscripts indicate the velocities above and below the root zone.

Observed residence times for water can be quite long by hydrologic time scales. For instance, White et al. (2009) in a Mediterranean climate calculated 10–25 years based on infiltration velocities 0.06–0.22 m/year in a 15 m thick soil profile.

Finally there is the question of whether a uniform flow through the profile is an appropriate approximation when there is ample evidence of heterogeneities, preferred flowpaths and macropores, so that infiltration is not uniform (Beven and Germann, 2013). Numerical studies coupling the chemical reaction principles in this chapter with preferential flowpaths are beginning to appear

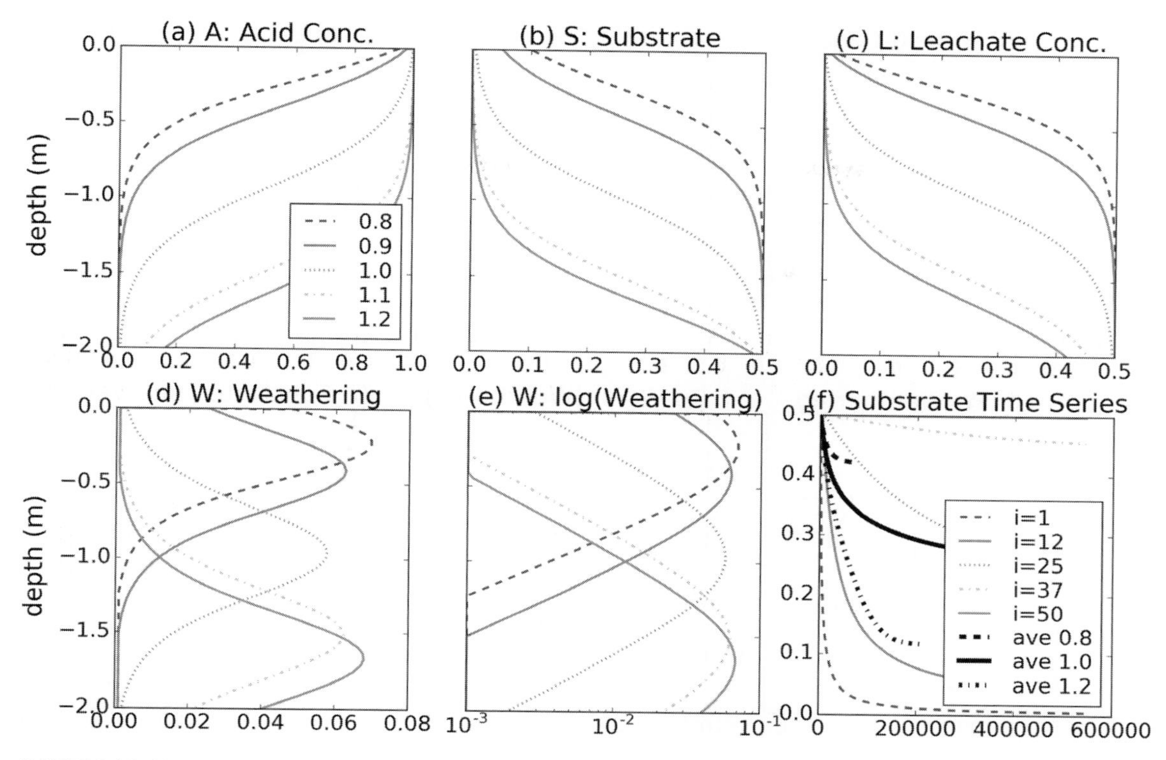

FIGURE 8.15: Velocity sensitivity analysis for the weathering model with the full depletion parameters showing the generation of a humped weathering profile. This figure is identical to Figure 8.14 except that it shows greater detail for the parameter values around the value that generates the mid-profile hump in the weathering profile. For a detailed explanation of the panels see Figure 8.4.

(Pandey and Rajaram, 2016; Rajaram and Arshadi, 2016) but experience from the hydrology community on macropore flow would suggest it will be a hard issue to resolve. However, the complexity of explicitly modelling macropores may not be necessary because Pandey and Rajaram (2016) present horizontal averaged results for the weathered substrate and weathering front, and these averaged results are qualitatively similar to the one-dimensional model in this chapter. For instance, their Figures 6 and 7 are remarkably similar to Figure 8.3 here, despite their results showing strong spatial heterogeneity in weathering due to the preferential flowpaths. This suggests that the simple one-dimensional model in this chapter might be a useful horizontally averaged effective representation of weathering even in the presence of preferential flowpaths.

A further complication regarding macropores and preferential flow is the possibility that the chemical weathering process may coevolve with the soil heterogeneity so macropore density and pathways evolve over time (and thus the parameters of an effective parameterisation will also evolve over time). For instance, a minor initial perturbation in soil or infiltration may result in higher flow at one location, and this higher flow leads to greater chemical weathering, which in turn increases porosity and permeability. This higher permeability then allows more water to flow through this localised highly weathered region, with a positive feedback on chemical weathering and water flow so that a preferred flow pathway is created. This is mathematically similar to the positive feedback between water flow and erosion that leads to channel networks formation in catchments (Willgoose et al., 1991a, b).

8.4 Multicomponent Chemical Weathering

Clearly the single-component model above is very simple, but even this simple model shows a spectrum of soil profile dynamics. We will not discuss in detail the extension of this model to multicomponent problems. This is the realm of sophisticated geochemical modelling tools like CrunchFlow (Steefel, 2009), and PHREEQC

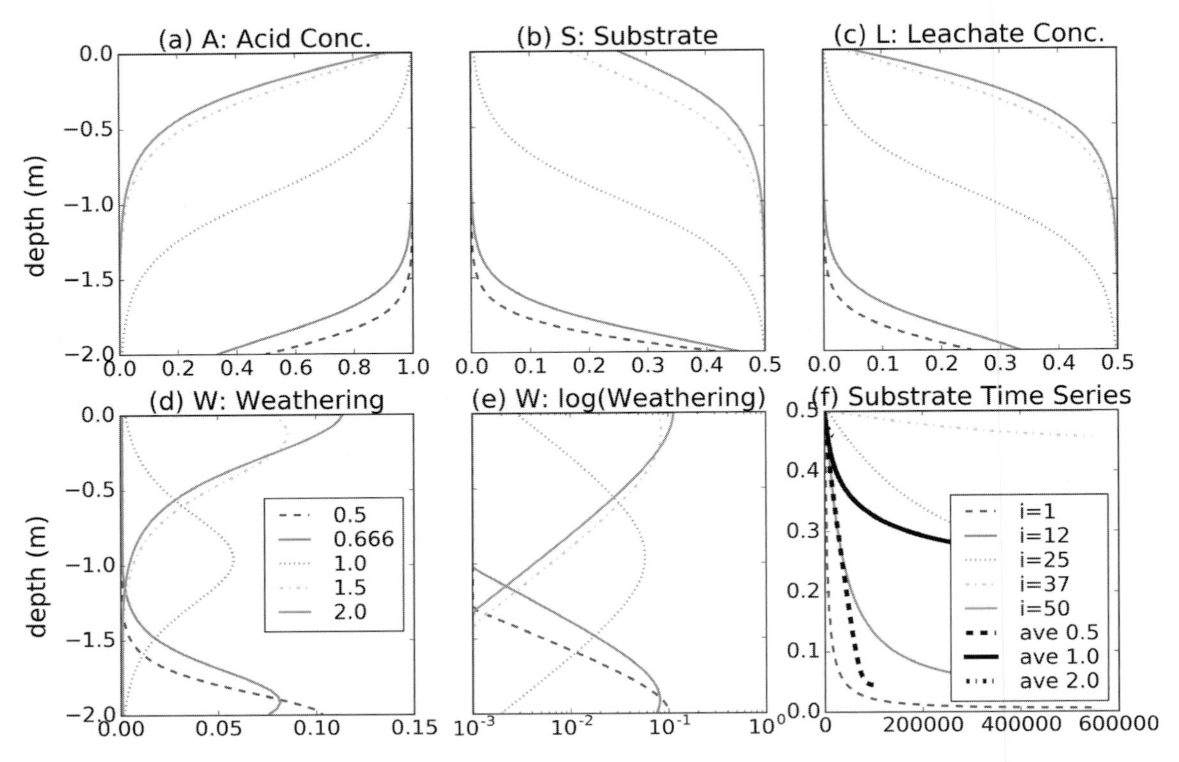

FIGURE 8.16: Erosion sensitivity analysis for the weathering model with the full depletion parameters for changes in the erosion rate parameter E showing the generation of a humped weathering profile. Compare this figure with that for the nominal parameters in Figure 8.10. For a detailed explanation of the panels see Figure 8.4.

(USGS, 2015). However, it is worthwhile to qualitatively consider some of the complexities introduced by multi-component modelling, many of which are still debated in the literature.

The most important complexity is that the weathering reactions from the different minerals may interfere with each other. If two mineral weathering reactions generate the same product, it is possible for a leachate limitation to inhibit one or both of the reactions. For instance, for two minerals

$$
\begin{aligned}
a_1A + S_1 &= b_1P_1 + c_1L_1 \\
a_2A + S_2 &= b_2P_2 + c_2L_2 \\
L_1 + L_2 &= L
\end{aligned}
\tag{8.21}
$$

the leachate product L is produced by both reactions, and if the $L = L_{\text{equilib}}$ then both reactions are inhibited compared with the situation if only one reaction was occurring. There are also issues of the rates of reactions with the faster reaction potentially interfering with the slower reaction. The mathematics of this is relatively straightforward, but is dependent on the geochemical model used and their databases of reaction rates for all the constituents in the reactions. This is the key capability of the multicomponent geochemical models mentioned above. Figure 8.26 shows calculations for the full depletion parameter set where all the weathering is done by a single infiltrating acid acting on two minerals with different reaction rates. Figure 8.26 can be compared with Figure 8.5, which has the same parameter sets (i.e. both have the multipliers 0.5 and 2.0), but where each weathering rate occurs independently of each other's rate in Figure 8.5. While the observation for Figure 8.5 that the front becomes sharper with increasing k_1 is still true, the details of the depth dependence of the weathering are different. When weathering occurs across several minerals simultaneously (and where the minerals are differentiated by their reaction rate), the peak depth of weathering rate is lower in the profile.

If on the other hand there is *no* interaction between the various minerals being weathered in the profile, then all

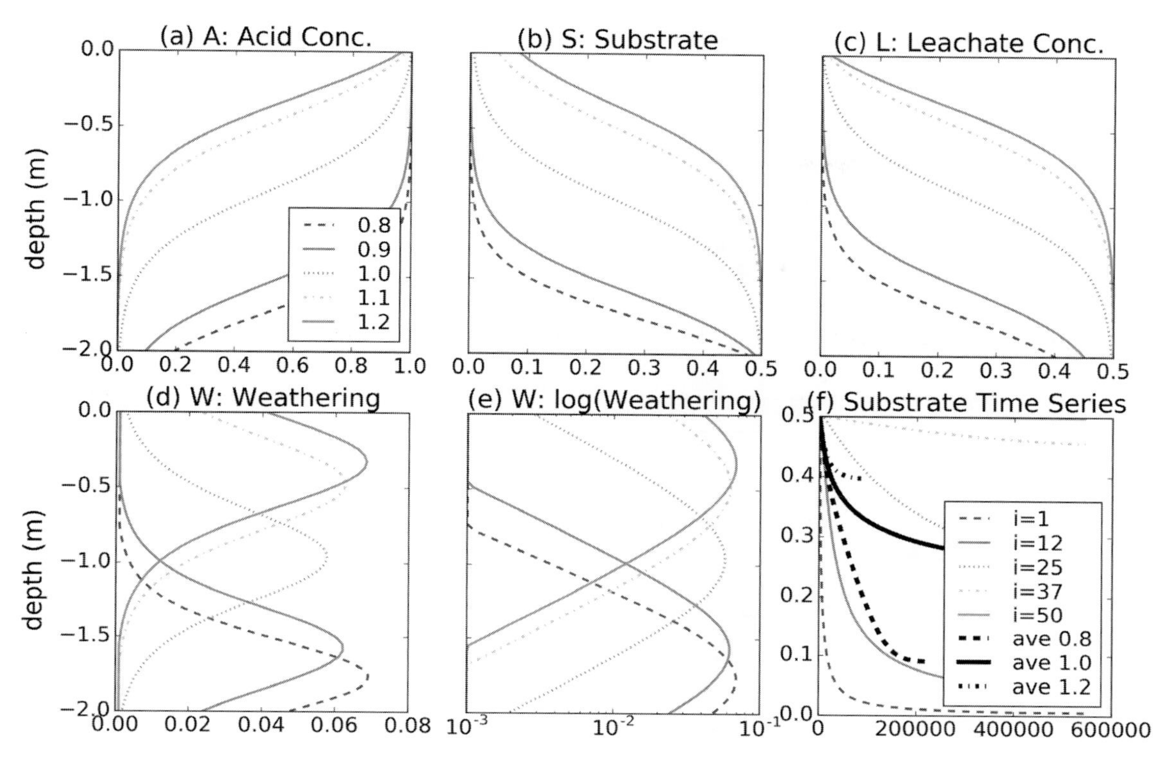

FIGURE 8.17: Erosion sensitivity analysis for the weathering model with the full depletion parameters showing the generation of a humped weathering profile. This figure is identical to Figure 8.16 except that it shows greater detail for the parameter values around the value that generates the mid-profile hump in the weathering profile. For a detailed explanation of the panels see Figure 8.4.

that is necessary is to run the single-component equations (i.e. Equation (8.6)) in parallel with each other and just sum up all the components to determine the cumulative impact of weathering (e.g. Yoo and Mudd, 2008). The only interaction is that the acid is consumed by all of the processes so there is the possibility of slower reactions being inhibited if the acid is preferentially consumed by the faster reactions. This would be consistent with applying the calculations leading to Figures 8.4 and 8.5 to each of the mineral constituents, respectively.

Another complexity is the surface available for weathering. The rate of the reaction depends on the surface area of the mineral exposed to the reaction. This surface area is more than the relative percentages by volume in the rock because the weathering will occur on the most reactive minerals (and mineral boundaries, defects, etc.) first. For instance, Hodson (2006) found biotite weathered about 100 times faster on grain boundaries than the grain interior. Accordingly at any time the specific surface area of each mineral component may deviate considerably from its percentage by volume in the rock. Predicting the time evolution of surface area due to chemical weathering is difficult, with no consensus on approach and with conflicting results from different experimental methods, rocks, and acids. Even with the use of the sophisticated geochemical models above this is an area of considerable predictive uncertainty.

Finally, Maher et al. (2009) noted that in her field study of a leachate-limited problem that her field data on reaction rates required either unrealistically high infiltration velocities, or a secondary reaction that removed leachate products from solution (i.e. a precipitation reaction). This requires the modelling of (at least) two chemical reactions (the original weathering reaction, and the secondary precipitation reaction involving the dissolved weathering products) and is beyond the scope of the model presented here. White et al. (2005) found that the solubilities of weathering products gibbsite and kaolinite changed with changing pH (which depends on the soil atmosphere CO_2 concentration) and the change in solubilities depended on the aluminium hydroxides present in the

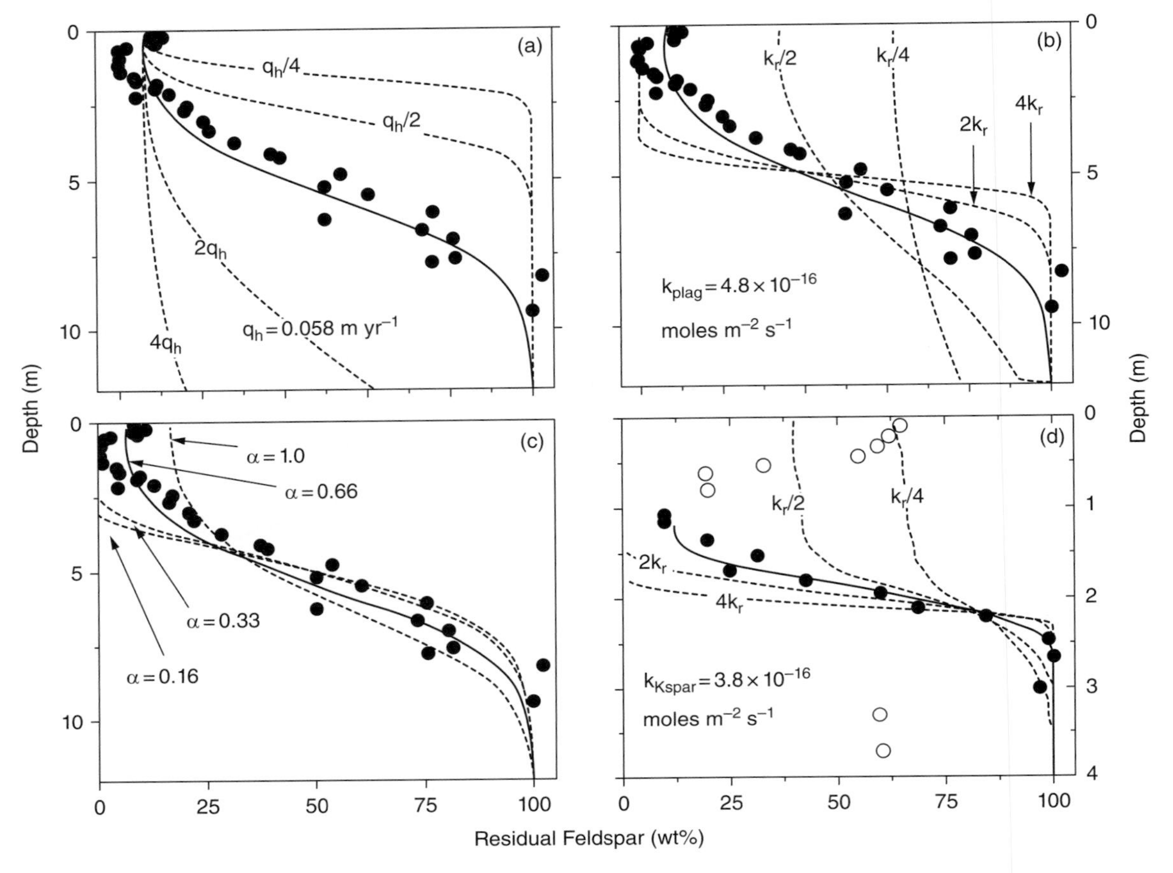

FIGURE 8.18: Data and model fits to a series of chronosequences in Santa Cruz. The data are the residual feldspar left in the rock after weathering, so are equivalent to substrate. The model is a multicomponent version of the model in this section. The lines are different parameters in the model for (a–c) plagioclase and (d) K-feldspar: (a) infiltration velocity, (b) reactivity, (c) surface area-diameter relationship and (d) K-feldspar reactivity (Figure 14 from White et al., 2008).

soil. Conversely, other mineral solubilities were independent of pH and aluminium. Again this suggests the need for multicomponent modelling in some situations.

Lebedeva et al. (2007) modelled multicomponent weathering with a model very similar to that proposed here but where transport in the vertical was only by diffusion and there was no advection. They showed the erosion is key to developing a steady state chemical profile because of the new material that is introduced at the base of the profile. This is consistent with the single-component model results in this chapter. A notable aspect of the multicomponent aspect of their modelling was that the different components had different stable weathering profiles through the soil, so that different weathering products dominated different depths within the profile. In Figure 8.6 the substrate profile varies differently with depth depending on the reaction rate, so for a number of components with different reaction rates the percentage concentration of each component will vary with depth. Lebedeva found layers of kaolinite that they postulated created porosity changes down the profile. It is tempting to speculate that these results hint at the possibility of horizon development within the soil profile.

8.5 Hillslope Modelling

The model framework and applications in the previous sections are for one-dimensional flow vertically down the soil profile. If we refer back to our discussion of dominant type of hydrology, this is consistent with arid zone hydrology, where there is no permanent water table within the soil profile (infiltration excess runoff generation; see Section 3.1.3.2).

In more humid climates there is typically a permanent water table within the soil and water flows laterally in the

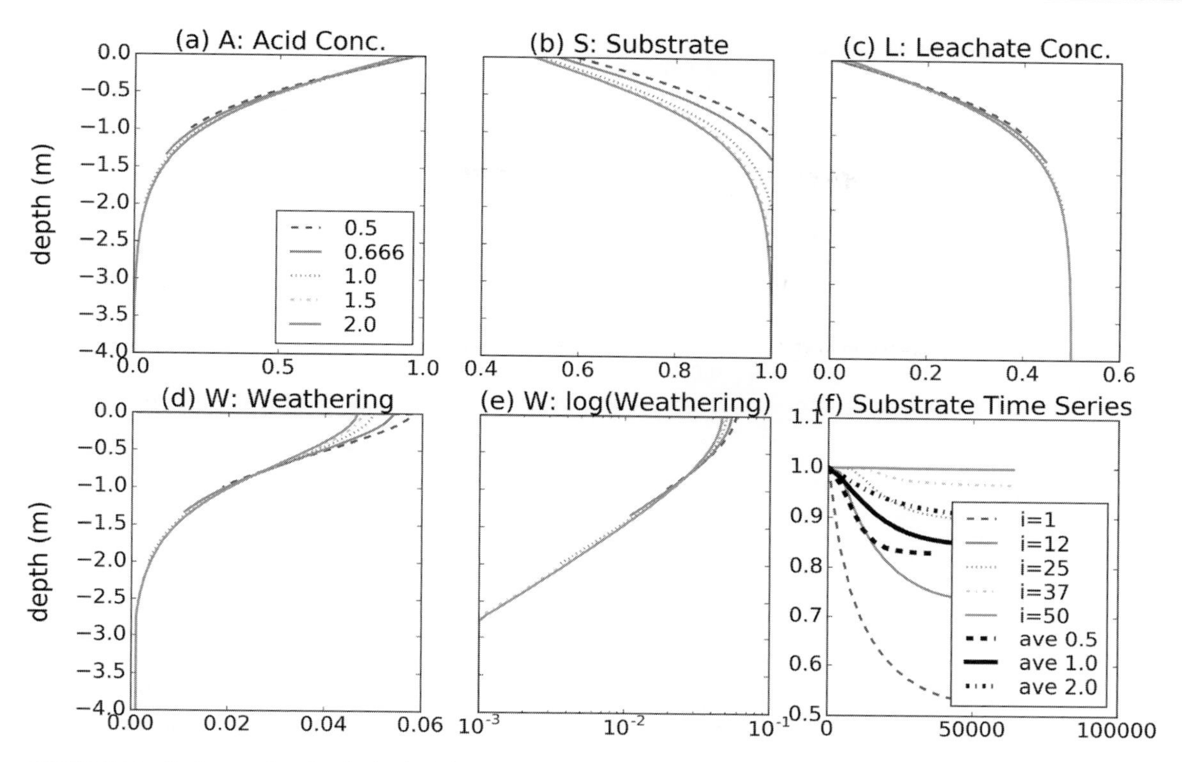

FIGURE 8.19: The impact of varying the depth of the soil profile from 1 m deep to 4 m deep. The multipliers in the legends indicate the depth relative to the nominal depth of 2 m. All other parameters are the nominal parameters. For a detailed explanation of the panels see Figure 8.4.

steepest downslope direction down the hillslope (i.e. saturation excess runoff generation; see Section 3.1.3.1). Above the water table infiltration will flow vertically down the profile until it hits the water table. Thus above the water the one-dimensional vertical flow model in the previous sections governs, but below the water table the dominant flow direction is near horizontal rather than vertical. This horizontal flow case has not been as well studied as the vertical flow case.

Some simple models are possible for transport-limited weathering. For transport-limited weathering the concentration of the leachate is determined by the equilibrium chemistry (i.e. Equation (8.5)) so that the rate of dissolved load transport is

$$q_d = qL_{\text{equilib}} \tag{8.22}$$

where q_d is the dissolved load carried by the groundwater (mass/time), q is the discharge (volume water/time) and L_{equilib} is the equilibrium concentration (mass/volume water) of the leachate. The cumulative amount of leachate carried by the groundwater can be used as a measure of the progress of weathering. Braun et al. (2016) used this observation to develop a model for the dynamics of hillslope soil weathering. The rate of change of elevation of the soil-saprolite interface is

$$\frac{\partial S}{\partial t} = -F\left(\frac{\partial q_{d,x}}{\partial x} + \frac{\partial q_{d,y}}{\partial y}\right) \tag{8.23}$$

where S is the elevation of the soil-saprolite interface (see Equation (2.7)), and F is factor that relates the amount of loss of mineral mass in the soil (which equals the amount of dissolved load in the groundwater) to the rate of change of the elevation of the soil-saprolite interface. The factor F depends on the mineralogy of the rock and how a kilogram of loss from weathering equates into the transformation from rock into soil. One way of characterising this is by the change in porosity of the saprolite to the porosity of the soil above the zone of active weathering.

A particularly simple result can be derived for a one-dimensional hillslope with parallel flow lines downslope, so that $q = (I - \text{ET})x$ where I is the infiltration rate (depth/time), ET is the evapotranspiration rate (depth/time) and x is the distance downstream from the hillslope divide. Substituting Equation (8.22) into Equation (8.23)

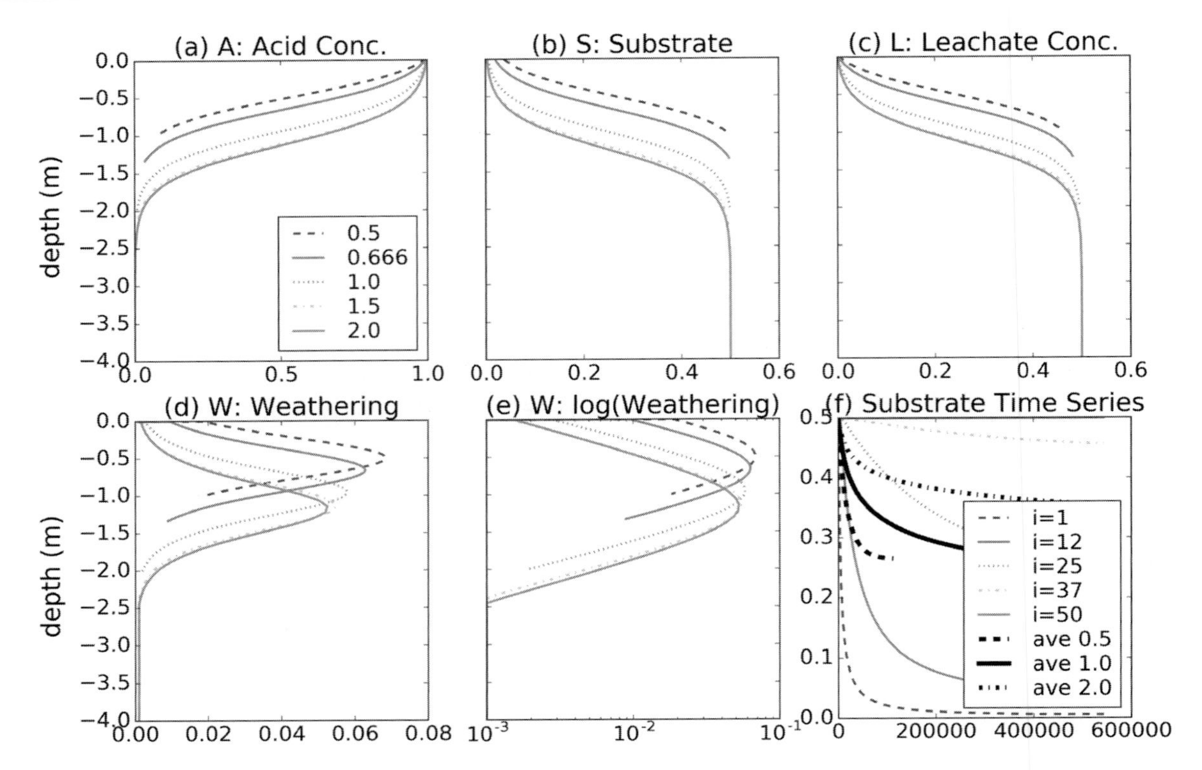

FIGURE 8.20: The impact of varying the depth of the soil profile from 1 m deep to 4 m deep. The multipliers in the legends indicate the depth relative to the nominal depth of 2 m. All other parameters are the full depletion parameters. Compare this figure with the nominal parameters (Figure 8.19) and note that the full depletion parameters generate a humped weathering profile. For a detailed explanation of the panels see Figure 8.4.

leads to an equation that estimates the change in elevation of the saprolite-soil interface for a hillslope in the direction of steepest descent

$$\frac{\partial S}{\partial t} = -FL_{\text{equilib}}\frac{\partial q}{\partial x} = -FL_{\text{equilib}}(I - \text{ET}) \tag{8.24}$$

The implication of this is that if the discharge is increasing linearly downslope, then the rate of transformation of saprolite to soil is constant across the landscape, or as some may prefer to express it the rate of elevation change of the saprolite-soil interface. This rate of soil formation is independent of depth, in contrast to, for instance, the exponential soil production function in Equation (6.2) which has the rate of saprolite to soil transformation as a declining function of the depth of the soil.

Braun et al. (2016) then concluded that a relationship for dynamic equilibrium of soil was possible. For landform topography the balance between a spatially constant tectonic uplift and the exactly balancing erosion is simply the argument leading to the classical slope-area relationship (Section 3.3.3). For a constant elevation saprolite-soil interface the rate of decline of the interface must exactly balance the rate of uplift U so that

$$U = -\frac{\partial S}{\partial t} = FL_{\text{equilib}}\frac{\partial q}{\partial x} = FL_{\text{equilib}}(I - \text{ET}) \tag{8.25}$$

It should be noted that everything of the right-hand side of this equation is a function of the geology or the climate, so there is nothing constraining the right-hand side to be equal to the uplift. Accordingly if the right-hand side is larger, than the uplift then the soil will deepen without limit, while if it is smaller it will decrease until (presumably) there is no soil. Thus it seems unlikely that this model will generate equilibrium soil depths that are constant with time for a hillslope at dynamic equilibrium.

In conclusion it is should be clear that there are a number of conceptual difficulties with this model including the inconsistency with the exponential soil production function, and the inability to generate an equilibrium soil depth. Those difficulties aside the approach of Braun et al. (2016) is novel and is consistent with the hydrology of hillslopes in humid regions. Accordingly this approach

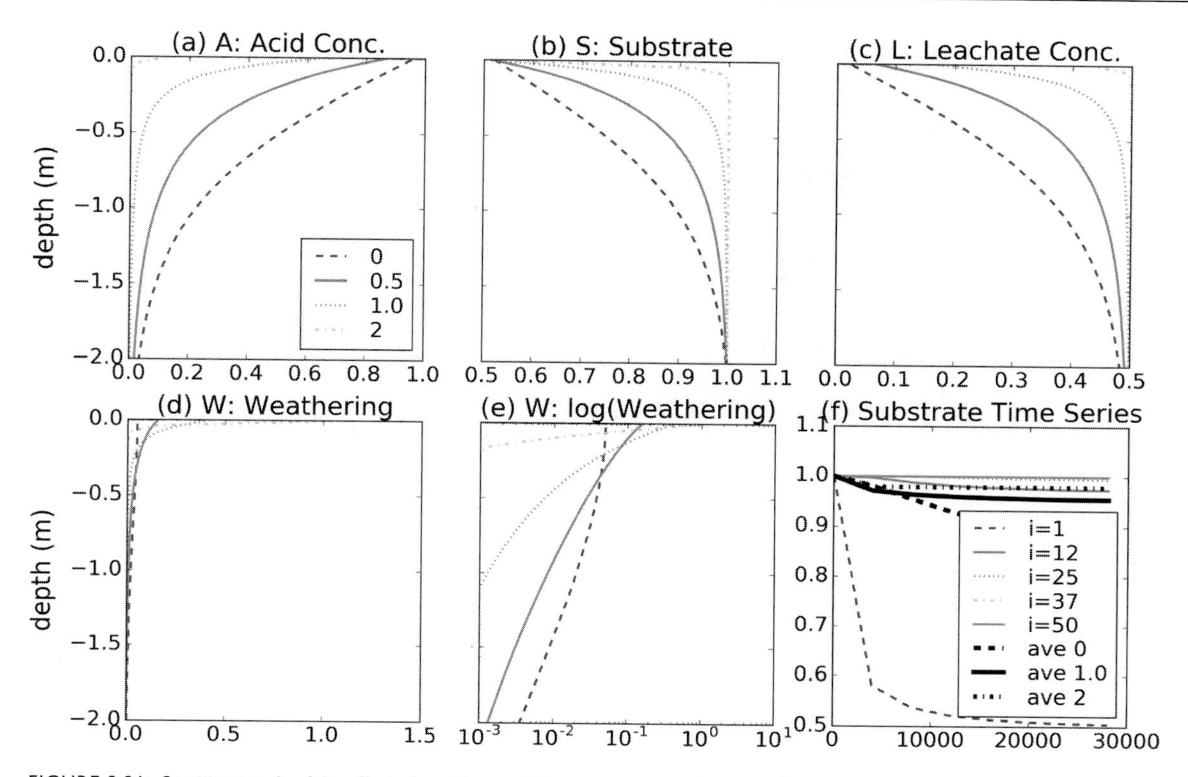

FIGURE 8.21: Sensitivity study of the effect of specific area. The parameters are the nominal parameters. A value of $\varepsilon = 0$ is equivalent to being independent of specific area (i.e. equivalent to the nominal parameters with multiplier = 1 in the sensitivity studies of Figure 8.4 to 8.11). For a detailed explanation of the panels see Figure 8.4.

seems a fruitful starting point for a physically realistic model for soil formation on hillslopes in humid regions.

8.6 Some Final Observations on Chemical Weathering

The discussion below returns to discussion of the vertical one-dimensional model for chemical weathering that has been widely discussed in the literature.

8.6.1 Integrating over the Profile for a Lumped Model

Transport limitation leads to a simple lumped model for chemical weathering in the soil profile. If the leachate can be used to uniquely identify the mineral that is weathering, then we can do a mass balance on the profile so the mass of leachate exiting the profile can be used to calculate the total amount of weathering of that mineral in the profile using the chemical mass balance in Equation (8.6). Taking a unit plan area soil profile, then

$$vL_{\text{equilib}}\Delta t = \frac{1}{c}\left(\int \frac{\partial S}{\partial t}\partial z\right)\Delta t \tag{8.26}$$

where Δt is the period over which the mass balance is performed, c is the mass balance parameter from Equation (8.6) and the integral inside the parentheses is the total rate of weathering of the profile. This leads to an expression for the total weathering rate averaged over the profile, W, per unit volume (i.e. unit plan area and unit depth of the profile)

$$W = \frac{cvL_{\text{equilib}}}{D} \tag{8.27}$$

where D is the total depth of the soil. If L_{equilib} cannot be attributed to the weathering of just one mineral, or if there is a secondary weathering reaction (e.g. the leachate may be involved in a precipitation reaction within the profile), then the value of c may not be known, so this equation, while correct in principle, may not have a value of c that can be calibrated.

The dependence of this result on infiltration is well supported by field evidence (e.g. Dessert et al., 2003).

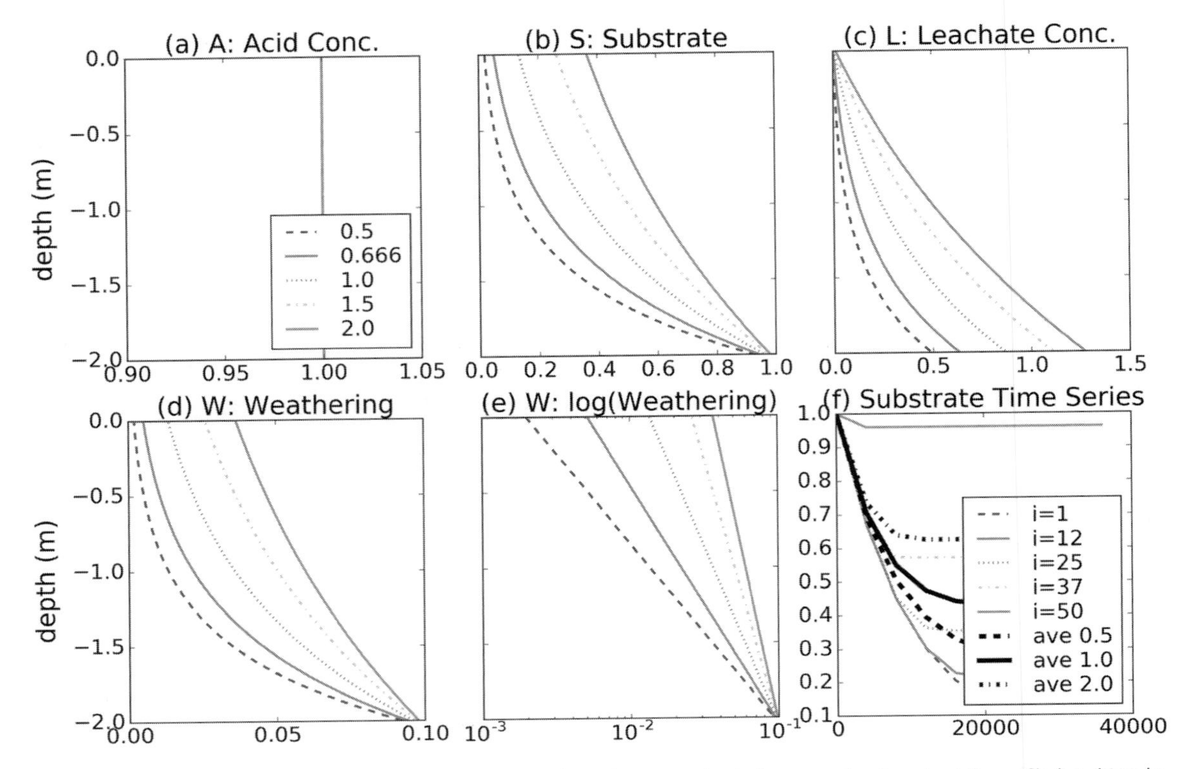

FIGURE 8.22: The weathering profile for varying erosion when the acid concentration is the same value throughout the profile (e.g. driven by equilibration to a soil atmosphere with constant carbon dioxide concentration throughout the profile depth). All parameters are the nominal parameters. Note the semi-log linear weathering profile in panel (e) ('reverse exponential'). For a detailed explanation of the panels see Figure 8.4.

Dessert also found a temperature dependence that is related to the leachate equilibration concentration, $L_{equilib}$.

8.6.2 Conceptual Lumped Models

Most existing soilscape models do not explicitly model the chemistry and the physics discussed in the depth explicit model above (with the exception of Finke's SOILGEN model; Finke, 2012). Typically they rely on conceptualisations of the rate of the dissolution over time from the entire profile, implicitly considering that the soil profile and the processes operating in it are the same from top to bottom.

Yoo and Mudd (2008) is a good example of the approach where they use the idea of mineral dissolution in Equation (8.6) to write the equation for dissolution loss of a mineral j as

$$\frac{\partial m_j}{\partial t} = (R_j A_j) m_j = k_j m_j \tag{8.28}$$

where m_j is the mass of mineral j (equivalent to S in Equation (8.6)), R_j is the reaction rate of that mineral per unit surface area (k_1 in Equation (8.6)) and A_j is the surface area of mineral j. The parameter k_j is the rate of the reaction ($k_1 A^a k_5$ and $S_{min} = 0$, and k_j should not be confused with k_1 in Equation (8.6)). Solving this with time, assuming k_j is constant with time and mineral mass, gives

$$m_j = m_j(0) e^{-k_1 t} \tag{8.29}$$

This equation is similar to the time component used by Minasny and McBratney (2001) but the latter authors added a humped function to simulate a soil depth dependency (their equation (4)).

Some of the assumptions built into Equation (8.29) are that (1) there is no addition of new saprolite to the profile, so eventually the mineral concentration will decline to 0 (i.e. a nonzero equilibrium is not possible), (2) the reaction rate does not change with time (i.e. either reactivity and surface area do not change or they compensate each other to keep k_j constant) and (3) it is possible to sum up the profile mineral mass over the profile and treat it as a lumped value. Yoo and Mudd (2008) do have a separate soil production function, so by using the SPR

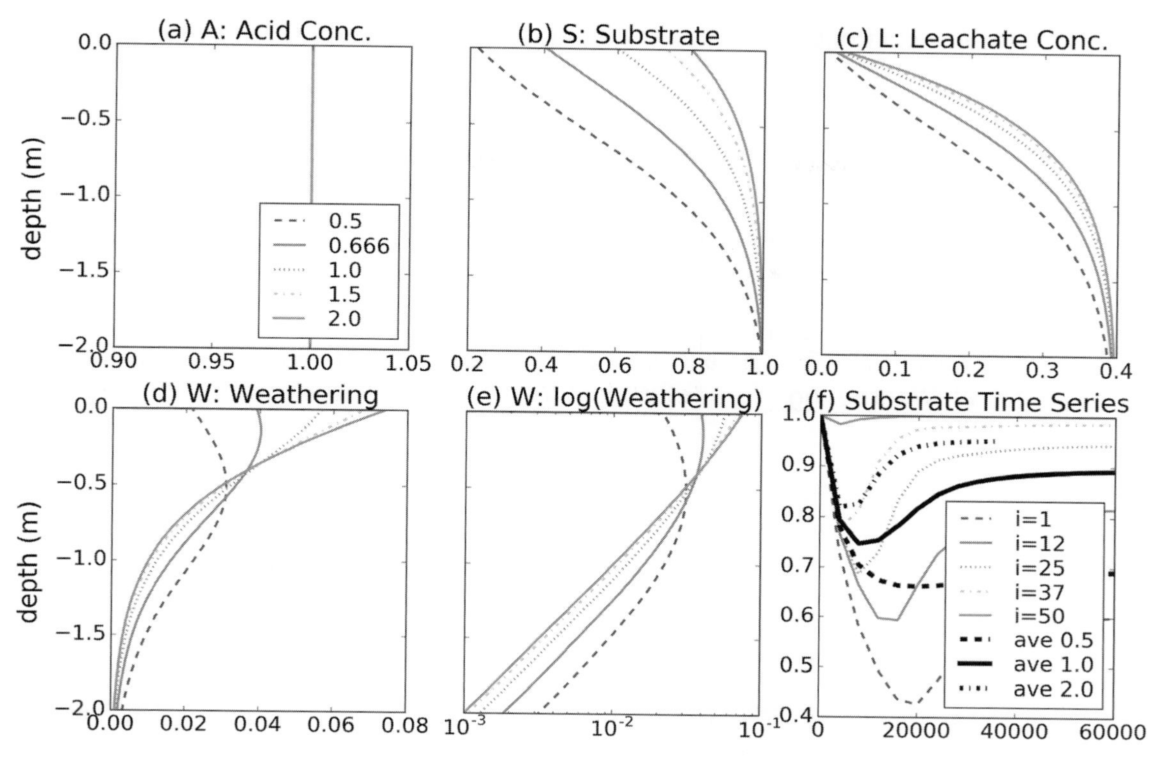

FIGURE 8.23: The weathering profile for varying erosion when the acid concentration is the same value throughout the profile (e.g. driven by equilibration to a soil atmosphere with constant carbon dioxide concentration throughout the profile depth). This figure is identical to Figure 8.22 but with leachate control with a maximum leachate concentration = 0.4. Note that the reverse exponential weathering profile no longer occurs because the leachate limitation at the bottom of the profile limits (panel (c)) the weathering rate rather than the acid concentration. The initial dips in the time series of the substrate in panel (f) are the initial transients when weathering rate is unlimited and before leachate limitation begins to limit (compare the early time evolution with that in Figure 8.22 without leachate limitation). The subsequent rise at later times is a result of new substrate being entrained (and not being highly weathered because of leachate limitation at the soil profile base) from the underlying saprolite by exhumation. This is also the reason why the dip occurs at earlier time for the soil profile base than at the soil surface (it takes more time for the surface material to be refreshed than at the base of the soil-saprolite interface). For a detailed explanation of the panels see Figure 8.4.

they are able to inject new mineral mass into the profile. By combining the SPR and their lumped model, Yoo and Mudd can potentially attain a nonzero equilibrium mineral content. If Equation (8.28) is used in the model, the time variation of k_1 is less problematic because it can be changed with time during the simulation, just as in the fully depth-explicit model. The third assumption is more problematic. In the sensitivity studies above for the depth-explicit chemical weathering model, the (f) panel of all the figures in this section are the depth-integrated values for the S, which is equivalent to m_j here. It is quite clear that in all the simulations an S-shaped curve is almost always generated that starts concave down, transitions to concave up, and then levels out at the equilibrium nonzero value. This is not the exponential in Equation (8.29). All the sensitivity simulations above have a nonzero saprolite exhumation in them because of the parameters chosen.

In Figure 8.27 the time series of substrate with zero exhumation is shown for the nominal and full depletion parameter sets. The deviation from the exponential with time in Equation (8.29) is related to the exhumation rate. While the exact trend with time is slightly different for the two plots, the initial concave section followed by a concave up period remains. If Equation (8.29) was a good approximation, then the lines for the three different velocity parameters should be a straight line on the semi-log plot. It is interesting to see that the substrate trends with time for the five layers within the profile do show exponential behaviour, after some time lag. This lag is greater the deeper the layer is within the profile. This lag is because the weathering front takes some time

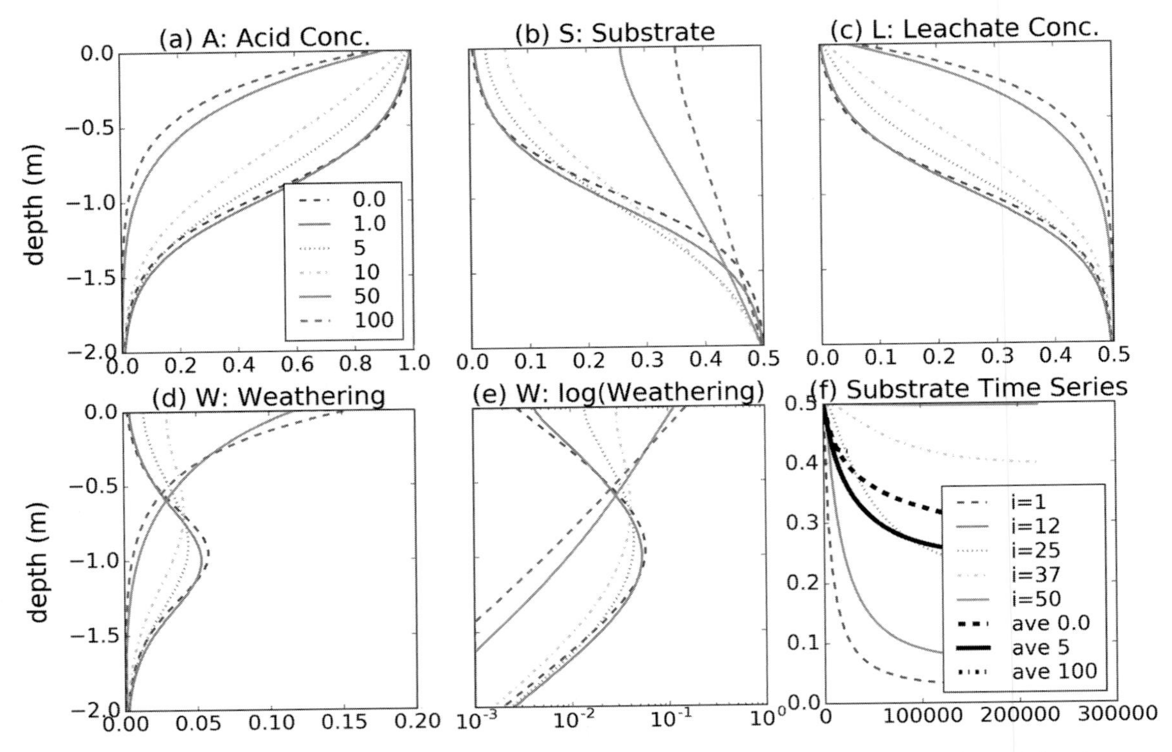

FIGURE 8.24: The weathering profile for varying bioturbation where the bioturbation is modelled with Fickian diffusion of the substrate through the profile for the full depletion parameters. The no bioturbation case is for the sensitivity multiplier = 0.0. For a detailed explanation of the panels see Figure 8.4.

to propagate down the profile. It is this depth dependency of the lag that creates the concave down section in the lumped weathering plots.

8.6.3 A Simple Steady State Humped Chemical Weathering Model

One of the conclusions of the sensitivity studies in the previous sections is that it is possible to have a steady state humped weathering profile if (1) there is active erosion so that new substrate is always being exhumed from the saprolite, (2) all the acid is depleted by the bottom of the soil profile and (3) all the substrate is weathered by the time it is exhumed to the surface. The governing Equations (8.6) can be rewritten as

$$\begin{aligned} \frac{\partial A}{\partial t} + v\frac{\partial A}{\partial z} &= -k_2 R \\ \frac{\partial S}{\partial t} - E\frac{\partial S}{\partial z} &= -R \\ R &= k_1 k_5 A S \end{aligned} \tag{8.30}$$

assuming that diffusive/dispersive transport is zero. For steady state the time derivatives are zero so that

$$\begin{aligned} v\frac{\partial A}{\partial z} &= -k_2 R \\ -E\frac{\partial S}{\partial z} &= -R \end{aligned} \tag{8.31}$$

and eliminating R from these equations yields

$$\frac{\partial A}{\partial z}\bigg/\frac{\partial S}{\partial z} = -k_2\frac{E}{v} \tag{8.32}$$

We can now see that the slopes of the weathering fronts are related to the exhumation rate and the infiltration velocity. It is worth noting that this result is independent of the depth of the soil profile, or the depth of the soil relative to the depth of the weathering front, is true for every point in the soil profile, and is independent of whether the weathering function is humped or not.

An intriguing result is possible for the humped weathering function. Note that the acid and substrate fronts are approximately linear (Figure 8.28). Over most of their range we see that the linear part of the curve fits Equation (8.32), so that the acid and substrate profiles can be approximated by piecewise linear function as shown in Figure 8.28. The approximate equations are

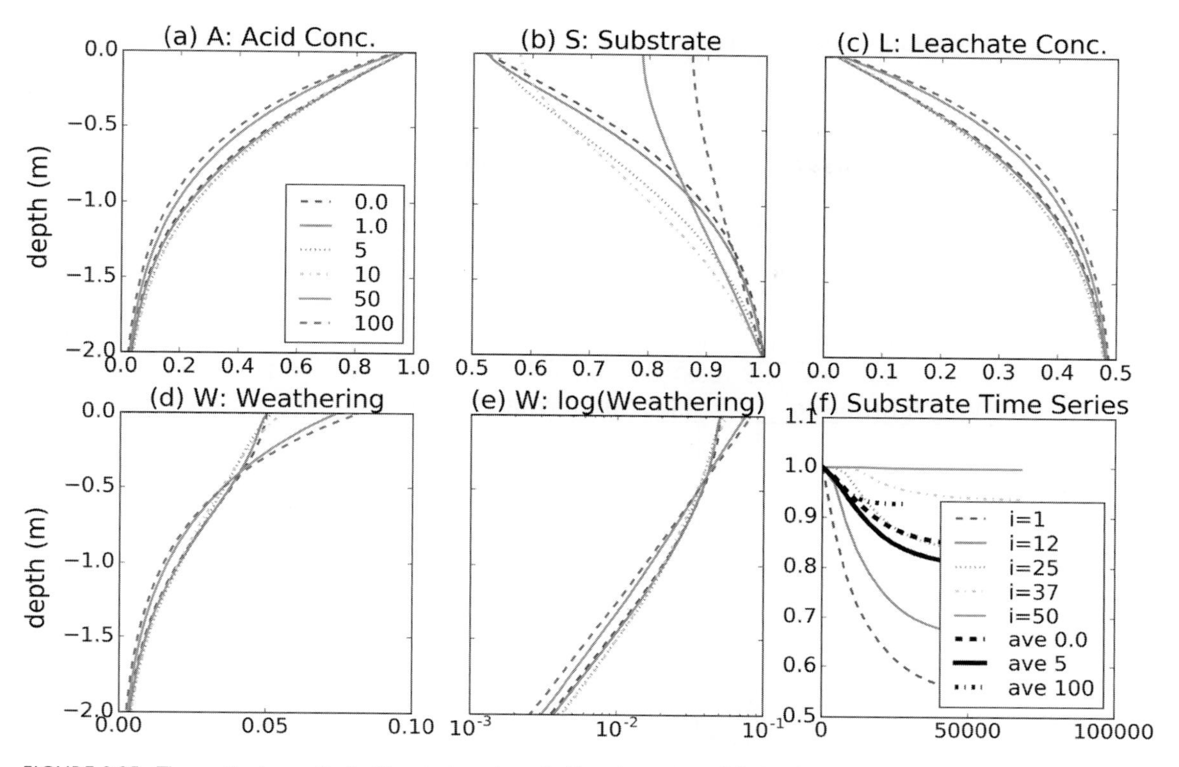

FIGURE 8.25: The weathering profile for bioturbation where the bioturbation is modelled with Fickian diffusion of the substrate through the profile for the nominal parameters. The no bioturbation case is for the sensitivity multiplier = 0.0. For a detailed explanation of the panels see Figure 8.4.

$$A = \begin{cases} A_0 & z < z_1 \\ A_0\left(\dfrac{z_1 - z}{z_2 - z_1}\right) & z_1 \le z \le z_2 \\ 0 & z > z_2 \end{cases}$$
$$S = \begin{cases} 0 & z < z_1 \\ S_0\left(\dfrac{z - z_1}{z_2 - z_1}\right) & z_1 \le z \le z_2 \\ S_0 & z > z_2 \end{cases} \tag{8.33}$$

which yields a humped function approximation to that observed in the sensitivity studies with zero value for $z < z_1$ and $z > z_2$. If we substitute the weathering front part of the piecewise approximation into Equation (8.32), we get

$$A_0/S_0 = -k_2 \frac{E}{v} \tag{8.34}$$

There are several notable aspects to this solution. This gives a condition within which a steady state humped weathering function is possible, and the combination of environmental conditions under which it is possible. The second is that this solution is independent of z_1, z_2 and $(z_1 - z_2)$. This says that provided the condition in Equation (8.34) is satisfied, then the weathering front can be in any location within the profile, and that it is independent of the slope of the front. This suggests that the equilibrium in Equation (8.34) is unstable so that the hump will migrate to be attached either to the top or to the bottom of the profile. This is consistent with the observation in Figures 8.15 and 8.17, which shows that it is a very small range of the erosion and velocity parameters that leads to a humped profile where substrate is near zero at the surface and acid is near zero at the base of the profile. Outside this narrow range the humped weathering profile has a significant nonzero value at either the top or bottom of the profile.

Equation (8.33) is an approximation to the weathering function and has nonzero values in the transitions at the boundaries of the linear part of the weathering front at z_1 and z_2. To examine the impact of this approximation one can imagine moving the actual weathering function and its associated weathering fronts up in the profile so part of the nonzero part of the function lies above the soil surface. In Figure 8.17 this is like changing the parameter multiplier from 0.9 to 1.2. The cumulative weathering on the substrate is the area under the weathering function, so any part of the weathering function that is cut off by the soil

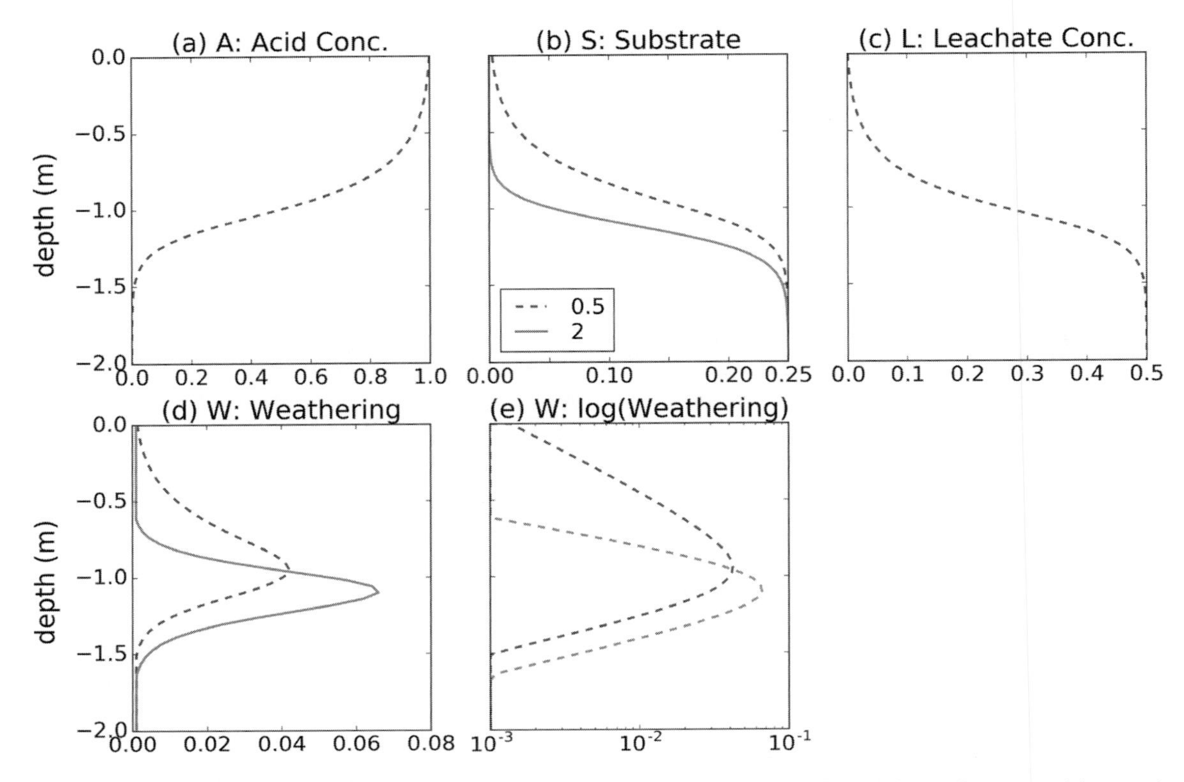

FIGURE 8.26: The weathering profile where there are two minerals being weathering by acid and where the weathering rate of the second mineral is 0.5 and 2.0 times that of the first mineral. For a detailed explanation of the panels see Figure 8.4.

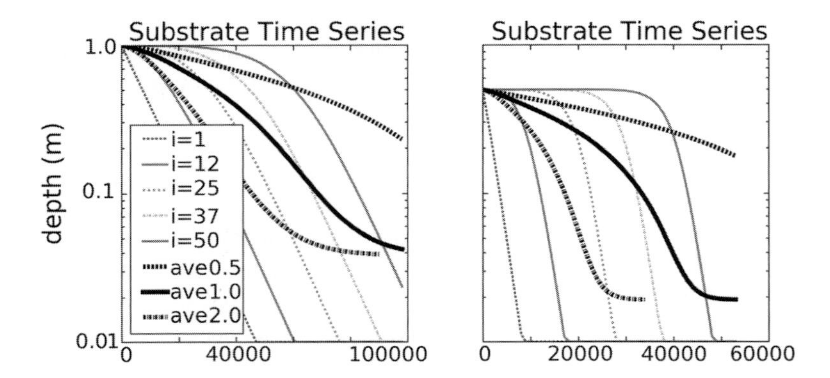

FIGURE 8.27: Our distributed depth model results showing how poorly the Yoo and Mudd (2008) lumped model approximates the depth-averaged weathering from the full chemical weathering model. For the Yoo and Mudd model to be a good approximation to the depth-averaged results (labelled in the legend in both panels 0.5, 1.0, 2.0; the thick lines) the decline in substrate should be linear on this semi-log plot (Equation 8.24): (a) nominal parameters, (b) Fully depleted parameters. The legend is the same for both panels. Note that the thick lines do not level off to an asymptotic value, rather they continue to decline to zero at a much slower rate off scale to the right. The lines labelled $i = 1, 12, 25, 37, 50$ are the weathering of substrate for five layers down the profile ($i = 1$ is the soil surface, $i = 25$ is mid-profile, $i = 50$ is the soil-saprolite interface) for the 'ave.1.0' run. The runs in these plots are equivalent to (a) Figure 8.8 and (b) Figure 8.14 but with zero erosion and exhumation. For a detailed explanation see Figure 8.4 for panel (f).

cosmogenics the channel sediments will differ from the soil profile average and the velocity profile will be one of the drivers of the difference.

Finally, Anderson ignored surface fluvial erosion and its impact on the age characteristics of the channel sediments. Fluvial sheet erosion will preferentially remove soil from the surface (i.e. the oldest particles), and in the extreme case where soil flow is zero all sediment in the channel will have the age distribution of the soil surface, weighted by the spatial distribution of the erosion. In the case of gully or rill erosion, the soil eroded will be averaged over the depth of the rills/gullies, so the sediment supplied to the channel will reflect the age distribution of the soil surface down to the depth of the rills/gullies.

Having now exemplified the importance of the soil velocity distribution we will now discuss various proposed physical mechanisms for soil creep.

9.2 Creep as a Slow Flow Process

In this section we will discuss the basics of soil flow. A good analogy for soil flow is honey flowing down a flat surface; the honey is the soil and the flat surface is the underlying saprolite. Figure 9.3 shows a schematic of the process showing the force balance for a small section of the soil down the hillslope. If the forces are in balance, then the soil will move down the slope with a space-invariant velocity and will be in steady state. If the forces are not in balance, then the soil will either accelerate or decelerate. We will focus on the steady state velocity case.

The relationship between the applied shear stress and the resulting shear strain per unit time of a fluid is called its rheology. Soil is commonly modelled with a viscous or viscoplastic rheology. A Newtonian fluid is an example of a viscous fluid and is where the rate of shear strain is proportional to the applied shear stress. A viscoplastic rheology is one where the soil does not flow until a threshold shear stress is reached (the plastic shear stress threshold) and when the applied shear stress exceeds that threshold the soil flows in response to the excess shear stress above the threshold. A specific form of a viscoplastic rheology is a Bingham fluid, where the relationship between stress and strain exceeding the threshold is linear. These rheologies are shown in Figure 9.4. Other types of fluids are commonly studied in the rheology literature (e.g. thixotropic/shear thinning, shear thickening, Herschel-Bulkley), but Bingham and Newtonian fluids are the ones commonly discussed for modelling slow soil flow (e.g. van Ash and van Genuchten, 1990; Ancey, 2007; Balmforth et al., 2007). We will start our discussion using Newtonian rheology and then extend this discussion to Bingham rheology.

9.2.1 Constant Depth, Newtonian Rheology

The first case to consider is the case of a Newtonian fluid with a constant soil depth down the hillslope (Figure 9.3). Summing the forces parallel to the soil-saprolite interface yields

$$T = \frac{\tau \Delta x}{\cos \theta} = W \sin \theta + (F_u - F_d) \cos \theta$$
$$= \rho_b g (D - z) \Delta x \sin \theta + (F_u - F_d) \cos \theta \qquad (9.3)$$

where ρ_b is the bulk density of the soil, g is the acceleration due to gravity, D is the depth of the soil, θ is the angle of hillslope and T and τ are the shear force and shear stress, respectively, acting on the saprolite at the soil-saprolite interface. If θ is small, then $\sin \theta = \tan \theta = S$ where S is the slope (in m/m). Since the soil flow direction is parallel to the soil surface, then the pressure distribution

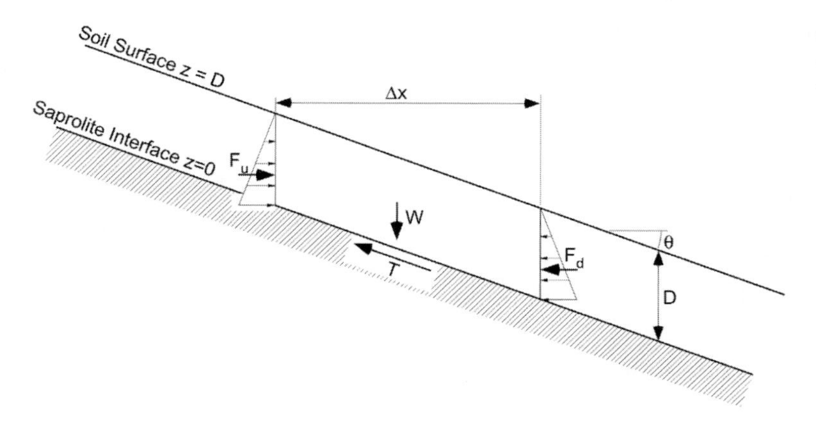

FIGURE 9.3: Schematic of downslope soil flow for viscous or viscoplastic flow.

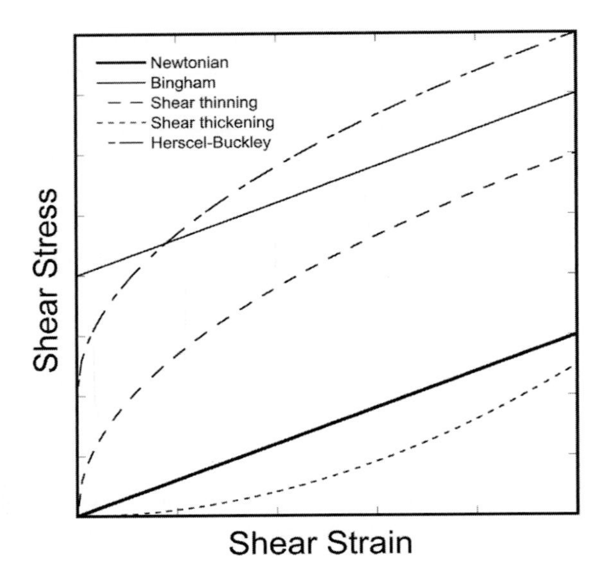

FIGURE 9.4: Shear stress versus strain rate showing some of the viscosity relationships (the slope of the curve) for rheologies that are commonly used for fluid flow.

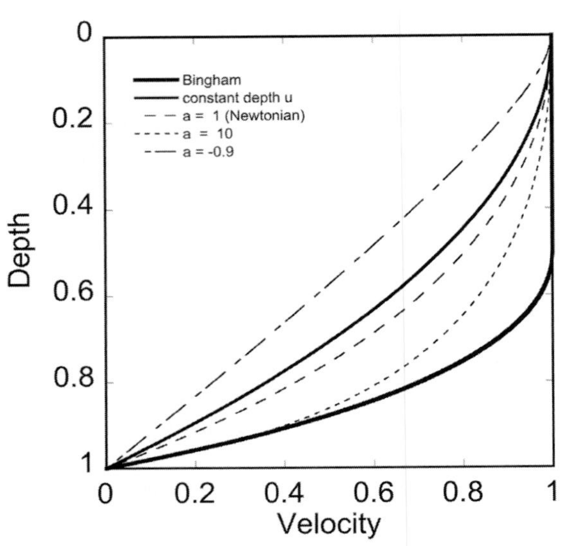

FIGURE 9.5: Velocity profiles for different rheologies for viscous flow.

down the soil profile is hydrostatic, the pressure at height above the saprolite z is $p = \rho_b g(D - z)$ and $F_u = F_d = \frac{\rho_b g D^2}{2}$. Accordingly, since the soil depth is constant downstream, $F_u - F_d = 0$. The shear stress parallel to the soil-saprolite interface, at any depth above the soil saprolite interface, z, is then

$$\tau = \rho_b g(D - z)\sin\theta\cos\theta = \frac{\rho_b g(D - z)}{2}\sin 2\theta \qquad (9.4)$$

The appearance of $\sin 2\theta$ indicates that the shear stress increases until the slope is 1:1, and for slopes steeper than 1:1 the shear stress decreases with increasing slope. Since most natural soil mantled hillslopes have slopes that are less than 1:1, the shear stress increases with an increase in slope. For a linear viscous fluid (a Newtonian fluid) $\tau = \mu\frac{dV}{dz}$ where μ is the absolute viscosity, and V is the downslope velocity at depth z so that

$$\frac{dV}{dz} = \frac{\rho_b g \sin 2\theta}{2\mu}(D - z) \qquad (9.5)$$

and integrating with depth (noting that $V = 0$ at $z = 0$) yields the velocity profile (Figure 9.5)

$$v = \frac{\rho_b g \sin 2\theta}{4\mu}\left(D^2 - z^2\right) = v_0\left(1 - \left(\frac{z}{D}\right)^2\right) \qquad (9.6)$$

where v_0 is the velocity at the soil surface. The depth-averaged velocity is

$$\bar{v} = \frac{\rho_b g D^2 \sin 2\theta}{6\mu} = \frac{2}{3}v_0 \qquad (9.7)$$

The displacement over time is obtained by multiplying Equations (9.6) and (9.7) by time. Equation (9.6) is controlled by two parameters

$$P_1 = \frac{\rho_b g D^2 \sin 2\theta}{4\mu} \qquad P_2 = \frac{z}{D} \qquad (9.8)$$

where P_1 is the parameter that determines the rate of flow of the profile, while P_2 determines how that velocity is distributed through the soil profile. Finally it is possible to recast the soil discharge calculated using Equation (9.7) into the form of classical creep (Equation (9.2)) where the diffusivity K is

$$K = \frac{\rho_b g D^3}{3\mu} \qquad (9.9)$$

if the slope is small, so $\sin 2\theta \simeq 2\sin\theta \simeq 2S$ and noting that $q_s = \bar{v}D$. The main conclusion from this formulation is that the diffusivity in Equation (9.2) is then dependent on the cube of the soil depth.

However, since the velocity profile in Equation (9.6) (Figure 9.5) does not match the observed soil displacement profiles in Figure 9.1, we conclude that this model does not provide a realistic model of creep.

This equation was derived by moving a particle upslope/downslope with a given energy and determining the travel distance of the particle sliding across the surface with a coefficient of friction S_t. The distance travelled upslope is less than the travel distance downslope, and Equation (9.19) is the net sediment transport downstream. As the slope approaches the coefficient of friction, the downslope particle will approach a travel distance of infinity. The power of 2 in the denominator arises naturally out of this calculation and is not a fitting parameter. Andrews and Bucknam fitted these data to the rounding of old fault scarps and found a better fit than using Equation (9.2). Roering et al. (1999, 2001, 2007) fit this equation to hillslope profiles and also found that it gave a good fit to hillslope profile, and that it fit better than assuming Equation (9.2) for slopes less than the threshold slope S_t, with infinite transport for slopes greater than S_t.

Roering (2008) extended Equation (9.20) by considering that the rate constant K was dependent on the depth below the soil surface, and by assuming that the main trigger for soil movement was the root density of trees, which decline exponentially with depth (Canadell et al., 1996). Equation (9.20) is then a depth-averaged equation after integration over the soil depth with depth-dependent root density

$$K = K'\left(1 - e^{-\beta D \cos\theta}\right) \tag{9.21}$$

where β is the length scale of the rate of decline of root density with depth and K' is the rate constant for root disturbance. Roering found that this equation gave a better fit to their study site than assuming a constant value for K with depth in Equation (9.20).

9.7 Conclusions

The main conclusion to this chapter is that if all that needs to be modelled is the mass flux to determine the soil surface elevation and the soil depth is the same everywhere, then most of the different processes can be reduced to the Culling Fickian diffusion model that has been traditionally used in landform evolution modelling. However, if the soil depth is changing, as is likely to be the case when modelling a coupled soilscape-landform evolution model, then the diffusivity in the Culling equation is dependent on soil depth. Not only that but the functional form of the dependence on the soil depth differs between the processes. Unfortunately, at this time there is no consensus on what is the appropriate physics. The problem becomes worse if we are interested in the velocity profile of the soil flow down the slope (Anderson, 2015) since most of the processes yield different velocity profiles, in addition to different relationships between the surface velocity and the profile average velocity.

One tentative conclusion can be reached. The representation of soil flow using methods developed in the geomechanics field for viscoelastic fluids, and which ignores field scale processes such as expansion and contraction, and bioturbation, appears to be a poor fit to observed displacement profiles in the field. This still leaves an open question over what is the soil depth dependency of the diffusivity in the Culling Fickian model.

9.8 Numerical Issues

For the classic Culling depth-integrated creep there is the choice of whether to use Equations (9.1) or (9.2). Equation (9.1) can be directly applied to the surface elevations but can be problematic if the planar boundaries of the domain are not rectangular. In this case applying the boundary condition (typically a zero slope, no flow, condition) can be messy. On the other hand, the flux form in Equation (9.2) when used in conjunction with a drainage analysis algorithm (e.g. D8; see Chapter 4 for a detailed discussion of drainage analysis algorithms) is easier to apply at irregular boundary conditions. The downside of a D8-based flux formulation is that normally diffusion dominated (i.e. creep-dominated) landforms tend to be convex down, and the flow directions are divergent. The D8 algorithm does not model this divergence since it allows flow only to one downslope node. The D∞ algorithm goes some way to addressing this by allowing two downslope flow directions. The main impact of the lack of divergence of D8 is that for short slopes (a few nodes long) generated landforms tend to look a little rougher than with Equation (9.1), and the catchment divides are sharper and less rounded. On the other hand, the total mass flux is the same for both D8 and D∞ (Willgoose, 2005a).

When using TINs instead of grids, the flux formulation of diffusion is also much easier to use, and this is the approach used in the CHILD LEM (Tucker, personal communication).

Perron (2011) proposed an implicit solver for Equation (9.20), which we discussed in Section 4.7.

To date no researcher has published a creep model that models depth-dependent variations of velocity (Anderson (2015) calculates analytical rather than numerical solutions), so we have no experience to share on numerical issues. We anticipate the main complexity here will be in the coupling of the layering in the soil profile at each node with the horizontal movement of each layer when the soil depth and velocity are varying downslope. As indicated when we discussed viscous flow for a varying depth soil, the thickness of the layers will vary downslope as the velocity downslope varies, so the depth boundaries of the layer discretisation of the soil profile at the upstream node will not match the layer discretisation of the soil profile at the downslope node. Furthermore, Anderson (2015) shows that even with constant soil depth, the combination of soil production and creep results in vertical movement of soil streamlines when moving downslope (and thus vertical movement and thickness change in the upslope layers as the profile moves downslope). Thus explicit accounting for material exchange between the layers (potentially defined differently at each node on the hillslope) as the soil moves downslope is required. In principle this is no different from the material exchange vertically within each individual profile that was performed in the physical weathering using mARM (Chapter 7), but it potentially adds an additional vertical numerical diffusion that may be an issue.

10 Soils: Colloids and Soil Organic Carbon

10.1 Introduction

This chapter focuses on (1) clay transport within the soil profile and (2) the dynamics of soil organic matter, of which soil carbon is the most important component. Clay and soil organic matter combine in the soil profile and are a major driver of the soil structure evolution (the 'clay-organic complexes'; Tisdall and Oades, 1982), and soil structure is a major influence on the infiltration rate of rainfall into the soil. There is a strong relationship between soil carbon content and soil water-holding capacity (Rawls et al., 2003). Soil organic matter consists of compounds containing carbon, phosphorus, nitrogen and sulphur. Operational models for the evolution of soil organic matter typically model the cycles of these elements (e.g. the CENTURY model; Metherell et al., 1994), but in this chapter we will consider only soil organic matter (SOM) in total, or the carbon component of it, soil organic carbon (SOC). SOM contains 58% SOC in units of mass C/mass soil. SOM contains particulate organics, humus, charcoal, living microbes, root exudates, fungi and fine plant roots. The living biomass (the vast majority of which is microbes and fungi) is a major influence on the decomposition of dead organic matter and the storage of carbon and nutrients (Wardle, 1992).

Jobbágy and Jackson (2000), Houghton (2007) and Stockmann et al. (2013) provide a summary of the fluxes, storage and importance of the SOM and SOC. Figure 10.1 summarises the carbon cycle. While fluxes and storage estimates are uncertain (Johnston et al., 2004) there is general agreement that the amount of carbon in terrestrial biomass (~550 Gt) is about equal to the amount of carbon in the atmosphere (~800 Gt), and that the amount of SOC in the top three metres of soil is about 5 times the terrestrial biomass (~2,300 Gt). Soil carbon is the second biggest store of carbon after the deep ocean. The vast majority of the terrestrial biomass is in the soil with only the aboveground biomass of vegetation not located in the soil. Total vegetation biomass is about 50% of the total biomass, and about 50% of vegetation biomass is aboveground (Larcher, 2003), so that about 75% of terrestrial biomass is in the soil. This chapter is focussed on this 75% of terrestrial carbon. Figure 10.1 also makes clear that the difference between carbon release and carbon sequestration is quite small, so errors in the calculation of the fluxes in the terrestrial carbon cycle will make large relative errors in the net balance of the fluxes and thus rates of change in carbon storage.

SOC is highly concentrated near the surface of the soil and declines with depth, with 54% in the top metre (~1,500 Gt) and 26% (~615 Gt) in the top 20 cm. In water-limited ecosystems SOC storage is also highly correlated in space with soil moisture (Hobley et al., 2015; Hobley and Wilson, 2016). This relationship is because plant growth and biomass are higher for higher soil moisture, so this source term for SOM is accordingly higher.

SOC is important because it is the dominant sink for carbon in the terrestrial environment (e.g. forests soils store up to 90% of global SOC; Peterman and Bachelot, 2012), and recent work with climate models suggests that it is the most sensitive sink to climate change because of the rapid change of the storage dynamics with changes in soil moisture (Friedlingstein et al., 2006). Stockmann et al. (2013) noted that a 10% change in the stored SOC equals about 30 years of current-day anthropogenic emissions.

10.2 Colloid Migration and Illuviation

Colloidal material moves down the profile through the pores in the soil carried by infiltrating water. The colloidal materials we are interested here are clay and biochar from wildfires. There is also evidence of microbial transport, but we will not discuss this here. The principle of clay migration is that clay is released at the surface of the soil profile due to rainsplash or as a result of chemical weathering

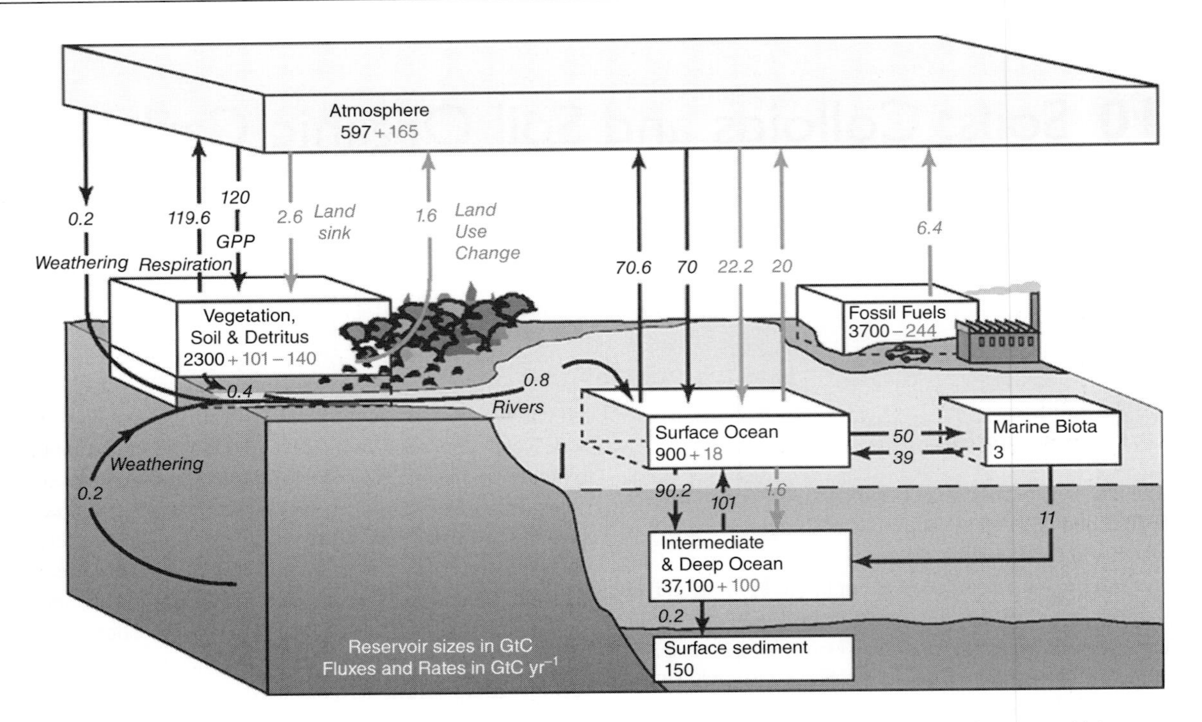

FIGURE 10.1: The global carbon cycle. Fluxes and storages in black are natural; fluxes and storages in grey are anthropogenic. Units are GtC or GtC per year. Houghton (2007) subdivides the vegetation, soil and detritus box into vegetation 550, detritus/litter 300 and soil 1,200 GtC so that soil carbon is about 60% of the terrestrial storage. For fluxes the soil litter flux into the soil approximately matches decomposition (from IPCC, 2007).

within the profile, and then is transported down through the profile by infiltration. It is then deposited within the profile by a number of mechanisms: (1) flocculation, which occurs if the salinity of the soil water increases down the profile, (2) particles get caught in small pores within the profile, (3) deposition in pores, which occurs when infiltration has reached its maximum depth in the profile for an individual rainfall event or (4) deposition on mineral surfaces (DeNovio et al., 2004; Bradford and Torkzaban, 2008). Similar processes drive biochar movement.

Rainsplash was discussed in detail in Chapter 4. The key points are that the low salinity of rainwater disperses clay, and the energy of the rainsplash detaches the clay particles from the surface, making them mobile. The mathematics of chemical weathering leading to clay formation was discussed in Chapter 8.

Jarvis et al. (1999) provides an empirical equation for the capture of clay particles F (g/m^3/hr) within the profile

$$F = f_{ref} v_{ref}^{n} v^{1-n} c\theta \tag{10.1}$$

where f_{ref} is a media calibration coefficient measured at a reference flow velocity v_{ref}, v is the specific discharge, c is the particle concentration (mg/l), n is an empirical coefficient approximately equal to 0.7 and θ is the soil moisture (volume m^3/m^3).

This equation models only mechanism (2) and does not account for the deposition of particles when infiltration stops, when water is transpired or evaporated from the profile, or flocculation.

Accompanying the transported clay is material transported adsorbed on the clay surfaces. In a typical soil these components are organic molecules, phosphate and metals. The amount adsorbed per unit surface area of clay is strongly pH dependent with adsorbed amounts decreasing with decreased pH (Dijkstra et al., 2004). If pH decreases, it is common to observe the release of these adsorbed constituents into the soil water solution.

10.3 Single Soil Layer Soil Organic Carbon Models

Figure 10.2 shows how most soil organic carbon (SOC) model conceptualisations are constructed. Generally the soil profile is considered as one mixed layer, with no distinction being made with the storage at different depths within the profile, so depth dependencies of SOC are

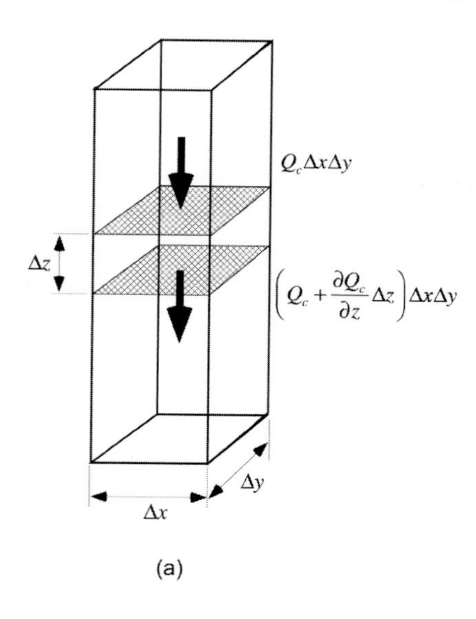

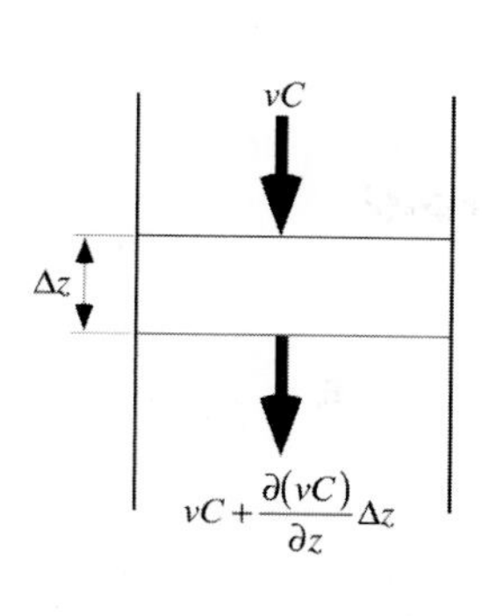

FIGURE 10.3: Schematic of SOC transport down the soil profile: (a) the profile and (b) advection down the profile.

mass(C)/mass(soil) so to convert the SOC concentration to a mass flux the SOC concentration must be multiplied by the soil bulk density ρ_b. If we define the decomposition rate of the SOC as before, then the mass of SOC degraded is $K\rho_b C\Delta x\Delta y\Delta z$, and the source of SOC, which we will assume is independent of the SOC in the soil, is $S\rho_b\Delta x\Delta y\Delta z$ where S is the rate of generation of SOC per unit mass of soil. Writing the mass balance equation yields

$$\frac{\partial(\rho_b C)}{\partial t}\Delta x\Delta y\Delta z = \left[Q_c - \left(Q_c + \frac{\partial Q_c}{\partial z}\Delta z\right)\right]\Delta x\Delta y - K\rho_b C\Delta x\Delta y\Delta z + S\rho_b\Delta x\Delta y\Delta z \tag{10.10}$$

where C is defined as the mass of SOC per unit mass of soil so that $C\rho_b\Delta x\Delta y\Delta z$ is the mass of SOC in the layer. If we expand the flux term inside the square brackets and divide the equation by $\Delta x\Delta y\Delta z$ we obtain the mass balance differential equation for SOC

$$\frac{\partial(\rho_b C)}{\partial t} = -\frac{\partial Q_c}{\partial z} - K\rho_b C + S\rho_b \tag{10.11}$$

where the first term on the right-hand side is the rate of change of SOC as a result of a changing SOC flux down the profile, the second term is the loss of SOC due to decomposition, and the third term is the increase in SOC from SOC production within the profile.

The decomposition rate for SOC K can be of quite general form, though it is common to assume that it is constant with depth. Some data suggest that K can decline with depth as result of a decline in biological activity with depth (Hobley et al., 2014) though the data are far from definitive, while some studies have assumed a decline with depth (Billings et al., 2010; Dialynas et al., 2016) with Dialynas adopting

$$K = K_0 e^{k_z z} \tag{10.12}$$

where k_z is the rate of decline of decomposition rate with depth, and K_0 is the decomposition rate at the soil surface.

A variety of constitutive equations are used for the flux term in Equation (10.11). Bioturbation (Chapters 6 and 7) can be modelled by Fickian diffusion so that

$$Q_c = -D\rho_b\frac{\partial C}{\partial z} \tag{10.13}$$

where the diffusivity D is a function of the rate of the bioturbation and will vary (typically declining along with declining biological activity) with depth. The flux term in Equation (10.11) then becomes

$$-\frac{\partial Q_c}{\partial z} = -\frac{\partial}{\partial z}\left(-D\rho_b\frac{\partial C}{\partial z}\right) = D\rho_b\frac{\partial^2 C}{\partial z^2} + \rho_b\frac{\partial D}{\partial z}\frac{\partial C}{\partial z} + D\frac{\partial \rho_b}{\partial z}\frac{\partial C}{\partial z} \tag{10.14}$$

In the case where SOC is transported down the profile by infiltrating water, it can be modelled with an advection term where SOC moves down with velocity v (Figure 10.3b). The mass entering from above the layer is

$$Q_c = v\rho_b C \tag{10.15}$$

where v is the velocity at which the SOC is transported down the profile. Note that this is not the velocity of the infiltration water carrying the SOC but is the average velocity at which particles of SOC move down the profile, which will be less than the infiltration velocity. During an infiltration event some particles will move but many will not, and the SOC velocity is the average across all SOC particles in the layer. The infiltration velocity may also decrease with depth as discussed in Chapter 3. The mass flux of SOC leaving the layer is

$$Q_c + \frac{\partial Q_c}{\partial z}\Delta z = v\rho_b C + \frac{\partial(v\rho_b C)}{\partial z}\Delta z$$
$$= v\rho_b C + \left(v\rho_b \frac{\partial C}{\partial z} + v\frac{\partial \rho_b}{\partial z}C + \frac{\partial v}{\partial z}\rho_b C \right)\Delta z \qquad (10.16)$$

so for advection the flux term in Equation (10.11) is

$$-\frac{\partial Q_c}{\partial z} = -\left(v\rho_b \frac{\partial C}{\partial z} + v\frac{\partial \rho_b}{\partial z}C + \frac{\partial v}{\partial z}\rho_b C \right) \qquad (10.17)$$

The main differences from the commonly used forms for advection are the terms in (1) $\frac{\partial v}{\partial z}$ which is required for cases where infiltration doesn't flow through the full soil profile every time it rains (see Chapters 3 and 7) so the average velocity will decline with depth, and (2) $\frac{\partial \rho_b}{\partial z}$ which is required for cases where changes in bulk density occur down profile (e.g. compaction). In the former case the top of the soil profile is wet more frequently than deeper down the profile, so there are more opportunities for SOC to be transported at the top of the profile than lower down, and the average velocity of the SOC particles at the top of the soil profile will be higher than it is deeper in the profile.

Combining these relationships yields the general equation incorporating bioturbation and infiltration for a multi-layer SOC model with one SOC compartment

$$\frac{\partial(\rho_b C)}{\partial t} = \left(D\rho_b \frac{\partial^2 C}{\partial z^2} + \rho_b \frac{\partial D}{\partial z}\frac{\partial C}{\partial z} + D\frac{\partial \rho_b}{\partial z}\frac{\partial C}{\partial z} \right)$$
$$- \left(v\rho_b \frac{\partial C}{\partial z} + v\frac{\partial \rho_b}{\partial z}C + \frac{\partial v}{\partial z}\rho_b C \right) - K\rho_b C + S\rho_b \qquad (10.18)$$

This equation simplifies if the soil bulk density is constant in both space and time so that

$$\frac{\partial C}{\partial t} = \left(D\frac{\partial^2 C}{\partial z^2} + \frac{\partial D}{\partial z}\frac{\partial C}{\partial z} \right) - \left(v\frac{\partial C}{\partial z} + \frac{\partial v}{\partial z}C \right) - KC + S \qquad (10.19)$$

The equation further simplifies if both the diffusivity and velocity are constant with depth so that

$$\frac{\partial C}{\partial t} = D\frac{\partial^2 C}{\partial z^2} - v\frac{\partial C}{\partial z} - KC + S \qquad (10.20)$$

In general these equations are complicated enough that they must be solved numerically. Wells et al. (2012) used Equation (10.20) to model not just the migration of SOC in soil but two radioisotope tracers (^{137}Cs and ^{210}Pb). They found that the velocity of transport down the profile for SOC was significantly slower than that for the radioisotopes, but subject to this discrepancy they were able to obtain excellent fits for the depth profiles of SOC, ^{137}Cs and ^{210}Pb. Wells et al. (2013) found a poor fit when using the same model at a site dominated by cracking soils, where they believed that the vertical cracks created preferential pathways for water and SOC. This preferential transport in cracks invalidates the one-dimensional vertical formulation of Equations (10.18) through (10.20).

There is one end member case for Equation (10.20) for which an analytic solution is possible. At equilibrium ($\frac{\partial C}{\partial t} = 0$) when bioturbation is small ($D = 0$), transport velocity is the same throughout the profile ($\frac{\partial v}{\partial z} = 0$), and all SOC is sourced from the surface and none is created in-profile (i.e. $S = 0$), then Equation (10.20) is

$$-v\frac{\partial C}{\partial z} - KC = 0 \qquad (10.21)$$

and solving this first order differential equation gives an exponential distribution of SOC down the profile

$$C = C(0)e^{-Kz/v} \qquad (10.22)$$

where $C(0)$ is the SOC at the soil surface. This shows that as the decomposition rate K increases and/or the velocity v decreases, the SOC declines faster with depth. For instance, Richter et al. (1999) attribute a rapid decline in SOC with depth to very porous soils that led to fast decomposition of SOC. An exponential profile of similar functional form to Equation (10.22) is often used to fit SOC data with depth.

If, in addition, SOC is generated within the profile at a constant rate (i.e. $S =$ constant), then the solution is

$$C = C(0)e^{-Kz/v} + \frac{S}{K} \qquad (10.23)$$

with the SOC declining to the constant value S/K rather than zero as in Equation (10.22).

O'Brien and Stout (1978) examined the case of steady state where D is constant with depth, and with no advection ($v = 0$) and found a near exponential decline of SOC with depth, and found that both experimentally and numerically the more stable SOC declined more slowly with depth as expected from Equation (10.22). Jobbágy and Jackson (2000) fitted a number of regression equations to SOC data with depth across a range of different biomes (Figure 10.4) and found Equation (10.22) gave a good fit; however, a slightly better fit, particularly for depths greater than 1 metre, was obtained with the equation $C = az^b$ where a and b were constants. Hobley et al.

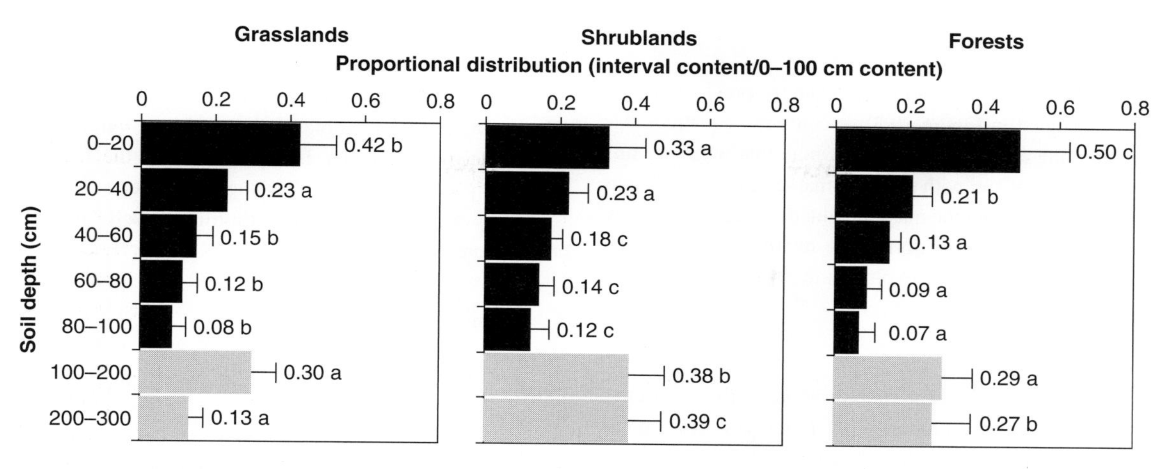

FIGURE 10.4: Global estimates of SOC content with depth for the three main vegetation types: grasslands, shrublands and forests showing the decline with depth. The error bars are one standard deviation. The horizontal scale is normalised so that they show the proportion of SOC with depth relative to the SOC content of the top metre of the profile, so that the scale shows the relative proportions of SOC with depth (from Jobbágy and Jackson, 2000).

(2013) found that an exponential curve best fit SOC associated with different particle size distribution fractions, while Hobley and Wilson (2016) found it satisfactory for most soil profiles across 100 profiles across contrasting soils and climates.

Returning to the single-layer SOC compartment models of the previous section and the different decomposition rates in each compartment we can easily extend the multilayer model to include the multi-compartment nature of the single layer models in the previous section. Using the terminology of the previous section a three-compartment version of Equation (10.19) is

$$\begin{aligned}
\frac{\partial C_1}{\partial t} &= D\frac{\partial^2 C_1}{\partial z^2} + \frac{\partial D}{\partial z}\frac{\partial C_1}{\partial z} - v\frac{\partial C_1}{\partial z} - C_1\frac{\partial v}{\partial z} - K_1C_1 + S \\
\frac{\partial C_2}{\partial t} &= D\frac{\partial^2 C_2}{\partial z^2} + \frac{\partial D}{\partial z}\frac{\partial C_2}{\partial z} - v\frac{\partial C_2}{\partial z} - C_2\frac{\partial v}{\partial z} - K_2C_2 \\
&\quad + \alpha_{12}K_1C_1 + S \\
\frac{\partial C_3}{\partial t} &= D\frac{\partial^2 C_3}{\partial z^2} + \frac{\partial D}{\partial z}\frac{\partial C_3}{\partial z} - v\frac{\partial C_3}{\partial z} - C_3\frac{\partial v}{\partial z} - K_3C_3 \\
&\quad + \alpha_{13}K_1C_1 + \alpha_{23}K_2C_2 + S
\end{aligned} \tag{10.24}$$

where the transfer from compartment 1 to compartment 2 ($\alpha_{12}K_1C_1$ in Equation (10.4)) is equivalent to an additional source term S in the equation for C_2, and likewise for transfers from compartments 1 and 2 to compartment 3 in the equation for C_3. Equations (10.18) and (10.20) can likewise be extended to multiple compartments.

Before we finish this section it should be noted that some portion of the passive/inert SOM in soils is charcoal/biochar resulting from fire (Rumpel, 2015). We will discuss fire in detail in Chapter 14, but we note here that charcoal is chemically very different from other forms of SOM, and it is highly resistant to decomposition. For a semiarid site Hobley et al. (2013, 2014) used spectrometry to determine the chemical composition of SOC down their profiles and found that they could distinguish charcoal from the other SOC, and that charcoal was up to 26% of the SOC in their profiles. Hart and Luckai (2014) also report significant components of charcoal in SOC in boreal forest soils, and related spatial variation in charcoal stability to different wildfire intensities.

Zhang et al. (2010) extended Equation (10.20) to model the downward movement of biochar. Their equation included terms for (1) linear reversible equilibrium deposition (of the same form commonly used for modelling first order sorption reactions) $K_e\frac{\partial C}{\partial t}$ and (2) irreversible deposition K_dC. They assumed that the in profile generation term S in Equation (10.20) was zero, and that there was no depth variation of infiltration and bioturbation, yielding

$$\frac{\partial C}{\partial t} = D\frac{\partial^2 C}{\partial z^2} - v\frac{\partial C}{\partial z} - \left(K_e\frac{\partial C}{\partial t} + K_dC\right) \tag{10.25}$$

Zhang et al. concluded that the parameters K_e and K_d were strongly dependent on the soil water pH and salinity, with retention of biochar particles increasing with lower pH and higher salinity, both of which are consistent with a first order sorption-like process.

10.4.2 SOC Age

Using the mass balance approach from the previous section and illustrated in Figure 10.3 we can write equations that describe the age of SOC. We first note that when we mix two samples of soil with different masses and ages, the age of the mixture is not the mass-weighted average of the ages.

The age A is a function of the relative masses of ^{12}C and ^{14}C where the time evolution of a packet of soil is

$$\frac{^{14}C}{^{12}C} = \frac{^{14}C(0)}{^{12}C(0)} e^{-\lambda t} \tag{10.26}$$

where ^{14}C and ^{12}C are the concentrations of the radioactive and stable carbon isotopes, $C(0)$ is the concentration at time zero (if all SOC is being input at the surface, this is also the concentration at the surface), λ is the decay rate and t is time. The decay rate is related to the half-life of ^{14}C as

$$\lambda = -\frac{\ln 0.5}{t_{1/2}} \tag{10.27}$$

where $t_{1/2}$ is the half-life of ^{14}C which is 5,730 years.

The mixing model for two packets of soil with different masses of carbon and ratios of ^{12}C to ^{14}C (and thus different ages) is

$$^{14}C_{\mathrm{ave}} = \frac{M_1 \cdot {}^{14}C(0)e^{-\lambda A_1} + M_2 \cdot {}^{14}C(0)e^{-\lambda A_2}}{(M_1 + M_2)} \tag{10.28}$$

and this can used to determine the age of the mixed sample

$$A_{\mathrm{ave}} = -\frac{1}{\lambda} \ln \left(\frac{M_1 \cdot {}^{14}C(0)e^{-\lambda A_1} + M_2 \cdot {}^{14}C(0)e^{-\lambda A_2}}{^{14}C(0)(M_1 + M_2)} \right) \tag{10.29}$$

This is different from what would be obtained if ages were simply a mass-weighted average, as is done for the mean residence time (MRT) method (Schaetzl and Thompson, 2015)

$$A_{\mathrm{ave}} = \frac{M_1 \cdot A_1 + M_2 \cdot A_2}{(M_1 + M_2)} \tag{10.30}$$

Accordingly, the best way to model age is to model the two carbon isotopes with two sets of coupled equations. Using Equation (10.19), then the appropriate equations are

$$\begin{aligned} \frac{\partial(^{12}C)}{\partial t} &= \left(D\frac{\partial^2(^{12}C)}{\partial z^2} + \frac{\partial D}{\partial z}\frac{\partial(^{12}C)}{\partial z} \right) \\ &\quad - \left(v\frac{\partial(^{12}C)}{\partial z} + \frac{\partial v}{\partial z}(^{12}C) \right) - K(^{12}C) + S \\ \frac{\partial(^{14}C)}{\partial t} &= \left(D\frac{\partial^2(^{14}C)}{\partial z^2} + \frac{\partial D}{\partial z}\frac{\partial(^{14}C)}{\partial z} \right) \\ &\quad - \left(v\frac{\partial(^{14}C)}{\partial z} + \frac{\partial v}{\partial z}(^{14}C) \right) - K(^{14}C) - \lambda(^{14}C) + S \end{aligned} \tag{10.31}$$

where the only difference between the equations for ^{12}C and ^{14}C is the addition of the radioactive decay term $\lambda(^{14}C)$ for ^{14}C. This formulation ignores fractionation effects, where the microbial decomposition slightly prefers to decay the lighter isotope ^{12}C over the heavier isotope ^{14}C so that K for ^{12}C is slightly higher.

At any time the age of the carbon in the soil can then be calculated by

$$A = -\lambda \ln \left(\frac{^{14}C/_{^{12}C}}{^{14}C(0)/_{^{12}C(0)}} \right) \tag{10.32}$$

Finally we can derive an analytic result for age with depth for the same assumptions that we used for the SOC concentration Equation (10.22). The depth profile for the isotopes is

$$\begin{aligned} ^{12}C &= \left(^{12}C_0\right)e^{-Kz/v} \\ ^{14}C &= \left(^{14}C_0\right)e^{-(K+\lambda)z/v} \end{aligned} \tag{10.33}$$

and when we substitute Equation (10.33) into Equation (10.32) we obtain the age profile

$$A = -\lambda \ln \left(\frac{e^{-(K+\lambda)z/v}}{e^{-Kz/v}} \right) = -\lambda \ln \left(e^{-\lambda z/v} \right) = \frac{\lambda^2 z}{v} \tag{10.34}$$

which is approximately consistent with measured age and concentration profiles, and the derivation is consistent with typical assumptions used in deriving the concentration profile (Rethemeyer et al., 2005). Note that the decomposition rate K divides out of the solution so this result is independent of the rate of decline with depth of the decomposition rate (e.g. Dialynas et al., 2016), so depth-related changes in K will not change the distribution of age down the profile. Note that any source of SOC within the profile (i.e. Equation 10.32) will result, on substitution into Equation (10.32), in the age growing more slowly with depth than Equation (10.34).

10.5 Erosion

Soil erosion and deposition is a component of the SOM cycle (Figure 10.2). We talked in detail about soil erosion in Chapter 4 (of particular relevance are Sections 4.1, 4.2.1 and 4.2.4). However, as a link to these sections a short discussion follows. Here we make a distinction between the mineral matter in the soil (i.e. the clay, silt and sand fractions), which typically has a density in the range 2,000–3,000 kg/m^3, the organic fraction (i.e. SOM), which typically has a density in the range 300–1,000 kg/m^3 and soil aggregates (see Section 10.6.1), which are about 1,300 kg/m^3.

The primary driver of erosion is the shear stress applied by the overland flow on the soil surface. Two factors that change the erodibility of the soil surface (it is good enough here to consider erodibility equivalent to the amount of erosion for a given discharge and slope, a more accurate definition is provided in Sections 4.2.1 and 4.2.4), and thus the erosion rate:

- The grading of the soil eroded from the surface into the flow. The finer the soil, the higher the erosion rate. This is not just the grading of the mineral matter in the soil but also the grading of soil aggregates (and specifically the water stable aggregates). The soil aggregates have a larger diameter than the mineral matter and the SOM in the aggregates.
- The density of the particles (e.g. Ahad et al., 2015). Less dense particles are easier to entrain into the flow and thus erode. They also settle more slowly and so deposit at a lower rate. Again this is not due to only the density of the mineral matter but also the density of the aggregates. The soil aggregates tend to be less dense then the mineral matter because they are bound together by SOM. Soil aggregates have a higher SOC concentration than the remainder of the soil, so erosion tends to have a higher concentration of SOC than the soil surface as a whole. This ratio of SOC concentration in the sediment in the flow relative to the SOC in the soil is called the enrichment factor and is typically greater than 1 (Figure 4.5) (Rumpel et al 2006).

Thus the SOM at the soil surface has two competing processes. The aggregates tend to be larger than the mineral matter on the surface (decreasing erosion), but they also tend to be less dense (typically around 1,200–1,500 kg/m^3, increasing erosion).

Ritchie et al. (2007) found patterns of SOC that were related to the regions of erosion and deposition (using caesium isotope analysis), with the locations of highest erosion having the lowest SOC percentage content, which was consistent with previous more qualitative observations (e.g. Gregorich et al., 1998).

Stallard (1998) estimated the global erosion rate and concluded that transport of SOC may be significant enough to have influenced the global carbon cycle. If this eroded SOC was sequestered by deposition, then it may be an important factor in terrestrial sequestration of carbon. However, for deposition to preferentially sequester SOM requires that the decomposition rate must decline with depth. Stallard had significant difficulties in estimating the net sequestration of erosion and deposition. Yoo et al. (2005, 2006) examined whether the trend in spatial distribution in SOC and erosion/deposition was simply a reflection of SOC enrichment, and thus preferential removal of SOC during erosion, or whether other sequestration processes were at work. They concluded, using a mixture of modelling and fieldwork, that there appeared to be a sequestration of SOC in the deposition regions and that the quantity of SOC sequestered was greater than that provided by the deposited sediment. They concluded that deposition enhanced the burial of vegetated material at the deposition sites (i.e. I in Equation (10.4) was increased by the act of deposition). Though Yoo et al. did not examine erosion regions, it might be inferred that in regions of erosion the value of I would be reduced, so the question of whether there was an enhancement of sequestration of carbon over the entire landscape including all regions of erosion and burial depends on the relative value of I in erosion and deposition regions, and the question remains open.

Billings et al. (2010) and Dialynas et al. (2016) coupled a model of erosion/deposition with a model for SOC decomposition to explore the role of deposition in carbon sequestration. Both used a single-compartment SOC model and assumed that the decomposition coefficient K declined with depth exponentially. Both showed a strong increase in sequestration as the SOC was deposited but that the results were strongly dependent on the exact form of the decomposition rate change with depth and that there were insufficient data to accurately define this rate of change with depth.

For burial of SOM to be a viable mechanism for sequestering carbon the decomposition rate after burial must be less than the decomposition rate at the surface, otherwise the generation rate of carbon dioxide per kilogram of SOM after burial is unchanged. While the idea of decline with depth of the decomposition rate (e.g. driven by reduced biological activity itself limited by reduced food and oxygen) is appealing, the field evidence for a decline in decomposition rates with depth (other than in permanently saturated regions, so that conditions are anaerobic and decomposition rates low) in the soil profile is ambiguous.

10.6 Soil Organic Matter-Hydrology Interactions

There are many ways that SOM interacts with hydrology. High values of SOM are associated with better soil structure, and higher water-holding capacities and infiltration rates. These relationships generate a positive feedback because when the SOM increases, both the infiltration rate and water holding capacity increase, so that the soil moisture content of the soil increases, and the SOM content of the soil increases as soil moisture increases. We will concentrate here on how SOM changes the hydrology relationships. Determining how the relationships below subsequently drive the hydrology and soil moisture (e.g. using soil moisture models such as HYDRUS) has been discussed in Chapter 3.

10.6.1 Soil Aggregation

Soil is not simply particles of mineral and SOM coexisting independently in the soils. A variety of biological, chemical and physical processes result in the particles combining to form aggregates. These aggregates are crucial in carbon storage, and the development of the soil structure that enhances infiltration rates. Furthermore different types of aggregates are formed as a result of interactions between different fractions of the soil. The smaller aggregates tend to result from chemical processes, while the larger aggregates tend to result from physical and biological processes. Some of these aggregates once formed are long lived, while others are more sensitive to disturbance and/or changes in the fungal biology of the soil. Six et al. (2004) reviews the processes that drive the formation of the aggregates.

Tisdall and Oades (1982) introduced a framework for how aggregates are formed by the clay and silt fractions combining, and how these aggregates are important for soil structure development. Aggregates tend to be unstable and disintegrate when wet unless they are also bound with the polysaccharide components of the SOM in the soil. The resulting clay-silt-SOM aggregates are then 'water stable', so the large pores between the soil aggregates (that enable high infiltration) do not collapse when the soil is wet. There is no single relationship for aggregate stability that has been found applicable for all soils, but Tisdall and Oakes discuss a number of the common relationships. For instance, Kemper and Koch (1966) (referenced in Tisdall and Oades, 1982) found

$$\text{aggregate stability} = 40.8 + 17.6\log(\%\text{SOC}) + 0.73(\%\text{clay}) - 0.0045(\%\text{clay}^2) + 3.2(\%\text{Fe}_2\text{O}_3) \quad (10.35)$$

where aggregate stability is defined by USDA-ARS (1999) and characterises the percentage by mass of aggregates greater than 0.25 mm in diameter that remains intact after soaking in distilled water for 5 minutes. Aggregates are classified on size as 0.002–0.02 mm, 0.02–0.25 mm and <0.25 mm, microaggregates, mesoaggregates and macroaggregates, respectively. Microaggregates are formed by flocculation of clays and organic molecules (particularly polysaccharides produced by root and fungal exudates). Mesoaggregates involve decaying particulate organic matter and biofilms. Macroaggregates involve physical aggregation of micro- and mesoaggregates by decaying plant roots and fungal hyphae (Voroney, 2007).

As well as influencing the infiltration rate, the presence of water stable aggregates increases the size of soil particles and thus reduces raindrop erosion (see Chapter 4). The effect of the change in soil grading, porosity and pore-size distribution due to soil aggregates can be estimated by pedotransfer function. For instance, the Kozeny-Carmen equation (Freeze and Cherry, 1979) relates the saturated hydraulic conductivity to the soil grading. Unfortunately, however, no pedotransfer functions provide guidance on how the grading and porosity of the aggregates changes with SOM (Nimmo and Perkins, 2002) though a number of fractal models have been suggested.

Because of the central role of SOM in forming soil aggregates, SOM content in aggregates can be much higher than in the finer, nonaggregate, fraction of the soil (e.g. SOC was 38% in particles greater than 250 μm, and less than 5% in particles less than 2 μm; Tisdall, 1996). Some preliminary work has linked SOM decomposition and aggregation (Malamoud et al., 2009).

10.6.2 Infiltration Rate

Barzegar et al. (2002) provides a direct link between SOM content and hydrology. They applied varying rates of organic material to an agricultural soil and found that the steady-state infiltration rate increased with increasing SOM and the infiltration rate was independent of the type of SOM (e.g. mulch, manure). While the data set was small and restricted to relatively high SOM contents ($8\% < \text{SOM} < 10\%$) they fit

$$I = 0.0064(\text{SOM})^{3.4} \quad (10.36)$$

where I is the continuing infiltration rate (mm/hr).

Arshad et al. (1999) present a comparative analysis of no-tillage paddocks with conventional tillage paddocks. Their no-tillage paddocks had a higher SOM content than conventional tillage by between 10% and 30% (the actual SOC content was between 2.2% and 4.5%), for the no-tillage sites the mean diameter of the water stable aggregates was larger by 20–100%, and the steady-state infiltration rate was higher by about 60%. They attributed the differences to the SOM content. Azooz and Arshad (1996) did not report SOC content but found that both saturated and unsaturated conductivities of no-tillage soils were 2–3 times higher than for a conventional tillage site.

In both cases the differences were attributed to soil structure changes and not to physical disturbance by tillage. These results are qualitatively consistent with Barzegar et al. (2002).

Dunkerley (2002) measured infiltration rates for a site with patchy groves of vegetation and compared the infiltration rates inside and outside the groves of trees. He found that the infiltration rate outside the groves (7.4–30.5 mm/hr) was significantly less than inside the groves (as much as 292 mm/hr), and that there was a significant correlation between infiltration rate and distance to the nearest tree stem. He did not measure SOM/SOC content and so was not able to distinguish whether this was a result of tree root macropores or SOM-soil structure effects, or surface sealing by rainsplash outside the groves (Wickens and Collier, 1971); however, others have found significant correlations between SOM content and distance from trees (Avirmed et al., 2014), suggesting that a link between SOM and infiltration rate was plausible.

Improved soil structure results in an increase in pore sizes and a reduction in bulk density, and it is likely that these are the fundamental drivers of the hydrology impacts of SOM. Thus models for the relationship between SOM, aggregation and porosity are key to predicting the influence of SOM on hydrology (specifically infiltration rates), but we do not have such relationships at the current time (Minasny, personal communication).

10.6.3 Water-Holding Capacity and Bulk Density

Another impact of SOM is that it increases the soil water-holding capacity and field capacity, and these are important for vegetation because they determine the amount of water available between rainfall events. Saxton and Rawls (2006) present various equations for the Van Genuchten unsaturated conductivity model that relate these parameters to percentages of sand, silt, clay and SOC content. He found the strongest effect of SOM was on the field capacity, FC, of the soil

$$\begin{aligned} FC &= FC_t + \left[1.283FC_t^2 - 0.374FC_t - 0.015\right] \\ FC_t &= -0.251S + 0.195C + 0.011\text{SOM} + 0.006(S \cdot \text{SOM}) \\ &\quad - 0.027(C \cdot \text{SOM}) + 0.452(S \cdot C) + 0.299 \end{aligned} \tag{10.37}$$

where S, C and SOM are the sand, clay and SOM proportion by mass, respectively. Other data suggest that FC is not as strongly impacted by SOM (Minasny, personal communication).

Hudson (1994) found when SOM increased from 0.5% to 3% by mass that the plant available water capacity was doubled across all soil types. Plant available water capacity is the difference between field capacity and the permanent wilting point.

Adams (1973) derived a relationship for predicting bulk density:

$$\rho_{b(\text{soil})} = \frac{100}{\left(\dfrac{\%\text{SOM}}{\rho_{b(\text{som})}}\right) + \left(\dfrac{100}{\rho_{b(\text{mineral})}}\right)} \tag{10.38}$$

where $\rho_{b(\text{soil})}$, $\rho_{b(\text{som})}$ and $\rho_{b(\text{mineral})}$, are the bulk densities of the soil, SOM and mineral respectively.

10.7 Catchment and River Network-Scale Carbon Cycling

This chapter has been focussed on soil and carbon processes on the hillslopes. However, given the book's overall focus on catchment and landscapes interactions, some discussion on the role of river networks in transporting and transforming organic carbon on its route to the ocean is warranted. For many years the view has been that rivers are simply the pipeline to transport eroded particulate organic carbon (POC) and dissolved organic carbon (DOC) to the ocean with some small decomposition of organic carbon while in transit. Recent years have seen a recognition that there is a significant mismatch between organic carbon creation on the hillslopes and the organic carbon delivered to the ocean, so that (1) significant transformation is likely occurring en route or (2) there is storage with long time scales within the catchment. While the data are by no means definitive, a recent review (Wohl et al., 2017) suggests that both processes are important. This discussion follows the interpretation presented by Wohl.

In Chapter 1 and Figure 1.2 the strings and beads conceptualisation of rivers was discussed. This conceptualisation divides the river into two types of reaches: (1) those with associated floodplains (the beads, typically alluvial river cross sections) interspersed with (2) river reaches without associated floodplains (the strings, typically bedrock river cross sections). Wohl argues that little carbon transformation and storage occurs in the bedrock river reaches because transit time are fast and there is little potential for storage of carbon within the reach. Accordingly all the transformation and storage must

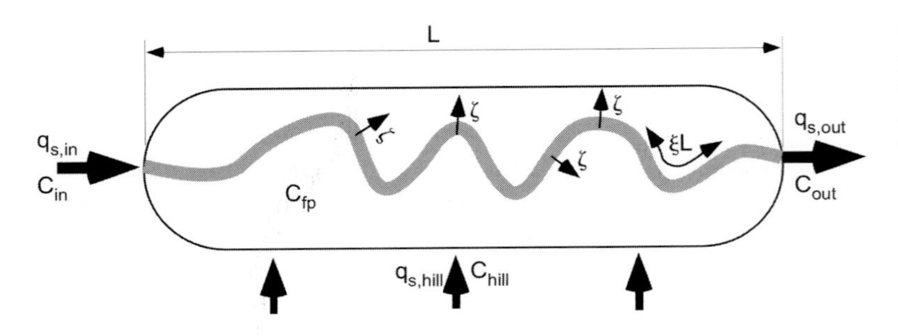

FIGURE 10.5: Schematic of the processes operating in the floodplain determining exchange and storage of organic carbon in the channel and the floodplain.

occur in the alluvial river reaches, and that the exchange processes between the channel and the floodplain (which includes the riparian or hyporheic zones) for both water and particulate matter are key to these processes. Because the floodplain is wet, it is an active zone for soil organic matter activity because vegetation growth and microbial decomposition are both high. Since SOM values are typically higher on floodplains than hillslopes, it seems reasonable to infer that the enhanced vegetation activity more than offsets increased decomposition. The exchange processes for POC are the same as they are for mineral matter (Figure 1.2; Sections 1.3.2 and 4.5) though organic matter tends to be more mobile because of its lower density. These processes are (1) deposition on the floodplain surface during flood events, (2) erosion on the outside of the meander bends and (3) deposition on the inside of meander bends. The DOC exchange is a result of groundwater exchange between the river and the groundwater table within the floodplain (the hyporheic zone).

The discussion below will focus on the POC. A simple model for the change of POC in the floodplain reach of the channel is a function of the exchange of sediment with the floodplain so that (using the terminology in Section 4.5; Figure 10.5)

$$
\begin{aligned}
q_{s,\mathrm{out}}C_{\mathrm{out}} - q_{s,\mathrm{in}}C_{\mathrm{in}} = {} & \zeta D_{\mathrm{fp}}\rho_b\left(C_{\mathrm{fp}} - \frac{C_{\mathrm{in}}+C_{\mathrm{out}}}{2}\right)\varepsilon L \\
& + E\rho_b\left(\frac{C_{\mathrm{in}}+C_{\mathrm{out}}}{2}\right)A_{fp} \\
& - K_{\mathrm{ch}}\left(\frac{q_{s,\mathrm{in}}C_{\mathrm{in}} + q_{s,\mathrm{out}}\mathrm{X}}{2}\right)\frac{\varepsilon L}{v} \\
& + (1-e_{\mathrm{fp}})q_{s,\mathrm{hill}}C_{\mathrm{hill}}
\end{aligned}
\tag{10.39}
$$

where q_s is the sediment transport in the channel (units mass/time, the subscripts in and out indicating at the upstream and downstream end of the floodplain, respectively), C is the concentration of POC in the sediment (units mass POC/mass sediment) and ρ_b is the average bulk density of the floodplain deposits. The first term on the right-hand side is the exchange of sediment between the channel and the floodplain by bank erosion as a result of meandering, and C_{fp} is the average concentration of POC in the floodplain sediments, ζ is the average rate of meander movement laterally, D_{fp} is the height of the floodplain above the channel bed on the outside of the meander, and εL is the effective length of the channel within the floodplain where L is the straight-line distance between the upstream and downstream ends of the floodplain so that ε captures the enhanced length of the channel from the tortuosity of the meandering. The second term on the right-hand side is the deposition of organic matter on the floodplain during high flow, where we use the notation of Chapter 4 that deposition is the negative erosion E (depth/time), and A_{fp} is the planar area of the floodplain. The third term on the right-hand side is the in-channel decomposition of the POC (assumed to be a first order decay process) and K_{ch} is the in-channel decomposition rate (units 1/time), and v is the velocity of the flow so that $\varepsilon L/v$ is the time spent in the floodplain reach. The fourth term is the POC delivered to the channel by hillslopes contributing to the floodplain, and e_{fp} is the efficiency of capture of POC from the hillslope by the floodplain ($e_{\mathrm{fp}} = 1$ is 100% capture of the POC by the floodplain with none reaching the channel).

This equation requires knowledge of the POC concentration in the floodplain sediments. We thus need an equation that predicts the dynamics of POC in the floodplain:

$$
\begin{aligned}
A_{\mathrm{fp}}D_{\mathrm{fp}}\rho_b\frac{\partial C_{\mathrm{fp}}}{\partial t} = {} & -\zeta D_{\mathrm{fp}}\rho_b\left(C_{\mathrm{fp}} - \frac{C_{\mathrm{in}}+C_{\mathrm{out}}}{2}\right)\varepsilon L \\
& - E\rho_b\left(\frac{C_{\mathrm{in}}+C_{\mathrm{out}}}{2}\right)A_{\mathrm{fp}} + e_{\mathrm{fp}}q_{s,\mathrm{hill}}C_{\mathrm{hill}} \\
& + SA_{\mathrm{fp}} - K_{\mathrm{fp}}C_{\mathrm{fp}}A_{\mathrm{fp}}D_{\mathrm{fp}}\rho_b
\end{aligned}
\tag{10.40}
$$

where the left-hand side is the rate of change of the total mass of POC in the floodplain sediments. The first two terms of the right-hand side come directly from Equation (10.39) and are the channel-floodplain exchange processes. The third term on the right-hand side is the hillslope POC transport that is captured by the floodplain. The fourth term is the source term for POC on the floodplain with S the rate parameter (mass/unit area of floodplain). The fifth term is a simple first order decay term as discussed in the SOM compartment models discussed in previous sections where K_{fp} is the decomposition rate for the floodplain.

These two equations are simple conceptualisations for the channel and floodplain processes, and it is likely they do not capture the full complexity of processes in floodplain reaches. However, given the extremely limited understanding of floodplain organic matter processes rate and storages (as discussed by Wohl et al., 2017), it is likely that more complicated modelling is not warranted by the data at this time.

Subject to that caveat some limitations might include that the decomposition terms in Equation (10.40) are likely to require the multi-compartment models discussed in previous sections. It may also be warranted to subdivide into the soils above and below the floodplain groundwater table because rates within the anaerobic groundwater table will be significantly less than the aerobic conditions above the water table (and this may be critical to long-term sequestration of POC within the floodplain sediments). The source term will also be a function of the vegetation growth models to be discussed in Chapter 14.

10.8 Conclusions

Before we finish, we note that the main source for SOM/SOC is vegetation leaf litter and dead roots, and the microbes feeding on them. Thus the terrestrial component of the carbon cycle has vegetation at its core: i.e. SOM → infiltration → soil moisture → vegetation growth → SOM source → SOM. Vegetation will be discussed in detail in Chapter 14.

10.9 Further Reading

The organic content of soil is a complex topic, and this chapter has been able to only scratch the surface of it. Many processes are so complex as to have defeated quantification to date (or even experimental study in some cases), and so fall outside the scope of this book. Schaetzl and Thompson (2015) provide a good general overview. Wardle (1992) provides an excellent review of the living biomass in the soil and the factors that influence it. Plante and Parton (2007) and Metherell et al. (1994) summarise the state-of-the-art for single-layer compartment models. Bondeau et al. (2007) present a modelling study using the LPJ model (see Chapter 14) of the impacts of agriculture on the global carbon cycle, including a 10% reduction in global soil carbon stocks, over the last century.

11 Soils: Constructing a Soilscape Evolution Model – Details and Examples

11.1 Soilscape Modelling

Before moving from soil evolution processes to landform evolution processes we will pause to consider how a standalone soilscape evolution model can be constructed from the components already discussed. Such a model can be used to explore the dynamics of soils on a fixed landform. This is not to deny that landform evolution may be important to the spatial distribution of soils but simply that soils may evolve more rapidly than the landform. In this case the soils will always be at or near equilibrium with the slowly evolving landform. Over the long term the soils may still evolve, but only because the landform itself changes. This simplification should be quite familiar because it is basically a restatement of the principle underpinning 'soil catena', where the position on the hillslope, but not the history of the hillslope, determines the soil properties. If the current-day soils were still responding to past landforms, then the link between the position on the current-day landform and soil would be less apparent.

11.2 Coupling Weathering Processes

11.2.1 What Is Soil?

Before outlining below how to couple all of the processes in this chapter to develop a general coupled model for soil dynamics we need to consider ongoing debate over 'what is soil'. At the beginning of the book we adopted a definition of soil as being all that material above the saprolite interface, where there was distinct change in the porosity and bulk density above and below the saprolite-soil interface. We discussed how this interface might arise from bioturbation as a result of tree throw where the tree roots rip rock out of the saprolite and mix saprolite material into the soil (and in the process mix the soil from top to bottom). However, other work suggests that this bulk density change may not form a distinct boundary if the bulk density change is due to chemical weathering processes (e.g. Graham et al., 2010).

At the heart of the soil definition problem is that we have four possible, but distinctly different, processes going on within the soil – (1) the soil production function, (2) physical weathering, (3) chemical weathering and (4) downslope soil flow – and the different intellectual communities that have worked in these four areas have each arrived at different definitions for 'soil'. Each of these processes contributes to the soil formation processes, but there is no consensus on their relative dominance. For example, Yoo and Mudd (2008, p. 249) state 'colluvial soil production and colluvial transport have not been considered in the geochemical evolution of the soils and chemical weathering has been neglected in understanding of the geomorphology of soil-covered hillslopes.'

For the soil production function the recent evidence supporting its existence and form is the rate of bedrock conversion as a function of soil depth determined using cosmogenic nuclides. Yet this same technique provides no insight into what the processes are that drive this bedrock conversion only that the rate of bedrock conversion decreases with soil thickness with a roughly exponential decline. Burke et al. (2007) showed a relationship between pH, cumulative chemical weathering and the soil production function, suggesting an exponential decline with depth for chemical weathering with a similar exponent as the soil production function.

On the other hand, for chemical weathering Maher (2010) suggests that the equilibrium soil stops at the depth where the acid driving the weathering is completely depleted. The argument is that if there is unused weathering potential (i.e. unused acid), then that acid will further weather the bedrock and the soil will become deeper as a result. Soil deepening stops only when the chemical weathering potential is depleted. This ignores the role of SOM decomposition and respiration in providing new acid potential in-profile. The philosophy behind Maher's viewpoint is that soil is all the material that is chemically weathered. On the other hand if there is a distinct change in bulk density at the saprolite boundary, it is hard to conceive of that being due to chemical dissolution, particularly if a humped chemical weathering profile has the maximum weathering rate in the mid-regions of the soil profile. There is evidence that dissolution is a significant cause of porosity development in soil profile (e.g. Brimhall and Dietrich, 1987; Brimhall et al., 1991; Anderson et al., 2002), but no credible mechanism for chemical dissolution to create an abrupt change in porosity has been proposed. This is particularly the case when the main indicator used for chemical weathering progress is the change in the ratio of mobile elements to immobile elements in the saprolite and the soil, and they typically do not show a step change at the saprolite boundary (e.g. Anderson et al., 2002; Riebe et al., 2003), only a gradual change from the saprolite to the soil surface.

There is thus the potential for conflict between the soil production function and chemical weathering definitions of soil. It is possible to have deep soils that are not fully chemically weathered to their full depth if the equilibrium depth from the soil production function is deeper than the maximum depth of chemical weathering.

There is also a geomorphic view of what defines soil. Geomorphologists (Anderson and Anderson, 2010) define it as that part of the regolith that is mobile, which is any material above the saprolite that can potentially move downslope or that can be mixed vertically within the soil profile. In this definition any material that is not physically 'locked into' the saprolite is soil. The movement downslope may be creep like processes, viscous or plastic flow of the soil, or mass movement.

The extent of these difficulties can be seen with a simple example. Consider a soil where the top half of the soil profile is subjected to the Gabet tree throw bioturbation mechanism: the underlying saprolite has not been excavated by tree roots, and the acid in the chemical weathering process has not been fully consumed at the bottom of the tree thrown layer. This can be modelled in the chemical weathering model discussed in previous chapters by considering that there is an abrupt increase in the porosity of the soil above the tree throw–saprolite boundary. Let's call the ratio of porosity above the boundary to the porosity below P_R. The impact of the porosity contrast is that the groundwater velocity below the boundary will be P_R times faster than above the boundary, and the dispersivity will likewise be P_R times higher below the boundary as dispersivity normally scales linearly with groundwater velocity. Figure 11.1 shows a simulation that is equivalent to Figure 8.16 but where $P_R = 10$ (e.g. porosity above the tree throw boundary is 0.3 and below the boundary 0.03). The main observation is for the weathering profile with multiplier = 1.0. Where it previously peaked in the middle of the profile (Figure 8.16) it now peaks at the bottom of the profile. There are now two distinct weathering regimes that do not overlap: the first where all the chemical weathering occurs above the porosity contrast boundary (for high erosion rates) and the second where it all occurs below (for low erosion rates).

11.2.2 What Soil Properties Should Be Replicated?

What we need to replicate will be different depending on the objective of the researcher. Data to test soilscape evolution models are rare. For instance, data from field studies where soil properties are available and can be dated, and where it is possible to definitively fingerprint the pathway and rates from saprolite to soil is rare. Beal et al. (2016) examines a volcanic field with a range of ages for lava flows and cinder cones up to about 12,000 years and were able to compare soil properties between sites with a good assurance of similar initial conditions. The results from previous chapters (e.g. soils fine with age, SOM increases with age) are consistent with their findings.

However, beyond the pragmatic issue of prioritising the modelling of those characteristics for which field data are available is the question of what characteristics of soil behaviour actually need to be modelled for any given application. It is rare that we can model everything simultaneously, so we prioritise those characteristics that are most important for our immediate needs.

Some geomorphology examples might include the following:

- Landslide researcher: Catchment wide soil depth so that they can determine volumes of source material for potential landslides, and potential for saturation to trigger a landslide.
- Landform evolution modeller: Grading of the soil surface and its erosion potential so they can determine future landform shape, and spatial distribution of rock outcrops (i.e. lack of soil).
- Ecologist: Water-holding capacity of the soil (depth, texture and structure), which limits plant growth during arid periods and nutrient availability from weathering.
- Water quality modeller: Leachate properties from weathering processes.

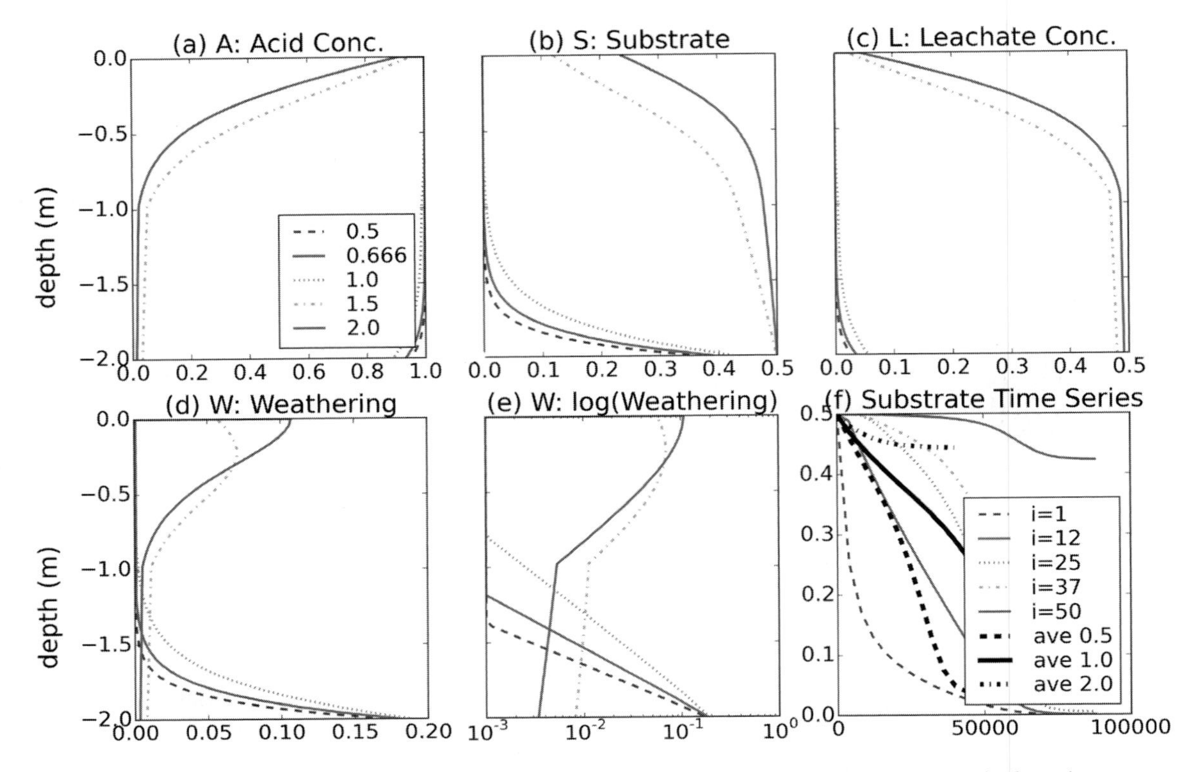

FIGURE 11.1: The chemical weathering model where there is a porosity change half way down the profile as a result of tree throw. This result has the same parameters as Figure 8.16 except for a 90% reduction in porosity in the bottom half of the profile.

In an attempt to answer this question from the perspective of a soil scientist, Minasny et al. (2015) surveyed 29 case studies involving a range of soilscape models. He listed 22 processes that potentially need to be included, of which only about 50% have ever been studied in a soilscape model (Figure 11.2). While it is a laundry list of processes, it does indicate the range of soil processes observed in the field. Taking a step back from the details of the processes, Minasny then suggested four criteria against which soilscape models should be evaluated:

1. Mass or solute fluxes along a soil catena (i.e. so fluxes match those observed)
2. Effects of slope along toposequences (i.e. so soil catena is correct)
3. Effects of climate variations in climosequences (i.e. so the trend with changes in climate is correct)
4. Effects of the duration of soil formation in chronosequences of soils (so modelled soil ages match field soil dating).

Minasny concludes with six challenges for the coupling of soils and landscape models:

1. Improve the ability of the models to respond to climate dynamics, climate change and landuse change
2. Balancing the desire for more process components (and the potential for over-parameterisation) against the need for model parsimony
3. Explicit representation of temperature and moisture dynamics
4. Optimise codes for high performance and reduced run times
5. Link models with global climate models and dynamic global vegetation models
6. Investigate the main drivers of soil spatial variability and scaling properties of soils in space.

Finally, from the soil science perspective McBratney (personal communication) asserts that a key objective for soilscape models is the reproduction of the soil horizons, without having explicitly incorporated soil horizon information into the model in its formulation. Soilscape models have not yet achieved this objective.

Hydrologists are typically interested in the catchment scale spatial trends in soil functional properties such as hydraulic conductivity. However, even in a relatively uniform soil, the grading and functional properties can

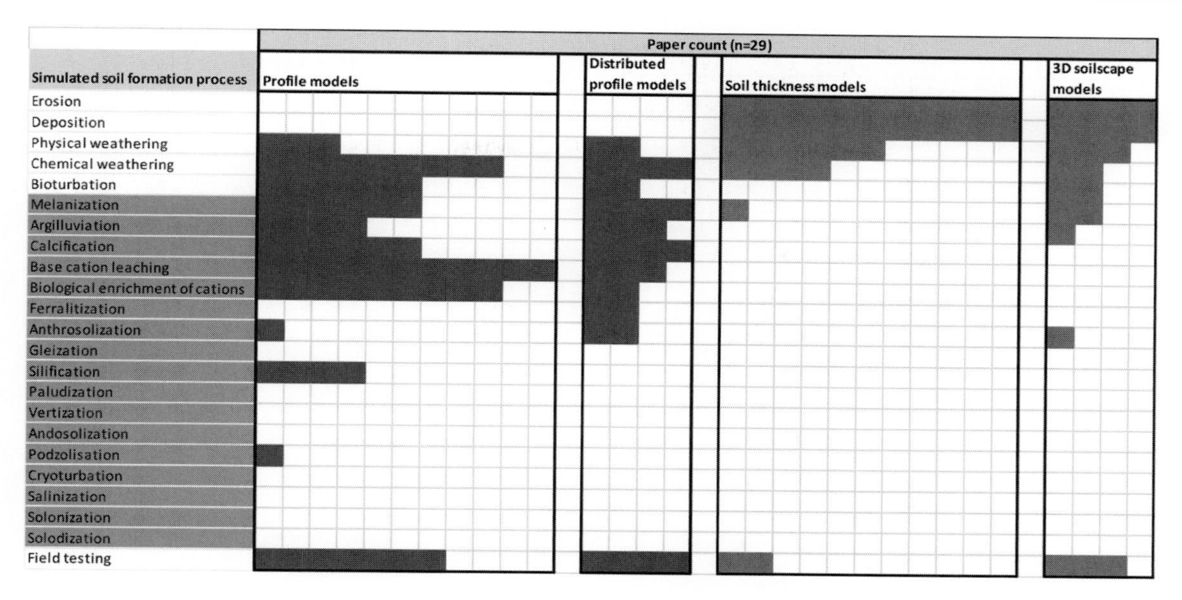

FIGURE 11.2: Presence (horizontal bars) of 22 soil processes in 29 case studies of profile models and distributed profile models (left two bar charts, black), soil thickness (two-dimensional) models and three dimensional soilscape models (right two bar charts, grey) (from Minasny, 2015).

vary substantially within a few metres. If the underlying source material is variable in space, then this may strongly condition this spatial variability. For instance, soil preferentially forms in cracks in rock as a result of feedbacks between water collecting in the cracks and enhanced weathering that results (e.g. Beal et al., 2016). Even with uniform source rock geology it is also possible to postulate a spatial positive feedback in soil formation where enhanced soil moisture enhances weathering, which enhances soil water-holding capacity, which enhances soil moisture. Thus locations where soil moisture has slightly increased for some reason might exhibit runaway soil weathering that results in spatially random, but persistent with time, patterns of soil moisture and weathering (e.g. the persistent but spatially random soil moisture patterns observed by Western et al., 1999 might result from this mechanism). This random patterning, in addition to any mean trends in space, of soil moisture and soil functional properties may then be an important property for hydrologists to be able to replicate.

11.2.3 Previous Modelling

This section is not intended as a comprehensive review of models, but rather as an outline of how previous modellers have constructed their models and place in context the components discussed in the previous chapters. Any review of a model will be out of date as soon as it is written, and interested readers are advised to download model manuals off the web.

It is generally considered that the modern era of quantitatively linking soil properties with pedogenic processes started with the empirical work of Jenny (1941, 1961), though it wasn't until the 1970s that quantitative dynamical models based on process began to appear in the literature. In parallel with the research motivated by Jenny, there was also the developing idea of soil catena, that a soil profile should not be viewed in isolation but as part of a (potentially evolving) hillslope system. For a discussion of these historic developments see, for example, Huggett (1975) and Hoosbeek and Bryant (1992) and the subsequent discussion of the latter (Amundson, 1994; Hoosbeek, 1994).

Kirkby (1977, 1985) presents one of the first models to attempt to model the evolution of saprolite weathering over a hillslope. While he recognised the different weathering characteristics of different minerals, he lumped all minerals into one component and modelled the degree of weathering of the saprolite into soil by a single variable, what he called the weathering deficit, ignoring the differences between the minerals. At any given time he considered that the solution and the soil have an equilibrium concentration as in Equation (8.15). However, he set the equilibrium leachate concentration as a function of the unweathered proportion of saprolite S (figure 4 of Kirkby, 1977)

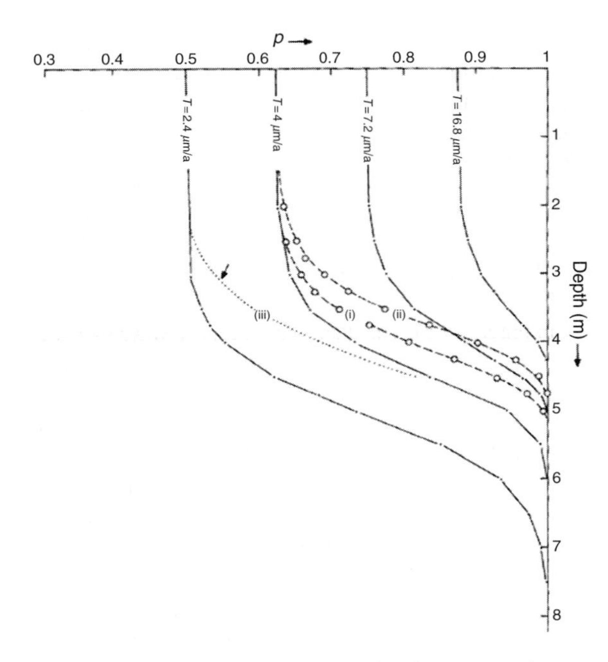

FIGURE 11.3: Equilibrium depletion profiles for a variety of different parameter sets (from Kirkby, 1985).

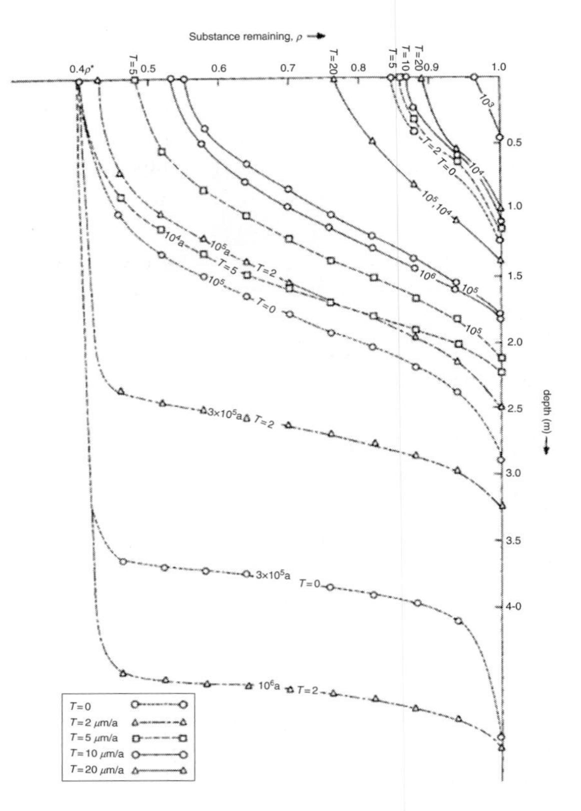

FIGURE 11.4: Transient depletion profile (from Kirkby, 1985).

$$R = k_1 k_5 (S - S_{\min}) \left(1 - \left(\frac{L}{S_{\min} + \alpha(S - S_{\min})}\right)\right) \tag{11.1}$$
$$A = A_{PCO2(soil)}$$

where he assumed that the leachate concentration was at equilibrium with the saprolite. The term $(S - S_{\min})$ meant that below $S_{\min}$ there was no weathering, so $S_{\min}$ was a lower bound on how much saprolite could be weathered. Unlike Equation (8.6), where the approach to zero reaction rate is asymptotic, Kirkby's was linear with an abrupt switching off of weathering at $S_{\min}$. He also assumed that soil atmosphere was driven by microbial degradation of the organic matter so that the acid content was the same throughout the profile. His assumption that the system was at chemical equilibrium is the same as assuming the reaction rate k_1 is very high in Equation (8.6). Kirkby coupled this chemical weathering model with a fixed rate of erosion as was used in the sensitivity studies in Chapter 8. He found depletion profiles (both the equilibrium profiles, Figure 11.3, and the transient approaching these equilibrium, Figure 11.4) that are very similar to those shown in Chapter 8 with two differences. The first difference is that the saprolite is never 100% weathered because of the threshold $S_{\min}$ below which weathering does not occur. The second, related, difference is that when he included a bioturbation term that was concentrated in the upper part of the soil profile, he found that the proportion of weathered saprolite in the top of the profile was fixed by the minimum value for weathering over the entire depth. Like all subsequent workers in this area he assumed the soil hydraulic conductivity was independent of the degree of weathering. Kirkby (1989) extended this model to examine the effect of climate dynamics on soil properties. Kirkby (1977) used the transition matrix formulation presented in Chapter 7 to simulate the interactive dynamics of four different types of soil organic matter (Section 10.3).

Minasny and McBratney (1999) implemented a coupled physical weathering and landform evolution model to predict the spatial distribution of the evolution of soil profiles on a hillslope. The physical weathering was the exponential soil production function (Chapter 6). Minasny and McBratney (2001) extended this work to implement both physical and chemical weathering models that were depth averaged over the soil profile, and coupled them with a simple landform evolution model to predict the evolution of the spatial distribution of soil depth. The physical weathering they modelled using a soil production function with an exponential decline with soil thickness and when soil was produced the bulk density was reduced so the volume increased. Their chemical weathering model was quite simple. The chemical weathering model was a modification of the exponential function and as the soil chemically weathered the thickness of the soil decreased. The function they adopted to calculate the rate of decrease in the soil thickness was a combination of (1) an exponential term that decreased with soil depth and (2) an exponential term that decreased with both time and thickness so that the rate of decrease decreased with time. The reduction of weathering rate with soil depth reflected a consumption of acid down the profile (and thus a weakening of weathering), while the reduction with time reflected a reduction of chemical weathering reactivity with time (the most reactive material reacts first leaving behind less reactive material).

Probably the simplest implementation of chemical weathering was a dissolution model of Mudd and Furbish (2004). In a mass balance model for hillslope soil mass loss they assumed that the loss by dissolution was constant down the hillslope. By assuming that bioturbation disturbed the weathering soil sufficiently so that the bulk density of the soil did not change with dissolution (Brimhall et al., 1992) they assumed that this dissolution led to a proportionate loss of soil thickness in a similar fashion to Minasny and McBratney (2001).

The matrix state-space approach for physical weathering described in this Chapter 7 was first described by Cohen et al. (2009, 2010) for his mARM soilscape model, who in turn had extended the work of Willgoose and Sharmeen (2006) that used their physically based ARMOUR fluvial erosion model (ARMOUR is discussed in detail in Chapter 4). Cohen modelled the soil by a large number (typically about 30) of size fractions from clay to large gravel, and used a variable number (typically about 20) of soil layers with depth. mARM was subsequently extended to examine the dynamics of soil profile adjustment due to climate variability (Cohen et al., 2013), and an aeolian deposition module was added to simulate soil development and spatial patterns of erosion at a field site in Israel (Cohen et al., 2015, 2016). Welivitiya et al. (2016) extended Cohen et al. (2010) to explore, using a parametric study, the generality, and process drivers, of scale-invariant behaviour found in the spatial organisation of soil grading generated by mARM across a landform.

One key outcome of Cohen and Welivitiya was the discovery of a spatially organised relationship between the area draining to that location, the slope at that location and the particle size grading at that location, the area-slope-diameter relationship (Figure 11.5). The trends of surface grading for five hillslope profiles are illustrated in Figure 11.5b:

1. This is a hillslope where the slope is increasing down the hillslope so is concave down in profile and looks like a rounded hilltop. The d_{50} increases down the hillslope (i.e. increasing area, moving from left to right in Figure 11.5). The diameter contours always increase from left to right and from bottom to top, so in general concave hillslopes always coarsen downslope.
2. This hillslope has constant slope downslope and, as for slope 1, always coarsens downslope.
3. This hillslope has slopes that are decreasing downslope and is concave up. The gradient of the hillslope in Figure 11.5 is less than the gradient of the contours so the hillslope coarsens downslope.
4. This hillslope is similar to profile 3 except that the rate of decrease of slope downslope is more severe so the gradient of the hillslope in Figure 11.5 is steeper than the gradient of the contours. This hillslope fines downslope.
5. This hillslope is a classic catena profile with a rounded hilltop and a concave profile downslope of the hilltop. Travelling down this hillslope it will initially coarsen. As the hillslope transitions to concave up, it will continue to coarsen until the rate of reduction of the hillslope slope is severe enough that is starts to fine downstream. Whether this latter region of fining occurs will depend on the concavity of the hillslope and whether the concavity is strong enough relative to the gradient of the diameter contours in Figure 11.5.

Figure 11.5 shows that the spatial distribution of soils, and any questions of downslope fining or coarsening of those soils, depends on the interaction between the pedogenesis processes that produce the soils (and thus drive the slope-area dependence of soil grading) and landform evolution processes that generate those profiles (and the slope-area relations for those slopes). Ultimately deeper understanding of these links will come only from a coupled landscape-soilscape evolution model (see Chapter 16 for examples).

Gasparini et al. (2004) used a landform evolution model that simulated sediment transport with a combination of sand and gravel, and that could armour the catchment surface by selective entrainment of the sand. She found for the landforms that she generated that the surface had a weak trend of fining downstream. Comparing her results with Welivitiya et al. (2016), he found the slope of

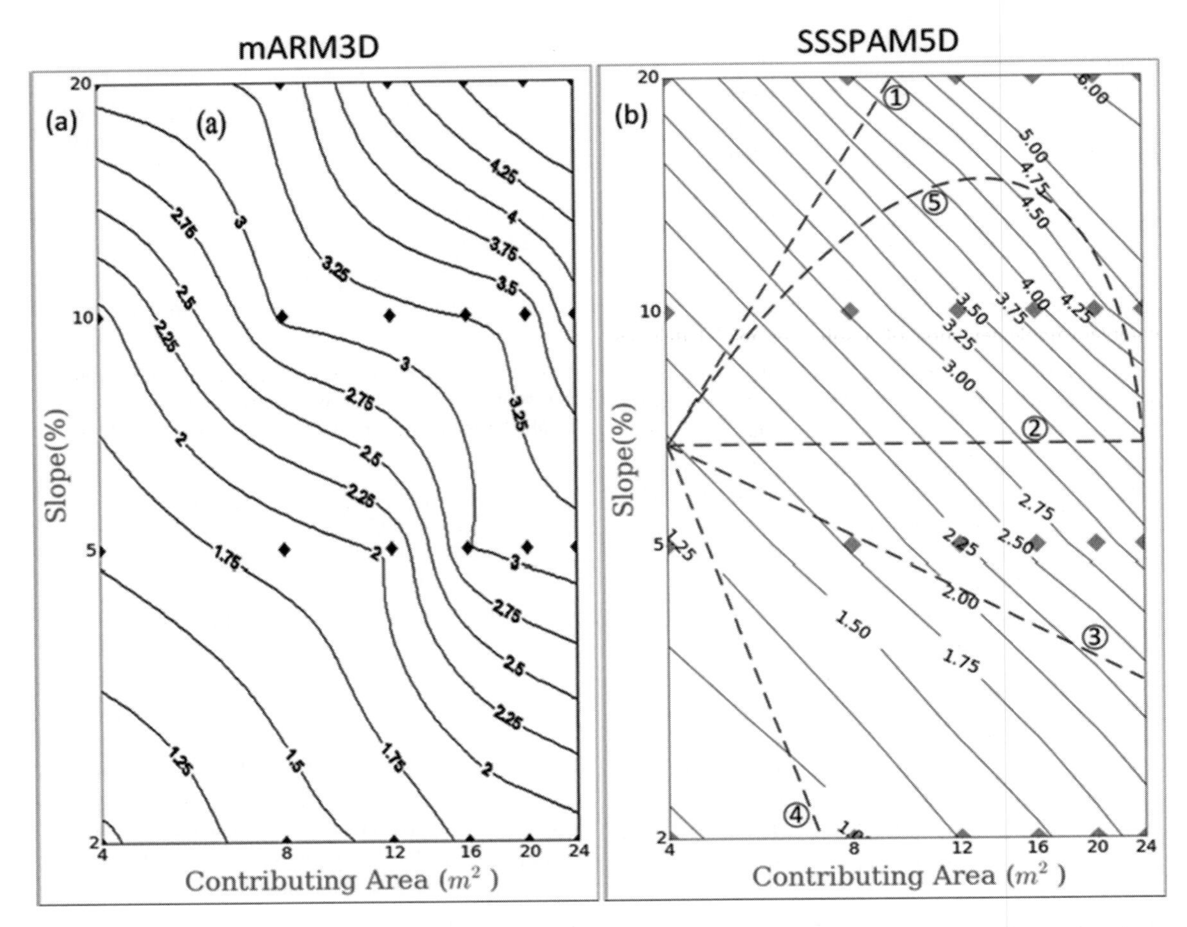

FIGURE 11.5: Area-slope-d_{50} plots from the (a) mARM and (b) SSSPAM models showing how the surface soil grading changes downslope. See the text for the explanation of the five lines in the SSSPAM plot (from Welivitiya et al., 2016).

the diameter contours was a function of the erosion exponents on discharge and slope. While the process formulations of Gasparini and Welivitiya are not directly comparable, the parameters of the erosion process used by Gasparini result in the slope-area relationship for the topography (the dotted lines in Figure 11.5) being approximately parallel to the contours for the d_{50} (the solid lines in Figure 11.5), so that if Welivitiya's relationship is used, then the grading will not change greatly downslope, which is consistent with Gasparini's results.

Finke (2012) presents the SoilGen2 model, which couples physical and chemical weathering for a single soil profile, but which did not couple the processes spatially. The chemical weathering component builds upon the LEACHC model. LEACHC couples Richard's equation for unsaturated flow down through the profile (so v is soil moisture dependent in Equation (8.6)) with a geochemical model following the principles outlined in Chapter 8 for four saprolite components (anorthite, chlorite, microcline, albite) generating four leachate ions (calcium, magnesium, potassium, sodium). It has a reaction rate that is a generalisation of Equation (8.6) where the exponent on the acid concentration, α, is calibrated, and where the equilibrium leachate concentration is calculated based on an equilibrium sum of the four ion concentrations and the dissolved aluminium concentration, which is pH (i.e. acid concentration) dependent. The physical weathering model is based on Salvador-Blanes et al. (2007), but instead of having 1,000 size fractions different by just 1 μm, he has size fractions that are powers of 2, so that a particle in one size fraction can be split into eight cubes and fall into the next size fraction smaller. This is similar to, but less general than, the size fraction approach of mARM, but Salvador-Blanes modelled the fragmentation as stochastic

soil chemistry in the absence of the geomorphic processes, geomorphologists discuss transport processes in the absence of pedogenic processes, and soil scientists do not quantify the long-term evolutionary processes leading to the soils observed. The best modern overviews are Anderson and Anderson (2010), and Minasny et al. (2015). Minasny and McBratney (2001) was the first (and very readable) paper to couple a pedogenesis model (albeit simple) to a landform evolution model. Brimhall et al. (1992) uses field studies to discuss the complexity of the field interactions between chemical weathering, physical weathering and bioturbation. Maher (2010) discusses the geochemical weathering model presented here in greater detail, while Kump et al. (2000) provides a good overview of geochemical weathering in the field. There is no equivalent paper to Maher about physical weathering other than the Sharmeen, Cohen, Wells and Welivitiya papers that were the basis for Chapter 7, though Wells et al. (2008) provides a quantitative interpretation of their experimental studies.

12 Tectonics and Geology

12.1 Introduction

One of the most important drivers of the movement of landforms is movement of the geology underlying the landform. Tectonic uplift provides the initial elevation that is then eroded away by the land-forming processes discussed in following chapters.

These tectonic forcings observed at and near the land surface are largely driven by interactions between the crust and the underlying mantle. The flow processes in the mantle lift, tilt, warp and translate the overlying geology, and the land surface responds accordingly. It is these latter land surface responses that are our primary interest in this book, but to understand them fully we do need to understand how the mantle drives them (even if we don't model the mantle explicitly, that being the primary domain of the field of geodynamics; e.g. Turcotte and Schubert, 2002). Our focus will not only be on how the mantle drives the landforms, but how the mantle responds to landform changes, and how that in turn influences the landform evolution into the future: i.e. the interaction between the landforms and the crust-mantle system.

A very brief overview of the crust-mantle system is warranted before we discuss details, and interested readers can find a more detailed overview in any modern introductory undergraduate geology book. The crust-mantle system consists of a thin layer (typically at most a few 10's of kms thick) of solid material (the crust) floating on top of a liquid layer underlying it (the mantle) (Figure 12.1). We will return to just how viscous the mantle is in a moment, but for the moment it suffices to think of it as a very viscous fluid (e.g. like refrigerated honey on steroids rather than water). This crust is not one continuous piece covering the whole planet, but a large number of interconnecting pieces (i.e. the plates in plate tectonics) that are floating around and interacting with each other at their horizontal boundaries. The plates that make up the land surface of the planet (called terrestrial or continental crust) comprise lower density rocks than the plates that underlie the oceans (oceanic crust). Both crust types are less dense than the mantle, which is why they both float on the mantle. This will be crucial later. Because of their lower density the terrestrial plates float higher in the mantle than oceanic crust. This is why the upper surface of the terrestrial plates is above sea level. At their horizontal boundaries the plates can be moving apart creating rift valleys or regions of normal faulting, on land, and spreading centres with undersea volcanic activity in the oceans. If the plates are converging, then mountains and subduction zones (where one plate slides under another, typically oceanic crust underneath continental crust because of the density differences) are created. Finally there is the possibility that the plates are moving laterally relative to each other, creating active strike-slip fault zones (e.g. the San Andreas fault in California).

The drivers of the plate movement are the convection currents within the mantle where hot liquid rises (upwelling) and cool liquid falls (downwelling). Each pair of rising and falling currents is called a convection cell (or in other fields a Hadley or Rayleigh-Benard cell). The schematic in Figure 12.1 suggests a convection schematic that is rather more neat and tidy than the current understanding of mantle dynamics which involves multiple internal layers (e.g. the mantle is demarcated into two regions, upper and lower mantle, by a change in seismic velocity at a depth of 660 km), isolated plumes and superplumes, and geochemical separation as a result of localised internal heat generation and density differences (Bercovici, 2003; Ogawa, 2008). It is believed that the driver for these convection currents is radioactive decay within the mantle and the core underlying the mantle. At the top of the convection cell the flow direction is predominantly horizontal, and it is the friction between the bottom of the crust and the horizontal mantle flow that drives the horizontal movement of the plates. Recent work also suggests that downwelling actively drags the crust

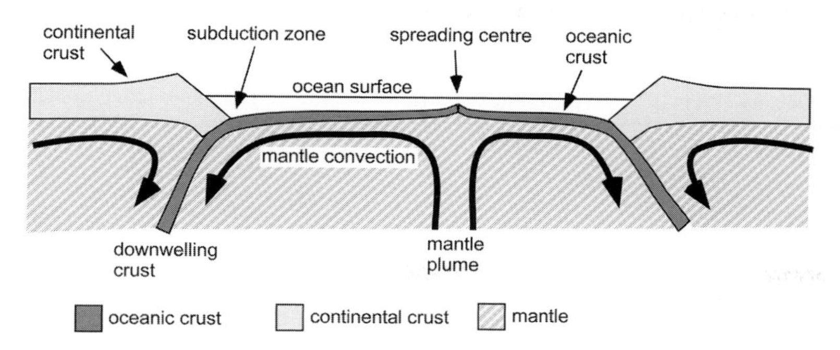

FIGURE 12.1: Schematic of mantle convection cells and crustal response. Note (1) the potential existence of mountain ranges at the subduction zone, (2) the increasing thickness of the oceanic crust with distance away from the spreading centre and (3) the relative thicknesses of the oceanic and continental crust.

down into the mantle in subduction zones (Conrad and Lithgow-Bertelloni, 2002; Mitrovica and Forte, 2004). These mantle convection velocities are typically less than 100 mm/year (Montelli et al., 2004). Some tectonic uplift is generated over the regions of upwelling (e.g. Yellowstone National Park) to balance the forces resulting from the change in the direction of the momentum of the upwelling convection current at the top of the plume, though whether this uplifting is ongoing (i.e. the topography is continuing to rise or fall) will depend on whether there is a force balance between the extra weight of the uplifted crust and mantle (acting downward), and the momentum change (acting upward).

Finally there is a slight complexity which we will not address here. This is that when the mantle is loaded, two processes happen. As mentioned above, over the long term the mantle flows viscously until equilibrium occurs. However, in the first few years of the loading the dominant cause of deflection is the elastic compression of the mantle fluid. It's only after a few years that the viscous flow has become sufficient that it dominates the elastic response. Since we are interested in landform evolution over thousands to millions of years, we will ignore this elastic response.

12.2 Isostasy and Isostatic Compensation

12.2.1 The Isostatic Compensation Principle

The most fundamental concept in tectonic-landform interactions is that of isostatic compensation. The basic idea of this is that the crust (both marine and terrestrial) floats on the mantle below (Figure 12.2) much as an ice cube floats in water. If we consider a unit area of crust of thickness d with density ρ_{crust} floating in a mantle with density ρ_{mantle} and the pressure in the mantle is hydrostatic (so the pressure is $p = \rho_{\text{mantle}} g h$, where g is the acceleration due

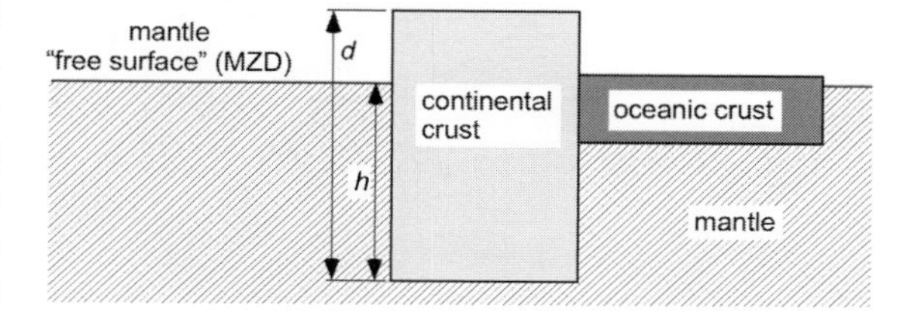

FIGURE 12.2: Schematic of isostatic compensation for continental and oceanic crust. Note that (1) the oceanic crust is thinner than continental crust and (2) the proportion of the oceanic crust above the MZD is less than continental crust because the oceanic crust is higher density.

to gravity, and h is the depth below the surface of the mantle), then by a force balance between the weight of the crust and the pressure on the bottom surface, the height of the crust above the mantle surface is

$$(d - h) = d\left(1 - \frac{\rho_{\text{crust}}}{\rho_{\text{mantle}}}\right) \tag{12.1}$$

so that the proportion of the crustal thickness that is above the mantle ($(d-h)/d$) is given by the difference between the density of the mantle and crustal density. When the crust obeys Equation (12.1) it is referred to as being in isostatic equilibrium or simply 'equilibrated', and h is called the 'compensation depth'. Landforms do not instantly achieve Equation (12.1) because of the high viscosity of the mantle, and prior to equilibrium the landforms will tend to evolve toward equilibration with some response time, as will be discussed in following sections.

Figure 12.2 is simplistic in three ways:

- The figure shows a mantle free surface against which height is measured. Since the entire planet is covered by crust, then a free surface does not actually exist, so it is not possible to measure heights against it. However, as a conceptualisation, this figure suffices. It is convenient below to refer to an equivalent of the mantle free surface

TABLE 12.1 Typical crustal properties

	Density (kg/m^3)	Typical thickness (km)	Heights relative to MZD (km)
Upper mantle	3,200[2]		
Terrestrial/continental crust	2,600–2,800[2]	15–75[1,2]	7.1–14.1
Marine/oceanic crust	2,800–3,000[2]	5–10[1,2,3,4] isolated locations to 35 km[4]	0.6–2.4

[1]Hammer et al. (2013); Klepeis et al. (2003), [2]Rogers (2007). [3]Canales et al. (2002), [4]Sallares and Calahorrano (2007).

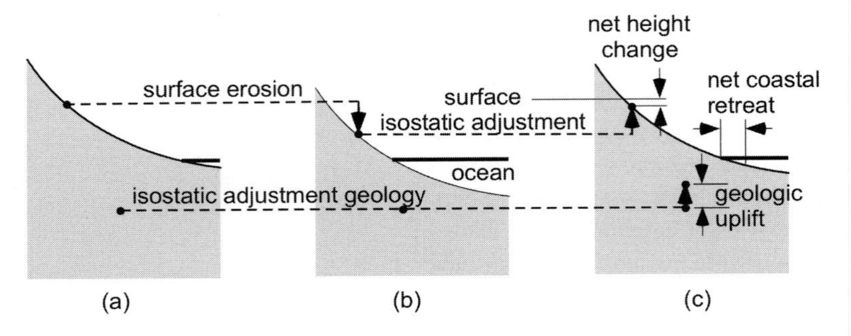

FIGURE 12.3: Schematic showing the net effect of the combination of erosion and isostatic compensation. (a) The original catchment showing the base level provided by the ocean height, (b) the change in land elevations if the crust is eroded instantaneously and (c) the adjustment provided by isostatic compensation over the long term. Note that in (c) not only is the net elevation lower by the amount shown, but also if the ocean level has not changed, then the base level provided by the ocean has retreated inland.

height and here we will label it the 'mantle zero datum' (MZD) as an analogy to the zero pressure datum provided by the free surface of a fluid (e.g. water in a lake).

- The figure does not consider mass continuity within the mantle. If the crust mass sinks into the mantle due to loading, then the displaced mantle material must move elsewhere. At equilibrium the mantle everywhere outside the location of the load must rise to compensate (and the MZD must rise) even if only by a small amount.
- Finally, as we will see below, it may take some time for this equilibrium to be attained, and in the meantime the displaced mantle will travel out in a slowly dissipating wave from the load, which may create wavelike uplift transients in the MZD (Lambeck and Johnston, 1998).

As noted above, the entire surface of the planet is covered by crust and thus there is no free surface, so the height in Equation (12.1) is the height above the MZD. Table 12.1 lists approximate densities for mantle and crust, typical thicknesses for continental and marine crusts and heights above the MZD to which the surface of crust can rise.

The heights above MZD in Table 12.1 are an approximation because they ignore the weight of the ocean water on the oceanic crust (so the marine crust will not rise quite as high as Table 12.1 suggests), but this calculation shows the difference in the maximum height of the terrestrial and marine crust. These depths are approximately consistent with the difference in the maximum depth of the ocean (Mariana Trench, 10,971 m) and the land (Mt Everest 8,848 m), and the average depths of the ocean (4,000 m) and the height of the land (840 m). The table also explains why marine crust underlies the ocean: it is more dense and thinner, and so forms the low spots on the surface of the crust, which is where the water collects.

Equation (12.1) is important in determining how the terrestrial crust responds to erosion and deposition of material. From this equation and Table 12.1, 1.0 m of erosion (Figure 12.3) will yield an isostatic rise in the crust of 0.76 m so that the net reduction in the elevation about the MZD is 0.24 m. This calculation ignores many important factors that will be discussed below but illustrates the hydrostatic balance that the terrestrial crust exhibits with the underlying mantle. The important factors that are missing from this calculation are the following:

1. The mantle is extremely viscous, and while in the example above the hydrostatic balance produces a 0.76 m isostatic uplift, this uplift may take tens of

thousands of years to reach equilibrium. For instance, regions that were covered in ice kilometres thick at the end of the last ice age (about 10,000 years ago) are still uplifting as a result of the isostatic adjustment to the loss of ice (Section 12.2.3).
2. The stiffness of the terrestrial crust laterally (i.e. east and north) means that any localised reduction in the load of the crust due to erosion will be spread over a larger area than the area of erosion so local adjustments will typically be less than predicted by Equation (12.1) but spread over a larger area.
3. The lateral stiffness of the plates combined with the buoyancy force means that there are some subtle far field effects from the local isostatic readjustment.

We will now discuss the details of these additional complicating factors. These factors form the core of models collectively called Glacial Isostatic Adjustment (GIA) models (e.g. ICE-3G, -4G, -5G; Tushingham and Peltier, 1991; Peltier, 1999, 2004). As their name suggests, they have been developed to understand crustal movements since the end of the last ice age and to reconstruct past ice histories from existing crustal movements. Insights from them provide a deeper understanding of crust-mantle interactions. At the simplest level these models couple the motion of a thin elastic crustal plate floating on a viscous mantle (Turcotte and Schubert, 2002). They model (1) the response of the crust to the removal of ice (up to 4 km thick) from the northern extremes of North American, Europe and Russia, (2) the extra loading on the crust of the resulting global rise in sea levels of about 130–135 m over the last 21,000 years (the end of the last glacial maximum, Lambeck et al., 2014) and (3) the (mostly) coastal retreat due to the combination of isostatic adjustment and sea level rise.

12.2.2 Spatial Linkages

The first of the complicating factors we will discuss is the lateral stiffness of the crust, which means that vertical deformations do not just occur locally near the load but also at a distance away from the load. A common representation is to model the crust as a thin elastic plate floating on a viscous fluid(s). The earliest models assumed that the mantle was uniform throughout with one viscosity. Recent models use multiple layers with depth and temperature-dependent variation in viscosity (e.g. Figure 1.12 in Rogers, 2007). The layer structure has been inferred from (1) seismic data, (2) better fits provided by the multilayer mantle models within GIA models to spatial patterns of uplift and subsidence around the planet and (3) data from satellite gravity missions (e.g. GRACE).

However, there is explanatory benefit from considering a simple single-layer mantle model. To illustrate the effect of the lateral stiffness consider a two-dimensional problem with a local out-of-plane (i.e. at right angles to the crust) load (Figure 12.4a), but ignore for the moment the buoyancy forces resulting from floating on the mantle. This problem can be a model for a volcano supported by the crust. If we assume that the crust is the same thickness and elasticity everywhere (Li et al., 2004), then

$$w = \frac{PL^3}{192\,EI} \tag{12.2}$$

where w is the deflection relative to the unloaded case, E is the elasticity of the plate, P is the point load, L is the distance between supports and I is the moment of inertia of the plate. For a unit width plate the moment of inertia is $I = T^3/12$ where T is the thickness of the crust. From Equation (12.2) it can be seen that the maximum deflection is reduced by either a stiffer crust (E high) or a thicker crust (I high). A doubling of the thickness of the crust reduces the deflection by a factor of 8.

The incorporation of the buoyancy force to this problem means that instead of the point load being supported at the edges of the plate as in Figure 12.4a the load is supported by the mantle nearer the load (Figure 12.4b). A good analogy for how the loads are distributed is obtained by noting that the buoyancy force is linear with deflection below the MZD so the buoyancy force can be replaced by a linear spring. The load on the spring is linearly related to the deflection of the spring, so that a classic formulation for this looks like Figure 12.4c. In the case of an infinitely flexible crust the entire load is taken by the spring directly under the load (this is the simple isostatic problem discussed at the start of this section). As the crust becomes stiffer, more of the load is carried by the springs to the left and right. The general two-dimensional solution for the deflection is (Ventsel and Krauthammer, 2001)

$$\frac{\partial^4 w}{\partial x^4} + 2\frac{\partial^4 w}{\partial x^2 y^2} + \frac{\partial^4 w}{\partial y^4} = \frac{1}{D}\left[P(x,y) - q_a\right] \tag{12.3}$$

where P is the vertical load per unit plan area, q_a is the elastic reaction force of the mantle per unit plan area $q_a = kw$, $k = \rho_{\mathrm{mantle}} g$ and D is the flexural rigidity of the plate

$$D = {}^{\mathrm{ET}^3}\Big/_{12(1-\nu^2)} \tag{12.4}$$

and ν is Poisson's ratio (typically in the range 0.2–0.3 for uncracked rock; Zhang and Bentley, 2005; Gercek, 2007).

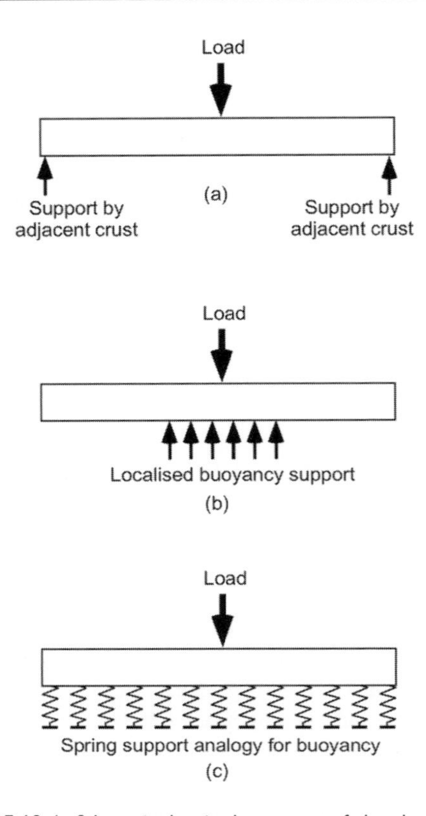

FIGURE 12.4: Schematic showing how an out-of-plane load can be supported by the crust in two end member conditions, and a third more realistic support condition: (a) if the crust is supported only by the end of the crust and there is no buoyancy, (b) if the crust is very flexible, then all the load is supported by buoyancy in the locality of the applied load and (c) a spring support condition for an elastic plate where the springs represent the support provided by buoyancy.

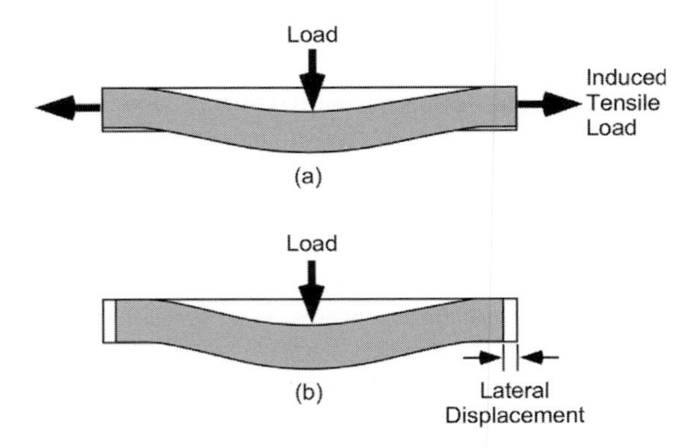

FIGURE 12.5: Two lateral constraint conditions for the elastic plate representing end members for the lateral movement of the crust under loading. The unshaded region shows the plate before loading, and the shaded region shows the plate after loading. (a) The case of an infinite plate where the out-of-plane load induces both bending in the plate as well as tensile forces that results from the plate being constraining from moving inward. Note that that plate is slightly thinner after loading because it is stretched over a slightly longer length than before loading, (b) The case where the ends of the plate can move laterally freely so that no tensile stresses are induced in the plate, and consequentially the deflection of the plate is slightly larger than the constrained case. In comparison with (a), in case (b) the plate does not thin after loading.

The elastic reaction force q_a is a good approximation to the buoyancy force if the density of the upper mantle is constant with depth.

Turcotte and Schubert (2002) consider a number of variants of the formulation of Equation (12.3). They are (1) where the deflected region is filled with water (e.g. the crust is underwater both before and after the load is applied) and (2) where the deflected area is filled with crustal material (e.g. where the deflected area is filled with eroded sediment and any uplifted area is eroded back flat). In the first case the new loading is $P_0 = P + \rho_{\text{water}} g w$ while for the latter case the elastic reaction force is $P_0 = P + \rho_{\text{deposited}} g w$ where $\rho_{\text{deposited}}$ is the density of the deposited material, noting that sedimentary rocks typically have a lower density than igneous and metamorphic rocks, so the deposited density is likely to be lower than the crust as a whole.

Equation (12.3) is an infinite plate solution and is derived assuming an elastic plate that is infinite in both horizontal directions and thus implicitly assumes that the elastic plate is horizontally constrained from shrinking laterally as the out-of-plane load is applied. This constraint means that a tensile load is induced in the plate with increasing load (Figure 12.5). As an approximate analogy, consider a high-wire walker on a cable between two trees. The load of the high-wire walker induces a tensile load in the cable. This tensile force is a direct response to the deflection of the wire. However, if the lateral constraint on cable fails (i.e. the wire breaks), then the tensile force and stress will be released, and the high-wire walker will fall on the ground (rather embarrassingly). This horizontal constraint results in the appearance of Poisson ratio (which describes the shrinkage of the elastic material at right angles to the tensile load in the plate) in the denominator of Equation (12.4), and the tensile forces induced by lateral constraint make the plate slightly more rigid than it would otherwise be (due to the high-wire walker effect). If this lateral constraint is zero, then the rigidity is

$$D = {}^{\mathrm{ET}^3}\!\big/_{12} \tag{12.5}$$

which is the rigidity used to derive the deflection in Equation (12.3). The key implication of this is that the lateral constraint on movement of the plate will reduce the magnitude of the deflection slightly (typical values for ν of 0.2–0.3 suggest by about 5–10%). This lateral constraint is provided by (1) subduction zones at the plate boundary and (2) frictional resistance to horizontal movement of the crust provided by the mantle. The lateral constraint is also less if the crust is vertically cracked. Most studies that use Equation (12.3) assume perfect lateral constraint and use Equation (12.4). The small difference between Equations (12.4) and (12.5) can reasonably be ignored in the field, but the differences may be important if using Equation (12.3) to validate the accuracy of computer models, where lateral constraints may be modelled explicitly, e.g. through the drag force of the mantle on the overlying crust, downwelling crust at subduction zones.

Finally it can be shown that the isostatic compensation equation (Equation (12.1)) can be derived from Equation (12.3). If the rigidity of the plate is zero ($D = 0$), then Equation (12.3) can be written

$$P(x, y) = q_a = (\rho_{\text{mantle}} - \rho_{\text{crust}})gw \tag{12.6}$$

and if the load P is the weight of the crust $P = \rho_{\text{crust}}gd$ then $w = \frac{\rho_{\text{crust}}d}{(\rho_{\text{mantle}} - \rho_{\text{crust}})}$ which yields Equation (12.1).

12.2.2.1 One-Dimensional Example: Line Loading

To demonstrate some of the characteristics of the physics of Equation (12.3) we can solve some simple one-dimensional problems. More complex spatially distributed two-dimensional solutions can be found in Turcotte and Schubert (2002) and elsewhere, but the underlying concepts and consequences are similar. For one dimension, Equation (12.3) simplifies to

$$\frac{\partial^4 w}{\partial x^4} = \frac{1}{D}[P(x) - q_a] \tag{12.7}$$

For the case where the load is independent of x (the crust thickness and density is the same everywhere), then for an infinite plate (i.e. the boundary conditions are at $\pm$ infinity), then the deflection w is also independent of x, and $P = q_a$. This is the isostatic equilibrium condition where the buoyancy force q_a equals the weight of the crust P. Equations (12.3) and (12.7) are linear in deflection and load, so superposition can be used to separate the various loads on the crust, calculate the deflections caused by each of the loads, and then sum up the deflections for the total deflection. For the examples below we will only consider the effect of an additional load over and above the weight of the crust and ocean, and consider only the incremental deflection as a result of this additional load.

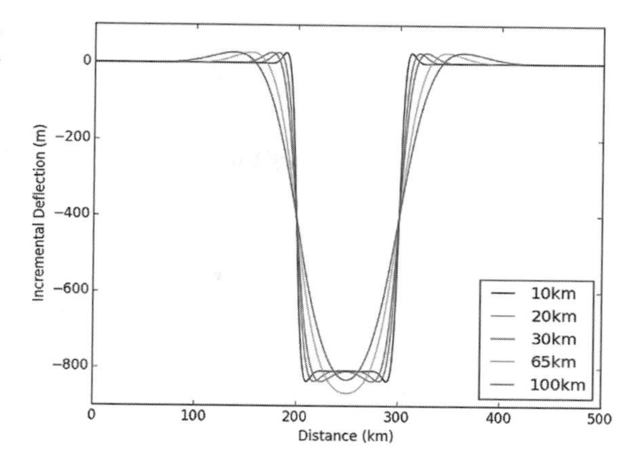

FIGURE 12.6: The elastic deformation of the crust for a uniformly applied distributed load between 200 km and 300 km. The curves are for different thicknesses of crust varying from 10 km to 100 km.

Figure 12.6 shows the deflection of the crust under a distributed load. The distributed load is assumed to be over 100 km wide in the x direction. The five different deflection curves are for a range of crustal thicknesses. The nominal thickness is 30 km, with the thinnest thickness being 10 km and the thickest 100 km, which covers the range of crustal thicknesses observed. The elasticity of the crust is 5×10^{10} Pa, which is a typical value for uncracked granite. The load is 10 km^3/km length with specific gravity of 2.65, which is about the line loading of the Hawaiian island chain (note: making this example a parallel to Hawaii assumes that the plate is continuous under Hawaii, which recent studies suggest may not be the case; Klein, 2016). Several features in Figure 12.6 are of note. The first is the existence of close to constant deflection under the load. This value of deflection (about 800 m in the figure) is the isostatic compensation depth for this problem. Outside the loaded area there is a small upward deflection (about 25 m), which is called the forebulge. This forebulge is accompanied by an equal and opposite increase in deflection just under the edge of the load. As the thickness of the crust increases, its rigidity increases, the deflection curve is smoother, and the forebulge moves further out from the edge of the loading. Also the absolute maximum deflection is not when the rigidity of the crust is lowest (as might be expected), but at a midrange rigidity where the distance to the forebulge is equal to the width of the loading (called the 'flexural parameter'). Figure 12.7 shows the same loading and crustal thickness cases when the same load is applied as a point in x (called a point load), not spread over 100 km. In this case the maximum deflection under the load is much higher than the distributed case (even though the

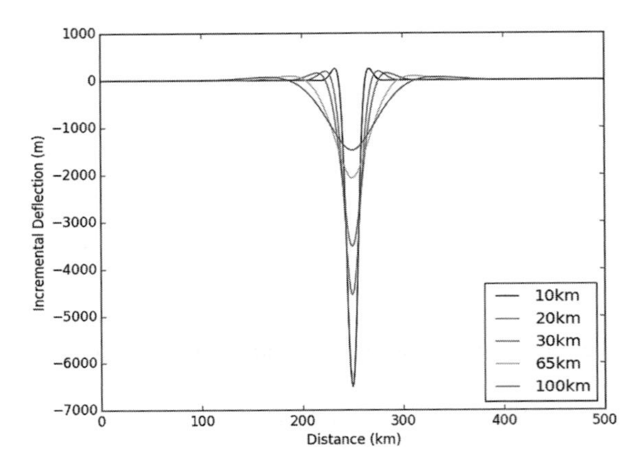

FIGURE 12.7: The elastic deformation of the crust for a point load at 250 km. The total load applied is the same as applied in Figure 12.6, only it is applied at a point rather than being distributed over 100 km. The curves are for different thicknesses of crust varying from 10 km to 100 km.

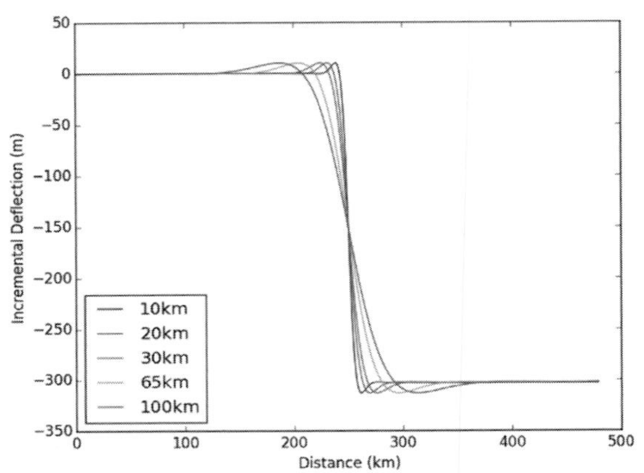

FIGURE 12.8: The deflection of the crust with an ice load of 1 km thick applied at 250 km and greater. The different curves are different thicknesses of crust varying from 10 km to 100 km, with the nominal thickness of 30 km as for the other examples in this chapter.

total loading is exactly the same) and as the rigidity increases the maximum deflection decreases (and is significantly higher than Figure 12.6), but the region over which the deflections occur becomes much larger. As the rigidity increases, the height of the forebulge decreases, the distance to the forebulge increases and the maximum deflection decreases. An analytical solution for a point loading case follows below which explains the results in Figure 12.7.

Equation (12.7) can be solved analytically, and the analytic solution yields insights into the form of the deflections in Figures 12.6 and 12.7. The deflection is (Turcotte and Schubert, 2002)

$$w = \frac{P\alpha^3}{8D} e^{-x/\alpha} \left(\cos \frac{x}{\alpha} + \sin \frac{x}{\alpha} \right)$$
$$\alpha = \left(\frac{4D}{\rho_{\text{mantle}} g} \right)^{1/4} \tag{12.8}$$

where P is the line load at location $x = 0$, and α is the 'flexural parameter' and is the length scale that determines the scaling in the x direction of the deflections. If α is larger, then the deflections spread further. The forebulge seen in Figure 12.7 is at location $x_{\text{fb}} = \pi\alpha$ and the height of the forebulge is $w_{\text{fb}} = -0.0432 w_0$. The maximum deflection that occurs directly under the load (setting $x = 0$ in Equation (12.8)) is

$$w_0 = \frac{P\alpha^3}{8D} \tag{12.9}$$

Equations (12.8) and (12.9) show that the shape of the deflection is always the same and that the only change is the scale in the x direction, given by α, and the magnitude of the deflection, determined by the load P and the plate rigidity D. This is consistent with Figures 12.6 and 12.7, which show the forebulge moving outwards as the rigidity D increases. The ratio of the height of the forebulge to the deflection under the load is fixed. For typical properties (i.e. the nominal parameter set in Figure 12.7) $\alpha \approx 33$ and the distance to the forebulge is about 100 km from the point loading.

While the height of the forebulge is quite small relative to the primary deflection under the load in situations where slopes in the region of the forebulge are very low, it is still possible for it to be important. Driscoll and Kerner (1994) suggest that drainage patterns on the coast near the mouth of the Amazon result from the forebulge from the loading of the Amazon delta. Coastal streams in the area drain inland rather than to the coast, and, given the low slopes in the area, this is consistent with the forebulge forming a barrier to seaward drainage. Thus, even though small, it may be important to model this forebulge feedback between deposition and tectonic response in landform evolution models.

12.2.2.2 One-Dimensional Example: Ice Sheet Boundary

Another interesting example is that of the deflection of the crust as a result of loading from an ice sheet, and estimating the deflections near the edge of the ice sheet. Figure 12.8 shows the case of 1,000 m of ice on a 10–100 km thick crust with a Young's modulus of 5×10^{10} Pa (the same parameters used in the previous example). The deflection of the crust is 300 m, and the height of the forebulge is about 11.2 m (or about 3% of the fully compensated

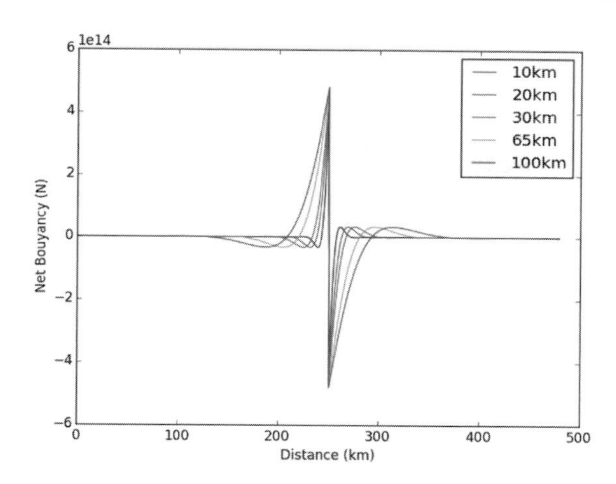

FIGURE 12.9: The net force balance of the downward weight of the ice minus the upward buoyancy force, showing the net forces applied by the elastic deformation of the crust. This is for the same case as in Figure 12.8.

deflection under the ice). For a crust that is 100 km thick, the forebulge is about 60 km in front of the ice front. The slight additional deflection just behind the front of the ice sheet is exactly symmetric with the forebulge, as was the case of the distributed load in Figure 12.6.

Figure 12.9 shows the vertical force balance for the crust relative to isostatic equilibrium at that location. The ice covers the right-hand side of the figure from 250 km to 500 km, and 250 km is the location of the ice front. In the 50 km in front of the ice front (200 km to 250 km) there is net upward force, which reflects that the crust is deflected but does not have any load applied to it (Figure 12.8), so there is a net buoyancy uplift that is balanced by the downward force of the elastic deflection of the crust. Moving further away from the front into forebulge, there is a net downward force, which indicates that the crust has been uplifted by the elastic forces of the crust (i.e. the forebulge in Figure 12.8). If on the other hand in the region immediately underneath and behind the ice front, there is a large net downward force, because the crust just behind the ice front has not deflected enough to balance the ice load and is being held up by the crust immediately adjacent and in front of the ice front. Moving further away from the front under the ice, there is a negative equivalent to the forebulge. Note that the net effect of these force imbalances is to create a clockwise moment, which results in the crust rotating clockwise at the ice front, which is consistent with the deflections in Figure 12.8. In Figure 12.9 for a vertical force balance the area above the zero line must exactly equal the area below, which is ensured by the symmetry in this example, and this is consistent with the symmetric forebulges in Figure 12.8.

12.2.3 Time-Varying Processes

The second complicating factor impacting on isostatic compensation is the rate at which the compensation occurs. The elastic plate response in Equation (12.3) above does not occur instantaneously, because for this to occur the liquid supporting the crust would need to be inviscid (i.e. zero viscosity) and for it to move instantaneously. Field data show a long-term post-earthquake adjustment of the gravity field to single earthquakes. These post-event changes are attributed to redistribution of mantle fluid (Han et al., 2008).

The mantle has very high viscosity ($\sim 10^{20}$–10^{21} Pa·s; Peltier, 1999; Mitovica and Forte, 2004; Paulson et al., 2007; Ivins and Wolf, 2008; Klemann et al., 2008) so that the rate of adjustment of elevations to changes in crustal loading is of the order of thousands of years. For instance, areas of terrestrial crust near the poles are still rising at up to 20 mm/year in response to the end of the last glacial max 21,000 years ago (Peltier, 1999; Klemann et al., 2008). This elevation adjustment varies markedly with distance from past glaciation, with locations just outside the glaciation showing subsidence as the former ice-age forebulge declines (e.g. Sella et al., 2007). Figure 12.10 shows histories of relative sea level for the last 20 kyr for three locations in Canada and one in Sweden. In Canada the relative sea level combines the effect of (1) sea level rise due to ice melting and (2) crustal rise due to ice removal, but since they are both driven by the same crustal response, they give a good indication of response time (Peltier, 1998, Figure 36). The solid line is an exponential fit to the field data with response times (i.e. time when the response is $1/e$ the full response, the e-folding time) of 3.3–4.7 kyr (Table 12.2) with an average of 3.7–4 kyr. These numbers agree well with the relaxation times derived by direct application of the ICE-4G GIA model for Angerman River, Sweden, of 4.2 kyr (Peltier, 1998), 1.7–3.3 (McConnell, 1968; presented in Peltier, 1998) and other independent analyses of Mitrovica and Forte (2004).

The relevance of the exponential time scale fitting is that it potentially provides an alternative less complex model for viscous impacts on isostatic adjustments than direct use of GIA models. Moreover, there is considerable uncertainty in the parameterisation of GIA models, specifically mantle physics (e.g. the type of viscosity (Maxwell versus Newtonian), and the pattern of viscosity both with depth and globally within the mantle, and the absolute values of mantle viscosity), and geodynamics researchers have found that the parameters for the viscosity parameterisations in GIA models are poorly

TABLE 12.2 The e-folding response times for GIA

Site	Response time (kyr)		
	Peltier (1999)	Peltier (2004)	Mitrovica and Forte (2004)
Richmond Gulf, Canada	4.7	-	5.3 ± 1.3
James Bay, Canada	3.3	3.4	2.4 ±0.4
Hudson Bay, Canada	3.9	3.4	-
Angerman River, Sweden	4.7	4.4	4.9 ±0.9
Average	4.1 (4.0 excl. Richmond)	3.7	4.2

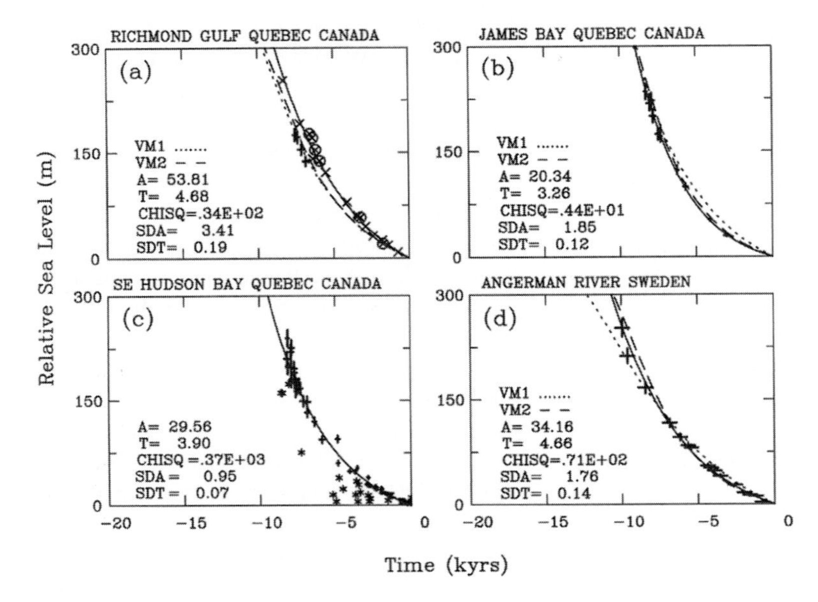

FIGURE 12.10: Eustatic sea level rise for four sites in Canada showing the initially fast response from the ice unloading, which gradually decreases, approximately exponentially, with time (from Peltier, 1999). The time axis is years before present.

identifiable (e.g. Peltier, 1998; Paulson et al., 2007). Finally, loads that cover a larger area mobilise deeper mantle material during the adjustment and have a different response time to more localised loads (e.g. Figures 2 and 4 in Peltier, 2004). Thus even with state-of-the-art GIA models there are uncertainties about the physical mechanisms driving mantle viscosity, and the inferred isostatic response times vary by a factor of 2 or more. Consequently, in the absence of better information (or a lack of desire to couple the landform evolution model directly to a GIA model), an alternative and conceptually simpler approach would seem to be to assume an exponentially declining isostatic adjustment response with a specified response time

$$z(t) = z_\infty e^{-(t/t_r)} \qquad (12.10)$$

where $z(t)$ is the amount of isostatic elevation adjustment at a given time t after the load is applied to the crust, z_∞ is the isostatic elevation adjustment that occurs at equilibrium (or, equivalently, would occur instantly if the mantle was inviscid) and t_r is the response time of the isostatic response, discussed above. This response time is likely to vary only slightly globally because it is primarily driven by mantle viscosity and does not interact with the ocean and continental crust thickness so that it is independent of the elastic plate spatial response (Equation (12.3)). Accordingly, the response time t_r and the equilibrium isostatic adjustment z_∞ are independent. Comparing a coupled lithosphere-mantle model with uniform lithospheric thickness with one with realistic global variation in thickness, Zhong et al. (2003) showed that the response time t_r to glacial rebound is relatively insensitive to lithospheric thickness if mantle viscosity is unchanged (though the magnitude of rebound z_∞ was sensitive because of the changing stiffness of the elastic plate). Moreover, while Peltier (1998, 1999) does not directly address the question of global variations in response times, his data, which include sites in both the northern and southern hemispheres and sites inside and outside glaciation in the last glacial maximum, are consistent with there being little variation (a maximum of a factor of 2) in response times globally. Thus to first order a starting point for the response time in Equation (12.10) is the response time in Table 12.2.

The final complicating factor is that horizontal flow of mantle material is required to move mantle material from areas that are subsiding to areas that are uplifting. This flow induces friction stress on the bottom of the crustal plates and results in horizontal motion in the plates. This motion is in addition to any mantle plume–induced motions (James and Morgan, 1990). Klemann et al. (2008) and Argus and Peltier (2010) suggest rates of GIA-induced horizontal movement of the order of 1 mm/year and that this rate of movement is typically less than 10% of the total observed horizontal motion of plates globally. Argus and Peltier (2010) highlight the difficulty in using GPS data in GIA models to infer horizontal movement rates. This suggests that, given the parameter uncertainty bounds in the other parts of the GIA model, these induced horizontal flows are second order and can be ignored.

12.3 Mountain Building

The previous section talked about the movement of the crust under loading but didn't address the question of evolution of the crust itself. The main topic of interest is mountain building as a result of convergence of two sections (typically two plates) of crust toward each other (Figure 12.11). As a result of continuity of mass the crust must thicken and mountains are formed. Isostasy is still active, but because of the dynamics of the crustal thickening the crust may not be equilibrated. As a result of isostasy there is a thickening of the crust under the mountains (called a 'root'), but the dynamics of this process cannot be described by the methods in the previous section.

The main issue is determining the dynamics of the crustal thickening that occurs at the collision point. For the case where one plate subducts under another the motion of the underlying plate is normally modelled with an elastic-plastic model where the plate softens as the temperature increases with depth. This is the realm of geodynamics and will not be discussed further here.

The subject of this section is what happens with the material that does not get subducted down into the mantle. Figure 12.11 shows a conceptualisation of two end-member cases where the converging material piles up to create a mountain range. In this conceptualisation the left-hand crust is moving toward a stationary crust on the right-hand side. A fixed proportion of the crust is subducted and forms the root (not shown in the figure), and the remainder piles up on the surface. It is how this material piles up that defines the differences between the models. The main method is critical wedge/taper theory (Dahlen, 1990), where the material is assumed to be a Coulomb material where movement is instantaneous when stresses exceed the yield stress (e.g. a granular material like sand).

Willet et al. (2001) summarise the concepts underpinning the competing approaches:

- *Frontal accretion* is simply the piling up of material against a resisting wall, which is the stationary crust on the right-hand side of Figure 12.11. Over time as new material is accreted at the mountain range front, two things occur: (1) the mountains get higher and steeper as more material is contained within the horizontal width of the mountain range (L in the x direction in Figure 12.11) and (2) the width of the range gets wider so that L increases and the range front moves to the left.
  - In case (1) any given vertical cross section through the mountain range (the vertical cross sections in the figure) over time is (a) compressed (i.e. becomes thinner in the x direction) horizontally, (b) to maintain mass continuity the same cross section gets higher and (c) the cross section itself moves from left to right.
  - In case (2) and if the prevailing surface slopes are unchanged, then the (a) left-hand side of the mountain surface will rise parallel to existing surface, (b) the mountain divide will rise and move to the left and (c) the point L will move to the left. As in case (1) the vertical cross section will be compressed, higher and moved to the right, but the effect will be less dramatic than in case (1).
  - The dotted lines with arrows show typical paths for a particle of crust. Once it reaches the surface, it is removed by erosion. At the land surface of the mountain range the surface velocity has both horizontal and vertical components, with the surface moving from left to right, as well as upwards.

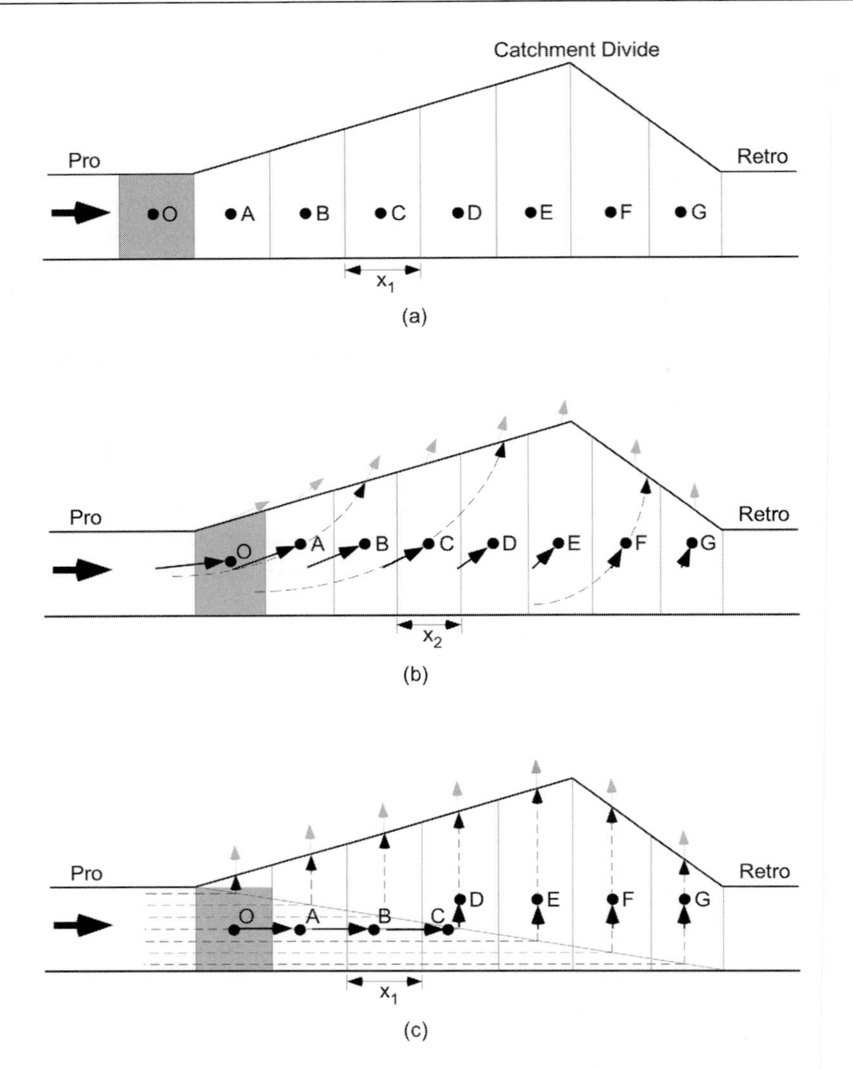

FIGURE 12.11: The two mountain range building conceptualisations of Willet et al. (2001). (a) The initial condition. The dots are labelled points, and the columns finite volumes used to show movements in the bottom two panels. The heavy arrow on the left-hand side is the direction of movement of the crust. (b) Frontal accretion, where material is pushed from the left into mountain range equally from the top to bottom of the crust (akin to a bulldozer blade). The black arrows show the movement of the labelled points from (a) to (b), and the light dotted lines show trajectories from beginning to end of rocks passing through each labelled point. The light grey arrows at the surface show the direction of the movement at the surface (note that for dynamic equilibrium where erosion equals uplift in the vertical direction, then the vertical component of these grey arrows are all equal). Column width x_2 is less than x_1 because the column is compressed in the horizontal direction and the ratio $x_2 : x_1$ can be calculated from volume continuity. (c) Underplating, where the material being pushed in from the left is preferentially pushed underneath the mountain range. All the arrows have the same meaning as in (b). Note that the columns are not compressed in the horizontal direction, so the surface movements do not have a horizontal component.

- *Underplating* is conceptually simpler than frontal accretion and is the case where the crust moves under the right-hand side and simply lifts the existing mountain wedge upwards. In this case the motion of the material as it reaches the surface is vertical and the motion of the surface of the mountain range is solely vertical with no horizontal motion. The dotted lines with arrows show typical paths, and what is important is that the particle moves horizontally in the underplating region before it finally moves vertically. The different paths for frontal accretion and underplating may have an impact of thermochronology dating techniques because the cooling profile of the rock will be different from frontal accretion where the pathway is always moving closer to the surface.

- The shape of wedges in the cross section is a direct result of the flat base below the horizontal crustal movement and the assumption of a Coulomb material. If the base is not flat, then the geometry is more complicated. The general form of the wedges is also well supported by sand box experiments, sand also being a Coulomb material.

Note that in both cases there was no interaction between crust and the mantle, so isostatic buoyancy processes play no part in this conceptualisation. Willet et al. (2001) coupled both models to a landform evolution model to identify unique geomorphic signatures to these processes so that they can be distinguished in the field. Their main finding was the resulting mountain ranges created were symmetric for underplating, but asymmetric for frontal accretion. In the latter case the divide of the mountain range was pushed from left to right, so the cross section conceptually looked like their schematic for the process (i.e. Figure 12.11b).

Whipple and Meade (2004) explored the effect of the coupling between erosion and accretion and allowed the width of the orogen to adjust to these processes. They found for steady state that an erosion increase (e.g. by climate changes) reduced the width of the orogen and increased the uplift rate. This increased uplift rate is required by the balance, at steady state, between erosion and uplift (see the discussions about slope-area analysis in Chapter 3). A consequence of this balance between erosion and uplift is that the rock uplift rate is relatively insensitive to the convergence rate and is most sensitive to the erosion rate. Roe et al. (2008) extended the underplating approach by postulating a precipitation variation with elevation, and leeward and windward sides of the range. They performed a scaling analysis to show the impact of the precipitation feedback on the equilibrium height, width and shape of the mountain range. Recent work has focussed on collecting experimental evidence for this erosion driver (e.g. Whipple, 2009).

Roe and Brandon (2011) extended this work by looking at the effect of different forms of rheology governing the way the crust flows in the convergence zone where the mountain range is being formed. They focussed on the relationship of width to height of the range, and found the range characteristics to be relatively insensitive to the type of rheology used. This is consistent with erosion, rather than convergence dynamics, being the primary driver of orogen characteristics.

12.4 Extensional Settings

The discussion of the previous section considered settings where the crust was in compression. In this section we will consider settings in tension. The main way that tensional strains can be accommodated is through faulting. Figure 12.12 shows how a single extensional fault and the crust interact. The subsidence and erosion in the region of the fault are a result of the stiffness of the elastic plate, which can be modelled with the thin plate techniques previously discussed, and far from the fault, elevations are unchanged from that predicted by isostasy. The failure surface is predicted by the Mohr-Coulomb failure surface assuming that the material is a Coulomb material (Olive et al., 2014).

When looking at a large area under tension, it has been found that thick crust tends to extend with a number of smaller offset faults, while thinner crust tends to break with a lesser number of higher offset faults (Buck, 1993; Lavier et al., 2000; Lavier and Buck, 2002) (Figure 12.12).

The angle of the fault is determined by the requirement that the two sides of the fault do not separate but slide relative to each other. The failure criterion is that the frictional shear strength of the failure plane per unit area in the horizontal direction (which is a function of the normal stress perpendicular to the fault) times the area of the fault is equal to the tension force in the crust (Figure 12.12). The angle that results in this strength being least is the angle of the fault. The shear strength of any surface is

$$\tau = \mu\sigma_n + \tau_c \tag{12.11}$$

where μ is the coefficient of friction of the fault surface, σ_n is the stress perpendicular to the fault surface and τ_c is the cohesive shear strength of the fault surface. For a friction coefficient of $\mu = 0.6$ the optimal angle to the horizontal of the fault is about 60° (Buck, 1993). Whether the fault rises to the left (as shown in Figure 12.12a and b) or to the right is a result of random chance and is likely driven by minor random variations in the rock mass that bias the fault in one direction or the other.

It should be clear from Figure 12.12 that the faulting leads to higher elevation on one side of the crust. Potentially this could form a mountain range. Buck (1993) calculated that the upwarping at the fault exposure could be as much as 1–2 km. Sachau et al. (2013) propose an additional way that faulting may result in mountain building. If the faulting does not fully penetrate the crust from top to bottom but only penetrates the top part of the crust while the hotter, softer, underlying crust undergoes

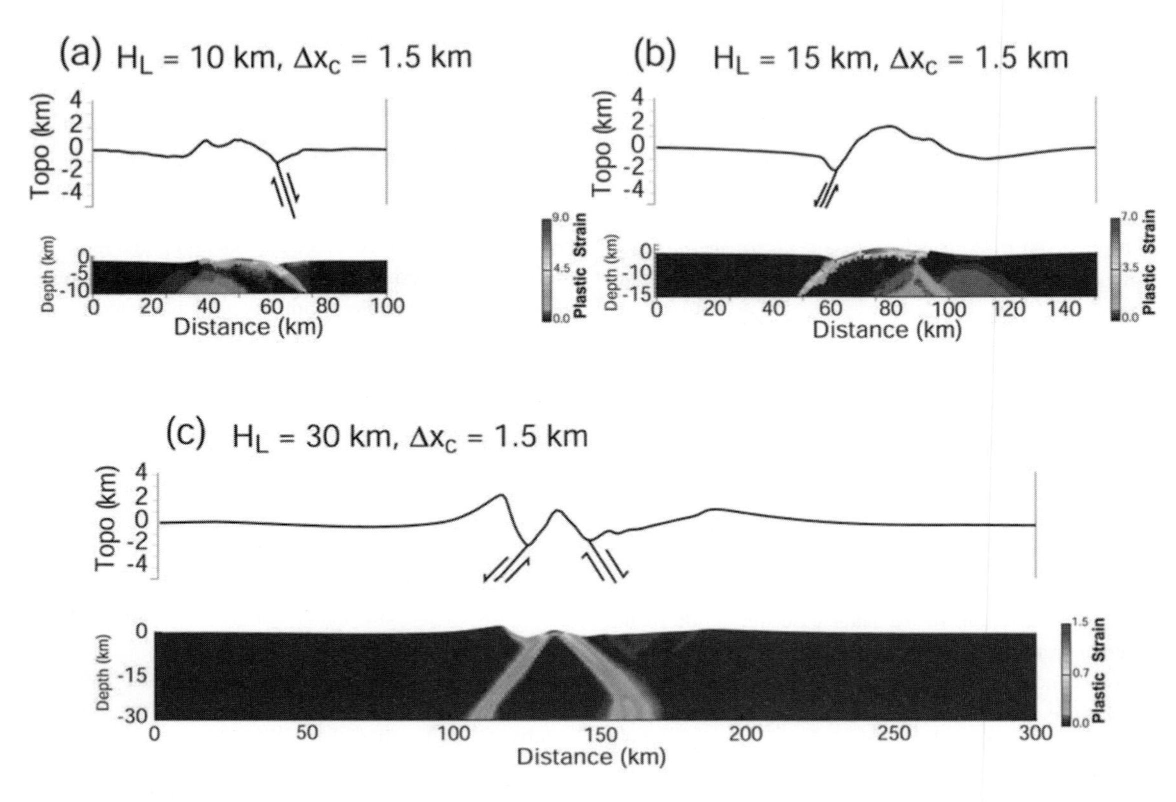

FIGURE 12.12: The form of elevation change resulting from tensional faulting in the crust as a function of thickness of the crust: (a) crustal thickness 10 km, (b) crustal thickness 15 km, (c) crustal thickness 30 km. (a) and (b) generate a single fault, while (c) for the thicker crust generates multiple faults and complex topography (from Lavier and Buck, 2002).

viscous flow, then the crustal plate will upwarp because of the parasitic moments induced in the crust by the tensile load that is no longer being applied through the centroid of the crust. Sachau proposes this as an explanation for crustal warping of 2,500 m and a gravity anomaly in the Alvertine rift system in Africa.

12.5 Renewal of Terrestrial Crust

If the terrestrial crust is always eroding and this eroded sediment is deposited in the ocean, an important question is how the continents still exist. Perhaps they should have all eroded away by now. Portenga et al. (2011) estimated global erosion rates using ^{10}Be. For drainage basins he estimated the mean erosion rate to be 218 m per million years. This suggests that a 30-km-thick continental crust will be eroded completely away in just 137 million years, which is significantly less than the measured age of continental rocks.

If erosion is thinning the continental crust and the rate of thinning is high enough to destroy the crust within geologic time, then there must be a process that is rejuvenating crust. For landscape evolution studies of less than a few million years this process is unlikely to be an important variable to model, but over longer time scales it is a consideration, though one that is poorly understood (O'Reilly et al., 2001; Bishop, 2007). A number of processes are discussed in the literature, and most evidence comes from geochemical analysis of rocks brought to the surface by igneous process, or by seismic studies. The possible processes include the following:

1. Buoyancy of mantle materials. Lighter melted rocks will float to the surface of the mantle as a part of a natural differentiation of materials. The specific gravity differences required for this process may be due to temperature differences or differences in geochemistry. However, if it is geochemistry, then there must be another parallel process that ensures that the lighter materials collect preferentially under the continental plates, otherwise there must be another process that maintains a difference in the densities between the mantle underlying oceanic and continental plates.

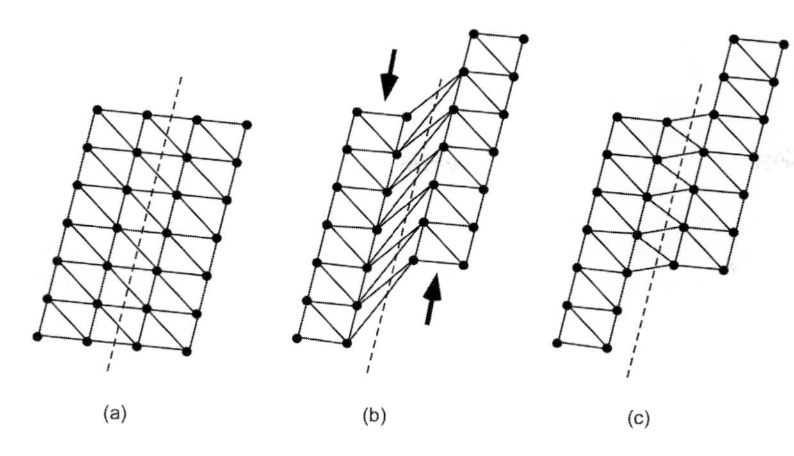

FIGURE 12.13: Schematic of the how a triangulation deforms across a strike-slip fault (a) the initial triangulation, (b) the deformed triangulation and (c) the retriangulated nodes. The dotted line shows the strike-slip fault. The arrows in (b) show the direction of movement on each side of the fault.

2. Heat flow. If there is a heat flow balance in the crust, then the solidification of mantle rocks always occurs when the crust reduces below an equilibrium thickness. As material erodes on the surface an equal amount of mantle material solidifies on the base of the continental crust. This will be the lighter mantle rocks as they will have floated to the mantle surface. This will not occur under the oceanic crust because material is not being eroded, so the insulating blanket of the lithosphere is not being thinned under the ocean. In fact, under the ocean with the deposition of eroded sediment, then the lithosphere will be thickening, and material may melt and detach from the oceanic crust base.
3. Preferential recycling of crustal material. O'Reilly et al. (2001) propose a mechanism of recycling where continental lithosphere is delaminated from the lithosphere and recycled back into the mantle, and then back into the lithosphere. Recent geochemical studies appear to support this hypothesis (Kelemen and Behn, 2016).

Along with many other questions about the behaviour of the mantle the mechanisms driving crustal rejuvenation appear to be unresolved.

12.6 Numerical Issues

The main feature of tectonic problems compared with other processes in this book is the possibility for features to move laterally (e.g. strike-slip faults, convergence, spreading) rather than movement being predominantly up and down. In this case a discretisation of the elevations that can move with these features is desirable to ensure that they are kept intact under lateral movement. In this case a TIN discretisation has significant advantages over a grid because at each time step the nodes at which the elevations are discretised can be moved with the lateral movement. For instance, points either side of a strike-slip fault can be moved relative to each other, and any features (e.g. catchment divides) will be kept intact (Figure 12.13). If this strike-slip behaviour is modelled with a fixed grid, then at each step there will be a smoothing of the terrain as the divide moves between grid points and is interpolated between the grid points and smoothed as a result. The reader might then legitimately ask why TINs are not used more frequently in landform evolution models. The reason is that TINs add complexity in a number of different ways, which may be unnecessary if lateral movement is not significant in the problem being studied. Specifically

1. The representation of the landform in TINs requires not just the easting and northing locations of the nodes in the DEM, but the geometry of each of the triangles joining these nodes (called the triangulation), and the geometry of how each of the triangles is joined to its adjacent triangles. As a result the calculation of flows across a TIN is more complicated than across a grid.
2. How the triangulation changes from timestep to timestep requires a degree of care, otherwise mass balance errors may enter the calculations. Normally the triangulation is done using a Delaunay triangulation algorithm (e.g. Sloan, 1987), which ensures that the triangles are close to equilateral in shape, which is optimal for numerical reasons. The area associated with each node is then determined using a Voronoi tessellation (e.g. Tucker et al., 2001b). When the triangles are long and skinny, which over time is inevitable for triangles across a strike-slip, then periodic

retriangulation is required (Figure 12.13). But if the nodes move relative to each other and the nodes are simply retriangulated, then there is no guarantee that the mass balance of the landform is maintained after retriangulation.

3. There is additional bookkeeping involved with the triangles and retriangulation. If the triangle's plan area changes area, this indicates lateral compression of the crust, so that the height of the elevations at the triangle corners must be increased to maintain the same mass within the triangle. But this will also change the mass in adjacent triangles because changing the elevation at a corner changes the triangles that also have this node as a corner. The Voronoi tessellation gives the area associated with each node, and the expansion and contraction of the nodal area in the Voronoi tessellation that occurs with node movement can be used to do this efficiently.
4. If new nodes are added, or old nodes removed, than mass balance checks need to be done before and after the addition/removal.

None of these issues are insurmountable but they do add to the complexity of the computer code. Accordingly many computer codes are gridded, which is less flexible, but these bookkeeping issues do not arise.

The REAR model (Melini et al., 2015) calculates the elastic response of the crust and the mantle to time- and space-varying ice load. This model does not model the viscous response of the mantle but only the short-term elastic response of the mantle. This elastic response has not been discussed in this chapter. Spada et al. (2012) used this approach to examine the tectonic effect of changes in ice thickness over the last decade in Greenland as measured by satellite instruments. To do longer term modelling (i.e. centennial to millennial time scales) requires one of the GIA models that simulate the viscous flow of the mantle.

Wickert (2015) provides a set of open source tools in Python, gFlex, for solving elastic plate flexure problem with isostasy. It uses the finite difference approximation to the flexure equation (Equation (12.3)) and allows the crustal rigidity to change laterally. The codes can be used standalone or integrated into the GRASS GIS package.

12.7 Further Reading

For a review of the research questions around the implications of the coupling of tectonics, erosion and climate the reader is referred to Bishop (2007). The classic text discussing the details of isostasy is Watts (2001). For greater depth on the elastic plate modelling of the equilibrium crust-mantle interaction see Turcotte and Schubert (2002). For the dynamics of the crust-mantle interaction the reader is best served by reading some of the work on glacial isostatic rebound modelling or GIA modelling. The details of these GIA models can be quite complex, but the modelling is also an area of active interest outside of earth sciences because of its time-varying impacts on satellite trajectories and in the field of geodesy (e.g. Han et al., 2008). Accordingly it is difficult to identify a single publication that neatly summaries the state-of-the-art that won't be out of date very quickly, but Lambert and Johnston (1998) provide a readable overview of the issues of GIA model formulation and impacts of glacial rebound. Gerya (2010) has a specialist discussion of mantle rheology and how to model various aspects of mantle geodynamics, and significantly extends the discussions here on mantle dynamics.

13 High-Slope Gravity Processes

This chapter is about processes that are primarily driven by gravity. These processes are typically also only active in steeper regions because they involve some form of strength failure in the hillslope. These processes are dominant on the steep hillslopes in the upper reaches of catchments. They may also be active in regions of the landscape's lowlands if localised geology is such that cliffs and scree slopes occur.

There is some overlap with previous chapters, particularly the chapter on soil flow and creep (Chapter 9) where slow-moving processes without soil or saprolite failure were discussed. In this chapter the distinguishing features of the processes are (1) the process begins only after failure of the strength of the material, the trigger mechanism and (2) the subsequent transport of the material is fast. These two features are linked because if there is a strength failure, then immediately after the failure there is a significant force imbalance (typically the force constraining the movement has disappeared as a result of the failure) and the soil/rock mass is rapidly accelerated by this force imbalance. While these trigger and transport processes are conceptually linked it is convenient below to treat them separately. In contrast the processes discussed in Chapter 9 do not involve a failure trigger mechanism, and consequently the dramatic and sudden force imbalance does not exist so that transport rates are much lower and less abrupt.

This chapter is organised into main parts. The first part discusses mechanisms that trigger failure of the hillslope (most of the processes in this chapter start on hillslopes rather than in the channels). The second part discusses the movement of this material downslope (both on the hillslope and in the valleys/channels) once this failure has occurred.

This chapter will not concern itself with deep-seated, multi-cubic-kilometre, landslides, which tend to be driven by very site-specific geology. This site-specificity makes them ill-suited for long-term evolution modelling approaches, and they are better addressed with site-specific models that explicitly account for local geologic effects. Rather the focus here will be on shallow mass movements, which tend to be driven by processes in the top few metres of the soil-saprolite profile, and which are largely contained within the surface soil. It is thus unsurprising that the soil-saprolite interface figures heavily in the sections that follow. As noted in Chapter 6 a number of processes and properties change either abruptly or very rapidly at this interface (e.g. porosity, permeability, mobility) so that the soil-saprolite interface tends to be either a plane of weakness, or a plane either side of which there are abrupt changes in transport and hydrology processes.

13.1 Trigger Criteria

In this section we will discuss the mechanisms that start gravity-driven mass transport. The components of this are the processes that trigger the slope failure, and that determine how big the resulting slope failure will be.

13.1.1 Empirical Relationships

Before we discuss physically based models of landslide triggering we will look at empirical relationships that have been observed in the field or derived from remote sensing studies. The typical methodology for these studies is that there has been some macroscale trigger process such as heavy rain, earthquake or snowmelt that has triggered a large number of microscale landslide events in the region affected by that macroscale event. These studies provide insight into what has been observed in the field. More importantly they may also provide a simple basis for a stochastic model of landsliding where landslides occur

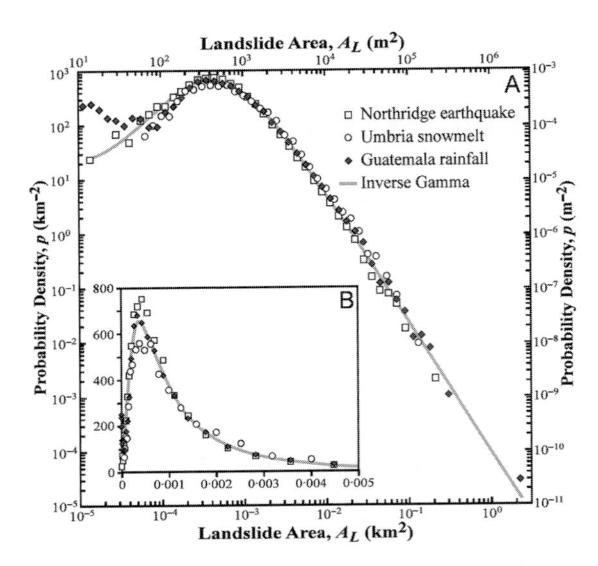

FIGURE 13.1: Dependence of landslide probability density function p for landslide area A_L (from Malamud et al., 2004).

randomly in space and with random mass, where the landslide properties can be simulated using Monte Carlo sampling from these empirically derived probability distributions.

For instance, Stark and Hovius (2001) and Malamud et al. (2004) examined landslide area data for a number of macroscale events worldwide. Malamud et al. (2004) examined published studies of landslide trigger rates (probability per unit area) and the volume of the landslide that resulted for three locations affected by three different macroscale trigger processes (rain, snowmelt and earthquake). Malamud found that the microscale data for three macroscale events were well fitted by a three-parameter inverse gamma distribution (Figure 13.1)

$$p(A_L) = \frac{1}{a\Gamma(\rho)}\left[\frac{a}{A_L - s}\right]^{\rho+1} e^{\left[\frac{-a}{A_L - s}\right]} \tag{13.1}$$

where $\Gamma(\rho)$ is the gamma function of ρ with parameters $\rho = 1.4$, $a = 1.28 \times 10^{-2}$ km², and $s = -1.32 \times 10^{-4}$ km². For large areas this is well approximated by

$$p(A_L) = \frac{1}{a\Gamma(\rho)}\left[\frac{a}{A_L}\right]^{\rho+1} \tag{13.2}$$

Both of these equations are for large areas of more than a few hectares. Equation (13.2) shows that as the landslide area A_L increases the likelihood of a landslide occurrence decreases with a power law with exponent $-(\rho + 1)$ and $\rho + 1 = 2.4$ for Malamud's data. Stark and Hovius (2001) examined landslides in Taiwan and New Zealand, and, while they fitted a double Pareto distribution on the entire dataset, for large areas they also found a power law relationship with landslide area with an exponent of -2.11 for Taiwan and -2.44 to -2.48 for New Zealand.

Using Equation (13.1) the average area of landslides $\overline{A}_L$ is

$$\overline{A}_L = \frac{a}{\rho - 1} + s \tag{13.3}$$

These relationships give the distribution of landslide areas for specific macroscale trigger events, but they lack information about (1) how often the macroscale trigger events occur, (2) the relationship for the absolute number of landslides as a function of the magnitude of the macroscale trigger event and (3) the volume of sediment moved given a landslide area. We will return to these power law relationships with area in Section 13.2.2.

Turning now to the volume of the landslide, if the landslide removes all soil down to the soil-saprolite interface, then the volume moved is $V_L = DA_L$ where D is the soil depth. However, studies have shown that the volume scales nonlinearly with area, so that events with larger areas remove a greater depth of material. Simonett (1967) and Hovius et al. (1997) found relationships of the form

$$V_L = \varepsilon A_L{}^{\gamma} \tag{13.4}$$

For landslides in Papua New Guinea Simonett (1967) fitted $\varepsilon = 0.024$ and $\gamma = 1.37$ while for landslides in New Zealand and Taiwan Hovius et al. (1997) fitted $\varepsilon = 0.05 \pm 0.02$ and $\gamma = 1.5$. Larsen et al. (2010) report $\varepsilon = 0.15$ and $\gamma = 1.1 - 1.3$ for their entire data set, but when they separated soil and bedrock landslides they found $\gamma_{soil} = 1.145 \pm 0.008$ and $\gamma_{rock} = 1.35 \pm 0.01$. The value near 1.0 for the soil landslides is consistent with the idea that their depth is independent of the area of the landslide, and that all soil down to a threshold layer (i.e. soil-saprolite interface) is removed regardless of the area of the landslide. For instance, for a landslide area of 1 ha Simonett's parameters give an average depth of the landslide of 4.3 m, while Hovius's parameters give 5 m. These depths are both larger than typical hillslope soil depths suggesting that large landslides (see Figure 13.1) will remove some of the saprolite as well as the soil, while smaller landslides (e.g. an area of 0.01 ha yields average depth of 0.8 m and 0.5 m for Simonett and Hovius, respectively) might be confined within the soil. Larsen et al. also noted that the depths of their small bedrock landslides tended to be indistinguishable from soil landslides, both being constrained by the soil availability.

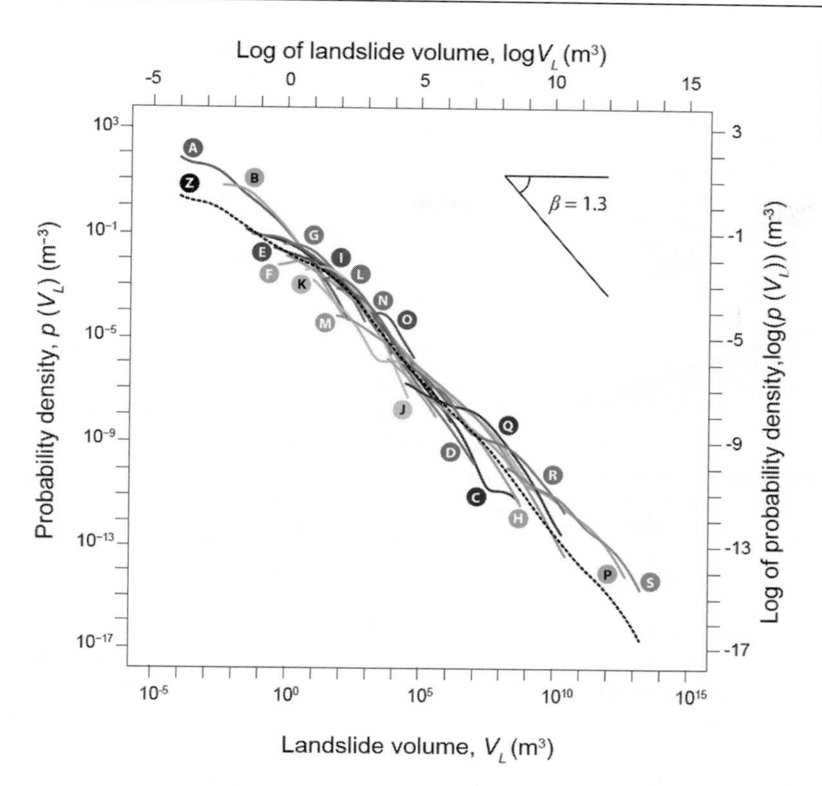

FIGURE 13.2: Dependence of landslide probability density function p for landslide volume V_L (from Brunetti et al., 2009).

Equations (13.2) and (13.4) can be combined to create a relationship for the probability distribution of landslide volumes

$$p(V_L) = \left[\frac{a^{\rho+1}}{a\Gamma(\rho)} \left(\frac{1}{\varepsilon} \right)^{-\left(\frac{\rho+1}{\gamma}\right)} \right] {V_L}^{-\left(\frac{\rho+1}{\gamma}\right)} \tag{13.5}$$

which for Hovius's parameters yields an exponent on the volume V_L of -1.6.

Brunetti et al. (2009) examined 19 datasets of landslide volumes to determine a probability distribution for landslide volume and found (Figure 13.2)

$$p(V_L) \propto {V_L}^{-\beta} \tag{13.6}$$

where $\beta = 1.1 - 1.4$ for rockfalls and rock slides, and $\beta = 1.5 - 1.9$ for soil slides, so that the scalings with volume in Equations (13.5) and (13.6) are approximately consistent even though they were derived from different datasets.

Warburton et al. (2008) mapped widths and lengths of landslide scars (their maximum size of landslide scar about 1 ha, so their landslides were much smaller than the studies discussed above) and found that length increased with width but there was significant scatter around this trend. Milledge et al. (2014) used these same data together with a number of other published studies of similarly smallish landslide scars and found that the mean of the length:width ratio was about 1–2, but with a standard deviation in log-space of about 2–3 around that mean. They also found, by plotting depth against area, that γ in Equation (13.4) was about 1.4, consistent with the studies of the larger scars discussed above.

13.1.2 The Physically Based Infinite Slope Model

The most commonly used physically based criterion for shallow slope failure has been the 'infinite slope' model. This model examines the stability of the soil for a hillslope that is infinite in both downslope and across slope directions, and where the hillslope is planar with no divergence or convergence (i.e. no ridges or valleys). The soil is assumed to be uniform in strength and depth everywhere. The slope is assumed to fail everywhere at the same time, so there are no lateral boundary effects influencing the failure scar. The rationale for ignoring the lateral boundary, finite size, effect is that the area associated with vertical edges of a landslide is much smaller than the basal area. For instance, for the example in the previous section, of a 1 ha scar with a depth of 5 m, the

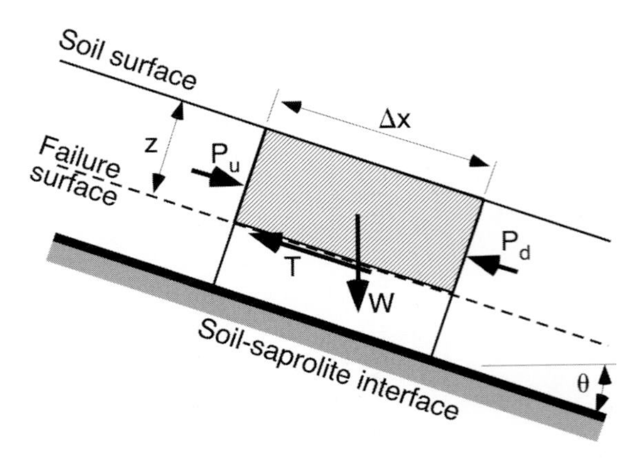

FIGURE 13.3: Schematic of the force balance for a control volume of failed soil on a hillslope where the cross-slope direction is considered infinite, where the failure surface occurs within the dry soil profile.

area of the vertical edges for a square landslide is 20% of the basal area, so to first order the edge area is small. Including the edge effects will be discussed in the next section when we extend this model.

Ignoring edge effects allows us to examine the force balance stability of a small element of the hillslope, independently of surrounding elements (Figure 13.3). Note that in the figure we have drawn a failure plane that does not necessarily coincide with the soil-saprolite interface, so that the discussion below is also valid if the slope fails along a plane that is contained within the soil (for instance, as discussed in the previous section for the depth of landslide scars where the equations can predict a depth less than the depth of the soil).

There are three main forces acting on the element. They are the weight force W acting vertically downwards, and two shear forces at the failure plane acting to resist movement, collectively labelled T. The first resisting force T_f is a frictional resistance force that is a function of the overburden pressure on the failure plane, and the second T_c is the cohesive shear strength of the soil at the failure plane surface and $T = T_f + T_c$. There are also two pressure forces at the upslope and downslope ends of the element, P_u and P_d, respectively, but they balance (because the soil and the failure plane is assumed the same in every direction) and so can be ignored until we consider boundary effects in the next section.

The cohesive shear strength force is active until it is exceeded and then decreases to zero post-failure. It can be considered to be a force resulting from a 'glue' between the overlying soil and the underlying soil/saprolite. Once the glue is broken it is no longer active. The two main cohesive forces are the cohesive strength of clays and the existence of vegetation roots that penetrate through the failure surface (e.g. Gorsevski et al., 2006; Meisina and Scarabelli, 2007).

The frictional shear force is a function of the weight of the overlying soil and the roughness of the interface between the soil and the saprolite, is active both pre- and post-failure of the soil, and is expressed as

$$T_f = \tau A = \mu \sigma A = \mu W_n \tag{13.7}$$

where T_f and τ are the shear force and shear stress acting at the failure surface, respectively, A is the area of failure surface that the shear stress is acting on, W_n and σ are the components of the vertical force and stress acting normal to the failure surface and μ is the friction coefficient. Pre-failure (when the soil is stationary) $\mu = \mu_s$, the static friction coefficient, while post-failure (when the soil has started moving relative to the underlying soil/saprolite) $\mu = \mu_d$, the dynamic or kinematic friction coefficient. For solid to solid sliding (e.g. rock fragment sliding against a rock surface) it is widely accepted that the dynamic coefficient is less than the static friction coefficient (Meriam et al., 2012). For granular material this also appears to be true (Géminard et al., 1999; Géminard and Losert, 2002) though there are fewer experimental data for granular material. Thus once movement starts, the frictional shear force resisting movement is reduced, and Geminard and co-authors found this reduction to be of the order of 20%. Post-failure only the frictional force is active.

Summing the pre-failure forces in the direction parallel to the slope, positive in the downslope direction

$$\begin{aligned} \sum F &= W \sin\theta - T_f - T_c \\ &= \rho_b g z \Delta x \sin\theta - \mu_s \rho_b g z \Delta x \cos\theta - \tau_c \Delta x \end{aligned} \tag{13.8}$$

where ρ_b is the bulk density of the overlying soil, z is the depth of the failure surface below the soil surface and τ_c is the cohesive shear stress strength of the soil at the failure surface (the shear strength of clays is normally expressed as a shear stress, while the shear strength due to roots will be a function of the number of roots and/or the total area of roots per unit area of the failure surface). If $\sum F$ is greater than zero then the slope fails because the destabilising forces (the weight) exceeds the resisting forces. This equation is commonly expressed as a factor of safety, FOS,

$$\text{FOS} = \frac{\mu_s \rho_b g z \cos\theta + \tau_c}{\rho_b g z \sin\theta} \tag{13.9}$$

where the destabilising forces are in the denominator and the stabilising forces are in the numerator. If FOS > 1 then the slope is stable, while if FOS < 1 then the slope

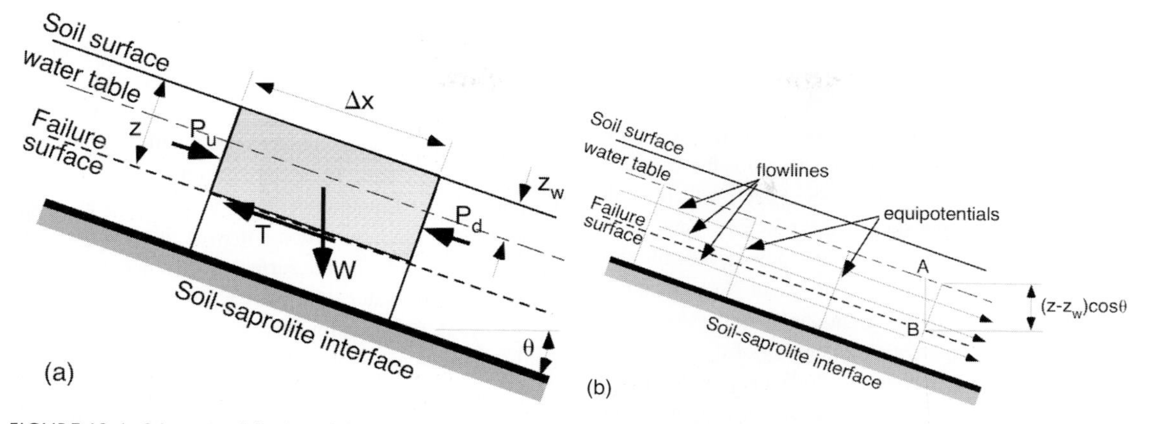

FIGURE 13.4: Schematic of the force balance for a control volume of failed soil, where the failure surface occurs within a soil profile with a water table above the failure surface so that some of the failed soil is saturated with water: (a) the force balance, (b) the equipotentials of the water table down the slope showing the pressure distribution down the slope.

fails. One end member is when the cohesive strength is zero and

$$\mathrm{FOS} = \frac{\mu_s \cos\theta}{\sin\theta} = \frac{\mu_s}{\tan\theta} \tag{13.10}$$

so that at the failure threshold, when FOS = 1, $\tan\theta = \mu_s$, where $\theta = \theta_r$ is the angle of repose and the internal friction angle for a granular material (i.e. a material with no cohesive strength). For nonzero cohesion

$$\mathrm{FOS} = \frac{\mu_s + \dfrac{\tau_c}{\rho_b g z \cos\theta}}{\tan\theta} \tag{13.11}$$

which shows more explicitly that cohesion always improves the stability of the slope, but that this stabilisation effect is reduced as z increases, so that if τ_c is constant with depth, then the plane most at risk of failure is the deepest plane in the soil, the soil-saprolite interface. In fact, even if the cohesion strength increases with depth, if the cohesive strength varies as $\tau_c \sim z^{\varphi}$ where $\varphi < 1$ then the soil-saprolite interface ($z = D$, where D is the depth of the soil) is the weakest failure plane.

Finally in the situation where the soil depth is increasing with time due to pedogenesis (see, for instance, the soilscape evolution chapters), then for some depth D^* the soil-saprolite interface will become unstable (provided only that the hillslope slope is greater than the internal friction angle), and from Equation (13.11)

$$D^* = \frac{\tau_c}{\rho_b g \cos\theta(\tan\theta - \mu_s)} \tag{13.12}$$

This suggests a mechanism whereby a dynamic balance can develop between soil thickening by pedogenesis and soil thinning by landsliding. Note that this result does not mean that the depth of the soil will be D^* everywhere, because once the soil is thicker than D^* the entire profile will be removed by landsliding because the plane of failure is the soil-saprolite interface, resetting the soil depth to zero at the failure location. Vanwalleghem et al. (2013) presents a simulation model that coupled a pedogenesis model with stochastic landslides and found that this combination of mechanisms acted to limit the maximum depth of the soil, and gave a better prediction of landslide initiation locations than assuming a constant depth soil. Montgomery and Dietrich (1994) found that for one steep study site that they classified as unconditionally unstable (i.e. $\tan\theta > \mu_s$ and $\tau_c = 0$) the site was dominated by exposed rock, consistent with Equation (13.12), and the soil was being stripped off by landsliding before being able to accumulate.

The main limitation of the infinite slope model as described above is that it is incapable of modelling hillslope response to a macroevent. In the model above, a hillslope is either stable or not, and the model cannot simulate time-varying instability and FOS (e.g. landslides after a heavy rainfall event).

The main extension to the model is to simulate the existence of a time-varying water table within the soil. Figure 13.4 extends Figure 13.3 to include a water table at depth z_w below the surface. As previously there are pressure forces at the upslope and downslope ends, P_u and P_d, respectively, which are now combinations of soil and water pressures, but they cancel out as before. We also note that the bulk density of the dry soil is $\rho_b = \rho_s(1-n)$ where ρ_s is the density of the soil particles and n is the porosity of the soil. The total weight of the soil (noting that below the water table the pores in the soil will be filled with water) is

$$W = \rho_s(1-n)gz\Delta x + \rho_w ng(z-z_w)\Delta x \qquad (13.13)$$

where ρ_w is the density of water. However, the weight that is effective on the failure plane to generate friction (the movement of the soil particles above the failure against the soil particles below the failure plane) is the weight of the wet soil minus the buoyancy effect, which is the vertical upwards force due to the pore pressures at the failure plane. Figure 13.4b shows a flow net for the flow down the hillslope in the ground water. In particular it shows the equipotentials at right angles to the flow direction. Along an equipotential the energy of the flow is the same so that $\frac{p}{\rho_w g} + z =$ constant. Since the pressure is zero at the groundwater table the pore pressure at the failure plane is $p = \rho_w g(z - z_w)\cos\theta$. For those with an interest in fluid mechanics this is not the pressure resulting from the hydrostatic pressure distribution from A to B (which would be $p = \rho_w g(z - z_w)/\cos\theta$) in Figure 13.4b. The difference is because there is a vertical component to the velocity from A to B, violating the hydrostatic assumption. Thus the effective weight to determine the frictional resistance is

$$W_{\text{eff}} = [(\rho_s - \rho_w)(1-n)z + \rho_w(1-n)z_w]g\Delta x\cos\theta \qquad (13.14).$$

The force balance is then

$$\sum F = (\rho_s(1-n)z + \rho_w n(z-z_w))g\Delta x\sin\theta - \mu_s[(\rho_s-\rho_w)(1-n)z + \rho_w(1-n)z_w] g\Delta x\cos^2\theta - \tau_c\Delta x \qquad (13.15)$$

and the FOS is

$$\text{FOS}_{\text{wet}} = \frac{\mu_s[(\rho_s-\rho_w)(1-n)z+\rho_w(1-n)z_w]g\cos^2\theta+\tau_c}{(\rho_s(1-n)z+\rho_w n(z-z_w))g\sin\theta} \qquad (13.16)$$

which can be rearranged into a form that is more easily compared with the dry soil FOS:

$$\text{FOS}_{\text{wet}} = \frac{\mu_s\rho_b gz\cos^2\theta+\tau_c-\mu_s\rho_w(1-n)g(z-z_w)\cos^2\theta}{\rho_b gz\sin\theta+\rho_w ng(z-z_w)\sin\theta} \qquad (13.17)$$

The first two terms in the numerator and the first term in the denominator are almost the same as in the dry soil FOS; the only difference is the $\cos^2\theta$ term instead of $\cos\theta$ which slightly reduces the FOS. The groundwater table (1) subtracts a term (which is always positive) from the numerator that reduces the FOS, this effect being the reduction of the weight of the solid component of the soil acting on the failure surface as a result of buoyancy of the soil particles and (2) adds a term to the denominator (which is always positive) that also reduces the FOS, this effect being the extra weight of the soil after it is wet. Any reduction in the cohesive shear strength of the clays in the soil due to the increased water content is in addition to these changes. Thus we can see that a water table in the soil always reduces the stability of the soil relative to a dry soil. There are several slightly different formulations of Equations (13.16) and (13.17), but they differ only in the assumptions of what the pore pressure distribution is at the failure surface, which depends on the assumed groundwater flow pattern (Iverson, 1990), and, in any event, pore pressures are difficult to predict in the field (Montgomery et al., 2009).

Montgomery and Dietrich (1994) called hillslopes that were always unstable even when dry as 'unconditionally unstable', and those that are stable even when fully saturated as 'unconditionally stable'.

The final observation about the effect of the water table is that this reduction in the FOS is a function of how fast the water table varies. As we noted in Chapter 3 the soil water table for saturation excess hydrology (Section 3.1.2.1) tends to vary slowly, yet it is common to see heavy rainfall trigger landslides very quickly. Several explanations for this apparent contradiction have been proposed including the following:

(1) Preferential flow paths (e.g. macropores) so that the soil-saprolite interface becomes very quickly saturated even though the soil profile itself may only be partially saturated. This allows the buoyancy term to be operative but without the increase in the soil density due to wetting (i.e. the third term in the numerator of Equation (13.17) changes, but the second term in the denominator does not change),
(2) Hillslopes are at the edge of their stability threshold, and even a small change in the soil wetness is enough to trigger landslides. This does not appear to explain observations that rapid changes in groundwater levels trigger landslides, but equivalent, but slower, changes do not.

The concepts above form the basis of the SHALSTAB (Dietrich and Montgomery, 1998), dSLAM (Wu and Sidle, 1995) and SINMAP (Pack et al., 1998) landslide stability models that have been incorporated into GIS packages. Early work did not use the general solutions above, but assumed that only one of either friction or cohesion was the stabilising force. Later work has recognised that both may be important. Dietrich et al. (1992, 1993), Montgomery and Dietrich (1994) and Borga et al. (2002) coupled this shallow landsliding model (using only frictional resistance, no cohesion) and a model for saturation excess runoff (to predict the depth of the water table based on catchment geomorphology) to predict regions of low stability, and then compared this spatial distribution with field observations with a good qualitative match. Dietrich et al. (1995) extended this work to include a spatially variable cohesion that was a function of root density. Ho et al. (2012) were able to improve their predictions of landslide scars by incorporating measured soil depth data rather than assuming soil depth was the same everywhere. Roering et al. (2003) found landslide scars concentrated in areas with less root strength, confirming the importance of the spatial distribution of root strength in promoting cohesion in forested areas and the impact of forestry clear-cutting (Wu and Sidle, 1995; Schmidt et al., 2001). Gorsevski et al. (2006) used a continuous soil moisture accounting model to simulate time-varying landslide susceptibility and coupled it with a model for forest clear cut. They found a strong link between landsliding and the dynamics of root strength in space and time.

Griffiths et al. (2011) constructed a two-dimensional finite element hillslope model (the axes were downslope and down the soil profile with no cross-slope component) and used it to validate the results of the analytic infinite slope model discussed above. By varying the length of the modelled hillslope, they showed that the infinite slope model gave good FOS estimates for length:depth ratios more than 15. They extended this model to examine cases where the soil strength increased with depth down the soil profile. They were able to generate failure surfaces that were fully contained within the soil profile. This was because the increasing strength of the soil with depth offset the decrease in stability with increasing depth in Equation (13.12).

Most field studies have compared model predictions for the spatial distribution of landslides after a single event, and relatively few studies have looked at the temporal behaviour at the same site for a series of rainfall events. Shuin et al. (2012) examined 13 events using a model that coupled a two-dimensional groundwater model with an infinite slope stability model and found that they could not satisfactorily predict landslide rate from event to event unless they allowed the soil cohesion to vary between storm events by an order of magnitude.

The main effect of vegetation is to provide shear strength across the failure surface (typically the interface between the bottom of the soil and top of the saprolite) reducing the rate at which slopes fail. It has been observed that following clear cutting of forests for silvaculture that for the first five years after clear cutting there is no change in rate of landsliding, but after that time there is marked increase (up to a factor of five times higher than pre-clearing), which stabilises back to pre-clearing levels after 10–20 years. The mechanism is that over the five years the strength of the older roots drops, and it is only after this drop that landsliding rates increase (Burroughs and Thomas, 1977; Sidle, 1992). We will return to this topic of root strength when we discuss the impact of fire on tree death and subsequent hillslope transport in Chapter 14.

13.1.3 Finite Slope Modelling

The major limitations of the infinite slope model are the assumptions that (1) slope and soils are the same in all directions and (2) the landslides are infinite in extent. Addressing these limitations results in formulations of the shallow landslide model that consider a landslide of finite extent. This means that in addition to the resisting shear forces on the base of the landslide (as for the infinite slope model, T_{basal}, Figures 13.3 and 13.4), there are now resisting shear forces due to friction and cohesion (T_{left}, T_{right}) along the sides of the landslide parallel to the downslope direction, a tension force at the upper boundary ($F_{upslope}$) and a resistance force at the lower boundary ($F_{downslope}$) (Figure 13.5). The side normal forces (F_{left}, F_{right}) cancel as they are equal if the soil depth doesn't change. The triangular volume at the upslope end is the wedge of soil applying the destabilising force at the upslope end, while the triangular wedge at the downslope end is the wedge of soil that is resisting the movement downslope. As noted at the start of the previous section, the main rationale for including boundary conditions is for when the area of edges is comparable to the basal area so the forces are potentially comparable, which occurs for the low length to depth ratios typical of small landslides.

Milledge et al. (2014) implemented a finite element (FEA) model for a long hillslope identical to that of Griffiths et al. (2011) (see the previous section) and generalised Griffith's results using a parametric study. Depending on the combination of parameters used, Milledge found that the infinite slope model was within 10% and 5% of the FEA model for length:depth ratios above 18 and 25, respectively.

Milledge et al. (2014) formulated a two-dimensional finite conceptual landslide model, called MD-STAB, where the two dimensions are the downslope and cross-slope direction (Figure 13.5) and which included a water table. An application at the Coos Bay landslide field site found that if the measured pore pressures were used, then the model gave better predictions of landslide occurrence than previous studies using the infinite slope model (e.g. see Montgomery et al., 2009 for some introspection about

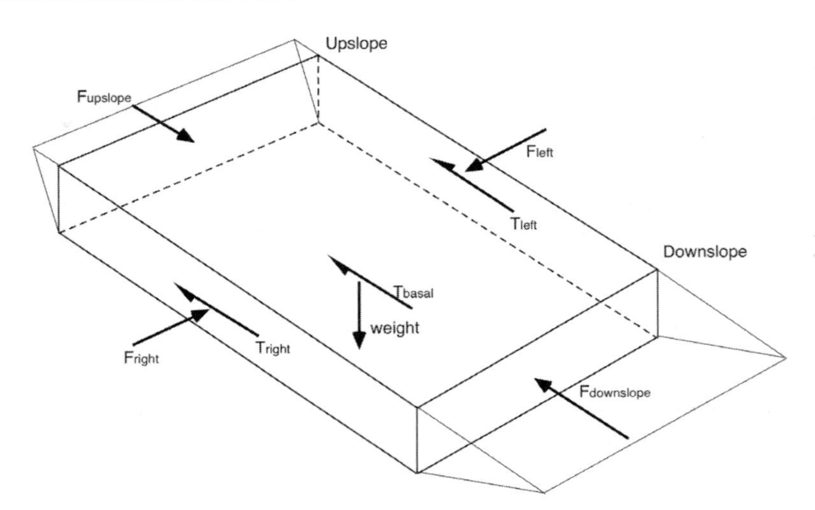

FIGURE 13.5: Schematic of the force balance on a three-dimensional control volume of failed soil.

the experimental and modelling work at this field site). One possible implication of Milledge's model is that for smaller landslides the FOS is increased relative to the infinite slope model, and the authors speculate that there may be a minimum area below which landsliding cannot occur. Milledge et al. speculated that this might explain the peak of the landsliding frequency curve in Figure 13.1 below which small events become less likely (Malamud attributed it to undersampling of small landslides due to the limited spatial resolution of the data he used).

The major difficulty with Milledge's model (and any finite area model for that matter) is the need to define the spatial extent of the landslide. This spatial extent can change depending on the spatial variability of all the inputs into the model including topography, soil depth, soil strength and pore pressure distribution. While it has been commonly assumed that pore pressure variations are the dominant unknown variable, it has been suggested that the spatial variation of soil depth and root strength distribution are equally important variables (Montgomery et al., 2009; Bellugi et al., 2015a). In principle, defining the failure surface for a natural hillslope involves searching through an infinite number of combinations of failure surface geometries and finding the least stable surface, which is clearly computationally intensive. Bellugi et al. (2015a,b) used Milledge's MD-STAB model and developed a search algorithm that starts with some region (typically a single pixel) and adds to that region adjacent pixels that are least stable and removes ones that are most stable. The coupling between the adjacent pixels occurs by the friction and cohesion forces on the vertical interface between the pixels (e.g. the edge shear and normal forces in Figure 13.5). In this way it searches for the least stable collection of pixels in the landscape and then assumes that these pixels fail as a single block. They calibrated the model to the Coos Bay field site and found it was able, after calibration, to predict 65% of the observed landslides with a relatively low false positive rate.

13.1.4 Soil Destabilisation by Dynamics at the Soil Surface

The sections above have dealt with the failure of an initially stationary soil. Another case that is important is where soil, mud or water is flowing over the surface, applying a shear stress to the surface of the stationary soil, and potentially destabilising the underlying soil as a result of this extra loading on the soil. This is an issue downslope of a failure along the path that the landslide takes downslope, and is the mechanism whereby debris and mudflows scour the hillslope downslope of a failure source area upslope. Hereafter we refer to this case as a debris flow, but the discussion is equally applicable to any geophysical flow that applies weight and/or shear stress to the soil surface.

Figure 13.6 shows a schematic of this process, highlighting the differences with the discussion in the previous section shown in Figure 13.4. The differences are the addition of a weight load W_d and a shear force T_d applied at the interface between the debris flow and the underlying soil. The shear force T_d is the component of the weight of the debris flow in the direction parallel to the surface. This has parallels to the shear force applied by the weight of the soil on the failure surface. However, unlike for the soil

motion can be of order of 20%, but the drop is highly sensitive to the grading of the material, which influences how much dilation is required before motion can occur.

The acceleration downslope can be calculated in two ways. The first using Newton's second law in a force balance downslope to do a momentum balance, while the second uses the kinetic energy conservation.

Writing a force balance in the downslope direction (downslope positive)

$$\sum F_{ds} = ma = mg \sin\theta - \mu_d mg \cos\theta \quad (13.29)$$

and solving for the acceleration a

$$a = g\cos\theta(\tan\theta - \mu_d) \quad (13.30)$$

We see that the acceleration is a function of the excess of the slope over the dynamic coefficient of friction (i.e. $\tan\theta - \mu_d$).

The alternative derivation based on energy conservation can be obtained by writing the energy balance, which requires calculating energy conservation over a short length of slope (Δx in the horizontal direction). The frictional work is (from Equation (13.23))

$$W_{\text{friction}} = \mu_d mg\cos\theta\left({}^{\Delta x}/{\cos\theta}\right) = \mu_d mg\Delta x \quad (13.31)$$

and the change in kinetic energy over Δx (noting that for a short length of hillslope, so the kinetic energy change is small, $v_2{}^2 = (v_1 + \Delta v)^2 = v_1{}^2 + 2v_1\Delta v + \Delta v^2 \approx v_1{}^2 + 2v_1\Delta v$ when Δv is small) is

$$\Delta\text{KE} = \frac{m}{2}\left(v_2{}^2 - v_1{}^2\right) \approx mv_1\Delta v = 2mg\Delta x(\tan\theta - \mu_d) \quad (13.32)$$

so $\Delta v = g\Delta x(\tan\theta - \mu_d)/v_1$. The acceleration is then

$$a = \frac{\Delta v}{\Delta t} \approx \frac{g\Delta x(\tan\theta - \mu_d)}{v_1} \cdot \frac{v_1\cos\theta}{\Delta x} = g\cos\theta(\tan\theta - \mu_d) \quad (13.33)$$

which is the same result as derived using the force balance method in Equation (13.30).

One point about the acceleration derived in Equations (13.30) and (13.33) is that it is independent of the depth below the surface and so is equally applicable to any interface at any depth below the surface, so there is no relative acceleration between soil at different depths below the surface. This is provided only that the dynamic coefficient of friction does not change with depth. The implication of the lack of relative acceleration with depth below the surface is that all the soil above the failure plane slides as a single block. Thus, in the case of Coulomb friction, we can conclude that the material will flow as a single block with all energy dissipation occurring at the base of the block and no energy dissipation occurs internally within the slide.

13.2.5 Debris Flows

The main type of flow we will talk about in this section is a debris flow. Classical flows have a no-slip condition at the base of the flow, and all the energy dissipation occurs internally within the flow. Debris flows are a little different because they have aspects of flows and slides simultaneously. They have some of the friction energy losses at the flow-soil interface of the slide with the energy losses within the flow as a result of turbulent and viscous losses. The discussion that follows is based on the general principles outlined in Iverson (1997) and Iverson and George (2015).

Debris flows are masses of rock, soil and water that travel downstream and deposit on flatter areas at the base of mountains. It is generally believed that a key condition is that the rock and soil mass must be saturated with water. Their behaviour is somewhere intermediate between dry granular flow (e.g. dry sand) and water flow. Their velocity is fast, commonly greater than 10 m/s. A key aspect of the debris in the flow is that the size of particles covers a large range from large boulders down to clay, and the shape of the particles is quite irregular. The fine particles are key because they form an important part of the 'lubrication' of the flow. There is a strong size dependence to debris flows with larger flows travelling longer distances (Johnson et al., 2016).

It is common to treat the debris flow as having only one component/phase (the mixture of rock, soil and water), and to mathematically formulate the problem as a fluid flow with laminar viscosity where the viscosity is described by a Bingham fluid (e.g. Whipple, 1997). We discussed viscous Bingham flow in the context of soil creep in Chapter 9. Iverson and George (2015) indicate that the calibration of a single-phase model results in viscosities that are too low so that runout distances are too high. They proposed that the physics of the debris flows is more clearly understood by differentiating two phases: (1) the granular rock phase and (2) the interstitial fluid between the rocky particles. The granular rock component is typically about 40–80% of the volume of the material, and the fluid component is a mixture of water and fine particles that form mud with about 10–20% solids by volume.

There are three stages on the temporal evolution of a debris flow: (1) the initiation of the debris flow movement through some form of instability or slope failure, (2) the flow of the rock, soil and water mass down the slope and (3) the slowing down and eventual stopping of the flow in what is normally a lobate-shaped deposit. In all three stages the fluid and its pore pressures in the interstices are key. Figure 13.8 shows how the rock and fluid potentially interact at the moment when movement is initiated.

As part of the debris movement process the material can either dilate (expand) or collapse (contract). Dilation occurs when the large particles are initially tightly packed and cannot move downslope unless they move up and over the particles directly underneath. Thus to move the material must expand/dilate. One consequence of dilation is that the volume of voids within the debris increases and the water level drops. If the material is initially fully saturated with no excess water, then upon movement the top layers of the pile will become unsaturated as the pore water flows into the increased void space underneath. The opposite process occurs when the material collapses and excess water appears on the surface after it is expelled from the decreased volume of voids. Water flow direction is in the direction of reduction in head (pressure p is sometimes used instead of head h but that is incorrect, $h = z + {}^{p}/_{\rho_m g}$ where h is head, z is elevation and ρ_m is the density of the mud), so for dilation and water drop the pressure at the base of the debris flow must decrease below hydrostatic pressure, while for collapse the pressure at the base of the flow must exceed hydrostatic pressure to expel that water to the surface. An important consequence of collapse is that if the collapse occurs because of movement and this movement was triggered by rises in pore pressures (e.g. as a result of infiltration during rainfall), then the pore pressures will rise further, further destabilising the material, leading to a positive feedback. On the other hand dilation will lead to a reduction of the pore pressure with the possibility of a stabilisation of the initial movement. Iverson and George (2015) talk about dilation leading to a gradual stop-start destabilisation of the slope, while collapse leads to a rapid destabilisation. One consequence of this is that the initial condition of the soil on the slope is important for how a debris flow commences since the initial consolidation state of the soil determines whether it will collapse or dilate upon movement.

The excess pore pressure at the base of the debris flow dissipates upwards and is driven by the vertical head gradient. The rate of dissipation of pore pressure is akin to a Fickian diffusion process for head h as can be seen from the groundwater flow equation in the vertical direction z:

$$\frac{\partial h}{\partial t} = \frac{\partial}{\partial z}\left(K\frac{\partial h}{\partial z}\right) = K\frac{\partial^2 h}{\partial z^2} + \frac{\partial K}{\partial z}\frac{\partial h}{\partial z} \approx K\frac{\partial^2 h}{\partial z^2} \tag{13.34}$$

where the diffusivity for pressure is the hydraulic conductivity of the flow in the vertical direction K, which is a function of the void geometry between the solid particles (e.g. the Kozeny-Carmen equation; Freeze and Cherry, 1979). Pore pressure adjustment occurs primarily in the vertical direction because the depth of the flow is much less than the length or width of the flow, so the vertical hydraulic gradient is much greater than the horizontal gradient $\frac{\partial h}{\partial z} \gg \frac{\partial h}{\partial x}, \frac{\partial h}{\partial y}$ where x and y are the downstream and across stream directions. Note that the hydraulic conductivity for the interstitial mud is not the same as the hydraulic conductivity of clear water, since the interstitial mud is more viscous than clear water (by 10^2–10^4 times so the conductivity will be 10^{-2}–10^{-4} times less for the mud).

One consequence of Equation (13.34) is that the response time of the adjustment of pore pressures scales with the (depth of the debris flow)^2 so that deeper (and bigger) flows will have excess pore pressures (and thus reduced frictional resistance) for longer periods. Using a similitude analysis Iverson and George (2015) suggest that large debris flows may take days to return to hydrostatic conditions, with the consequence that it seems reasonable to suggest that excess pore pressures from debris flow initiation may persist for the entire duration of the debris flow event. The increase in pore pressure relaxation times with size may also explain why larger debris flows appear to have lower frictional resistance and larger runouts than smaller debris flows (Iverson, 1997). This relaxation time scaling is consistent with field data that shows excess pore pressures occur in the main body of a debris flow deposit long after the event (Major and Iverson, 1999).

This scaling solution assumes that the only pathway for pore pressure relaxation is for water to flow from the base of the flow up through the layers of debris in the flow to the surface. There are other potential mechanisms (Figure 13.9), including (1) vertical mixing within the debris, (2) development of preferred flowpaths to the surface of the debris flow such as cracks and (3) pressure relief into the underlying soil material. All of these mechanisms would significantly reduce the relaxation times, but they were not considered by Iverson and George (2015).

For typical conditions, scaling analyses by Iverson and Denlinger (2001) and Iverson and George (2015) suggest that the dominant resistive force against flow is bottom friction and not turbulent dissipation, so by that definition

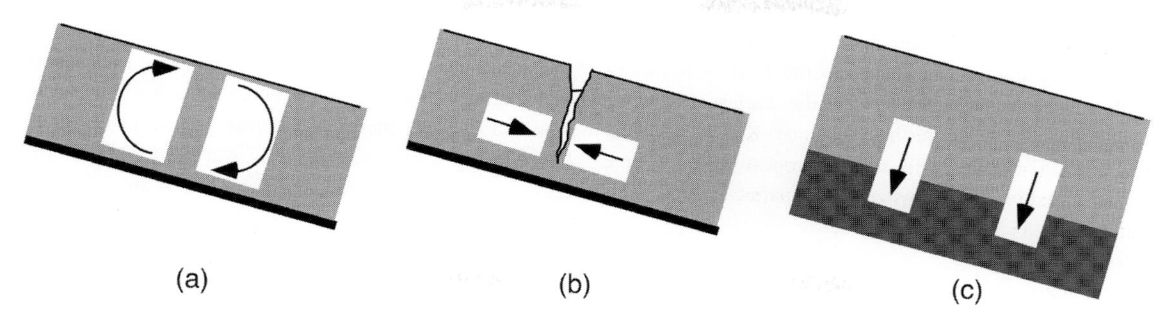

FIGURE 13.9: Schematic showing pathways for pore pressure relief other than as a result of flow of pore water flow through pores. The lighter hatched material is the moving debris flow material, and the darker hatching underneath indicates the natural surface the debris flow is moving on top of. The arrows indicate the direction of pore pressure relief. (a) Vertical mixing where the higher pore pressure materials are advected to the surface by vertical mixing, and the lower pore pressure materials from near the surface are transported downward, (b) pore pressure relief by pore water flow into a vertical crack in the debris flow, (c) higher pore pressure water flow into the underlying material (either into fully or partially saturated pores or cracks in the underlying material).

a 'debris flow' is not a flow but a slide, with most energy dissipation occurring at the bottom and sides of the debris flow. A simple model for debris flows is then the (1) slide model of the previous section (Section 13.2.4) coupled with (2) a pore pressure dissipation model based on diffusion (Equation (13.34)) and using debris flow depth, where the coupling occurs through (3) the reduction of the normal force in the friction equation as a result of excess pore pressure. The friction coefficient in part (3) should have an approximately log-log linear relationship with flow volume varying from about $\mu = 0.5$ for a volume of 10^6 m^3 declining to about $\mu = 0.1$ for a volume of 10^{10} m^3 (Campbell et al., 1995; Johnson et al., 2016).

The derivation of Iverson and George does not directly address what happens in the termination stage of the flow when the flow is slowing to a stop and the lobate-shaped deposits are being formed. However, it is quite simple, and as the slope reduces at the bottom of the hillside, the frictional resistance will be greater than the self-weight driving force and the flow will slow, eventually stopping.

Finally, one interesting consequence of this pore pressure diffusion model is what happens around the edges (i.e. the sides and front) of the debris flow. Because the edges are closer to the atmosphere any excess pore pressures are more readily dissipated than deeper within the flow. As a result the pore pressures are lower near the edges so that the frictional resistance will be greater. It is then possible that the moving material at the edges will fall below the undrained angle of repose and become stable. Thus at the edges of the debris flow there may be levees of deposited material formed, while at the front of the debris flow will be a pile of stable material potentially being pushed by the 'liquid' debris flow behind the front and a surface layer on top of the debris flow (Major and Iverson, 1999; Iverson and Denlinger, 2001; Figure 2 in Iverson and Vallance, 2001). The segregation that has also been observed in these levees (and on the surface of flows) where the levees are richer in coarse materials will be a positive feedback, since coarse material will result in faster dissipation of the pore pressures (because of its higher hydraulic conductivity in Equation (13.34)), enhancing stabilisation (Johnson et al., 2012).

Finally it should not be concluded from the discussion above that pore pressures from water/mud in the sediment is essential to debris flow behaviour. For instance, Campbell et al. (1995) notes that the trend of reduced frictional resistance for large flows has also been observed on Mount St Helens and Mars. For Mount St Helens it is unreasonable to expect that water played a role. Traditionally this was also the conclusion for Mars, but recent evidence of historical water processes may change that interpretation. This suggests that there are additional, as yet unresolved, mechanisms at work in large debris flows.

13.3 Rockfall Modelling

This section presents a modelling methodology for examining rockfalls, where the main transport process is individual rocks moving down a slope. Dorren (2003) reviews the dominant physics of rockfall, noting that for low slopes (less than about 45°) the dominant process is rolling of rocks down the slope and that for steeper slopes rocks bounce. Sliding of rocks typically occurs only for the initial and final stages of movement.

For rolling of rocks the main energy loss is rolling resistance, and the stopping of motion typically occurs quite suddenly as the rock is captured in the surface roughness of the slope. Smaller rocks tend to be captured more easily than larger rocks, so the final positions of rock fragments are sorted with the largest fragments being transported the furthest distance. Kirkby and Statham (1975) captured this sorting process with a simple model based on friction where the dynamic friction coefficient is

$$\mu_d = \tan\phi + \frac{kd}{2R} \tag{13.35}$$

where ϕ is the angle of internal friction or angle of repose (from 20° to 34°), k is a constant (between 0.17 and 0.26), d is the mean diameter of scree on the surface and R is the radius of the moving rock. This coefficient of friction is then used in the sliding model discussed in Section 13.2.3 to determine the travel distance. The larger the diameter of the rock, the lower the friction coefficient and the longer the travel distance. This results in the aforementioned downslope sorting on diameter.

For higher slopes the approach that is described in more detail in the following section is based on Dorren et al. (2005, 2006) and Bourrier et al. (2009) who present a feasible (in the sense of being able to be incorporated into a LEM) model for rockfall, called RockyFor3D. Their model focuses on the transport process and does not address the triggering process. A review of this model, and other physically based models, is provided by Volkwein et al. (2011).

13.3.1 Collision with Boundaries

When a particle collides with another particle, (1) kinetic energy (KE) is potentially transferred between particles and (2) during this transfer of energy some of the KE is lost. If the particle hits a deformable surface (e.g. soil), there will be no transfer on KE (unless the soil starts to move as a result of the collision), but some of the KE of the particle is lost and used in plastic deformation of the soil. Finally if either of these collisions above results in fragmentation of the particle, then some of the KE of the particle(s) is used in the fragmentation process. In the terminology of the fragmentation literature there is energy associated with the creation of the new surfaces resulting from the fragmentation, called 'surface energy'. The more the particles fragment, the greater is the surface area generated and the greater is the transformation of the particle's KE into surface energy.

All the energy transformation processes above are lumped into the 'restitution coefficient', which

$$\mathrm{KE_{bi}} = \mathrm{KE_{ai}} + E_l + E_f \tag{13.36}$$

where $\mathrm{KE_{bi}}$ is the KE of particle before the impact, $\mathrm{KE_{ai}}$ is the total energy of all of the daughter particles after impact and fragmentation, E_l are the energy losses during collision and E_f is the energy loss during fragmentation. If we first consider a particle that collides at right angles to the surface without fragmentation (i.e. $E_f = 0$), then

$$\mathrm{KE_{ai}} = \frac{1}{2}m(C_r v_{\mathrm{bi}})^2 = C_r{}^2\mathrm{KE_{bi}} \tag{13.37}$$

where C_r is the coefficient of restitution of the particle. For soil-covered slopes C_r is about 0.1–0.2, while for rock-covered slopes C_r is about 0.3–0.7 (Asteriou et al., 2012). For particle-particle collisions without fragmentation it can be as high as 0.65 to 1 (Durda et al., 2011). If in addition the particle is travelling downslope so that there is a component of velocity parallel to the collision surface as well as normal to the surface, then Cagnoli and Manga (2003) found that as the angle of impact decreased (i.e. the downslope velocity increased relative to the velocity perpendicular to the surface), then C_r increased (e.g. 0.15 for an impact at 70° to the surface, up to 0.6 at 30°, both figures for pumice particles). Asteriou et al. (2012) proposed a relationship between Schmidt hammer hardness of material and the coefficient of restitution for five different rock types with C_r varying from 0.3 to 0.8

$$C_r = 0.235e^{0.022R} \tag{13.38}$$

where R is the Schmidt hammer hardness value. They also proposed a relationship for the coefficient of restitution with impact angle, which was

$$C_{r,\theta} = 1 - \alpha\frac{\theta}{90} \tag{13.39}$$

where α defines the relationship between $C_{r,\theta}$ and the angle of impact θ (in degrees). The two relationships in Equation (13.38) and (13.39) are not necessarily consistent because Equation (13.38) is the result of lumping together all of their results for different impact angles (ranging from 10° to 90°).

Wyllie (2014) extended the definition of the coefficient of restitution to consider separately the effects on the component of the particle velocity normal to the surface that is being impacted and the component parallel to the surface so that

$$\begin{aligned} C_{r,N} &= \frac{v_{N,a}}{v_{N,a}} \\ C_{r,T} &= \frac{v_{T,a}}{v_{T,a}} \end{aligned} \tag{13.40}$$

where the subscripts N and T indicate velocities normal and tangential to the impacted surface, respectively. He noted that the energy loss in the normal direction is primarily a function of the plastic deformations of the surface and the energy losses resulting from this plastic deformation. The energy loss in the horizontal direction is a function of the friction between the particle and surface, which is a function of the friction coefficient between the particle and the surface, and the normal force of the impact (which in turn is a function of the normal impact velocity and the plastic deformation of the surface). By examining various field rockfalls he found a tangential restitution coefficient in the range 0.3–0.8, and a normal restitution coefficient

$$C_{r,N} = 19.5\theta^{-1.03} \tag{13.41}$$

which gives a normal restitution coefficient greater than 1 for angles of impact less than about 18°. The values greater than 1 were attributed to interaction between the particle rotation and its translation upon impact, with transfer between rotational energy and translational energy (see Equation (13.27)).

One complication of this discussion is that the definition of the coefficient of restitution above considers only linear velocity and not rotation of the particle (Chau et al., 2002). At low angles of impact the rotational energy of particles can interact with the rough surface being impacted with, to result in impacts where the coefficient of restitution as defined in Equation (13.37) is greater than 1. This is a result of some of the particle rotational KE being transformed into translational KE by the geometry of impact. For instance, Buzzi et al. (2012) were able to construct laboratory experiments using rotating particles that gave a coefficient of restitution that was greater than 1 and sometimes as high as 2 because the rotational energy was converted to translational energy by the impact, which is consistent with the findings of Wyllie (2014).

As a final note for cricket and tennis fans the interaction between rotational energy and translational energy upon impact is how applying spin to a ball can result in a ball 'jumping' with a higher speed after the impact with the ground than its speed before impact.

13.3.2 Particle-to-Particle Collisions

The discussion above considers collisions between a particle and a solid boundary. In the next section we will discuss situations where many particles interact as a result of particle-to-particle collisions. In this case both kinetic energy and momentum will be transferred between the particles. The pre- and post-collision (subscript b and a, respectively) energy (both linear and rotational) and momentum (in the x, y, z directions) for two particles 1 and 2 are

$$\begin{aligned} &m_1 v_{1b}^2 + m_2 v_{2b}^2 + I_1\omega_{1b}^2 + I_2\omega_{2b}^2 = m_1 v_{1a}^2 + m_2 v_{2a}^2 \\ &\quad + I_1\omega_{1a}^2 + I_2\omega_{2a}^2 + \text{losses} \\ &m_1 v_{1xb} + m_2 v_{2xb} = m_1 v_{1xa} + m_2 v_{2xa} \\ &m_1 v_{1yb} + m_2 v_{2yb} = m_1 v_{1ya} + m_2 v_{2ya} \\ &m_1 v_{1zb} + m_2 v_{2zb} = m_1 v_{1za} + m_2 v_{2za} \end{aligned} \tag{13.42}$$

where I is the moment of inertia of the particle. For a sphere $I = {}^2\!/_5 mr^2$ where r is the radius of the sphere. The transfer of linear and rotational energy occurs as a result of tangential frictional force between the particles at the time of collision, which in turn is a function of the normal force between the particles during the collision. Energy loss can occur as a result of these tangential friction forces if the surfaces of the two particles slide against each other. This can occur when the inertial forces required to change the rotation rate of the particle exceeds the tangential friction force. The energy lost is the work done by the sliding surfaces $W = Fs$ where F is the frictional force between the particle surfaces, which may vary during the collision as the normal contact forces change.

This equation considers only collision without fragmentation. If multiple particles are generated by the collision, then the energy and momentum of all the post-collision particles must be summed on the right-hand side. There will be additional energy losses as a result of fragmentation, which will be discussed in the next section.

13.3.3 Fragmentation

Returning to Equation (13.21) it remains only to characterise the energy lost as a result of fragmentation. Giacomini et al. (2009) performed a series of experiments to define the energy loss due to fragmentation, E_f, relative to the impact energy (i.e. the ratio ${}^{E_f}\!/_{\mathrm{KE_{bi}}}$) and found that this ratio ranged from 0.1 to 0.6, and that it was independent of the number of daughter particles generated by the impact so that they could express the energy lost in fragmentation as

$$E_f = C_f \mathrm{KE_{bi}} \tag{13.43}$$

where C_f varies from 0.1 to 0.6.

We are also interested in the grading of the daughter products from any fragmentation (e.g. see the discussion about fragmentation modelling in Chapter 7). Using experiment and discrete element modelling, Wang and Tonon (2011) tested the main distributions used to characterise fragmentation products (exponential, Weibull and Voronoi distributions) and found that the Weibull distribution fitted the data and model results best. They also found that mean particle size decreased as the strain rate (i.e. how rapidly the fragmentation occurred) increased.

13.3.4 Modelling

Bourrier et al. (2009) discusses an energy balance model as outlined above, but doesn't consider fragmentation of the rock. The general principle is that a rock bounces down the slope, and at every bounce energy is lost and this energy loss is parameterised by the restitution coefficient. However, instead of considering only the energy loss from that component of the velocity at right angles to the surface, Bourrier also considers energy lost from the tangential (i.e. parallel to the surface) and rotational energy. At each impact the velocities after impact are determined before impact by the matrix equation

$$\underline{v}_{\text{ai}} = \mathbf{A}\underline{v}_{\text{bi}} \tag{13.44}$$

where the velocity vector is the two translational velocities, and the rotational velocity and matrix **A** are

$$\underline{v} = \begin{bmatrix} v_{\text{tangential}} \\ v_{\text{normal}} \\ \omega \end{bmatrix}; \quad \mathbf{A} = \begin{bmatrix} a_{11} & a_{21} & a_{31} \\ a_{21} & a_{22} & a_{32} \\ a_{31} & a_{32} & a_{33} \end{bmatrix} \tag{13.45}$$

For the case where only velocities normal to the surface are considered, as previously discussed, then $a_{ij} = 0$ for all i and j except $a_{22} = C_r$. This matrix equation characterises both the transformation of KE from one form to another (the off-diagonal term) and the loss of KE for each bounce (mostly the diagonal terms in the matrix, though the off-diagonal terms can include a loss component in the transformation process). For instance, a bounce at a low angle will tend to make the particle want to rotate, thus transferring KE from the tangential component to rotational, while a particle that is rotating quickly prior to impact will tend to transfer rotational KE to tangential and normal velocities. Thus it should be clear that the idea that there is a single restitution coefficient for each of the three velocity components is too simplistic (Bourrier et al., 2009). In fact, as was observed in the Buzzi et al. (2012) experiments, the Bourrier model RockyFor3D predicts restitution coefficients greater than 1 for low impact angle.

One interesting aspect of the RockyFor3D model is that there is a random bounce component based on probability distributions presented by Dorren et al. (2005), who measured impacts in a forested hillslope. In principle, a similar probability distribution could be derived or assumed for rock-to-rock impact on rocky slopes, or on the impacts between the rocks and the topography from the model digital elevation model. RockyFor3D does not deterministically calculate velocities (i.e. the downslope and cross-slope components of the tangential velocity) but uses an empirical rule that is based on the probability distribution and the steepest downslope direction derived for the topography from the digital elevation model.

Potentially, the RockyFor3D approach could be extended by disaggregating the tangential velocity into its downslope and cross-slope components (or its (x,y) components if that is more convenient), and by disaggregating the rotational velocity into its three components. This would provide a greater physical basis for their post-impact velocities so that the velocity vector would then be

$$\underline{v} = \begin{bmatrix} v_{\text{x}} \\ v_y \\ v_{\text{normal}} \\ \omega_x \\ \omega_y \\ \omega_z \end{bmatrix} \tag{13.46}$$

RockyFor3D did a good job of matching an experiment on a hillslope for (1) the spatial distribution of the final location of the rocks (both distance downslope and lateral spreading) and (2) the trajectories of the bouncing rocks down the slope (compared with video imagery) except for two regions:

- The model underpredicted the number of particles close to the original release point. This was because many of those particles did not roll but slid down the slope under friction (because the experimental particles were not perfectly round but blocky) and as a result stopped their motion closer to the release point due to frictional energy dissipation.
- The model overpredicted the number of particles at long distances out in the runout zone. This was because a soil mantle at the surface dominated the runout zone in the experiment, rather than a rock surface as assumed by the model. As noted above, the restitution coefficient for soil is much less than rock, so the model underestimated the energy loss in these soil mantled regions, and the modelled particle moved further than the experiments.

Bourrier et al. (2009) does not document how they derived the coefficients of the matrix **A**, but Bourrier (2008) indicates they were derived from laboratory experiments where particles were impacted on a rough surface.

13.4 Discrete Element Models

An emerging modelling technique, which is implicitly Lagrangian and which follows directly from the discussion of the previous section, is discrete element modelling (DEM; not to be confused with Digital Elevation Models). The concept of DEM is that each rock fragment is modelled separately (i.e. mass, location, velocity etc.), and the interaction between the particles and between the particles and the boundaries are modelled. Typically many thousands of particles are modelled, and the average behaviour of the particles can provide a continuum representation of the flow.

The general physical principles of how the DEM particles move and interact (both with each other and with the boundaries) are as discussed in Section 13.3. To model the motion of individual rocks from rockfalls requires the user only to simulate a small number of DEM particles (e.g. Mitchell and Hungr, 2017). However, to model a debris flow, a typical simulation involves tens of thousands up to millions of computational particles. Even with these numbers of particles, the model is not simulating all of the particles in the flow, rather the idea is to model the aggregate behaviour of the flow, which is dominated by the interaction between the particles. Early DEM studies were limited to two dimensions (e.g. the downslope direction and depth), circular particles and relatively simple interaction dynamics.

Many DEM models follow the approach of Cundall and Strack (1979). They modelled a two-dimensional, slowly moving, geomechanics problem with a small number of different size circular discs. The interparticle forces were modelled by a damped spring, as were the tangential frictional forces. These interparticle spring forces were active only when the particle discs overlapped. The timestep used for the solution must be small enough to simulate the gradual overlap of the particles and the subsequent development of the interparticle normal and frictional forces, which results in the method being very compute intensive. Campbell et al. (1995) extended this approach to a highly dynamic situation by modelling a debris flow. Johnson et al. (2016) used Campbell's code to examine the enhanced runout with increased size dependence of debris flows and was able to successfully model the mechanism. They attributed this behaviour to acoustic fluidisation, which results from the dynamic interaction of the particles.

Computations of interparticle forces are significantly simplified if only circular or spherical particles are allowed. Accordingly many studies have adopted circular or spherical particles. One way to model irregularly shaped particles is as aggregates of these spherical particles. This is a common technique for rockfall analysis, as rock shape has been shown to strongly influence trajectory calculations as a result of the interaction between rotational and translational energy during impact with the ground (Mitchell and Hungr, 2017). The trade-off of this approach is that irregular particles can be modelled, but at the expense of using significantly more particles.

Three-dimensional simulations are less common because of the computational requirements. Mead and Cleary (2015) modelled a dry debris flow (comparing it against experimental data in Iverson et al., 2004) in three dimensions and with irregularly shaped particles. The particles were modelled to be superquadric in shape. Superquadrics are a generalised version of an ellipse. By changing the properties of the particles they were able to evaluate their effect on flow properties. They found particle aspect ratio, angularity and interparticle friction to be important to the flow dynamics.

The main limitation of DEM at the current time is that it is possible only to model the interactions between particles, so it is unable to model the interaction between particles and interstitial fluids like water and mud, and the resultant pore pressure effects. Thus they can model only dry flows (i.e. the Iverson et al., 2014 data in the previous paragraph), while wet flows and snow avalanches (for the latter air entrainment is important) cannot be modelled.

13.5 Escarpment Retreat

One recurring topic upon which LEMs have been applied is escarpment and cliff retreat. The questions around this topic are (1) what processes operate on or near the escarpment and how do they determine that an escarpment is maintained and (2) what determines the rate of retreat of an escarpment. The intent of this section is not to review all the work in this area, but to discuss how the process formulation has been found to have profound implications for whether cliffs are developed and/or maintained over the long term.

Tucker (1994), Kooi and Beaumont (1994) and Howard et al. (1994) carried out a series of one-dimensional simulations (a long profile that included a combination of plateau, cliff front and scree slope) using a variety of river erosion formulations (see Chapter 4) to model the evolution of the landform. The conclusion these papers reached was that the stream power detachment-limited model did a better job of maintaining the vertical cliff face than the transport-limited model. Figures 13.10 and 13.11 show similar simulations for similar starting landforms for the range of parameters that are expected in the field. All mechanisms show some flattening of the initially (near) vertical cliff, but the detachment-limited landforms show something that looks like parallel retreat,

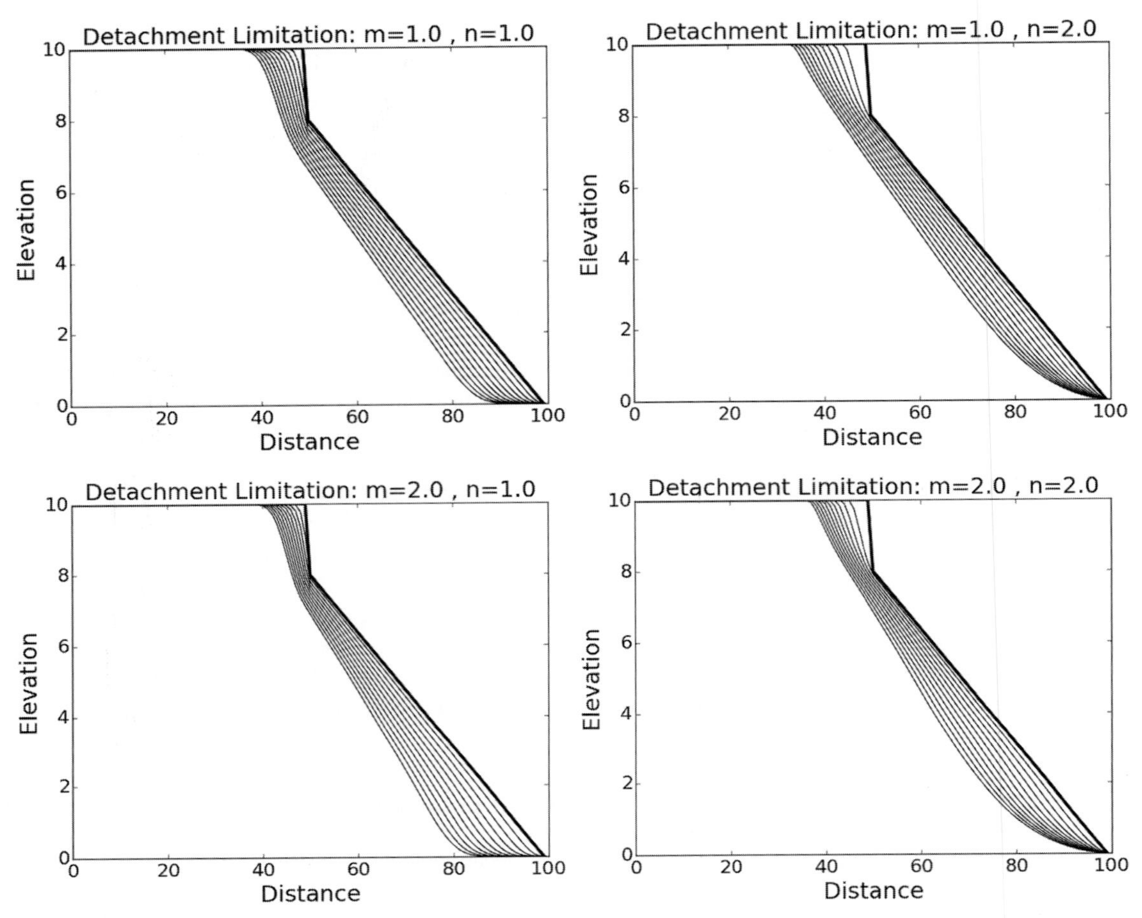

FIGURE 13.10: Cliff retreat modelled with the stream power detachment limitation equation (Equation (4.29)) for the parameters indicated and $\tau_c = 0$. The heavy line is the initial hillslope profile, and the end time for all the simulations is when the average elevation of the landform is 6.0, a convenient number that allows comparison of the final hillslope profiles with Figures 13.11 and 13.12.

whereas the transport-limited results show that flattening of the cliff dominates. The detachment-limited cases for $n = 1$ are particularly interesting because they are of the form of a pure advection equation with no diffusion, so that for our one-dimensional example

$$\frac{\partial z}{\partial t} = -\beta Q^m S = \beta x^m \frac{\partial z}{\partial x} \tag{13.47}$$

where x is the distance from the divide (the left-hand boundary of the figure). This is the classical advection equation where the velocity of movement of a disturbance in z (in this case the cliff line) is $-\beta x^m$ where the sign indicates the movement is from right to left in Figure 13.10. Thus provided that the cliff remains vertical (so that the top and the bottom of the cliff are both at equal distance from the divide), then the cliff will retreat at a velocity $-\beta x^m$ and the cliff profile will NOT change with time, only retreat, and the parallel retreat speed will decrease as the cliff approaches the catchment divide (i.e. x decreases).

Much of the smoothing of the cliff profile with time in Figure 13.10 is a result of numerical diffusion in the finite differences used to solve the detachment-limited equation. Lest the reader think that this smoothing in Figure 13.10 is a result of defective numerics, the algorithm used for the calculations is identical to that used in most existing landform evolution models, with the slope term being a forward finite difference (i.e. the slope is that slope from the current node to the next node downstream as done in D8) and the timestepping is explicit Euler. To further confirm that this smoothing of the cliff face is numerical diffusion, Figure 13.12 shows a pure advection solution with constant advection velocity to the equations ($m = 0$, $n = 1$ in Equation (13.47)), and this smoothing of the cliff face is still evident and of about the same magnitude as seen in Figure 13.10. Accordingly it is reasonable to

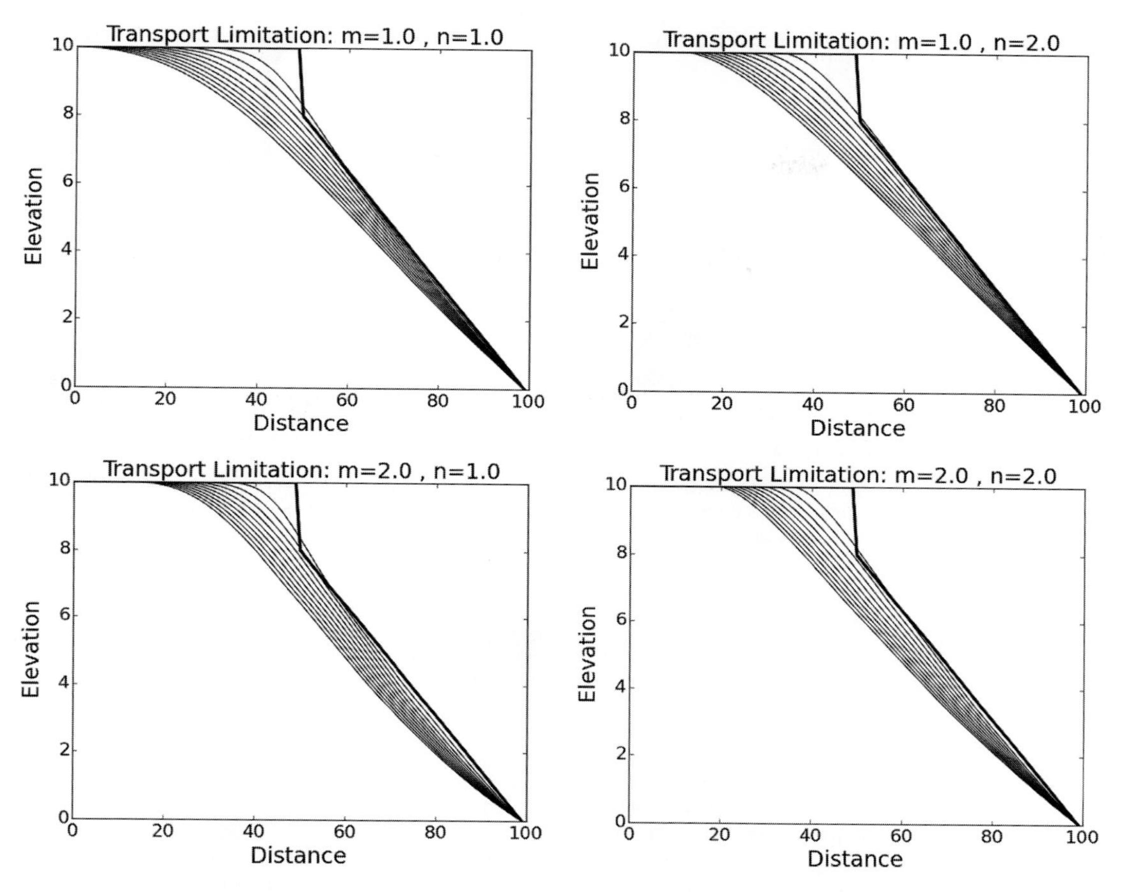

FIGURE 13.11: Cliff retreat modelled with the transport limitation equation (Equation (4.10)) for the parameters indicated. The heavy line is the initial hillslope profile, and the end time for all the simulations is when the average elevation of the landform is 6.0, a convenient number that allows the comparison of final hillslope profiles with Figures 13.10 and 13.12.

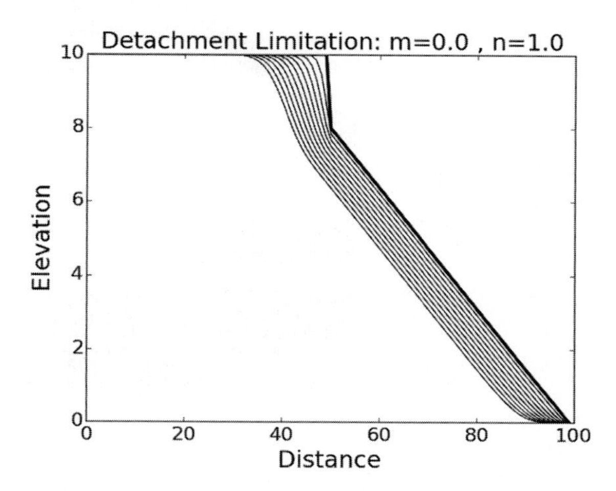

FIGURE 13.12: Cliff retreat modelled with a pure advection equation (equivalent to detachment-limitation $m = 0, n = 1$). The heavy line is the initial hillslope profile, and the end time for all the simulations is when the average elevation of the landform is 6.0, a convenient number that allows the comparison of final hillslope profiles with Figures 13.10 and 13.11.

assume that at least part of the smoothing of the cliff face that Tucker, Kooi and Howard observed in their detachment-limited simulations was a result of numerical diffusion.

Kooi and Beaumont (1994) also coupled their erosion with an elastic plate to model isostatic rebound (see Chapter 12). They found that the unloading of the plate as a result of erosion led to the uplifting of the plateau upslope of the cliff and warping so that flows upstream of the cliff reversed direction and flowed away from the cliff face. As a result, erosion of the cliff face was eliminated (since the catchment divide had moved to the cliff face) and the cliff face becomes stable. The cliff face only retreated as a result of erosion and undercutting on the slope downstream of the cliff face (for further discussion see Section 16.1.3).

The alternative simulation method for cliff retreat is to assume that the cliff face is more erosionally resistant than the geology upstream and downstream of the cliff face (Weissel and Seidl, 1997). Figure 13.13 shows

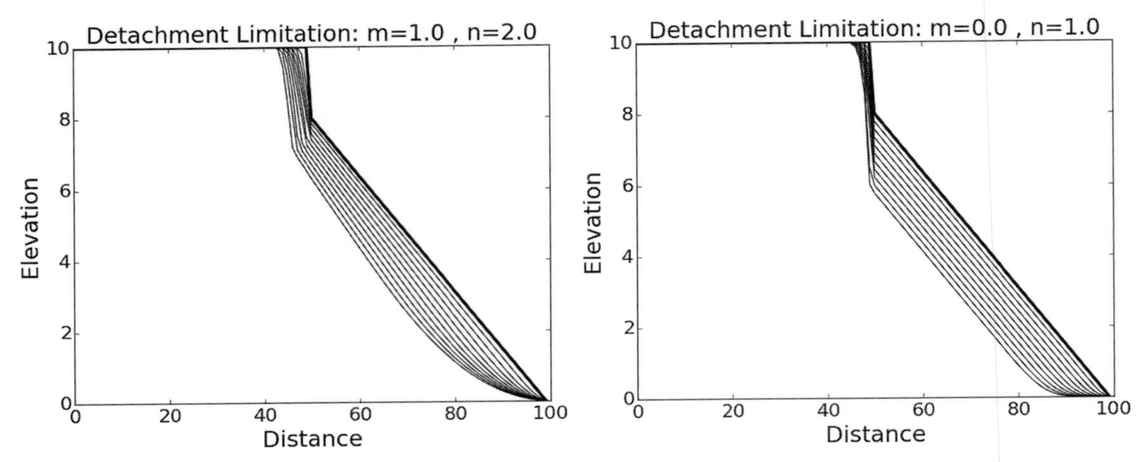

FIGURE 13.13: Cliff retreat modelled with the detachment limitation equation (Equation (4.29)) for the parameters indicated and where the cliff face has a detachment rate 0.1 times that of the material below and above the cliff face. The heavy line is the initial hillslope profile, and the end time for all the simulations is when the average elevation of the landform is 6.0, a convenient number that allows comparison of the final hillslope profiles with Figure 13.10.

simulations for a case where the cliff erodibility is 10% of the material above and below the cliff face, and this clearly shows the expected result of the maintenance of the vertical cliff face, and that the numerical diffusion, while still present, is also reduced.

The persistence of numerical diffusion in the simulations results suggests that custom numerical solvers might be advantageous for modelling cliffs. The main problem is that cliffs are vertical or near vertical and that the retreat occurs horizontally, and both of these properties are hard to resolve with a (fixed) horizontal grid of land elevations.

One solution would be to model the cliff as an explicit discontinuity in elevations and to move this discontinuity in elevations laterally across the grid with the rate of movement determined by some retreat mechanism. The physics that control the retreat mechanism would need to be parameterised. If the landform evolution model uses a TIN with nodes that can move in time (e.g. the CHILD landform evolution model), this discontinuity could be the boundary between the triangles of the TIN so that the cliff face is modelled as occurring along the boundary of the TINs, and as the cliff moves the triangle boundaries could also be moved. Such discontinuities along TIN boundaries are a common technique in structural and geomechanics finite element codes. The author's gridded SIBERIA model likewise uses discontinuities in the grid to model contour banks (a discontinuity in flow directions, albeit stationary) and channels (a discontinuity in elevations based on where channels occur) but is otherwise limited by its stationary computational grid. The challenge is then determining at what stage does a cliff become steep and high enough to require special treatment with a 'cliff model' as outlined above.

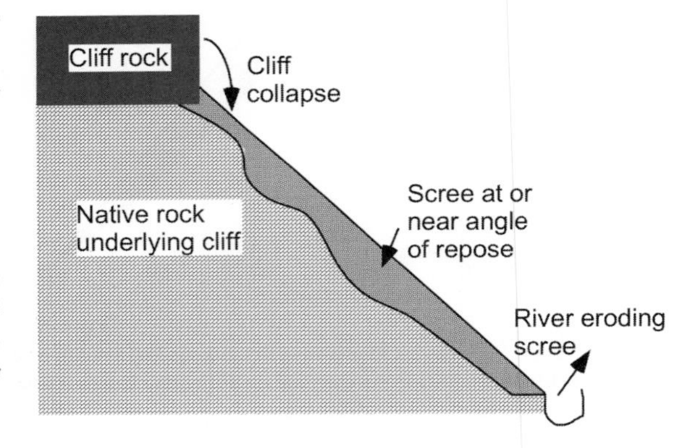

FIGURE 13.14: A schematic showing the mass balance for processes acting on a scree slope downslope of a stable retreating cliff face.

A simple conceptualisation of cliff retreat can be constructed for a cliff with a scree slope below. For cliff faces with scree slopes underneath them, the retreat rate is likely to be a balance between the clearance rate of scree at the base of the slope (e.g. by a river) and the rate of collapse of the cliff face (Figure 13.14). If the rate of removal of scree by the river is higher than the rate of retreat of the cliff, then eventually the river will remove all of the scree. If the rate of removal is lower than the rate of cliff collapse, then the scree slope will aggrade, push the river toward the other side of the valley and eventually cover the cliff. Thus cliff retreat rates may be a balance between the cliff source term and the river removal term. Rather than being the result of a lucky coincidence, this balance might be achieved by (1) if the river removal rate is too high, then the cliff face will be undercut increasing the rate of collapse due to instability of the cliff face and increasing the retreat rate and (2) if the river removal rate is too low, then the scree aggrades covering part of the cliff, reducing the cliff face instability and weathering, stabilising the cliff and reducing the retreat rate.

13.6 Further Reading

For energy dissipation and fragmentation resulting from rockfall and collisions, there is a body of work in the asteroid and comet collision community that, while it may not be directly relevant, may be of interest for this research community's different approaches (e.g. Durda et al., 2004, 2007). For greater detail on the theory and experimental evidence for debris flow processes, readers are referred to the extensive body of work by Iverson and his colleagues.

14 Vegetation and Wildfire

14.1 Introduction

This chapter is about the role of vegetation in the evolution of landforms and soils, and the potential feedbacks between the three. The chapter covers (1) vegetative effects on fluvial erosion rates, (2) vegetation and climate feedbacks including the development of patterned vegetation and hillslopes in arid zones and (3) wildfire. We have touched on other vegetative effects elsewhere in the book including (1) tree throw and its role in soil depth evolution (Chapter 5) and soil mixing (Chapter 6) and (2) the role of leaf litter as a source material for soil organic matter (Chapter 9).

The chapter will touch on the broader issues of ecosystems dynamics and dynamic vegetation models. Typically these models have been developed for other purposes such as climate change impact assessment and ecohydrology applications, and while they consider a changing climate, they mostly assume fixed soils and landforms. This chapter will not provide a comprehensive review of these models because that would require a book in its own right (e.g. for ecohydrology see Eagleson, 2002 or Rodriguez-Iturbe and Porporato, 2005). Rather we will examine these models to assess their suitability as models in a coupled soilscape-landscape evolution model.

In any discussion of vegetation fire needs to be considered. In many places in the world fire is an integral part of the ecosystem function, including vegetation, and it is an important process impacting on long-term average soil erosion rates and soil organic matter mass balances. In the context of soilscape and landform evolution research, it has often not been accorded the profile it deserves, and the discussion here aims to emphasize the impact of fire, and introduce techniques that might be suitable to study its impact.

We will not talk about some biotic factors that undoubtedly influence landform and soilscape evolution such as animal effects (e.g. grazing, beaver dams), biocrusts, fungal/mycorrhizal processes on soil weathering and large woody debris in rivers. The lack of coverage here is not to deny their importance. The lack of treatment simply acknowledges the lack of widely applicable models for them. See Viles et al. (2008) for a conceptual discussion of some of these processes.

14.2 Landforms, Soils and Vegetation

Landforms, soils and vegetation are intimately linked, and this linkage has significant impacts on other important environmental processes. Models for exploring these linkages and impacts are currently a work in progress, so this chapter focuses on the individual components of vegetation systems, with some examples of impacts from field and usage in models where appropriate.

As an example of these links, Peterman et al., (2014) examines the correlation between soil depth (Chapter 6) and soil carbon (Chapter 10) and found a strong correlation. The underlying driver of the correlation was that deeper soils provide greater water-holding capacity, and, at least in water-limited regions, this increases the rate of production of vegetative biomass. As discussed in Chapters 3 (Hydrology) and Chapter 11 (Soilscape Modelling), the distribution of both water and soils is driven by the landform. Increased biomass is an increased carbon source term in the models for soil carbon in Chapter 10 so soil carbon increases. Thus vegetation is the key link underpinning the empirical observation by Peterman and colleagues of a link between soil depth and soil carbon. They also observed a link between soil depth and fire, and as we will see later in the chapter, this also arises because of the vegetation biomass growth. Finally, Peterman and Bachelet (2012) note that soil nitrogen is a product of a biogeochemical cycle driven by vegetation, so without vegetation soil nitrogen has been found to decline sharply.

14.2.1 A Simple Model for the Effect of Vegetation on Erosion

A very simple way to allow for vegetation effects on landform evolution is for the user to define the vegetation and its characteristics. In this way the effect of vegetation on erosion rate can be estimated using one of the many ways that vegetation is accounted for in existing agricultural erosion models. For instance, for the Universal Soil Loss Equation (USLE) Table 10 from Wischmeier and Smith (1978) (Figure 14.1) gives tabulated values for the ratio of the hillslope fluvial erosion rate with and without vegetation, called the cover factor, C. To allow for vegetation the erosion for unvegetated conditions is multiplied by the cover factor for the vegetation, C, to calculate the erosion rate with vegetation. This table explicitly accounts for two factors: (1) the reduction of rainsplash detachment as result of the canopy intercepting raindrops (see Section 4.2) and (2) the reduction in the shear stress applied to the soil surface, and thus erosion by overland flow, as a result of the groundcover protection by grasses (see Section 4.2.3). What is not clear from the table is that it also implicitly includes the effect of vegetation on infiltration and thus the reduced runoff that occurs for highly vegetated areas. The RUSLE2 (Revised USLE, Version 2; USDA-ARS, 2008) provides an explicit equation for calculating the cover factor that includes (1) the canopy effects, as above, (2) the groundcover effect, as above, (3) surface roughness, (4) ridge/furrow height, (5) soil biomass, (6) soil consolidation, (7) water ponding and (8) antecedent soil wetness. These factors can also change over time, acknowledging, for instance, that it is possible for peak rainfall/runoff to occur at times when vegetation provides relatively poor cover. Vegetation contributes directly to factors (1) and (2), and indirectly to factors (5) and (8). The remainder are largely agricultural management factors. The cover factor is the multiple of the eight factors above, but since this section is about vegetation, then if we lump all the non-vegetation factors together, then the cover factor is

$$
\begin{aligned}
C &= C_{\text{canopy}} C_{\text{groundcover}} C_{\text{remaining factors 3-8}} \\
C_{\text{canopy}} &= 1 - f_c e^{-0.33 h_f} \\
C_{\text{groundcover}} &= e^{-b f_g}
\end{aligned}
\tag{14.1}
$$

where C is the cover factor where the canopy and groundcover effects are made explicit, and for the presentation here the remaining factors are simply lumped into one factor (in RUSLE each of the remaining six factors also have equations for their corresponding C factor, but they will not be discussed here). The parameter h_f is the height of the canopy (in metres), f_c and f_g are the fraction of the plan area covered by canopy and groundcover, respectively, and b is a parameter in the range 0.025–0.06. Thus the reduction in erosion due to the canopy is linear with the proportion of cover, while the reduction due to groundcover is nonlinear with the maximum rate of reduction with increasing groundcover occurring for low

TABLE 10.—*Factor* **C** *for permanent pasture, range, and idle land*[1]

Vegetative canopy		Cover that contacts the soil surface						
Type and height[2]	Percent cover[3]	Type[4]	Percent ground cover					
			0	20	40	60	80	95+
No appreciable canopy		G	0.45	0.20	0.10	0.042	0.013	0.003
		W	.45	.24	.15	.091	.043	.011
Tall weeds or short brush with average drop fall height of 20 in	25	G	.36	.17	.09	.038	.013	.003
		W	.36	.20	.13	.083	.041	.011
	50	G	.26	.13	.07	.035	.012	.003
		W	.26	.16	.11	.076	.039	.011
	75	G	.17	.10	.06	.032	.011	.003
		W	.17	.12	.09	.068	.038	.011
Appreciable brush or bushes, with average drop fall height of 6½ ft	25	G	.40	.18	.09	.040	.013	.003
		W	.40	.22	.14	.087	.042	.011
	50	G	.34	.16	.08	.038	.012	.003
		W	.34	.19	.13	.082	.041	.011
	75	G	.28	.14	.08	.036	.012	.003
		W	.28	.17	.12	.078	.040	.011
Trees, but no appreciable low brush. Average drop fall height of 13 ft	25	G	.42	.19	.10	.041	.013	.003
		W	.42	.23	.14	.089	.042	.011
	50	G	.39	.18	.09	.040	.013	.003
		W	.39	.21	.14	.087	.042	.011
	75	G	.36	.17	.09	.039	.012	.003
		W	.36	.20	.13	.084	.041	.011

[1] The listed C values assume that the vegetation and mulch are randomly distributed over the entire area.

[2] Canopy height is measured as the average fall height of water drops falling from the canopy to the ground. Canopy effect is inversely proportional to drop fall height and is negligible if fall height exceeds 33 ft.

[3] Portion of total-area surface that would be hidden from view by canopy in a vertical projection (a bird's-eye view).

[4] G: cover at surface is grass, grasslike plants, decaying compacted duff, or litter at least 2 in deep.
W: cover at surface is mostly broadleaf herbaceous plants (as weeds with little lateral-root network near the surface) or undecayed residues or both.

FIGURE 14.1: Cover factors for the relative impact of vegetation cover on erosion rate. Note the distinction between the relatively small effects of canopy versus the strong effects of ground cover. The differentiation between G and W ground covers reflects plants where the canopy is strongly attached into the soil surface and thus provides strong protection against flowing water (G) versus plants where the canopy simply lies on the surface and is easily moved or detached from the soil surface by flowing water (W) (from Wischmeier and Smith, 1978).

groundcovers. In the USLE (Figure 14.1) the cover and groundcover effects are lumped together in the C factor for the erosion calculation. In our LEM calculations, if we model rainsplash and rainflow separately from overland flow fluvial erosion, then the C_{canopy} term should be applied to the rainsplash transport term and $C_{groundcover}$ to the fluvial erosion term. While the details of the vegetation effects on erosion vary slightly between the various erosion models, the general principle of the cover factors of the form in Equation (14.1) is commonly used worldwide and is empirically well founded on field data. The manuals for the USDA agricultural erosion models USLE (Wischmeier and Smith, 1978), RUSLE2 (USDA-ARS, 2008) and WEPP (Laflen et al., 1991) provide more detail. It is important to note that these models have been developed for application to agricultural areas subject to low gradient fluvial erosion, and so do not represent processes that primarily occur on steeper slopes including rilling, gullying and debris flows.

The form of the groundcover correction is exponential, and the reason for this nonlinearity requires a brief explanation. If the sole effect of groundcover was to protect the surface by covering it, then the reduction with cover would be linear from 1.0 at 0% cover to 0.0 at 100%. However, vegetation also reduces the shear stress action on the soil surface by changing the soil surface roughness. The bottom shear stress equation (i.e. $\tau_0 = \gamma y S$ for a wide channel) is derived by a force balance between that proportion of the self-weight of the water acting downslope (the right-hand side of the bottom shear equation) and the resistance to the flow provided by the bottom roughness (the left-hand side of the bottom shear equation). In the case of vegetated surface there is the drag force from the flow acting on the stems of the groundcover, which provides a further resistance to flow so that the force balance equation is

$$(1 - f_g)\tau_0 + f_g \tau_v + \frac{\text{drag}}{\text{area}} = \gamma y S \tag{14.2}$$

where τ_v is the shear stress acting on the groundcover, and the drag force per unit area is a function of the number of plant stems per unit area, the average area of a stem at right angles to the flow direction and the velocity of the flow squared (Lopéz and Garcia, 1998). The drag on the stems results in a reduction in erosion over and above the groundcover effect (the second term in Equation (14.2)) leading to the exponential decline. Nepf (2012) notes that while Equation (14.2) is useful to explain the effect of vegetation, it is a poor tool for experimental determination of the soil bottom shear stress because the bottom shear stress term (the first term in Equation (14.2)) is typically a small term determined by taking the difference between two large numbers (the total shear stress, and the resistance to flow per unit area from the vegetation).

Foster (1982) and Foster et al. (1995) proposed a simpler version of Equation (14.2) that doesn't explicitly include the drag term, called the shear stress proportioning fraction, where

$$\tau_f = \tau_0 \frac{f_s}{f_t} \tag{14.3}$$

where τ_f is the shear stress acting on the soil surface, and f_s and f_t are the roughness of the bare soil and vegetated surface, respectively.

While the formulations in Equations (14.1)–(14.3) are different in detail, they all partition the shear stress between that applied to the bare soil between the vegetation, and that applied to the vegetation. They all then calculate the detachment from the soil by erosion as a function of the shear stress applied to the bare soil fraction. Moreover, all formulations have the steepest reduction in bare soil shear stress for low percentage cover by vegetation with the rate of reduction declining as the percentage cover increases.

The main conclusions from this section is that if the characteristics of the vegetation are known, then it is possible to estimate the effect of vegetation on erosion by using the correction factors from agricultural erosion models. Perhaps the main problem with using agricultural erosion models is that they are heavily (though not completely) focussed on agricultural applications, so the data for the cover factors are biased toward annual crops and agricultural management practices (e.g. ploughing). Accordingly, given the focus on agriculture, much of the data are not applicable to problems involving the modelling of the pre-human agricultural past, steep mountainous landscapes and geomorphic time scales.

Willgoose (1995) used the table in Figure 14.1 to estimate the effect of vegetation (relative to an unvegetated condition) on a rehabilitated mine waste rock dump for 1,000 year landform evolution simulations (Willgoose and Riley, 1998a, b). The main uncertainty in this approach was the ability to measure or predict the vegetation canopy and groundcovers. Willgoose had measurements of these from adjacent natural field sites and assumed that these were constant for the 1,000 year design life of the containment structure (despite the landform and soils being predicted to evolve significantly over the 1,000 years). Subsequent field experiments approximately validated the use of USLE cover factors for current-day conditions (Evans et al., 1996; Evans and Willgoose, 2000) but was not able to test the assumption of stationarity of the vegetation or the potential impact of fire over 1,000 years.

Temme and Veldkemp (2009) in their LAPSUS LEM also adopted a cover factor but where the erosion rate was a linear function of vegetation cover. Relative to the USDA factors this approximation overestimates erosion for midrange vegetation covers. They estimated vegetation cover as a function of soil thickness, and the rainfall and temperature relative to averages over the last 50,000 years. If soil was less than 0.5 m, then the vegetation was reduced linearly by the soil thickness so that at 0.0 m soil thickness the vegetation cover was 0. In Section 14.4 we will discuss models that exhibit a similar relationship with soil depth where the amount of water stored in the soil profile limits the rate of growth, and therefore vegetation density.

Yetemen et al. (2015a) use a conceptually similar approach to Equation (14.1), but they used it to predict the effective shear stress applied to the soil surface relative to an unvegetated surface (functionally equivalent to a cover factor) as a function of LAI. They used an empirical function to relate the LAI to the percentage cover of the surface. Like Equation (14.1) their function had the greatest rate of erosion reduction for low vegetation density and groundcover.

In cases with no ground-based data there has been a body of research, with mixed success, that has tried to estimate the canopy and groundcover directly from remote sensing (Vrieling, 2006), or by correlating estimates of aboveground biomass with cover (e.g. Schönbrodt et al., 2010). In the absence of defensible remote sensing correlations it is very common to simply use tabulated values relating cover factor with cropping type (e.g. Panagos et al., 2015).

Finally it is important to note that the cover factor conflates the combined effect of vegetation on erosion resistance and hydrology. Thus the cover factor decreases with increasing vegetation cover because of (1) the protection of the vegetation, (2) the increased infiltration rate and (3) the reduced soil moisture at the start of a rainfall event (Kinnell, 2010). In LEMS that model erosion resistance and runoff generation separately, in adjusting for vegetation it should be noted that the C factor captures the combined effect of vegetation on both hydrology and erosion, not just the erosion.

14.3 Dynamic Vegetation Models

Models of dynamic vegetation (DVM) (and their close relatives, dynamic global vegetation models, DGVM) aim to model not only the cycles of nutrients, climate and water but also the evolution of the vegetative ecosystem that responds to those cycles so that the vegetation characteristics (e.g. light use efficiency (LUE), water use efficiency (WUE), species structure, proportion of understorey to over-storey canopy) evolve in response to those cycles. These cycles and the vegetative response are quite complex and involve many processes (Figures 14.2 and 14.3). In general they have not been fully characterised. We will only be able to scratch the surface of these processes in this section. There is considerable interest in dynamic vegetation models to model the terrestrial carbon cycle in global climate models (e.g. Cramer and Field, 1999; Quillet et al., 2010), but the complexity of these models is beyond what will be discussed below, even though the principles are similar.

We start by outlining a simple mathematical framework within which the DVMs discussed in the remainder of the chapter can be viewed. Almost all DVMs and DGVMs model biomass, and if they are interested in factors that are related to biomass (e.g. vegetation cover), they use empirically derived relationships or another model to estimate them. Some models simulate carbon rather than biomass because they have a compelling interest in the carbon cycle, though as indicated in Chapter 10 there is a good empirical relationship between carbon and biomass. The discussion below uses biomass per unit plan area as the property to be modelled. Conceptually the biomass dynamics at a point can be formulated as a mass balance equation

$$\frac{\partial B}{\partial t} = P - R - D - H - F \tag{14.4}$$

where B is the live biomass (mass per unit area), P is the growth rate of live biomass, R is the respiration loss (mass per unit area per unit time), D is the death rate of live biomass and H and F are the biomass loss caused by herbivores and fire, respectively. We need to make a distinction here between live and dead biomass because while the total biomass is important for many processes (food for herbivores), only the live biomass grows, so D is the conversion rate of live biomass (e.g. plant leaves and flowers) to dead biomass (e.g. leaf litter and seeds). We will return to modelling dead biomass shortly.

A simple biomass production equation is

$$P = \alpha B\left(1 - \frac{B}{B^*}\right) \tag{14.5}$$

where B^* is the maximum biomass possible, and α is the rate parameter for biomass growth. For low biomass this term is linear in B, and this reflects that for scattered trees

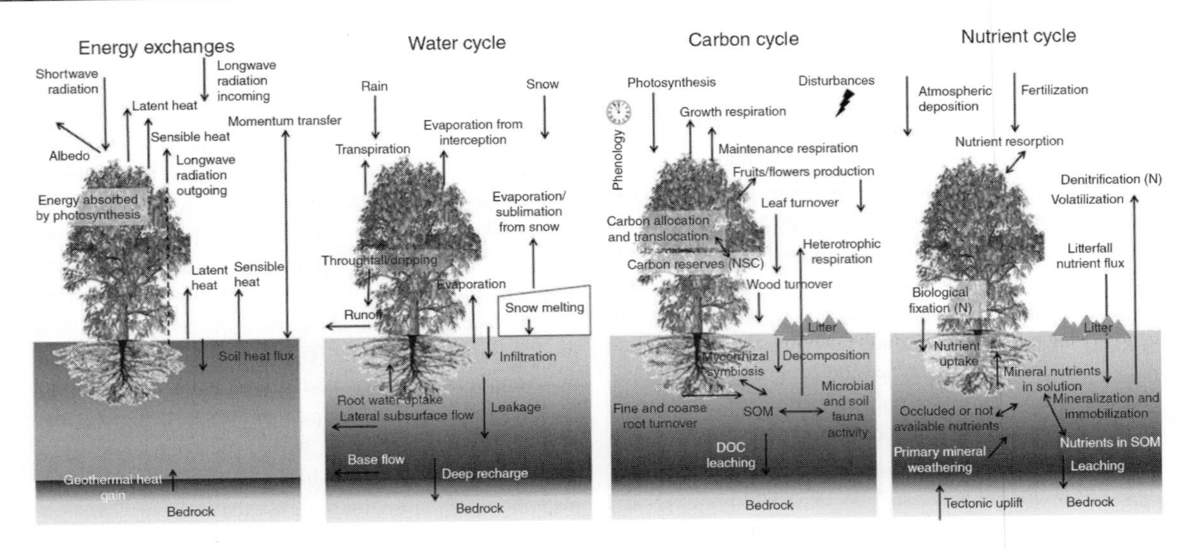

FIGURE 14.2: A summary of the energy, water, carbon and nutrients for a forested ecosystem. Note that the picture simplifies the typical modelling of canopy and rhizosphere processes (see, e.g. Figure 14.3) since many models (1) discretise the processes with elevation above/below the soil surface so that process rate varies with elevation, (2) allow a number of species to coexist and evolve as a result of competition for resources and (3) differentiate the canopy and rhizosphere properties of these species (e.g. grasses and shrubs versus trees) (from Fatichi et al., 2016).

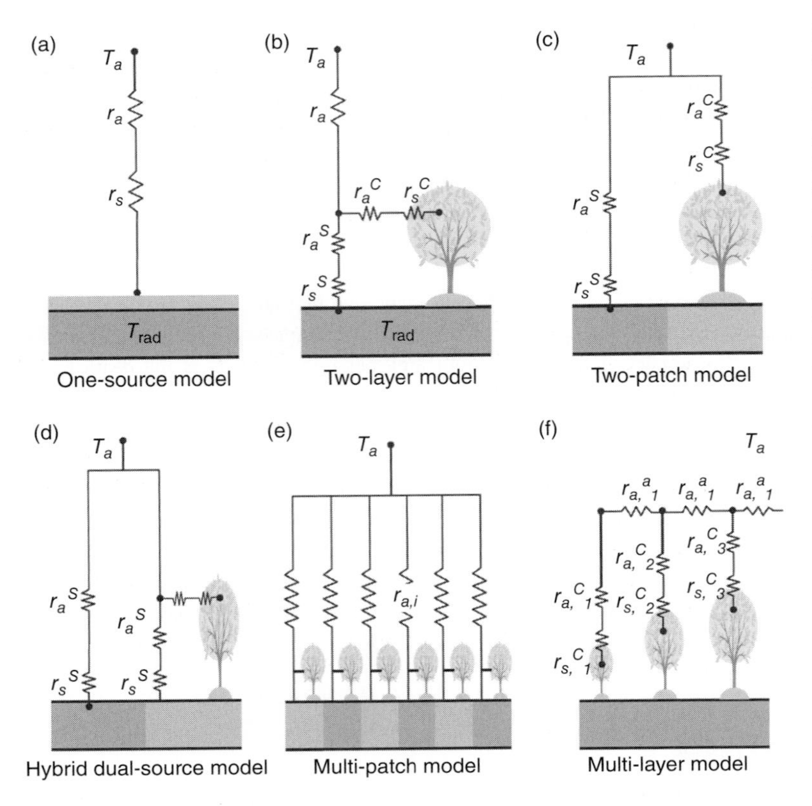

FIGURE 14.3: This is a schematic of some of the approximations used to model evapotranspiration in ecohydrology and land surface models. T_a is the total evapotranspiration, and the various r's are the resistances to flows across the atmospheric gradient in water content: (a) bare soil with no vegetation, (b) the single layer vegetation canopy which shades the soil, (c) where the soil and vegetation are independent, (d) a mixture of cases (b) and (c), (e) multiple independent patches of vegetation (each patch potentially different species) and (f) a multilayer canopy (from Zhang et al., 2016).

is short for forests where average tree lifespans are measured in decades or even centuries (van Mantgem et al., 2009). We can thus reasonably assert that (1) for forests tree death is not dominant and other processes likely dominate the loss of live biomass, (2) for grasses seasonal growth and death factors such as water availability might be dominant and (3) herbivory and fire are likely to be important in many locations.

The discussion above simplifies processes considerably and does not include some of the complexities built into existing DGVMs. Some of these excluded processes include (Quillet et al., 2010) (1) changes in plant functional type (i.e. ecosystem changes, such as changes in the type of forests or grasses) in response to evolving climate, soilscape and landform, (2) nutrient competition and nutrient cycling, (3) internal plant chemistry and (4) response to changes in temperature and atmospheric CO_2.

14.3.1 Agricultural Crop Models

There is an emerging interest in applying LEMs to agricultural settings either to look at the historical long-term impact of agriculture on soils and landforms (Schrool and Veldkamp, 2001; Baartman et al., 2012) or to investigate the future impacts of agriculture management practices (Barreto et al., 2013; Fleskens et al., 2016). LEMs have also been applied at mine sites with a range of different surface conditions and surface treatments, though to date they have not been coupled to vegetation models (Hancock et al., 2008).

LEMs provide a number of conceptual advantages over traditional erosion models including explicit gully/rill erosion (many traditional models have gully/rill capabilities but typically the properties need to be input, rather than develop naturally out of the landform evolution), soil depth changes due to erosion and the dynamics of erosion and deposition for erosion protection measures (e.g. sediment deposition in the drainage line behind contour/graded banks that result in the drainage line eventually choking and failing). It is common for traditional erosion models to be coupled with a crop model to allow the erosion model to assess erosion through the growing season. These crop models vary in complexity from (1) simply imposing an assumed growth cycle through to (2) fully coupled agricultural productivity models. In principle there is no reason that these same crop models cannot be coupled with a landscape evolution model. However, various considerations may be important, including the following:

- They assume fixed soil properties: soil depth, water-holding capacity, grading, SOC, chemistry and pH. Changes in these properties may or may not be important depending on the time scales of interest. For instance, depth, water capacity and grading may not change significantly at the decadal time scale unless erosion is particularly high (e.g. the author has seen Spanish olive groves on steep lands where erosion is high enough that significant soil depth loss has occurred), but SOC and chemistry and pH can change quite quickly.
- The crop models are highly specialised, focussing on crops of economic value, and so typically are not optimised for rangeland, grazing and natural lands. In particular, by focussing on annual crops they do not typically deal with recruitment and reproduction, nor with competition. This is not true of all models, with APSIM (Agricultural Production Systems sIMulator) having a rangeland option that allows a mixture of annual and perennial species (Johnson, 2008).
- In most cases fire is ignored.

Other aspects of these crop models are more sophisticated than what has been discussed here including the following:

- Nutrient cycles are modelled included N, P and C cycles. While the soil geochemistry is fixed, the nutrient cycles are dynamic responding to the geochemical reactions, and sources and sinks of the nutrients.
- Very sophisticated crop physiology parameterisations might be difficult to predict for an evolving ecosystem. Competition, if it is modelled at all, is modelled very simply.
- It is common to use daily climate data.

One crop model, widely used internationally, is APSIM (Keating, 2003), which has been developed by amalgamating a number of previous models customised to specific crop types. In particular its pasture model is derived from the SGS model (Johnson, 2008), and the points above capture its structure. It is a complex model and includes extra complexity to do with farm management and irrigation. Many crop model users recognise that the complexity in most crop models leads to 'low science transparency' (Wang et al., 2002), which is a shorthand for saying that it is difficult to understand how the model leads to the answers obtained.

Ramankutty et al. (2013) recognised that a simpler model based on APSIM would elucidate first-order dependencies in the model operation and would have value in other applications. They fitted regression models to the output of APSIM for a wheat crop in a water-limited environment and found that a model that used only mean rainfall, number of rainfall events/month, mean radiation and temperature provided a satisfactory fit ($R^2 = 0.9$). Thus one use for the agricultural models is in the development of reduced-order models where much of the complexity of the reductionist approach in the agricultural models is removed. The most important complexity removed by Ramankutty was the daily time series of rainfall, temperature and net radiation, which we do not typically have and cannot generate reliably (see Chapter 3), for landscape evolution applications. Brooks et al. (2001) presented a similar simplification of the SIRUS crop model for a site in England and found that the best predictors of crop yield were soil water capacity and temperature, though they did not attempt to simplify the leaf area, water balance and transpiration models.

14.3.2 Ecosystem Models

The ecosystem community has developed models of vegetation evolution that have been adapted by the global climate modelling community, who have used them to estimate the terrestrial interaction with the atmosphere. Originally this interest was simply to model latent heat (i.e. evapotranspiration) exchange at the land surface as this is the major energy interaction between the land and the atmosphere. In recent years, as noted in Chapter 1, there has developed an interest in the carbon assimilation capacity of the terrestrial biomass. These models now model the evolution of and competition between different phenologies and ecosystems and are collectively known as Dynamic Global Vegetation Models (DGVM). It is generally considered that the BIOME model (Prentice et al., 1992) was the start of this class of model.

The LPJ (Lund-Potsdam-Jena) model is typical of these DGVMs and draws heavily from BIOME. The discussion below follows Sitch et al. (2003) (Figure 14.4). The biomass model above generally follows the principles in LPJ though with a number of differences:

- LPJ does not distinguish live and dead biomass, but does distinguish between leaves and sapwood and roots (i.e. nonstructural versus structural biomass).
- LPJ divides vegetation into 10 ecosystem types (called plant functional types, PFTs), which are generally different types of forest (e.g. temperate versus tropical, evergreen versus deciduous), and each PFT has a separate set of processes and parameters.
- The mechanics of photosynthesis in LPJ are modelled explicitly at the daily time step. The production and respiration functions in Equations (14.5) and (14.7) are conceptually the same as LPJ, but the mechanics of the chemistry, water limitation and canopy effects are much more explicit in LPJ.
- A soil organic matter model is included in LPJ. It is a single-layer three-compartment model as discussed in Chapter 10, with an aboveground litter pool.
- LPJ explicitly models leaf phenology (e.g. evergreen and deciduous) with leaves responsive to budburst and temperature.
- C_3 and C_4 photosynthesis are explicitly modelled in LPJ.
- Respiration follows Equation (14.7), but LPJ has the respiration rate also responsive to the carbon:nitrogen ratio in the plant.
- Tree mortality occurs as a result of light competition, low growth (e.g. respiration greater than photosynthesis) and heat stress,
- LPJ uses monthly climate data, with some of the monthly data being disaggregated to daily resolution.

14.4 Soil Moisture and Water Limitation

The dynamic vegetation models of the previous section did not explicitly discuss the impact of water limitation on plant growth. About 50% of the planet is water-limited at some point during the growing season(s). To model this process explicitly we need to model the evolution of soil water content (SWC, measured as depth of water in the soil profile) over time.

In Section 14.2 we discussed a simple model used by Temme and Veldkamp (2009) that limited vegetation cover as a result of shallow soil depth and rainfall. While not explicitly derived for water-limited environments, the net effect of the model was to impose a water limitation effect on vegetation cover. In this section we look at a more explicitly physically based model structure, albeit more complex, for the effect of water limitation on vegetation cover.

14.4.1 The Physical Principles

If we model the soil as a single layer, then a mass balance equation for the change of soil water content, SWC (mm depth of water), per unit time for a pixel with unit plan area is

$$\frac{\partial(\mathrm{SWC})}{\partial t} = I - T - E - D_D - L \qquad (14.16)$$

where I is infiltration of water from the surface (in mm depth per unit time), T is transpiration consumption of

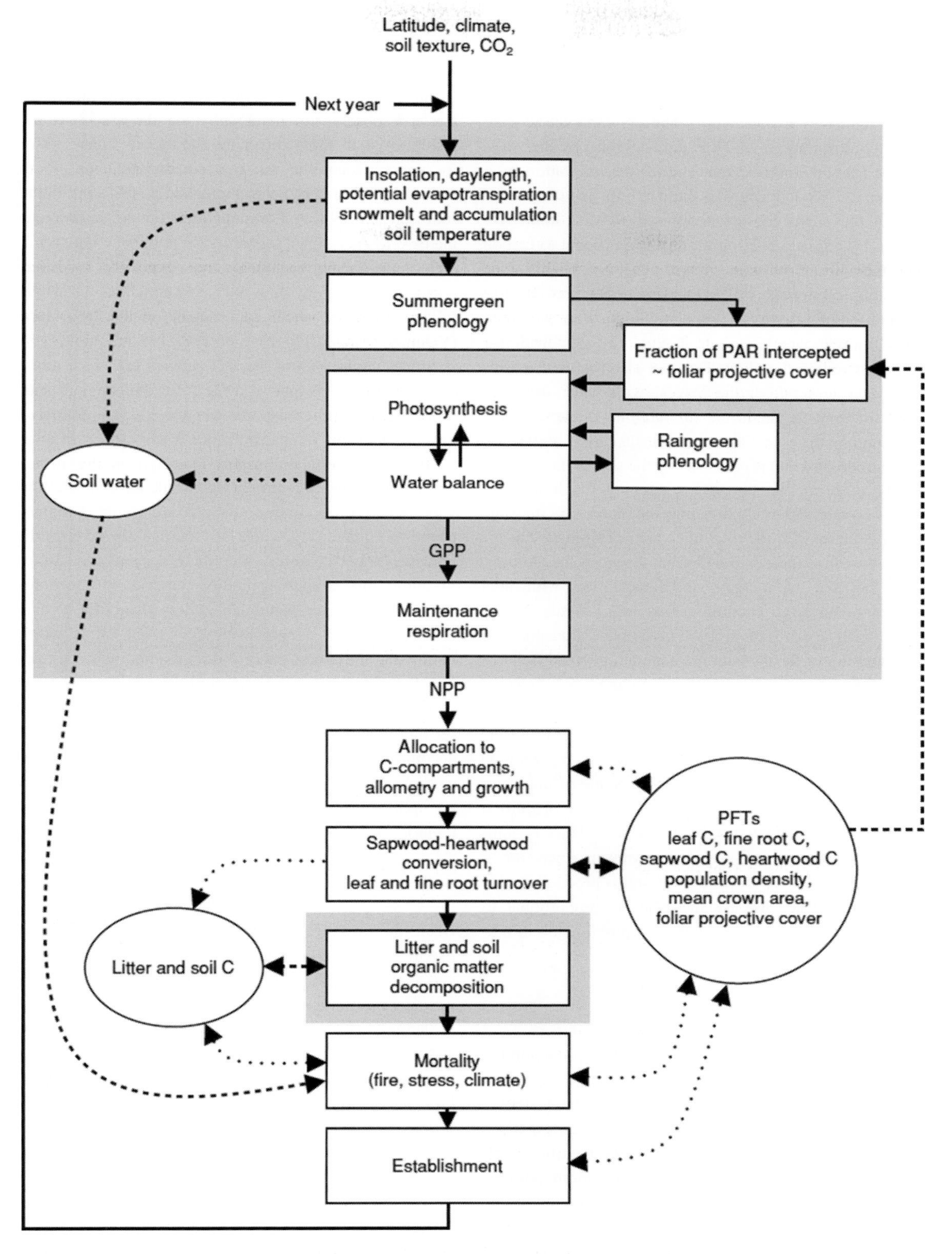

FIGURE 14.4: Flow chart of the calculation modules in LPJ (from Sitch et al., 2003).

water by the vegetation, E is bare soil evaporation of water from the near surface, D_D is deep drainage of water from the bottom of the soil profile and L is net lateral transfer of water down the slope within the soil profile (see Chapter 3 for more detail). The key variables here are (1) the transpiration, which is the water use by the vegetation, (2) soil water content and (3) the infiltration from the surface. We will consider these in turn, and then return to how the overall ecosystem responds to water limitation.

In dry climates transpiration is proportional to carbon assimilation and biomass growth. This is a result of the operation of stomata in leaves, which open and close as part of the photosynthesis process. Stomata are the openings through which carbon dioxide diffuses from the atmosphere into the plant. They are also the main source for water loss out of the leaves, with water diffusing outward from the air in the stomata into the atmosphere surrounding the plant. The rate of diffusion (of both water and carbon dioxide) is governed by the size of the opening in the stomata: the bigger the opening the higher the diffusion rate. For small openings the increase in transpiration is linear with the total open area of stomata, but for large openings there is interference between the stomata, and the rate of increase in transpiration with area decreases for large openings (Ting and Loomis, 1965). Since this is a diffusion process, then the net transport rate of water out of the stomata is a function of the difference in the humidity of the air in the stomata (typically saturated) and the air around the leaf, which for water-limited environments is typically low (mainly because in water-limited conditions the size of the stomata opening is partially closed to reduce water loss). Some of the water in the leaves is used in the transformation of carbon dioxide into carbohydrates, but the water balance is dominated by the water loss out of the stomata. Since carbon dioxide is the source material for biomass production, this means that water use by the plant is proportional to biomass production so that a good approximation to biomass production is

$$\frac{\partial B}{\partial t} = \mu T \tag{14.17}$$

where μ is the coefficient relating biomass production per unit area to transpiration.

The soil water content constrains how much water the plant can extract from the soil, and therefore how much water the plant can transpire. If there is plentiful water, then the plant can transpire at the maximum possible rate. In this unrestricted case bare soil evaporation and transpiration are lumped together as evapotranspiration, and the maximum rate is known as the potential evapotranspiration, PET. The PET is calculated from an energy balance at the surface with the energy supplied being the net incoming radiation, and this energy being consumed as the latent heat of vapourisation of water. PET has a feedback with topography via the aspect of the hillslopes, with more incoming energy per square metre for slopes facing the sun (i.e. equatorial facing; south facing in the northern hemisphere, and north facing in the southern hemisphere). For midlatitude (i.e. 45°N or S) the increase in energy per square metre at midday between a 10% slopes facing the equator and a horizontal surface is about 23%, and everything else being equal this means the PET will also be about 23% higher on the 10% slope (Yetemen et al., 2015a, b).

A key component of the PET process is that the moist air needs to be transported away from the bare soil and/or the leaves (otherwise the air will become saturated near the leaf/soil so that no more water can be evaporated), and this vertical transport up away from the surface into the atmosphere is driven by turbulence generated by wind, so PET equations (mostly based on the Penman-Montieth equation) typically also involve a wind term. If, however, water is limited, then there is insufficient water to allow this evapotranspiration process to act unrestricted, so evapotranspiration is less than PET and is called the actual evapotranspiration, AET. It is worth noting that the excess energy (from the net radiation) that is not used to vapourise water turns into heat energy and temperature rises (see Kalma et al., 2008 and Chapter 3 for more detail). The key concept here is that as the soil becomes drier it becomes more difficult for the plants and soil to transpire at the maximum rate. A common approximation that is used is (Rodriguez-Iturbe and Porporato, 2005)

$$\mathrm{AET} = \mathrm{PET}\left(\frac{\mathrm{SWC} - \mathrm{SWC}_{\mathrm{min}}}{\mathrm{SWC}_{\mathrm{max}} - \mathrm{SWC}_{\mathrm{min}}}\right)^{\beta} \tag{14.18}$$

where the subscripts min and max are the minimum and maximum amount of water that can be held in the soil, and β is a parameter that is commonly assumed to be 1 but is sometimes larger (Walker et al., 2001). It is common to assume that $\mathrm{SWC}_{\mathrm{max}} = (\text{soil porosity}) \cdot (\text{soil depth})$, and $\mathrm{SWC}_{\mathrm{min}} = (\text{field capacity}) \cdot (\text{soil depth})$ for bare soil evaporation and $\mathrm{SWC}_{\mathrm{min}} = (\text{wilting point}) \cdot (\text{soil depth})$ for transpiration. In Equation (14.17) we then substitute $T = \lambda \cdot \mathrm{AET}$ where λ is an empirical partitioning coefficient to determine how much of the AET is transpiration and $(1 - \lambda)$ for bare soil evaporation.

The drier the soil is at the start of a rainfall event, then the faster the rate of water infiltration at the start of the rainfall event (Chapter 3). Infiltration during a storm is divided into two parts: (1) the initial loss, which is the water that wets up the soil to saturation at the start of the event, and is a function of the porosity and grading of the soil and the initial soil wetness and (2) the continuing loss, which is the amount of water that can infiltrate through the profile under the action of gravity once the soil is saturated and is equal to the saturated hydraulic conductivity of the soil. A commonly used expression for these is the Philip equation (Philip, 1969) for the cumulative amount of water infiltrated I (or i the infiltration rate) at time t:

$$I = St^{1/2} + ct \qquad i = \frac{1}{2}St^{-1/2} + c \tag{14.19}$$

where S is the sorptivity (i.e. initial loss) and c is the continuing loss. In this form S is a (undefined) function of how wet the soil is before the rainfall event (note that Philip derived this equation assuming that the soil was dry at the start of the rainfall event). The Green-Ampt equation (Section 3.1.4) yields similar equations with appropriate soil property assumptions. This can be converted into a more useful form by substituting the expression for I into the expression for i to replace t (noting that in the initial stages of the rainfall event the sorptivity term dominates the continuing loss), which yields an equation between the amount of water in the soil profile and the infiltration rate

$$i = \frac{S^{3/2}}{2I^{1/2}} + c \tag{14.20}$$

and as the soil becomes wetter (i.e. I increases) the initial loss is reduced. The result of this transformation is that S is now an intrinsic property of the soil rather than also being a function of the antecedent wetness. It is common in engineering hydrology models to assume that SWC $= fn(I)$ since I is a function of how wet the soil is at the soil surface while SWC is the amount of the water in the soil profile (or at a minimum down to the base of the root zone).

As noted in Chapter 10 the infiltration rate is also a function of SOM. Thompson et al. (2010) developed an empirical relationship between infiltration rate and dead biomass for arid environments

$$i = 56 \cdot B^{0.43} \tag{14.21}$$

where i is in mm/hr and B is in kg/m^2. The infiltration rates from this equation are quite high relative to infiltration rates that are typical of sandy loams (i.e. their sites), so the results likely reflect macropore effects (see Chapter 3). The advantage of this equation is that it allows an estimate of infiltration rate directly from the biomass equations in this chapter (Equation (14.4)). The more complicated alternative would be to estimate biomass from Equation (14.4) and use this as input to the soil carbon model in Chapter 10, and then link the SOM estimate with infiltration rate.

HillResLambers et al. (2001) use a simpler approximation for the interaction between biomass productivity and aridity. While more conceptual, it still captures qualitatively the feedback between reduced water availability and reduced productivity

$$\frac{\partial B}{\partial t} = g_{max}\left(\frac{\text{SWC}}{\text{SWC} + k_1}\right)B \tag{14.22}$$

where k_1 is a parameter and g_{max} is the maximum biomass productivity when water is not limiting. Conceptually this relationship is similar to the term for SWC in Equation (14.18), but is not normalised by the maximum water-holding capacity of the soil.

From the discussion above it is possible to construct a simple biomass dynamics model in water-limited environments. What should be clear, however, is that this requires modelling at the rainfall event scale, which may be problematic because of the data and computations required. A considerable effort is currently being expended in the ecohydrology research community to simplify this by deriving temporally averaged equations. This averaging is technically quite challenging and is well summarised elsewhere (e.g. Eagleson, 2002; Rodriguez-Iturbe and Porporato, 2005).

Eagleson (2002) provides a simple conceptualisation of the complicated feedbacks involved in water-limited ecosystems (Figure 14.5). Eagleson presented this graph for a single leaf with light intensity on the horizontal axis and carbon assimilation (i.e. biomass production) on the vertical axis, but it is easily extended to the case of an ecosystem. Presenting Eagleson's interpretation first, the left-hand side of the graph is the utilisation of light by the leaf, and as the light intensity (the horizontal axis) increases, the carbon assimilation (the vertical axis) increases linearly until it reaches a threshold (labelled 'optimum bioclimatic state') beyond which the leaf cannot utilise additional light energy since it has reached its maximum assimilation capacity. This is simulated by Equation (14.5) though Eagleson has a discrete threshold at $B = B^*$ rather than the asymptotic approach to B^* in the equation. For increasing light intensity I_0 above the light threshold the biomass production rate remains the same but SWC decreases (because of increasing evaporation due to the increasing incoming energy that is not used by the transpiration) until another threshold light

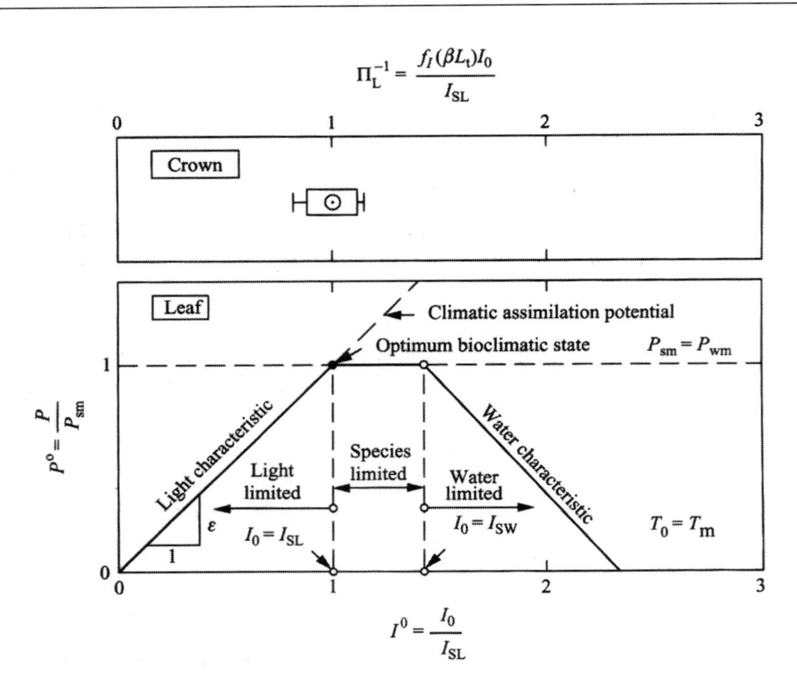

FIGURE 14.5: Schematic of light and water limitation effects on the production of biomass (from Eagleson, 2002).

intensity is reached (and its corresponding SWC) when there is insufficient water to allow the leaf to operate at maximum capacity. For increasing light intensity beyond this SWC threshold (and corresponding decreasing SWC), the stomata close to reduce water use and carbon assimilation drops accordingly. This 'water-limited' part of the curve is Equation (14.18).

The ecosystem analogy to Eagleson's curve is that the light-limited part of the curve is much the same, but the biomass generated (vertical axis) does not increase linearly because of shading of leaves as the leaf area index and biomass increases (the horizontal axis). The 'optimum bioclimate state' is then simply the biomass at which LAI saturates (i.e. LAI $= 4 - 6$), and no extra incremental biomass can be generated with an increase of light. The right-hand part of the axis is simply the response of the ecosystem to soil water limitation and AET. The horizontal middle part of the curve may still exist and reflects an ecosystem that can't grow any faster because it is already using as much of the light as possible and SWC does not limit growth.

Noy-Meir (1973) and Sala et al. (1988) found that for water-limited ecosystems the precipitation and soil water-holding characteristics of the soil interacted, what has subsequently become known as the 'inverse soil texture hypothesis'. This hypothesis is that sandy soils have higher biomass productivity than finer loam, or clay soils. The rationale for this is that water can infiltrate faster, penetrate deeper into the sandy soil reducing bare soil evaporation, with the result that more water is available to plants. This is despite loam and clay soils having better water-holding characteristics so that they hold more of the infiltrated water between infiltration events. Sala noted that the opposite occurs in humid areas, and high-water-capacity soils are preferred. The model structure presented above is consistent with this behaviour (see, e.g. sections 5.3 and 7.2 in Rodriguez-Iturbe and Porporato, 2005).

No discussion about water limitation and vegetation would be complete without a brief discussion of C_3 and C_4 plants. The difference between C_3 and C_4 plants is in the chemistry of the photosynthesis process. All plants are C_3, C_4 or (less often) CAM. Most grasses are C_4 and most other plants are C_3. The details of the chemical reactions are not important for this discussion, only that typically in conditions without water limitation C_4 plants use about half the water of C_3 for the same amount of carbon assimilation (and thus biomass production) (Larcher, 2003; Lambers et al., 2008), so that in Equation (14.17) the ratio $\mu_{C4} : \mu_{C3}$ is about 2:1. However, their response to water limitation differs, and the water use efficiency advantage of C_4 diminishes with increasing water stress (Nayyar and Gupta, 2006; Ghannoum, 2009; Taylor et al., 2011). Modelling studies with varying water stress (Vico and Porporato, 2008) suggest that β in Equation (14.18) is slightly smaller for C_4 relative to C_3, so that the water use

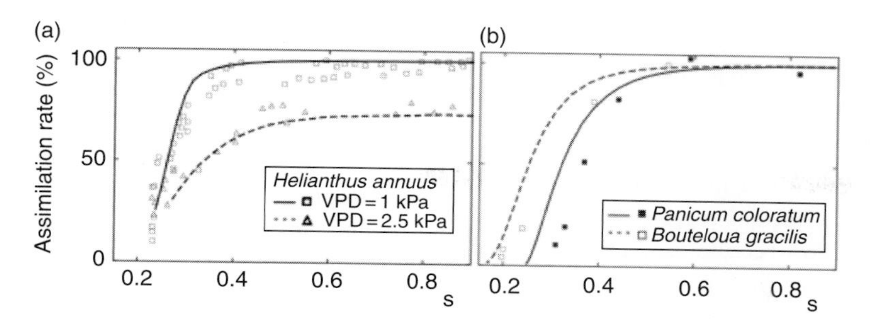

FIGURE 14.6: Soil moisture versus normalised biomass growth, where the growth rate is normalised against the growth rate for the non-water limited case: (a) C_3, (b) C_4, for two different vapour pressure deficits (VPD) (from Vico and Porporato, 2008).

efficiency of C_4 starts to decline more rapidly at a lower water stress than C_3 (Figure 14.6). Most studies involve experiments or models using a single species. In the field C_3 and C_4 coexist (and probably compete), and there is little information about these mixed cases. The approach adopted in most DGVMs, which is to weight behaviour by the percentage cover within the pixel, seems a sensible approach in the absence of more definitive information.

14.4.2 Spatial Self-Organisation of Vegetation Patches and Hillslope Profiles

One apparent conclusion from the preceding section is that as water becomes increasingly limiting, then the biomass density should decrease uniformly until there is not enough SWC to support vegetation. However, one intriguing aspect of low-slope arid and semiarid landscapes is the almost ubiquitous occurrence of banded and patchy vegetation where vegetation is concentrated into high-productivity areas surrounded by areas of low or no productivity (e.g. tiger bush). This suggests that arid ecosystems do not respond to increasing aridity by uniformly reducing biomass but by breaking up into a patchwork of high- and low-productivity zones (Dunkerley and Brown, 1995, 1999). The functioning of these ecosystems and exact mathematical formulation of a set of processes that create this patterning are still subject to debate, but some general principles are generally accepted (Tongway et al., 1990; Ludwig et al., 1999, 2005; Dunkerley, 2000, 2002):

1. In the downslope direction there are unvegetated patches separated by vegetated patches/bands.
2. The unvegetated patches are low in nutrients and organic carbon, and have a low infiltration rate so almost all rainfall on them runs off.
3. The vegetated patches are high in nutrients and organic carbon, and the infiltration rates within the band/patch are high.
4. Runoff from the unvegetated patches (and any suspended sediment and nutrients) runs into the vegetated patches and infiltrates into the soil in the vegetated patches. Thus the vegetated patches become sinks of water and nutrients, and unvegetated patches are sources.
5. Destruction of the bands/patches has a disproportionate impact on water capture and net primary production relative to the percentage of area covered by the bands/patches.
6. Many hillslope profiles show a slight rise in topography (or at least a lessening of the hillslope gradient) within the bands.

Various models have been developed to capture these principles (e.g. Dunkerley, 1997; Saco et al., 2007). Recent modelling work has suggested that coevolution of the landforms and vegetation is essential to create some of the observed patterns (Saco et al., 2007; Saco and Moreno de Las Heras, 2013). Saco et al. coupled a landform evolution model with a vegetation model and evolved the landform in response to the transfer of sediment from the unvegetated to the vegetated zones. The persistence of the spatial location of the patterns even after overgrazing and fire (Ludwig et al., 2005) suggests that a long-lived process is essential to the formation and persistence of these bands whether it be topography, organic matter changes in soil structure, nutrient storage or other aspects of soil evolution.

A more conceptual approach to the effect of banded vegetation was adopted by HilleRisLambers et al. (2001) and Rietkerk et al. (2002) where the erosion resulting from overland flow was not modelled, so it was not possible to couple the vegetation and topography. A consistent finding with this model and by various subsequent authors (e.g. Dagbovie and Sheratt, 2014) is that the bands of vegetation generated move upslope for almost all parameter values. The evidence for or against band movement in the field is weak, but a conclusion of the modelling studies is that to fix the bands in space requires a coupling between biomass and topography.

Some intriguing questions about the pattern formation remain open, including the following:

1. What drives the transition between uniform, patchy and banded vegetation? Some modelling suggests that bands capture more overland flow because they infiltrate all except the most extreme high flows, while patches allow short circuiting around the patches, particularly if the bands/patches have slightly higher topography as a result of erosion and deposition (Moreno de Las Heras et al., 2012). This is consistent with the field observation that patchy landscapes have slightly less improvement in biomass production relative to uniform vegetation than for banded landscapes (Ludwig et al., 2005).
2. What is the governing process that determines the spacing of bands/patches? It seems reasonable to conclude that for a given aridity, the percentage of cover on average will be determined by water availability and the redistribution from unvegetated patches, but to achieve that average biomass there is an undetermined degree of freedom between larger, more widely, spaced patches, versus smaller, more closely, spaced patches.
3. Climate appears to control the percentage cover, but how do the dynamics function in response to climate variability? As the climate dries, do patches get smaller, or does the spacing increase (and vice versa)? For the latter case patches will need to move with climate variability. What are the implications for soil evolution if the patches are fixed in space?

14.5 Wildfire

Wildfire is an integral part of natural vegetated systems by periodically removing vegetation, creating regions of new vegetation growth and ecosystem patchiness. It is key for soil carbon because fire is the main source of long-residence-time carbon in the soil, in the form of charcoal and other low-lability soil carbon. Fire is a significant process in the return of carbon to the atmosphere from terrestrial biomass (4.5 Gt/year, not listed separately in Figure 10.1) (Stocker et al., 2013). Finally fire creates patches on the landform where, in the aftermath of the fire and for some years afterwards, hydrology and erosion, and thus landform evolution, are changed. Istanbulluoglu et al. (2004) modelled the impact of post-fire erosion using the processes outlined below and found that even though fires at their field site were infrequent, the increases in erosion rates post-fire were large enough that the cumulative impact on erosion was very important, so the contribution of fire to average erosion rates cannot be ignored.

There are three main types of wild fire. These are ground fire (slowest moving and lowest temperature), surface fire and crown fire (fastest moving and highest temperature). Ground fires largely burn and smoulder in the soil and consume peat, roots and other buried organic matter. Surface fires burn surface litter and grasses. Crown fires burn the canopy of forests. Another form of fire, mainly in eucalyptus forests, is a ‘fire bomb’; this ignites organic vapours released into the atmosphere by heated vegetation and the fire front (in the atmosphere) can propagate much faster than the ground and crown fires. Finally very high-intensity fires can induce convection currents in the troposphere (resulting in a thunderstorm-like cloud called a ‘fire storm’; Sharples et al., 2016), and the induced winds can be important in increasing both the fire intensity and the speed of propagation of the fire front. Crown fires are further subdivided into (1) passive, where crowns of individual trees burn as a result of propagation up from surface fire and (2) active/continuous, where fire in the crowns propagates across the canopy from tree to tree.

Section 14.2 indicates the overwhelming influence of groundcover on the erosion rate, with canopy cover being secondary. This would indicate that the groundcover loss by fire is more important than the canopy loss for erosion-driven landform evolution. However, crown fires also burn at higher temperatures and induce greater pyrogenic changes in the soil (e.g. destruction of soil carbon, thermal transformation of clays, creation or destruction of soil hydrophobicity), though these pyrogenic changes have been less well characterised and appear to be somewhat site and/or soil specific. Crown fires are important, however, in driving tree mortality, and they have longer-term impact on catchment water yield as recruitment of tree seedlings (and the accompanying water usage that accompanies biomass production) occurs to replace the dead trees (see the Kuczera curve, discussed later).

The fire models that follow do not typically distinguish between these fire types and do not model atmospheric feedbacks such as fire storms. The models focus on the frequency and spatial extent of fire occurrence, and it is rare for models to simulate fire temperature. Most of the models that follow have a general structure with three components (Venevsky et al., 2002): (1) fire ignition, a process for triggering the start of a fire, (2) fire spread, determining how fast a fire propagates once triggered and (3) fire termination, a process that suppresses a fire and that is typically weather related (e.g. rain, change of wind direction). For ecosystem modelling there is a fourth component, the impact of the fire on ecosystem mortality, which is related to fire type and temperature.

The discussion below will begin with physically based models that simulate the three fire components explicitly. These physically based models might be too complex and computationally infeasible in a SLEM but will highlight the physical processes. We will then discuss simpler stochastic models that might be adequate in other applications. There is a range of other modelling approaches, such as cellular automata, percolation networks and self-organised criticality, but to date the evidence is that, despite their conceptual appeal, they perform poorly, and the reader is guided to Sullivan (2009c) for a discussion of these approaches.

14.5.1 Physically Based Modelling

A number of deterministic fire models are available. They are mostly used for modelling of fire propagation during a fire, that is, how a single fire, once started, propagates. We will focus on one model that is typical of these models, which will enable the highlighting of generic physics that are required to model fire, and aspects of the input data and parameters that are required to run these models. It is the detail of these requirements that may make these models difficult to use in the context of hundreds to thousands of years of landscape evolution simulation. The model discussed below is FARSITE, developed and used by the USDA and US Forest Service (Finney, 1999, 2004).

FARSITE models the propagation of a fire front, since most of the dynamics of the fire occur at the fire front (the interface between the burnt area and the surrounding unburnt material). The area burnt is then simply the region traversed by the fire front during the simulation. The methodology is that at every timestep the fire front is advanced. The rate and direction of the fire front movement are primarily a function of (1) wind speed and direction and (2) the intensity of the energy release by the fire at the fire front. Many of the underlying physical principles in FARSITE, and a number of other models, are based on Rothermel (1991).

The spreading of a fire from a localised point is well approximated by an elongated ellipse. The initiation location of the fire is at the rear foci of the ellipse, and the rate at which the long axis of the ellipse extends is a function primarily of wind. The faster the wind, the more elongated the ellipse (Figure 14.7), with a length-to-breadth ratio of

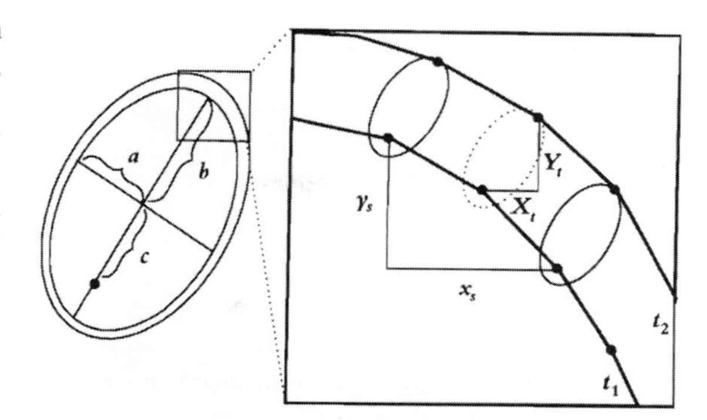

FIGURE 14.7: Elliptical spread of fire (from Finney, 1999).

$$\mathrm{LB} = \begin{cases} 0.936e^{0.2566U} + 0.461e^{-0.1548U} - 0.397 & \mathrm{LB} < 8 \\ 8 \end{cases} \tag{14.23}$$

where U is the velocity of wind at the height of the middle of the flames. The 'rate of spreading' (interchangeably R or ROS in the literature, the rate at which the ellipse increases in size with time) is given by

$$R = \frac{I_R \xi (1 + \phi_w + \phi_s)}{\rho_b \varepsilon Q_{\mathrm{ig}}} \tag{14.24}$$

where I_R is the fire reaction intensity and is a function of the amount of energy released by the fire per unit time and per unit length of the fire front, ρ_b is the bulk density of the fuel, Q_{ig} is the heat of pre-ignition and characterises how much energy per unit mass must be input into the fuel to ignite it, and ξ and ε are fitting parameters. The effect of wind and slope is characterised by ϕ_w and ϕ_s, respectively and are (Rothermel, 1972)

$$\phi_w = C(3.281U)^B \left(\frac{\beta}{\beta_{\mathrm{op}}}\right)^{-E}$$
$$\phi_s = 5.275\beta^{-0.3} \tan^2\phi \tag{14.25}$$

where ϕ is the slope in the direction of the wind, and B, C, E and β are fitting parameters (all typically positive). Note that Finney (1999, 2004) incorrectly quotes the slope term in the slope equation as $\tan\phi^2$ and fail to note that Rothermel derived this only for fire fronts travelling upslope. For slopes less than 10%, the slope correction is minor (Rothermel, 1991). The velocity of the front of the ellipse is

$$V_{\mathrm{front}} = \frac{3}{2}\frac{R}{\mathrm{HB}} + \frac{R}{2}$$
$$\mathrm{HB} = \left(\mathrm{LB} + \left(\mathrm{LB}^2 - 1\right)^{0.5}\right) \Big/ \left(\mathrm{LB} - \left(\mathrm{LB}^2 - 1\right)^{0.5}\right) \tag{14.26}$$

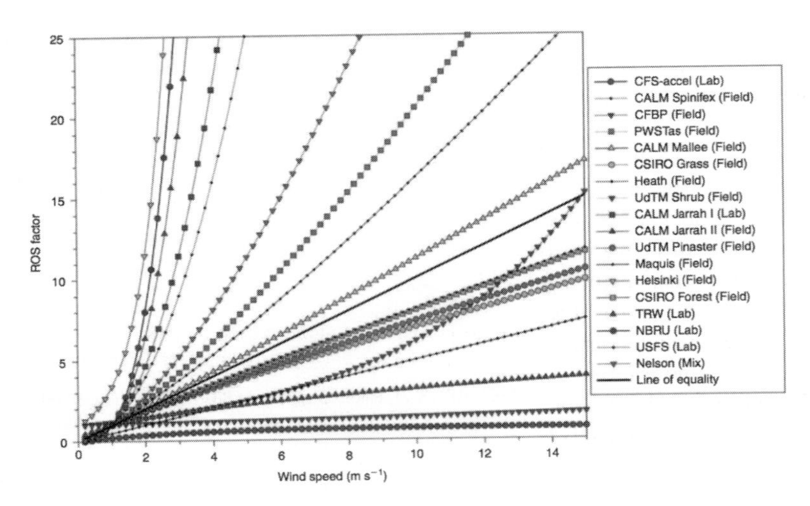

FIGURE 14.8: Rate of fire spreading as a function of wind speed (from Sullivan, 2009b).

Some key concepts embedded in these equations are common to many other models even if their mathematical formulations are different:

1. As the wind speed increases, the ellipse of burnt area becomes increasingly elongated (Equation (14.23)), ϕ_w becomes larger so that the rate of spreading of the ellipse increases, so more area is burnt per unit time both in the direction of the wind and at right angles to it (Equation (14.24)), and the velocity of advance of the fire front increases (Equation (14.26)).
2. The parameter B has been found to be in the range 1–2 (Sullivan, 2009b), so the rate of spreading increases nonlinearly, and more quickly than the wind speed (Figure 14.8).
3. Since wind speed increases with height above the ground, the speed of the fire front will increase for higher flames because of the dependence on the mid-flame height U.
4. As the slope of the topography increases then the velocity of the fire front increases (Equation (14.25) and (14.24)).
5. As the amount of energy released increases (I_R in Equation (14.24)), the velocity increases. Since wet fuel releases less energy than dry fuel (because the water has to be driven off before it can ignite), the fire front propagates less rapidly in wet conditions. If there is more fuel (e.g. there has been a good vegetation growth season leading into the fire season), then the fire front will move more quickly.
6. As the fuel becomes harder to ignite (higher Q_{ig} in Equation (14.24), typically because it is wet), the fire front slows (this effect of wet fuel is in addition to point 5).

Two broad conclusions can be reached here. The first is the fire will interact with the topography because of the slope term in Equation (14.25). The second is that climate data are needed, including (1) wind speed and direction and (2) inputs for a water balance model to determine how dry/wet the fuel is before the fire. The biomass model to determine the quantity of available fuel from the previous growing season(s) will also require pre-fire climate. Point 5 highlights the important issue of the time sequencing of climate variations. The worst fire conditions occur as a result of a wet growing seasons in winter and spring followed by a dry hot summer because there is plentiful biomass and it is dry (Bradstock, personal communication; O'Donnell et al., 2011). Finally, FARSITE is a fire propagation model and does not simulate (1) when and where fires begin and (2) fire termination. While we won't discuss it further, FARSITE has criteria for surface fire and any transition to a crown fire, but the speed of propagation is assumed to be the same for both types.

FARSITE makes it clear that a considerable number of spatially distributed data are required to describe the pre-fire conditions, including elevation, slope, aspect, the fuel (type, rates of accumulation, wetness), canopy cover, canopy height, canopy base height and canopy bulk density, albeit the latter four are required only for crown fire modelling. These properties can be spatially variable, and propagation of parts of the fire may be slower or faster as a result of local spatial variation. In addition the climate data (which may also be spatially variable though that type of data is rarely available) required are (1) a wind speed and direction time series and (2) minimum and maximum daily temperature, humidity and precipitation to determine fuel wetness. Note that the precipitation is not used in FARSITE for fire termination.

These equations and their data requirements are typical of deterministic physically based fire models. We can now consider how this model and others like it might be used in a landscape evolution simulation, which may be simulating thousands to millions of years. Clearly we will have the topographic data required, and if we have a coupled vegetation model, we will likely be able to simulate the fuel properties required. The difficult issue is the simulation of the climate variables. There are three alternatives:

1. Use a historic climate record and repeat this record multiple times until the total simulation time is complete. This is a common solution (e.g. Willgoose and Riley, 1998a) and ensures that all the different climate series are correctly sequenced and the correlations between the time series (e.g. relationships between temperature and rainfall which will drive fuel wetness) are correct. This of course requires the climate series data be available, and that they be consistent with the varying climate over the period being simulated (e.g. does current-day data simulate what happened in an ice age?). One important consequence is, however, that the input data cover only the extremes in the recorded data, and mega-events that might have occurred in the pre-instrumental record cannot be simulated.
2. Randomly simulate synthetic times series to use. This needs to be done with extreme caution. The generation of multiple randomly varying climate time series (i.e. temperatures, rainfall, evaporation) must be done so that not only do they vary in time correctly (e.g. wet winters, dry summers) but also that the temporal relationships between the climate variables are also correctly maintained (e.g. wetter than average periods typically have lower than average evaporation, and reduced temperature extremes). The dangers multiply considerably if you want to simulate random spatial variability as well. Techniques to do this exist but typically rely on the variables being distributed with a Gaussian distribution, and this can be problematic, particularly for rainfall, even after transforming the data (e.g. using the Box-Cox transformation; see Chapter 3).
3. Generating high-resolution climates using regional climate models (RCM) such as WRF (Weather Research and Forecasting Model). It is early days for these climate models, and the author's experience with hydrology applications using this type of data suggests that the RCM model's ability to preserve relationships between climate variables is still patchy (Lockart et al., 2016; Chowdhury, 2017; Parana Manage, 2017).

The complexity of the modelling task using a physically based model like FARSITE means that researchers have tended to use simpler models that attempt to capture conceptually the processes above but without the detailed modelling of every process. All the approaches in the points above are suitable for inclusion in a LEM, the only constraint being data availability. If these data are not available or cannot be generated, then the approaches in the next two sections, though not as strongly physically based, are preferred.

14.5.2 Conceptual Modelling

Venevsky et al. (2002) present a conceptual fire model, Reg-FIRM, designed for incorporation into a global climate model (GCM), and the model captures some of the physical processes from the previous section. Their model determines the average burnt area arising from a stochastic representation of the processes above, and from this determines the average percentage of a GCM cell (typically about 300 km $\times$ 300 km) that is burnt. They do not recommend the model for grid cells less than 100 km^2, which is larger than normally used in LEMs; however, we will discuss later how Reg-FIRM is the basis of another model, PESERA (Fleskens et al., 2016), which was designed for typical LEM resolutions.

The number of fires N_{fire} in a grid cell per year is estimated using two ignition potentials: (1) human-caused ignitions and (2) lightning strikes

$$N_{\text{fire}} = CN_d\left(\delta P_D^\theta + l\right) \tag{14.27}$$

where C is a Fire Danger Coefficient (FDC) averaged over the daily values for the year, N_d is the number of days with high fire danger per year, P_D is the population density, l is the number of fires started by lightning strike, θ is a coefficient that has been found to lie between 0.43 and 0.57 and δ is a coefficient. Venesky defined FDC using the Nestorov Index, which is a function of the temperature on preceding days and whether rainfall is less than 3 mm on those preceding days. They note that a range of fire danger indices are used worldwide that relate fire climate with the rate of fire ignitions that could also be used. They also note that the number of human initiated fires is much greater than natural fires in recent decades (e.g. 97% in Spain, 86% in Russia and 87% in southeast Australia of total fires; Venevsky et al., 2002; Collins et al., 2015). For Russian boreal forest l is 0.015–0.028 fires per day per million hectares. The lightning ignition rate l increased with increased number of hot days, elevation, terrain slope and percentage vegetation cover (Collins et al., 2015). Penman et al. (2013) found that the rate at which a lightning strike caused ignition increased with number of years since the last bushfire, provided that the last bushfire occurred more than 15 years previously, and for less than 15 years the ignition rate was independent of the time since the last fire (and thus accumulated fuel/biomass). To yield l this ignition rate needs to be multiplied by the lightning strike rate, which is largely a function of incidence of thunderstorms, and at the global scale is strongly positively correlated with increases in elevation and terrain slope and is higher over forest than bare soil (Kotroni and Lagaouvardos, 2008; Albrecht et al., 2016).

Lynch et al. (2007) and Verdon et al. (2004) used paleoclimate data and ocean teleconnections to estimate the temporal variability of forest fire risk. Verdon found an increased risk of fire in eastern Australia during El Niño episodes and IPO negative (i.e. dry hot periods). Proxy records based on ice cores and other paleo-data (see Chapter 3) provide a possible means to develop millennial scale paleo-fire records.

The area burnt by the fire once initiated is a function of the rate of spread of the fire and the duration of the fire. Venevsky et al. (2002) used an exponential probability distribution for the duration of the fire ($p(d) = \mu e^{-\mu d}$ where d is the random duration) where the average duration of the fires was one day, which was based on Russian fire statistics (Korovin, 1996 in Venevsky et al., 2002). There is some rationale for the mean fire durations being about one day because many fires burn during the (hot) day but then extinguish in the cooler conditions overnight (Penman et al., 2013). The rate of spread is a function primarily of the wind speed using the elliptical fire burn area, as discussed above for FARSITE. The effect of fuel load and soil moisture (a proxy for fuel moisture content) modifies the rate of fire spreading. Topography and any overlap between the burnt areas of separate fires is ignored, so wind direction data are not required. This leads to their equation for the fraction of a grid cell burnt in one year, PB,

$$\mathrm{PB} = \frac{N_{\mathrm{fire}} \pi \overline{U}^2}{8 \mu^2 A_{\mathrm{grid}}} \tag{14.28}$$

where $\overline{U}$ is the annual average rate of fire spreading, which is the average over the daily values, and the daily values are a function of average wind speed, fuel load and soil moisture on that day, μ is the exponent in the exponential probability distribution for fire duration, and A_{grid} is the area of the grid cell. Spatial variations in PB were then driven primarily by spatial variations in the rate of spreading and population density, while temporal variations were a function of temporal variations in climate (and thus the rate of spreading $\overline{U}$).

As noted above, Reg-FIRM has been used as the basis of another model, PESERA, which is designed for use at high spatial resolutions (Fleskens et al., 2016). They calculate a Fire Danger Index (FDI) for a given month that is a simplified version of Venevsky's FDC as

$$\begin{aligned} \mathrm{FDI} &= 1 - \left(1 - e^{-\lambda N}\right)/\lambda N \\ N &= \overline{T}(T_E/2 + 4)D_{>3} \end{aligned} \tag{14.29}$$

where $\overline{T}$ and T_E are the mean monthly temperature and mean monthly temperature range over a day and $D_{>3}$ is the number of days per month with more than 3 mm of rain. The mean number of fire starts per month is calculated with Equation (14.27), and the number of wildfires started in that month is FDI $\times N_{\mathrm{fire}}$. The area of fire spread R (m/s) is calculated using a simplified form of the equation used in Reg-FIRM

$$R = \frac{0.9(1 + U)}{B\left(16 + 100\mathrm{AET}^2\right)\mathrm{PET}} \tag{14.30}$$

where AET and PET (mm/month) are actual and potential evapotranspiration, respectively, B is the biomass (kg/m^2), U is the average wind speed (m/s), and fires are assumed to last one day and are circular in plan. A fire severity, based on biomass and dryness, is then used to estimate the loss of biomass and subsequent increase in erodibility (Kirkby, personal communication). To generate a time series over a region, fires are assumed to randomly occur in time and space based on the expected number of fires per month in a grid cell.

The models above do not simulate fire intensity. A more advanced model that follows the approach of Reg-FIRM is LPJ-SPITFIRE. In addition to modelling fire extents it also uses the biomass and climate models of LPJ to simulate the intensity of the fire (e.g. surface fire versus crown fire), and thus the proportion of the biomass burnt by the fire (Thonicke et al., 2010).

14.5.3 Stochastic Modelling

Martin (2007) and Willgoose (2011) present simple stochastic models with strong similarities that build upon empirical observations of fire occurrence and spatial distribution. The simulation method is that a fire initiation time and location is randomly simulated, the burnt area

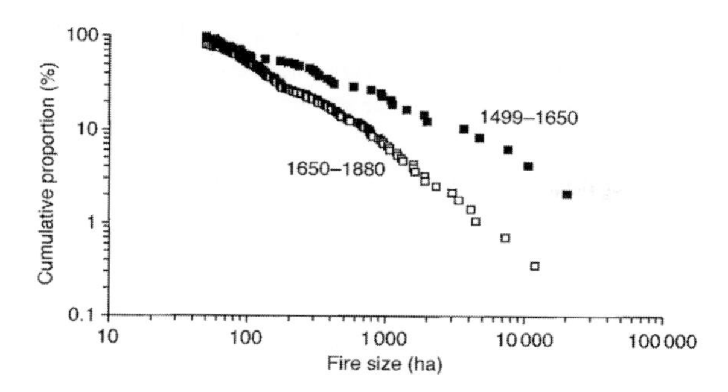

FIGURE 14.9: Cumulative size distribution of all detected fires for the two periods 1499–1650 and 1650–1880. Fires smaller than 50 ha have been lumped together (from Niklasson and Granström, 2000).

from the fire is determined and the burn area placed at the fire initiation point. A key input is the size of the fire that occurs. Malamud et al. (1998, 2005) and Nikalsson and Granström (2000) analysed historical fire data and found that the frequency of fire and the area of the fire were inversely related (Figure 14.9) by a relationship of the form

$$p = \beta A_F^{-\alpha} \tag{14.31}$$

where p is the probability of a fire larger than area A_F (per unit time and per unit area) and α was in the range 1.3–1.8. This general trend of decreasing frequency with increasing area has been widely found (Hantson et al., 2016). As a result, fires of small area are more frequent than fires with large area. Malamud's data were for a number of different locations worldwide and so are applicable worldwide. O'Donnell et al. (2014) found a similar relationship but with change in α for large areas, which they interpreted as being a size limitation based on ecosystem fragmentation due to land use. Fire suppression in the last 100 years has changed this relationship in some areas (Steel et al., 2015). Nikalsson and Granström (2000) attributed a change in α pre- and post-1650 to human occupation. Malamud et al. (2005) calculated frequency rates for the United States and Europe and found that the number of fires per year with area greater than 10 km^2 that occurred in a 1,000 km^2 area was about 1–10 for the western half of the United States and about 0.01 for the eastern half, while in the Mediterranean it was about 2 per year. These rates were for recent years and so will likely have a significant human initiation component, as in Equation (14.27).

Malamud et al. (2005) also calibrated their recurrence model to individual decades and found that there was a decade-to-decade difference in β but not in α. While they did not show that this decadal variability resulted from climate variability, other than changes in fire management policies (e.g. Dennison et al., 2014), no alternative explanation seems plausible. Verdon et al. (2004) found a relationship for eastern Australia between fire occurrence and ocean teleconnections (El Niño and IPO). These results are suggestive of a means of using their model to simulate natural climate variability and anthropogenic climate change: simply change the value of β so that fires of a given area become more or less frequent, as was done in Willgoose (2011). Malamud did not link his results to climate data, so an empirical link between frequency and climate needs to be developed for any given application site.

Willgoose (2011) calibrated his model to fire recurrence data of Russell-Smith et al. (1997) and found that a range of area-scaling models, including that of Malamud, and fire shapes (square, round, triangular) could be successfully calibrated to the Russell-Smith data, and that the only difference was that Malamud's area scaling model yields a spatial pattern of burning that had more fine detail (because of the high frequency of small-area fires) and was less spatially clumped than models that had a more uniform distribution of fire areas. Neither Martin nor Willgoose examined the spatial statistics of the burn patterns that might be ecologically significant, for example, the connectivity of unburnt regions. Rather the focus of their studies was how a change in fire frequency (e.g. due to climate change) might change (1) landslide and sediment delivery impacts (Martin, 2007) and (2) runoff yield (Willgoose, 2011).

14.5.4 Recovery from Fire

One important variable for estimating the impacts of fire is how long it takes for the ecosystem and soils to recover from the fire, and how these impacts decrease with time. For instance, the vegetation models in previous sections can be used to model biomass recovery, but what about other impacts such as on soils? Some impacts will have longer recovery times (e.g. loss of soil carbon) than others (e.g. hydrophobicity in the soil), but in general it would be expected that there would be a gradual decline with time (e.g. exponential with time). This recovery process has rarely been quantified beyond a simple decline with time lasting for a few years for small surface fires to decades for larger more intense crown fires (e.g. stand-clearing fires that kill all or most trees). It is also important to distinguish between biomass lost (mostly leaves and ground litter) and tree mortality. Tree mortality effects

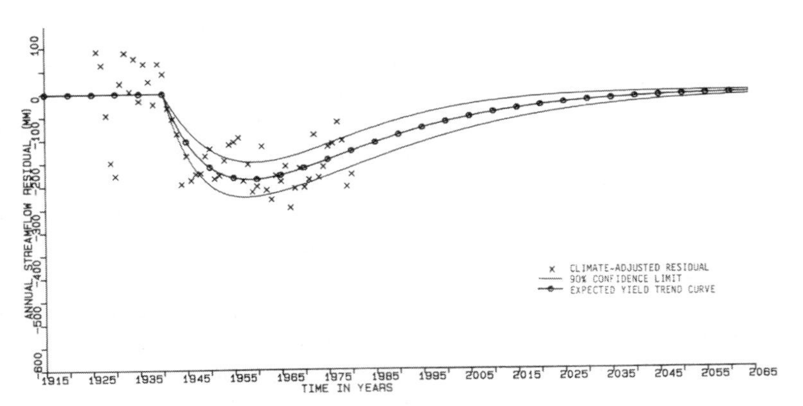

FIGURE 14.10: The Kuczera curve showing the impact on water yield from a catchment as a function of years from fire (from Kuczera, 1987).

will have longer recovery times because they involve the regrowth of fully grown trees (typically decades), while biomass recovery will be faster because it is recovery of vegetation on existing surviving trees (typically only years).

Kuczera (1987) fit a curve (the Kuczera curve) to catchment runoff yield after a stand-clearing (i.e. widespread tree mortality) fire in water supply catchments in southeast Australia (Figure 14.10). Immediately post-fire the runoff yield was not significantly impacted (though some authors have noticed an immediate small increase in water yield post-fire; Watson et al., 1999), then a short time later the runoff yield of the catchment dropped as the enhanced biomass growth, and consequent enhanced water use, resulting from the forest recovery occurred.

Forest water use peaks at the time that the biomass growth peaks and then asymptotically declines to the pre-fire condition some decades later. Since water use is strongly linearly correlated with biomass growth, it is reasonable to expect a yield reduction to occur in all catchments post-fire though the yield decline will vary with the amount of biomass lost (and thus the amount that will be required to recover over and above normal plant growth, which can be measured by NDVI remote sensing post-fire; e.g. Katagis et al., 2014) and the time scale of recovery. Nolan et al. (2015) estimated evapotranspiration (ET) using sap flow measurements in a mixed species burnt forest over time and found a result consistent with the Kuczera curve but with ET peaking at three to four years in her case rather than the 30–40 years found by Kuczera. The fires studied by Nolan were for fires involved biomass growth but low mortality, while the fires studied by Kuczera involved widespread tree mortality. Thus the recovery time for Nolan's fires was much less than Kuczera's.

If the loss of biomass is small, then the impact on runoff yield will likely be less, and thus it may be difficult to separate it from the other effects discussed below. The Kuczera curve mechanism will only occur in situations of water excess where water availability does not limit growth.

Much post-recovery work has used remote sensing and NDVI to determine when the NDVI, reflecting photosynthesis rates, has recovered to adjacent unburnt areas. Rodrigues et al. (2014) estimated the time for vegetation in Spain to return to its pre-fire canopy height and canopy cover and found that grasses recover faster (two to five years) than shrubs (10–25 years) and trees (20–50) years, and that if the vegetation recovered by sprouting on surviving vegetation (the numbers quoted above) then recovery was about 2 times faster than if recovery was predominantly from seed. These results are consistent with Polychronaki et al. (2014) who found in Greece that 20 years post-fire the grass and shrub areas had recovered their pre-fire NDVI, while forest areas were still recovering.

Finally the question of how likely a previously burnt area is to burn at some later stage is partly a question of recovery and particularly the recovery of the flammability of the vegetation (which may not correlate exactly with the amount of biomass; McCarthy et. al., 2001). Moreover it is also dependent on what percentage of the area is burnt each year and how quickly it recovers. Low burnt area percentages mean that the impact of previously burnt areas on fire occurrence and area will be small (Price and Bradstock, 2011; Price et al., 2012).

14.5.5 Impacts of Fire

This section can provide only a brief overview of the impacts of fire, and accordingly the focus here is on those with a first-order impact on soilscape and landscape evolution. Unfortunately the literature in this area is incomplete with apparently contradictory conclusions for different case studies, and, at least in part, these differences in responses are claimed to be a result of different responses to different severity fires (Keeley, 2009). Moreover as the previous section indicates, there may be time-varying impacts during the post-fire recovery. However, in general, impacts are greatest for the highest intensity fires, which are generally correlated with those areas of highest biomass (Pierson et al., 2002). Shakesby and Doerr (2006), and Moody et al. (2013) discuss many of the points below and note that fire effects on soils are relatively poorly understood compared to biomass and vegetation impacts. Nyman et al. (2013) review some of the modelling approaches used for modelling fire impacts on hydrology and geomorphology, while Langhans et al. (2016) and Sidman et al. (2016) propose detailed modelling approaches for hydrology and water quality impacts. Some of the impacts of fire are as follows:

- *Hydrology:* Hydrology impacts can be categorised as those that impact on event scale runoff that in turn drives erosion processes, and longer-term-yield impacts that impact on the average runoff on the catchment over the longer term. As discussed in the previous section, some impacts on runoff yield may last from years to decades:
  - *Event-scale hydrology:* Runoff response times tend to decrease as a result of the loss of leaf litter, but with smaller fires post-fire leaf fall offsets this. The ash plays an important role in retarding runoff initiation (Woods and Balfour, 2008). There is sometimes a period of increased/decreased hydrophobicity due to volatile by-products of burning organic matter (lasting several years to decades; DeBano, 2000; Cerdà and Doerr, 2005; Tessler et al., 2013; van Eck et al., 2016), which must be seen relative to any pre-fire hydrophobicity, but which results in changed runoff rates post-fire. Higher runoff rates with increased fire severity can be attributed primarily to surface cover decreases, not hydrophobicity (Hosseini et al., 2016).
  - *Yield hydrology:* Higher soil moistures at depth post-fire have been reported, and they appear to be related to the reduction of water use within the soil profile as a result of vegetation removal/damage. The higher soil moisture has also been linked to increases in groundwater flow (Silva et al., 2006). Nolan et al. (2014a) found that in the first three years after forest fires that AET decreased by 41% in areas of high severity fire but then increased above pre-fire rates at three years. Decreased AET was driven by tree death, while increased AET was driven by seedling water demand (Nolan et al., 2014b). A longer-term study suggested that AET of eucalyptus forests have recovered by eight to 12 years post-fire (Nolan et al., 2015). Destruction of organic matter destroys soil structure and decreases the hydraulic conductivity of the soil. Recovery times for vegetation can be variable from years to decades depending on the severity of the fire and percentage of tree mortality.
- *Erosion:* The main impact of post-fire erosion is the loss of litter and groundcover protection of the soil and the resulting increased erosion (up to 25 times higher, but more typically 5–10 times; Johansen et al., 2001; Llovet et al., 2009; Smith et al., 2011; Hosseni et al., 2016), and the recovery follows from reinstatement of the vegetation groundcover. The recovery to near pre-fire response is highly variable (from 1 to 10 years) but typically takes about two years (Cerdà and Doerr, 2005; Lane et al., 2006; Sheridan et al., 2007; Ryan et al., 2011). There is evidence for dry ravel and debris flows post-fire, and, together with fluvial erosion, a significant amount of removed hillslope sediment can be stored elsewhere within the catchment in the immediate aftermath of the fire. This stored sediment is then removed from the catchment over the longer term (Shakesby and Doerr, 2006; Santi et al., 2008; Nyman et al., 2011). The consumption of the organic-rich surface O layer in the soil can destabilise any established surface armours (Shakesby, 2011), and provide loose sediment, ash and charcoal that is readily available for transport by overland flow (Nyman et al., 2013).
- *Water quality:* The main impacts are an increase in suspended sediment (up 1,400 times higher), total N (specifically nitrate) increases in the first year, and dissolved organic carbon being unchanged while particulate organic carbon increases markedly if rainfall occurs soon after the fire (Smith et al., 2011). Significant impacts on the N and P cycles have been measured up to 15 years post-fire (e.g. 50% increase in N fluxes 13 years after fire has been observed). The suspended sediment is high in ash, and ash chemistry varies with fire temperature. For temperatures $< 450°C$ the ash is rich in organic carbon, for temperatures $> 450°C$ most organic carbon is volatised and carbon is in the form of carbonates, and for temperatures $> 580°C$ ash is dominated by oxides (Bodi et al., 2014) (see also ash effects under 'Soils' below). There have been some reports of elevated cyanide immediately post-fire as a result of aerial deposition.

- *Mass movement:* Dry ravel, debris flows and mini-landslides are common in the immediate aftermath of a fire (Ryan et al., 2011). This creates loose material that is then available for transport during subsequent runoff events (Cawson et al., 2012; Nyman et al., 2013). As discussed in Section 13.1.2, rates of landsliding show a marked increase after clear cutting of forest (up to a factor of 5), which is attributed to death of fine roots and a reduction of strength in the older roots. Istanbulluoglu and Bras (2005) used clear cutting as an analogy for the effect of fire on the rate of landsliding, though this is probably valid only for stand-clearing fires. For smaller fires the effect on landsliding rate is unknown. The reduction in strength of roots will occur only for tree death, so it is only in ecosystems with fire-sensitive trees that post-fire root weakening will be observed. Another consideration is how quickly the forest recovers. If forest reestablishment is fast, then new roots may be established by the time the dead trees roots weaken (Sheridan, personal communication).
- *Soils:* Fire consumes soil carbon (SOC). SOC consumption starts when soil temperatures reach 200–250°C and is complete at 460°C (Certini, 2005). Low-severity fires may heat the soil only to 60°C, while high-severity fires may heat the soil to 800–900°C (Cawson et al., 2012). Consumption of SOC breaks down soil aggregates that are held together by SOC, destroys soil structure (down to about 10 mm below the surface; Nyman et al., 2013) and thus reduces infiltration rates. Surface ash/char that may cover the surface post-fire is generally lost from the surface within a year or two, either by erosion or by filling up soil macropores, with higher carbon loads in runoff post-fire primarily ash from the fire and not soil carbon (Reneau et al., 2007; Hosseini et al., 2016). Rumpel et al. (2009) found that about half of the surface ash/char was eroded, while the remainder was transported down into the soil profile. Soil aggregate stability was reduced by the thermal alteration of the organic matter cementing particles at soil temperatures around 380–460° (Greene et al., 1990; Garcia-Corona et al., 2004), while Mataix-Solera et al. (2002) found greater SOM loss and soil structure breakdown after surface fires than crown fires. Fires return nutrients from the aboveground biomass to the soil and enhance the cycling of nutrients (De Marco et al., 2005). Belowground vegetative biomass and microbial activity is relatively unaffected by fire (for low fire severity), and belowground productivity is more affected by grass species and drought than burning (Burnett et al., 2012; Fontúrbel et al., 2012).
- *Ecosystem:* Greene et al. (1990) found that cryptogam crusts were destroyed by fire resulting in increased erosion, but cryptogam recolonisation occurred within four years. A number of authors have noted that the balance of grass and trees in savannah ecosystems is difficult to maintain without regular fire (Sankaran et al., 2005; Bowman et al., 2008; Lehsten et al., 2016; Murphy et al., 2010). Patches of unburnt canopy and wet gullies provide refuge for tree-dwelling mammals (Chia et al., 2015).

14.5.6 Fire and the Long-Term Erosion Rate

This section is rather more speculative than the previous sections and takes a long-term geomorphic perspective on the impact of fire on erosion and landform evolution. Many of the ideas below have been triggered by conversations and debates with my colleague Gary Sheridan.

In the previous section typical values quoted for fluvial erosion are that it increases by at least 10–15 times the unburnt erosion rate (sometimes orders of magnitude more) and that impacts may last for two to 10 years. Let us consider a thought experiment. If the increase is a factor of 10 for two years, both midrange values, then one fire will have the same amount of erosion as 20 years of unburnt erosion. Thus if an area burns every 21 years, then the post-fire erosion will be the same magnitude as the inter-fire unburnt erosion. In many fire-prone areas a return period for fires of 21 years is not abnormally long, suggesting that post-fire erosion may be a significant, if not dominant, component of the long-term erosion rate for a catchment. The thought experiment shows that as the fire impact on erosion and the recovery time for that impact increases, and the recurrence interval decreases then the relative importance of post-fire erosion in the long-term geomorphic erosion rate will increase.

Recent work has also highlighted other processes, even though quantification has still not been performed. In the immediate aftermath of a fire dry ravel and small debris flows move loose sediment (recall that fire can destroy soil structure and make soil more erodible by volatising the SOM that binds soil into aggregates) from higher parts of the catchment to lower parts of the catchment. This loose material is temporarily stored on the hillslope and valleys, and is available for movement in subsequent fluvial erosion after the catchment vegetation has recovered. The evidence suggests that this mobile material is then eroded over the long term after the catchment has 'recovered'. The quantities of sediment that may be temporarily stored in this way on the hillslope and in the valleys are still unclear.

Let us return to our thought experiment. Let us assume 50% (this is simply assumed for the thought experiment; I have no evidence to support this value) of the sediment eroded off the hillslope leaves the catchment during post-fire runoff (and is thus the 10–15 times increased erosion that is observed in experiments at the catchment outlet) while the other 50% is moved from the hillslope into areas of temporary storage elsewhere downstream on the hillslopes and in hollows. If this temporary storage is then eroded over the long term after the catchment has 'recovered', then this material may then be equal to 10–15 years of unburnt erosion. We then speculate that there are three stages of erosion in a fire-prone catchment:

1. *Stage 1:* In the year or two after the fire, the erosion rate (as observed at the catchment outlet) is elevated by 10–15 times. At the same time dry ravel and debris flows are mobilising soil, and this loose sediment is stored lower down on the hillslopes and in the hillslope hollows.
2. *Stage 2:* For a period of a decade or more (recall there are observed impacts on nutrient fluxes 15 years afterwards, so a decade or more to recover does not seem unreasonable) fluvial erosion is removing the loose sediment temporarily stored on the hillslope and in the valleys during Stage 1. Stage 2 ends when the stored sediment has all been removed. Nyman et al. (2013) suggest that the erosion rate post-fire declines exponentially, so there may be a gradual transition from Stage 1 to Stage 2, and thence to Stage 3.
3. *Stage 3:* After a decade or more of removal of loose sediment created and stored on the hillslopes in Stage 2, the erosion drops to a source-limited condition where the erosion is determined by the recovered vegetation cover and the equilibrium unburnt soil conditions. This might be called the true unburnt erosion rate. Stage 3 continues until the next fire.

There are two implications of this conceptualisation of erosion impacts into three stages.

The first implication is that if a catchment burns very frequently, then the catchment may never reach Stage 3. In this case the long-term erosion rate is solely determined by the post-fire transport processes. The burn frequency for this case, while short, is not outside the bounds of observed burn frequencies. What should be clear is the impact of fire on erosion will be a function of the fire frequency (how long the catchment remains in Stage 2), and the severity of the fire and the recovery time (i.e. the duration of the Stage 2). Thus the geomorphic impact of fire is likely to vary markedly with climate and dominant ecosystem.

The second implication is that the burn frequency in populated areas is a function of population density, so that it is possible to conceive that in areas of low burn frequency (where the catchment spends significant time in Stage 3) that if the population density increases and the burn frequency also increases, then the average erosion rate can increase because a greater proportion of time is spent in Stages 1 and 2. Thus it is possible for the geomorphically effective average erosion rate to increase as a result of human impacts without there being any obvious direct link between human activities and erosion (e.g. a direct impact would be farming, land management changes or climate variability).

Much of the discussion above depends on the relative magnitude of burnt and unburnt impacts, most of which are poorly quantified at this time, so the discussion above should be considered somewhat speculative. However, typical numbers from the literature provide support for the relative impact of burnt and unburnt erosion rates.

The mechanisms of dry ravel and debris flows that are the sources of the temporarily stored sediment are specific to high-slope environments. However, the destruction of soil structure by the burning of SOM is equally applicable to high, and low-slope environments, and it is possible that there are also mechanisms that provide temporary storage for mobile sediments on low-slope hillslopes.

We thus conclude that it is possible that in fire-prone areas the average erosion rate may be determined by the sourcing of sediment as a result of fire, and that the fluvial erosion between fires is simply the mechanism to deliver this sediment to the catchment outlet.

14.6 Other Forms of Vegetation Disturbance

Fire and herbivore grazing are only two forms of disturbance of vegetation that might feed back to erosion rates. A number of other disturbances were briefly discussed in Chapter 7 in the context of soil bioturbation in Section 7.3.5.

Tree throw excavates a hole in the ground when the tree roots are pulled up. There are two aspects to the post-tree throw behaviour. The soil in the root ball is now above the ground and is loose and can potentially be more freely transported, leading to higher erosion rates. On the other hand, there is now a hole into which water and sediment can flow and collect, potentially capturing erosion. The post-tree throw erosion is the net effect of these competing processes, and at least in the short term the roughening of the hillslope surface leads to a reduction in sediment delivery (Hancock et al., 2012).

Fauna can also influence erosion by excavating around trees to feed on the roots. In Section 7.3.5 we discussed rates of soil turnover from wombats digging for roots to eat. Wild pigs have also been observed to be an active soil turnover mechanism. As for tree throw, the net effect of this digging is to roughen the surface and enhance the capture of runoff from the hillslope, reducing delivery of sediment to the catchment outlet (Hancock et al., 2015b, 2017b).

14.7 Conclusions

Vegetation and fire impact on almost all the processes in the preceding chapters. Vegetation

- Increases the infiltration rate of the soil by the creation of macropores (Chapter 3), and influences soil moisture dynamics and water flow down the soil profile by transpiration (e.g. flow velocity down the soil profile in Chapter 8).
- Protects hillslopes from the full effects of shear stress arising from overland flow, thus reducing sheet and gully erosion (Chapter 4).
- Appears to be the main driver of soil depth via tree throw, with roots ripping rock fragments out of the saprolite (Chapter 6).
- Provides tree roots that strengthen the soil profile and protect the soil against gravity processes such as creep (Chapter 9) and landslip (Chapter 13).
- Is the main source of soil organic matter for the soil by providing leaf litter to the soil surface, and dead roots and exudates within the soil profile (Chapter 10).

The complexity of a general vegetation model that addresses all these issues simultaneously means that we can only scratch the surface of vegetation representation in a LEM in this chapter, even if the science actually existed to attack all of these issues simultaneously. One of the problems with the literature is that the treatment of vegetation is rather fragmented, with little linkage between the various approaches. Inevitably this chapter in part reflects this fragmentation.

14.8 Further Reading

For those with an interest in the range of agricultural crop simulation models used internationally, a good place to start is the 2003 Special Issue of the *European Journal of Agronomy*, 18(3–4). Research in Dynamic Global Vegetation Models (DGVM) is extremely active and fast moving because of global climate modelling imperatives, but the BIOME model (Prentice et al., 1992) and its descendants (BIOME2, BIOME3, BIOME4, LPJ) are seminal founding works. The 1999 Special Issue of *Global Change Biology*, 5(Supplement 1) compares DGVM model characteristics (Cramer and Field, 1999) and while it is somewhat dated, it overviews many aspects of DGVM construction and response at that time, and provides a grounding for understanding more recent work. Bondeau et al. (2007) used LPJ to examine the impact of vegetation (and specifically agriculture) during the 20th century on the global carbon cycle and the reduction in soil carbon stocks. Eagleson (2002) provides a very detailed discussion of vegetation adaptation in water-limited environments (Chapter 1 is a particularly good overview of the principles). For those interested in a broader range of forest fire models than discussed here, and the scientific principles underpinning them, see the review papers by Sullivan (2009a, b, c) and Hilton et al. (2015).

15 Constructing a Landscape Evolution Model – Details

15.1 How to Couple the Processes

The preceding chapters have focussed on how to solve the science of the various components of soilscape and landform evolution. In Chapter 2 we had a brief discussion of some concepts of how to model these processes and how to couple them. This coupling is key because many of the interesting aspects of landscape evolution arise as a result of the temporal or spatial coupling between processes. Having covered the science details in the previous chapters, this chapter dives into the detail of how to couple these processes in a model, calibrate it and test it against field data.

The core concept used to couple processes in time by most models is operator splitting, discussed in Section 2.2.2. At each timestep the change due to each process over that timestep is calculated, and the impacts of all the processes are summed and applied for that timestep. At the next timestep the results of the previous timestep are used as the starting point. This simple idea hides some subtle issues that may need consideration. The use of the results from the previous timestep as the starting point for each process means that during that timestep the processes are acting independently, and that the coupling occurs through the summing of the processes at the end of the timestep. This leads to a criterion on the maximum size of the timestep, which is that the timestep should be small enough that the fastest changing interaction is captured. If the timestep is too big, then (1) the interactions will not occur as fast they would in the field system and the model will model the interaction as acting too slowly and (2) instabilities may be created in the states that are interacting, leading to model mass balance errors or in the worst case a model crash.

A simple example may suffice to explain this principle in practice. To model erosion on a grid you solve the elevation change for a node; i.e. a mass balance is calculated on the inflows and outflows of sediment at that node. A maximum step size can be used to obtain the desired numerical accuracy for the elevation change, and that timestep size depends on the numerical solver used. It is straightforward to write a Courant number numerical constraint for this timestep size when an Euler solver is used. However, another factor is if the elevation change is large enough, then there is a possibility that the drainage directions in the code may be changed (i.e. the steepest slope direction is changed) and the inflows and outflows into that node are also changed (potentially dramatically) because of the change in drainage area, leading to large instabilities and model crashes at the next timestep. The experience of many LEM developers is that this latter drainage direction timestepping constraint is often the limiting constraint on timestep size when fluvial erosion is the dominant process. In this case the Courant number constraint is not applicable, and using it typically leads to model crashes.

These latter timestepping constraints have some interesting consequences.

The first consequence is that advanced solvers may not yield the performance speedups conventionally seen in other fields. A common method to speed up timestepping in traditional finite difference codes (e.g. a groundwater model) is to use an implicit solver instead of an explicit solver. Most LEMs use an explicit solver because it is significantly easier to implement. However, the solver instability that the Courant limit is designed to address occurs only for explicit models, and implicit methods allow a larger timestep. However, implicit methods are significantly more complex to code and require the inversion of a banded matrix that is $N \times N$ where N is the number of nodes in the domain. Domains of $1{,}000 \times 1{,}000$ nodes are not uncommon, leading to the need to invert a $10^6 \times 10^6$ banded matrix. If, however, for fluvial erosion the limit on the timestep size is the change in drainage directions as a result of erosion and NOT the stability of the erosion solver at any node, then there may be no advantage in using a more sophisticated solver because the problem that the solver is addressing is not the critical limit on the timestep size.

The second consequence is how to write the computer code to couple the processes. It should be clear from the previous chapters that for some of the processes discussed significant codes already exist to model them, and the LEM developer is left with three alternatives as to how to incorporate their capabilities in their LEMs. If we refer to the established process model as CodeA, then we can (1) recode CodeA from scratch so that they fit within an existing landscape evolution framework, (2) refactor CodeA so it can be made to work within an existing landform evolution (typically by writing some wrapper subroutines around CodeA) or (3) run the two codes (the LEM and CodeA) in parallel and transfer the output between codes using some intermediate computing infrastructure (e.g. file transfers).

The first option is a major waste of time. Recreating CodeA from scratch, or translating CodeA from one language to another (if you are lucky enough to have access to the source code for CodeA), and then debugging it, is potentially a major task. However, it is common practice. It also means that as developers update CodeA, then whether the LEM developer updates their LEM is a function of the availability of funds, time and energy, and it is common not to have the most up-to-date science in the LEM.

The second option is better but depends on (1) the CodeA having a structure that is amenable to having wrappers written around it that can be used by the LEM and (2) the language of CodeA and/or the wrapper subroutines being compatible with the language of the LEM. For instance, both Tellusim (Willgoose, 2009) and the CSDMS BMI modelling framework (Peckham et al., 2013) require four subroutines that can be called (a) program initialisation, (b) a computation loop, which may be called repeatedly, (c) program finalisation and (d) data transfer routine(s) between the two programs. For both Tellusim and BMI, if the model cannot be cast into this four subroutines structure, then the coupling may not be possible. The main difference between Tellusim and BMI is that Tellusim is designed to couple with models in the languages Fortran and Python/Cython, while BMI provides support for a broader range of computer languages.

The third option is to simply run the LEM and CodeA in parallel and have some infrastructure to transfer data between the codes. At the simplest level this may involve running both the LEM and CodeA for one timestep, writing results to files, and then before running the next timestep reading the output files of both codes, extracting out those data required as input for the next timestep and constructing input files for the next timestep. While writing/reading files to/from hard disks is slow, if RAM disks are used, then this slowness can be reduced to some extent. Solid state drive speeds are intermediate between hard drives and RAM and are an increasingly cost-effective intermediate technology. This doesn't require much software infrastructure other than scripts to run the codes and set up the files for each timestep. However, the two executable programs do need to be controlled by scripts. The scripting languages in Python or MATLAB both have the tools required to control executables that use a command line interface and do the ancillary file modifications. However, the control by scripting generally requires that the executables not be run from a graphical user interface (GUI) (e.g. controlled by a mouse) since the current generation of Python and MATLAB scripting tools cannot control a GUI. This may be a problem if the GUI is heavily integrated into the computational engine of the codes and so is difficult to isolate from the science engine. Fortunately most science codes have fairly rudimentary GUIs, if they have a GUI at all, so this is rarely a problem. Commonly science codes consist of two separate programs, one which is the C or Fortran computational engine, and then a separate graphical interface that may or may not be a separate program. In this latter case the graphical interface's sole role is to provide the easy to use tools to create input files and interpret output files from the underlying computational engine.

We now return to the issue of what timestep to use. The temptation is to use as large a timestep as possible (particularly for the third option) simply because the data transfer between codes is likely to be the limit on model performance. However, this may result in instability in the interaction between processes. For instance, Temme et al. (2011) coupled their LAPSUS SLEM with an agricultural management model and used it on two case studies to study the interaction between landform evolution and agricultural management options. They found that if the timestep was less than one year, then the interaction between the processes in their model was satisfactorily modelled, but for longer timesteps some of the interactions became unstable, even when the processes, when used on their own, were stable for timesteps of 50 years (water redistribution within the landscape) and 2,500 years (tillage effects) (see their Figure 10). Thus the interactions forced the reduction of the timestep from 50 years to 1 year.

Finally one advantage of having independent codes for each process is that it addresses one major issue with widely shared codes; the problem of code forking. Code forking is where we encounter multiple versions of a 'single' code, each of which has some unique capability that a researcher has added to the code, but there is no longer one single master code that contains all the capabilities that all researchers have added. Thus when somebody speaks about a model X, they may be referring to any number of multiple variants of a code, X1, X2, X3 etc. Thus if a user needs the capabilities of X1 and X2, they need to manually merge those capabilities themselves. Nothing is stopping the 'owner' of the code from converging the various forked codes, but (1) this assumes the 'owner' has access to all the variants of the code and that the developers are happy to share the source code for their variants, (2) the extensions in the variants may be incompatible and so cannot be converged into one master code and (3) most research funding is to achieve insights and results, and it is rare that funding support is available for code maintenance after a project is finished. However, if a standardised interface/wrappers allows any extension to be used by the landscape evolution model without needing to modify the underlying landscape evolution code, then it is easier to manage the incipient 'chaos' that can occur when multiple researchers across multiple projects and research groups are extending a code simultaneously. This is a nontrivial management problem for any successful open source computer code and is commonly commented on by developers of open source environmental codes that are widely used (e.g. Holzworth et al., 2015). The plug-and-play and plug-in architectures in Tellusim, CSDMS/BMI and Landlab are designed to begin to address this issue.

This discussion of how processes are coupled, and the various technologic ways that this may be done, may seem like overkill for many readers who are only interested in developing a small standalone landscape evolution model to be used by themselves or at most a small group of researchers. However, once the landscape evolution model escapes the lab environment and is used by outsiders, the demands for project-specific extensions can be overwhelming for the code developer (who no doubt has his or her own projects to worry about). Much of the technology mentioned above is designed to address this conflicting need to extend the model capabilities (by multiple groups) and the need to ensure that the basic code is not subject to forking, while trying not to introduce too much overhead into the code development process. For instance, we don't want graduate students, who can be struggling to implement their own research code as it is, to have to struggle with another layer of intellectual difficulty in the form of the coupling infrastructure.

Tellusim (Willgoose, 2009), SIGNUM (Refice et al., 2012) and Landlab (Hobley et al., 2017) are all designed to address this conflict between code forking and active code development by multiple research groups. The differences are in the paths taken to achieve that objective. SIGNUM is a modelling system in MATLAB that modularises the interacting processes but where each process is a conventional subroutine. In many respects SIGNUM is akin to traditional programming of landform evolution model, only it is done in MATLAB rather than Fortran or C. For example, both SIBERIA and CHILD are modularised in this way except they use Fortran and C, respectively, rather than MATLAB. On the other hand Landlab and Tellusim are designed around plug-in modules that have a standardised interface, and each module can be dynamically included or excluded without having to modify the underlying landform evolution model. This dynamic behaviour requires an amount of overarching infrastructure to support it as discussed above. Landlab is designed such that all modules are coded in Python, Cython or C, while for Tellusim they are coded in Python, Cython or Fortran. For both Landlab and Tellusim it is likely that other languages may be supported in the future because in both cases the overarching infrastructure is written in Python, and Python provides the capabilities to link to many other languages.

15.2 Model Testing

The objective of model testing is to provide confidence that a model is working as expected and/or designed, and that it can be used to describe and/or predict behaviour observed in the field. To provide this confidence the model must be subjected to tests that are capable of identifying that the model is not working correctly when it is wrong, and that does not have a high likelihood of rejecting a good model (i.e. saying it is wrong when it is in fact correct) or of accepting a bad model (i.e. saying it is right when it is in fact wrong). The rest of this chapter expands on these ideas and provides examples of methods that have been used, or could conceivably be used, to test landscape evolution models. Along the way we will discuss aspects of the landscape evolution problem that limit model testing.

The first testing concepts are those of validation and verification. The two concepts are often confused or conflated in the literature:

1. *Verification* is testing to ensure that the model solves the problem as designed. Normally this is testing that the mathematics of the model formulation is being

solved correctly by the numerical implementation. Verification is normally done by running the model against solutions for problems that are known to be correct (e.g. analytic solutions to equations for simple problems, other well-tested models, the same model with a different numerical solver) or convergence testing (e.g. multiple runs with increasingly smaller timesteps and/or finer spatial discretisation to see that the results converge to a stable answer . . . though this approach can be fooled when the discretisation error of the solver is a function of the ratio of the timestep to the grid size as is the case in advection problems; see Celia and Gray, 1991).

2. *Validation* is testing that the model correctly simulates the practical problem being simulated (e.g. field or laboratory experiments). This is a subtler task than verification because it involves construction of a model. A 'model' is the combination of the computer implementation of the mathematics, and the parameters and boundary conditions used in the computer implementation.
   a. Just how good this validation is, is then a question of how independent the data used in the construction of the model is from the data used in the testing of the fit. This latter point about independence is important because fitting a model to data (i.e. model calibration), and then asserting that this model is validated as a result, is incorrect. This just proves that the model can be forced, with some choice of parameters (realistic or otherwise), to fit a particular situation so that the model is *feasible*. If, however, the data were not used in determining the parameters, then this is an independent test, and more strongly validates that the underlying physics in the model and the process parameters leads to what is observed in the field. Thus calibration of the model (i.e. determining the model parameters and boundary conditions) is a separate task to validating the model, and so validation requires data independent of those used in the calibration. This independence can be quite a severe constraint in earth sciences, and it is common for independent validation to be only partly achieved.
   b. It is common to exercise only part of a model in one or both of calibration and validation so that the testing has validated only one aspect of the model not all of it. For instance, if all the calibration and validation data were obtained in the diffusive part of the landscape, you should not expect to be able to test and validate characteristics that are dependent on the nondiffusive components of the model. A second example is that calibrating the fluvial erosion model to data collected over a short period (say a few years) may not accurately reproduce the erosion over the long term (hundreds to thousands of years), because aspects of erosion arise only after a significant time (e.g. incision, which is a function of the feedback between erosion and landform evolution), and they are dependent on aspects of the erosion process not normally measured and calibrated over the short term.

Another important aspect of modelling is determining how reliable the model predictions are. If you make predictions (for a different site, or at some time in the future), an estimate of the reliability of the projections may be required (e.g. best estimate $\pm$ standard deviation). As we will show below the validation process is key to determining the properties needed to make these quantitative error estimates. For some engineering problems where a statutory requirement might be a key design constraint (e.g. incision less than the thickness of a protection cap on hazardous waste) it may be necessary to assess how likely the model projections are to meet this regulatory requirement. For example, if a waste containment structure must contain the waste against release by erosion for 1,000 years, what is the probability that this requirement will be breached during the design lifetime of 1,000 years?

15.3 Model Verification

Model accuracy reflects various aspects of model performance. The most obvious is the numerical representation of the processes accurately represents what is happening in the field. Since LEMs are discretised in both space and time. an important question is what resolution is required so that field processes are faithfully reproduced. All models are approximations of the real world, so this is a multistage question (1) how are the field processes converted into mathematics, (2) how is that mathematics spatially discretised and (3) how is that spatially discretised model then discretised in time so that the temporal evolution can be modelled. The previous chapters of this book have been focussed on the first of these stages, generating the mathematics that reproduces field behaviour. This section will focus on the latter two stages, which is how accurately is the mathematic solved. However, the central purpose of this section is to describe the methods that are used to assess the reliability of a model as a whole and how this model can be validated against field data.

15.3.1 Model Spatial Resolution

The main criterion for spatial resolution in an LEM is the transition of the landform from concave down, divergent flow (hilltops and ridges) to concave up convergent flow (valleys). In Chapter 3 we discussed how the slope-area relationship and cumulative area diagram both allow quantification of this transition area for catchments at dynamic equilibrium. In both cases we can calculate a catchment area at which this transition occurs. For the slope-area relationship, if we know the uplift (or the erosion rate ... equivalent to a negative uplift), we can estimate this area directly from the model parameters and the analytic expressions for slope area. Zhang and Montgomery (1994) analysed the saturation-excess hydrology of a catchment that had a transition distance of about 100 m and found no incremental benefit in a higher resolution of 10 m, suggesting a grid size of about 10% the transition distance. Claessens et al. (2005) also found that to model landslide hazard, a 10 m resolution DEM (the highest they tested) was required, and this appeared to be about 5% of the hillslope length (my visual estimate from their DEM). Hancock (2005) found that a grid size of 10 m was sufficient to capture hillslope shape and curvature in natural catchments (i.e. no imposed drainage structures such as contour banks), while Hancock et al. (2006) demonstrated the potential errors in using coarse data such as the 90 m SRTM DEM. Wechsler (2007) reviewed sources of errors in DEMs, including spatial resolution, and after reviewing the literature concluded that (1) higher resolution is not necessarily better and (2) the optimal resolution 'ultimately depends on characteristics of the study area such as topographic complexity.'

The limit of these works was that in many cases they did not have any higher resolution DEMs than 10 m, so all that can be concluded is that coarser resolutions than 10 m were less satisfactory. This limitation may be resolved with the more recent availability of high spatial resolution LIDAR DEM data. For instance, Thomas et al. (2017) found, after looking at LIDAR-derived DEMs of resolution ranging from 0.25 m to 5 m, that a 2 m DEM was optimal for capturing small microtopography effects on drainage paths.

One important issue in determining the optimal grid resolution is the need to determine drainage directions and slope, and the need to use a field DEM as initial conditions for the LEM or as a comparison with LEM predictions. The error in the slope is a function of both the elevation accuracy of the DEM and the horizontal resolution. For a given resolution, if the elevation accuracy is lower, then the slope accuracy will be lower, while for a given elevation accuracy, the finer the DEM resolution, the lower the slope accuracy. A simple example follows. Consider the calculation of slope between two adjacent nodes A and B, separated by distance equal to the grid resolution Δx, where the variance of the elevation accuracy of both A and B is σ_e^2 so that the mean $\overline{S}$ and the variance σ_S^2 of the slope estimate are

$$\begin{aligned} \overline{S} &= \frac{z_A - z_B}{\Delta x} \\ \sigma_S^2 &= \frac{2}{\Delta x^2}\left(\sigma_e^2 - \sigma_{AB}^2\right) \end{aligned} \tag{15.1}$$

where σ_{AB}^2 is the covariance of the error between points A and B. If the error structure of the DEM is such that the errors in the elevations at A and B are independent, then $\sigma_{AB}^2 = 0$. However, if the DEM is such that the error in the elevations is simply because the elevations have not been georeferenced to ground control, then it is likely that all elevations at all nodes will be either biased high or low, so that $\sigma_{AB}^2 = \sigma^2$ and the error in the slope is zero. For example, Moreno de Las Heras et al. (2012) found that a major component of the errors in the ASTER DEM was due to inaccurately ground-truthed stereo images, and found major jumps in elevations at image boundaries but little within one stereo image pair. Thus the correlation structure of the errors in the DEM is important.

By comparing DEMs generated by three different methods (ground survey, cartographic photogrammetry and digital photogrammetry) Walker and Willgoose (1999) showed that the derivation of the geomorphic properties in Chapter 3 could be extremely error prone, and that the method of DEM creation changed the spatial correlation of the error structure and thus the slope errors.

The derivation in Equation (15.1) is overly simple because as the errors in the slopes become larger, another possible outcome is that the drainage direction may also change (the location of the next downstream node B for any particular node A). Gyasi-Agyei et al. (1995) found that when the (elevation resolution or error)/(horizontal resolution) approaches 1, it becomes difficult to reliably extract the drainage network. Other issues arise when the only available data are those of coarse grid data. For example, the SRTM DEM allows unsurpassed global surface data; however, its coarse grid (at best 30 m) can provide poor catchment resolution (Hancock et al., 2006). There are two conclusions:

1. That slope error is a function of the combination of the elevation error and the grid resolution, and neither elevation error nor grid resolution should be considered alone if slope is important.

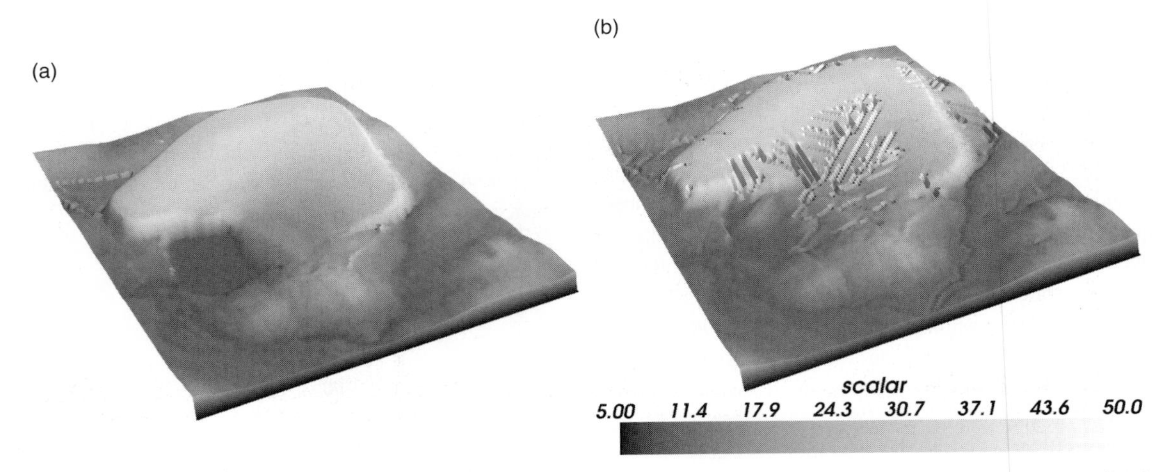

FIGURE 15.1: The Ranger landform at times (a) 0 years, (b) 1,000 years. Note that the landform is vertically exaggerated with the horizontal length being 1.1 km × 1.5 km, and the elevation range is about 40 m. What appear to be gullies at 1,000 years are in fact valleys 7–8 m deep by 60 m wide.

2. DEM error can be significant so that when we discuss model validation in a following section, it is important to remember that the DEMs, which are the data against which we compare our landform evolution models, have their own errors which need to be considered at the same time as model errors.

15.3.2 Model Timestepping Convergence Error

In this section we will demonstrate methods for quantitatively assessing model accuracy. For simplicity we will do this by using an example of the SIBERIA landform evolution model applied to a proposed rehabilitation of a mine site. This DEM has been extensively discussed in papers by the author and colleagues (e.g. Willgoose and Riley, 1998a, b), and it is a convenient example for the discussions that follow because in a single DEM it has an extensive range of features including (1) areas of flow convergence and divergence in both erosion and deposition regimes, (2) extensive areas of alluvial fan development on both flat and gently sloping terrain, (3) sudden changes in longitudinal slope that trigger gully development, (4) a mixture of man-made and natural terrain and (5) a defined objective for the solution of the problem, successful containment of the mining waste for 1,000 years. It is worth pointing out that this DEM is no longer the proposed rehabilitation strategy, and the design lifetime has recently been extended to 10,000 years, so no conclusions should be reached by the reader about the performance of this landform for the currently proposed rehabilitation strategy. The parameters used in the simulations are those determined in Willgoose and Riley (1998a, b).

One convenient way to characterise the accuracy of a model is the mass balance of the model, that is, how well

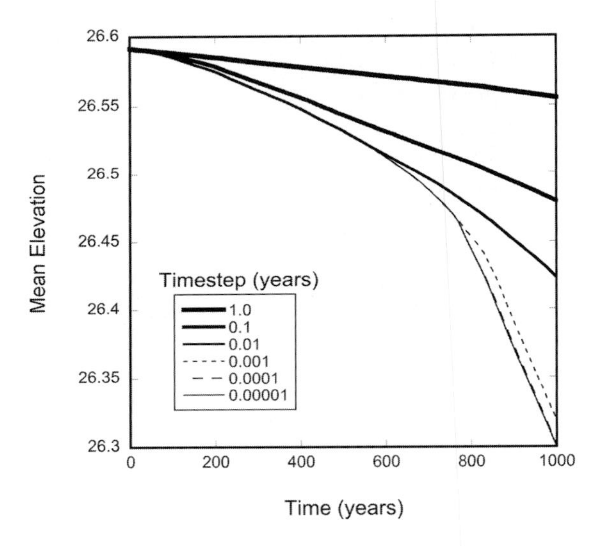

FIGURE 15.2: Time evolution of the mean elevation of the Ranger landform over 1,000 years against timestep sizes ranging from 10^{-5} year to 1 year.

does the mass change of the simulation match the exact mass change. For most of our applications we do not know what the exact change is because we don't have an exact solution. The major cause of inaccuracies in time-evolving models is the size of the timestep used, so to approximate the exact solution we use a model with a very small timestep, assuming that this model is a good approximation to the exact solution. We then plot the difference of the mass balance from the small timestep solution against solutions for various size timesteps.

Figure 15.1 shows the initial landform and the landform at 1,000 years. Figure 15.2 shows a plot of the mass

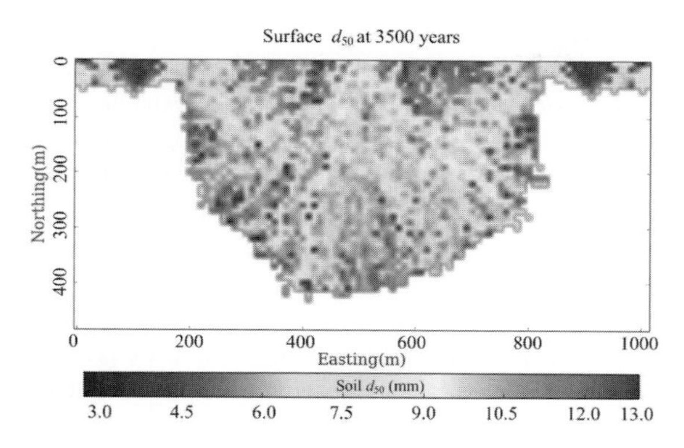

FIGURE 16.5: Planar view of the d_{50} of the surface grading of the alluvial fan. Note the linear filaments of coarse material propagating from the top of the fan at the centre top of the figure. The white area does not have significant deposition, and the localised rectangular areas at the top extreme left and top extreme right are artefacts of the design of the experiment (from Welivitiya, 2017).

On a related topic Cohen et al. (2015, 2016) used his mARM model to show how the deposition of a fine aeolian fraction can destabilise an armour by reducing the average particle size of entrained sediment and thus increase the sediment transport capacity of overland flow. Cohen's conclusion was that even though the overland flow could not carry the coarse armour by itself, if the armour particles are mixed with finer (aeolian) materials, then those coarse armour particles can be moved.

Finally, Welivitiya simulated the deposition of an alluvial fan (Figure 16.5) and found that his selective deposition (on particle diameter) model created banding in the particle size distribution down through the deposit, some in-situ weathering within the deposits and a spatial pattern of grading across the fan, driven by avulsions, that was consistent with experimental and field data for alluvial fans and modelling of submarine fans (Koltermann and Gorelick, 1992).

16.1.2 Coupled Vegetation and Landform Evolution Models

Moglen et al. (1998) examined the impact of climate change on drainage density using an empirical relationship between climate, vegetation and erosion that dates back to Langbein and Schumm (1958). This relationship (Figure 16.6) says that as the climate becomes wetter, the vegetation cover increases, offsetting the increase in erosion rate as a result of increased runoff. The result is a curve (Figure 16.6c) that has a low erosion rate for a dry climate (i.e. no runoff) and wet climate (i.e. heavy vegetation protection) and a high erosion peak at an intermediate rainfall (200–300 mm/year; Langbein and Schumm identified this rainfall as the transition between shrubland and grassland). They concluded, using their LEM, that for deserts, drainage density (or valley density) increased with increasing rainfall, for humid/wet climates, drainage density decreased with increasing rainfall, and for intermediate rainfalls, the drainage density deceased with both increasing and decreasing rainfall, and that the maximum drainage density occurred at the intermediate rainfall when erosion peaked.

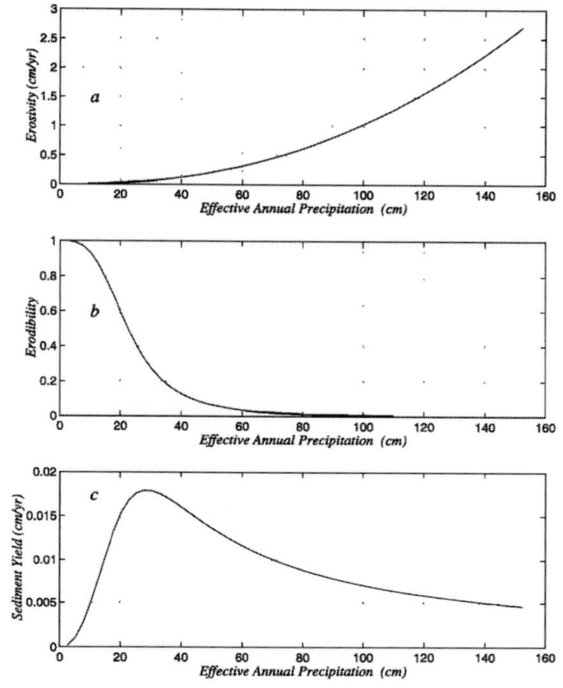

FIGURE 16.6: The Langbein and Schumm relationship between erosion rate and climate: (a) the erosivity, which is a function of runoff and is proportional to erosion rate in the absence of vegetation protection, (b) the erodibility of the landscape, which reflects the increasing protection of the surface by vegetation cover and (c) the actual sediment yield which is the multiple of erodibility and erosivity (from Moglen et al., 1998; after Langbein and Schumm, 1958).

Collins et al. (2004) and Collins and Bras (2010) examined a coupled model of erosion and vegetation where the vegetation growth model simulated vegetation cover (not biomass) in Equation (14.6). Collins and Bras revisited the work of Moglen by applying their vegetation dynamics model and showed that the coupled model could generate the coupling between erosion, vegetation and valley density in Figure 16.6. Moreover, they highlighted that there is a third regime that occurs for even

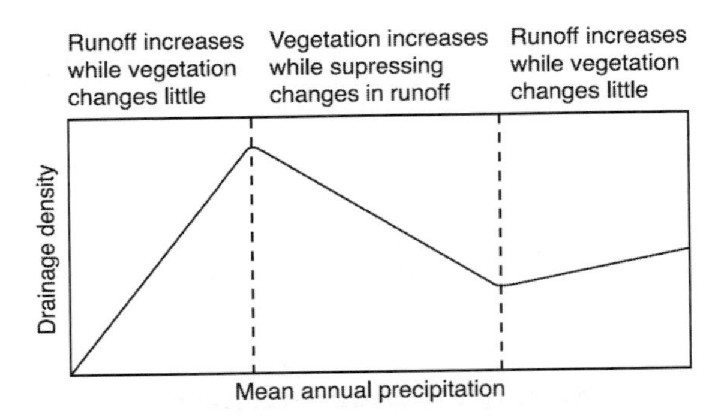

FIGURE 16.7: Schematic of the three regimes of erosion with the third region of increasing erosion rate with increasing rainfall on the far right of the figure (from Collins and Bras, 2010).

higher rainfalls than in Figure 16.6. In this third regime, as rainfall increases, the vegetation cover increases only slowly because it (1) is already quite dense and (2) the marginal benefit of increased vegetation cover decreases markedly for high vegetation cover (Figure 16.7). Collins et al. (2004) examined the effect of the competing processes of plant growth and death, and removal of vegetation by high shear stress during extreme runoff events. As vegetation became denser, the shear stress threshold increased. This had two consequences. The first consequence was that as the vegetation cover became denser, the average erosion decreased and the slopes in the catchment increased (this is consistent with the slope-area relationship and Equation (3.19)). The second consequence was as the erosion threshold increased, the size of the runoff event required to remove vegetation became higher. The larger size of the erosion event meant that the relative impact of the extreme rainfall was higher, so sediment output from the catchment, while it decreased on average, became more variable. Finally, Collins and Bras (2008) examined the impact of varying climate and found an asymmetry in vegetation recovery from climate variations. When the climate dried, the vegetation cover declines as expected, and vice versa for a wetter climate. But the recovery time from a drier period was slower than for a wetter period because the vegetation growth rate was smaller because it was limited by water availability. Moreover, the increase in vegetation density is driven primarily by the growth rate, while the decrease in vegetation density is driven primarily by the vegetation death rate, and these two rates are different.

Istanbulluoglu and Bras (2005) presented a landform evolution that was coupled with a vegetation biomass model (Equations (14.4), (14.5) and (14.8)), where the vegetation was subjected to a simple fire model that burnt the entire domain on average once every 200 years. The vegetation model was coupled with the erosion model so that erosion dropped with increased vegetation cover (Section 14.2.1), and landsliding rate increased after fire occurrence (see Sections 13.1.2 and 14.5.5). They performed a number of landform evolution simulations to explore the interactions between these processes. In the first they simulated the difference between an unvegetated and vegetated catchment for the case where the vegetation had no impact on the landsliding rate. Unsurprisingly (Figure 16.8), by reducing the erosion rate without impacting on the diffusive rate of transport (due to landsliding), they generated a catchment with much longer hillslope length and greater distance between valleys. This is consistent with the slope-area relationship (Chapter 3) for two competing processes where the reduction of fluvial erosion rate results in the inflection point in the slope-area relationship shifting to larger areas (i.e. reduced valley density). Most of the catchment gradients in Figure 16.8 are limited by the angle of repose, otherwise we would also expect the maximum gradients to increase with this shift of the inflection point. For the cases where the vegetation is periodically removed by landsliding and/or fire, the landform developed was intermediate between the two landforms in Figure 16.8. Collins and Bras (2010) also found that when the erosion was less (as a result of vegetation cover), the hillslope gradients increased.

Yetemen et al. (2015a, b) noted that in previous work (e.g. Istanbulluoglu and Bras, 2005; Collins and Bras, 2010; Saco et al., 2007) coupling vegetation to landform evolution assumed a 'flat earth' assumption for vegetation dynamics. In the flat earth, the land surface is assumed horizontal and no account is taken of the hillslope gradient and its orientation to the sun, so that all parts of the landscape are subjected to the same amount of incoming solar energy per unit area irrespective of the evolving landform and its aspect and gradient. Yetemen extended the Istanbulluoglu and Bras model to include the effect of hillslope aspect. Figure 16.9 shows how aspect changes incoming solar radiation with latitude, showing the significance of the differences in aspect for incoming solar radiation, and how for hillslope gradient in the N-S direction greater than 5° (about 9%) the effect of aspect is significant for midlatitudes. Yetemen also incorporated a soil water mass balance in his model so that as plants grew they consumed water at a rate determined by their growth rate. Thus faster growing plants consumed more water. For their study site they found that as a result of the soil water storage capacity that for the equatorial facing slopes (EFSs, in the northern hemisphere these are the south facing slopes), the soil water limited growth during the

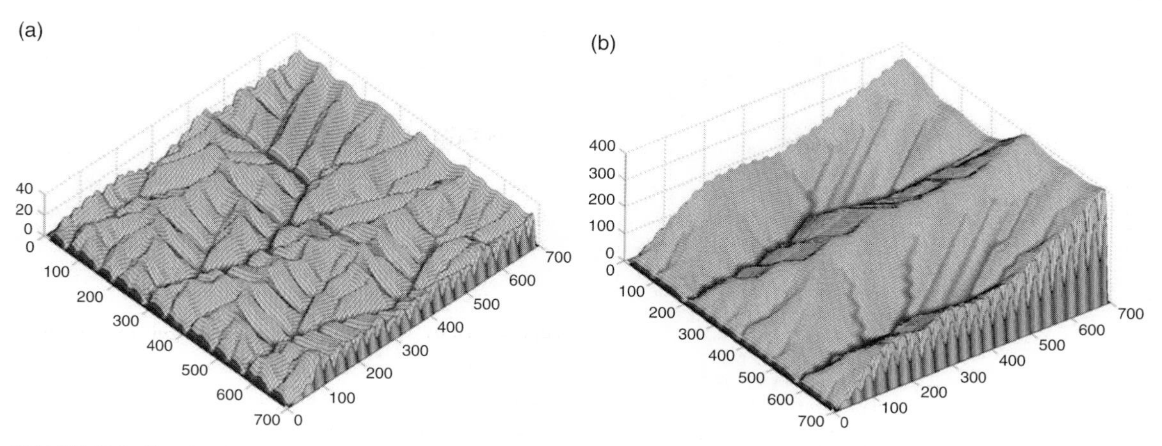

FIGURE 16.8: Two landform simulations showing the impact of (a) no vegetation and (b) fully vegetated on the valley density and hillslope gradients of the landform (from Istanbulluoglu and Bras, 2005).

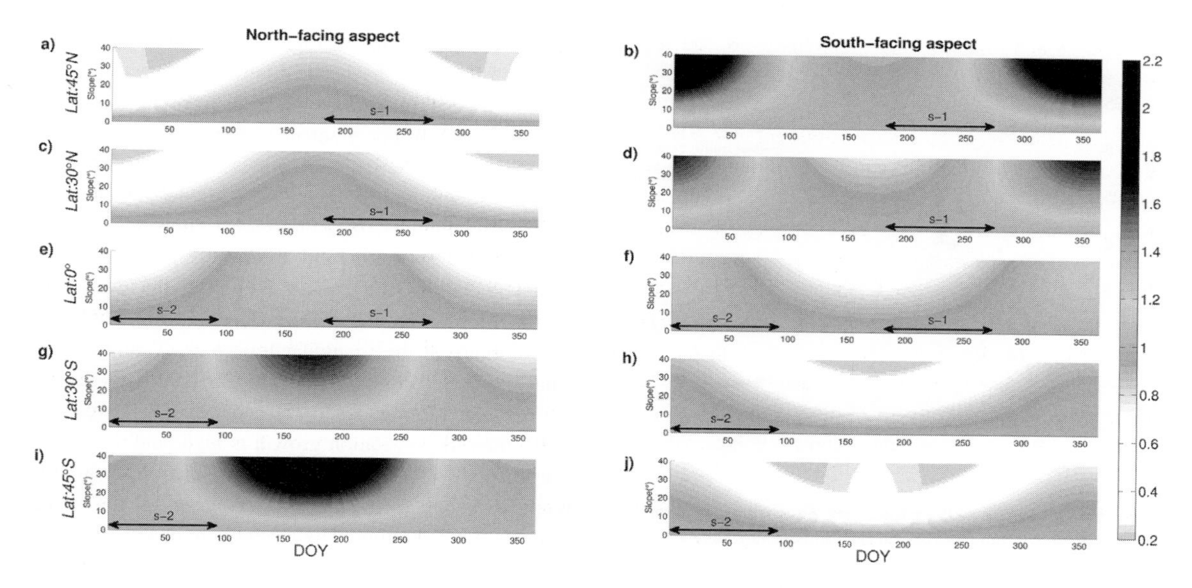

FIGURE 16.9: The change in the incoming solar energy at the top of the atmosphere (i.e. cloud free) versus latitude, normalised by the annual average solar energy at that latitude (from Yetemen et al., 2015a).

summer, while for the polar facing slopes (PFSs), growth was less limited by water availability. The consequence of this was that on average over the year the PFSs had higher biomass than the EFSs. Yetemen assumed that soils were the same for all hillslopes irrespective of aspect (i.e. flat earth for soil properties); however, the hydrology (the runoff mechanism used was saturation excess; see Chapter 3) varied with aspect as a result of the coupling with vegetation density and soil water storage. The higher vegetation cover on the PFSs relative to EFSs meant that erosion was least on the PFSs, leading to an asymmetry in the hillslope gradients where the PFSs were steeper and shorter than the EFSs (Figure 16.10). The hillslope gradient dependency of erosion is because of the erodibility dependency in the slope-area relationship (Section 3.3.3). This was consistent with observations at their field site.

Examining seasonal dynamics Yetemen et al. (2015b) highlights further important subtleties. The hillslope

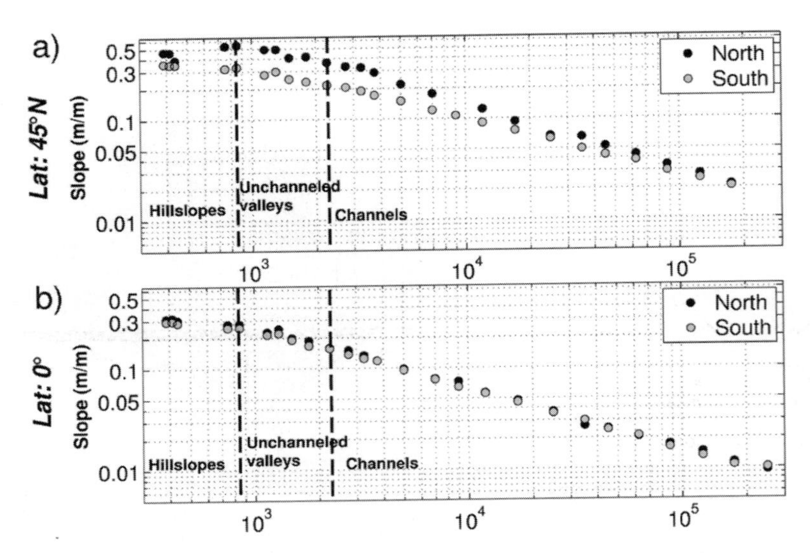

FIGURE 16.10: Slope-area plots for north (polar facing, PFS) and south (equatorial facing, EFS) facing slopes showing the steeper and shorter hillslopes for the PFSs: (a) 45°N, (b) equator. Note that this aspect effect on hillslope gradient disappears for a site at the equator (from Yetemen et al., 2015a).

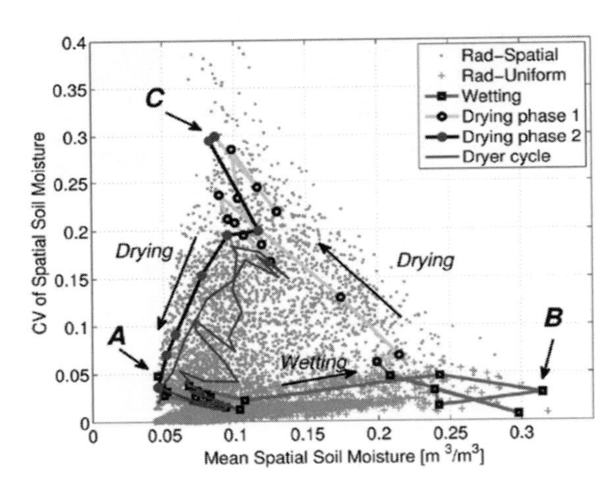

FIGURE 16.11: Plot of the trajectories of the catchment average soil moisture over time (horizontal axis) and the variability within the catchment of soil moisture (vertical axis). The dots are for the entire period simulated, while the lines are for two selected years. Points A, B and C, and the two years are for the same as in Figure 16.12 (from Yetemen et al., 2015b).

asymmetry they calculated was a function of the water limitation of the slopes. Consider the alternative case where nowhere was the vegetation water-limited (e.g. the soils were such that there was sufficient water to sustain vegetation through the summer on the EFSs). In this case the highest vegetation growth will be on the EFSs because of the higher incoming energy per unit area, so that erosion will be least on the EFSs and the slopes steepest. Thus for the non-water-limited case the hillslope gradients will be greatest on the EFSs, while for the water-limited case the hillslope gradients will be greatest on the PFSs. Thus the balance between water-limitation and non-water-limitation on the EFSs is key to the development of the hillslope gradient asymmetry. Figures 16.11 and 16.12 show the daily dynamics of the vegetation of the simulations, and highlight the seasonal dynamics of this process. Figures 16.11 and 16.12a show the spatially averaged soil moisture and biomass highlighting two annual trajectories (the times A, B and C in the two figures are for different days and were selected to highlight the three corners of the triangular trajectory; Yetemen, personal communication). As the soil moisture is increasing (i.e. non-water limited since water in the soil is increasing, A–B), then the vegetation growth is high, and as the soil moisture is decreasing (B–C), the vegetation is initially maintained but then starts to decrease as the water becomes more limiting. The spatial pattern illustrates the varying influence of hillslope aspect. For the non-water-limited part of the annual cycle (time B) the highest vegetation density is on the EFSs, showing the influence of the higher incoming solar radiation. As the EFSs become water limited (time C), the highest vegetation is now on the PFSs. Time A shows the persistence of this water limitation from the previous year with much lower vegetation densities on both the EFSs and PFSs (note the change in the colour bar units between times A and C), but the relatively higher density of vegetation on the PFSs remains.

The effect of aspect is greater at higher latitudes, but the effect levels off beyond about 30°. The effect is directly related to the relationship between solar radiation and latitude, and Figure 16.9 shows this increasing markedly up to about 30° and then increasing more slowly

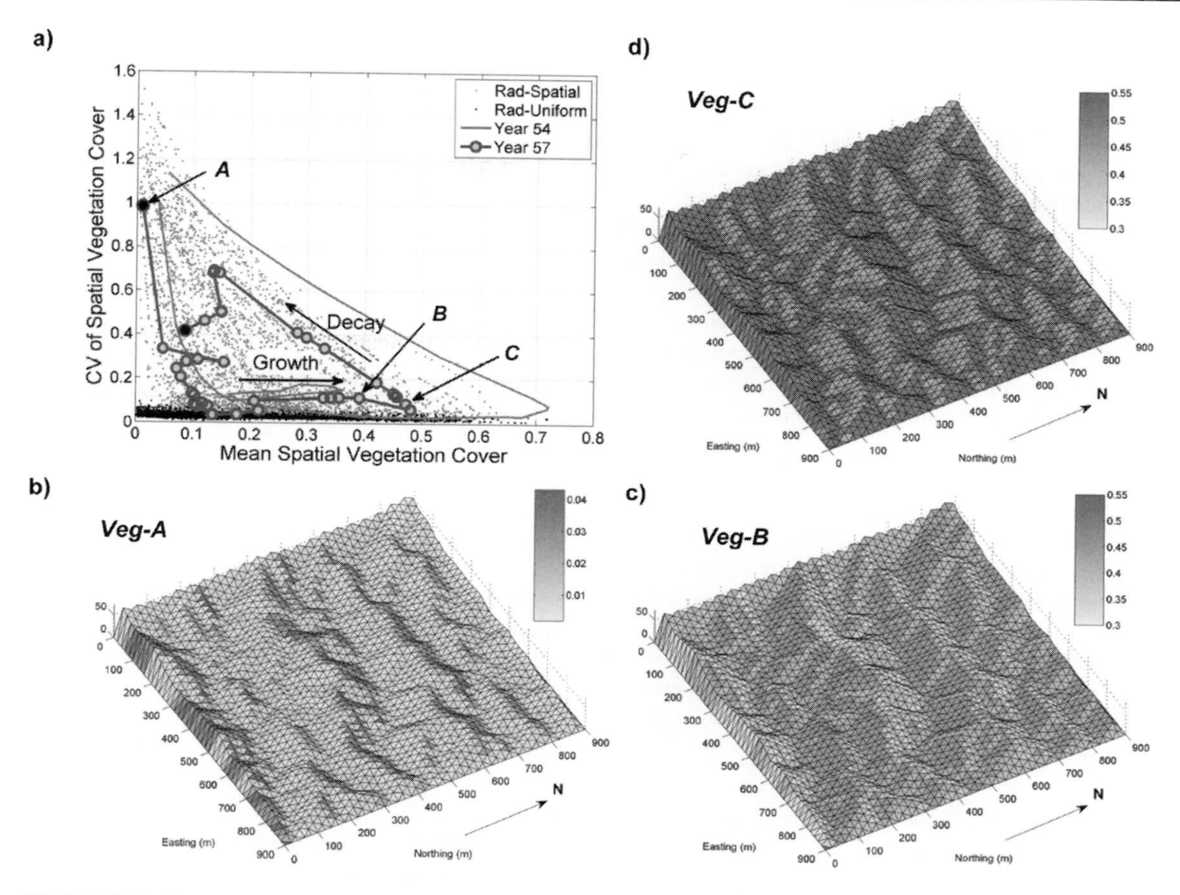

FIGURE 16.12: Vegetation cover for the catchment over time: (a) the catchment average vegetation cover (horizontal axis) and the variability of the vegetation cover over the catchment (vertical axis), (b)–(d) maps of the vegetation cover at the times A, B, C in plot (a). Points A, B and C and the two years are for the same as in Figure 16.11 (from Yetemen et al., 2015b).

thereafter. Yetemen et al. (2015a) confirmed that the simulated asymmetry in hillslope gradient increases with latitude up to about 30°, and then declines at higher latitudes because of a general reduction in water limitation at higher latitudes (Gerten, 2013), probably as a result of decreased evaporation rates at high latitudes (Yetemen, personal communication). Their modelling varied only the latitude and aspect, and so did not consider any latitudinal dependencies in altitude, soils, mean annual precipitation and potential evaporation. At any specific site an important variable is how quickly the EFS become water-limited, and this will be a function of the plant available water capacity, which will be a function of the depth of the soil (Chapter 6), the depth distribution of the water in the profile (relative to the depth distribution of the plant roots) and the ability of the soil to hold water between rainfall events (soil water-holding capacity), which is a function of the grading of the soil profile and its organic matter content (Chapters 7, 8 and 10). This interaction between water limitation and the soils raises the question of the landform response if the model of Yetemen et al. is coupled to a soil evolution model as discussed in this book. Some possible interactions include the following:

- SOM increases infiltration and water-holding capacity (Sections 10.6.2 and 10.6.3). If vegetation density is lower (due to water limitation), then SOM will also be lower, and infiltration and soil water-holding capacity will both be lower, reducing the stored soil water and providing a positive feedback enhancing the water limitation.
- Some researchers (e.g. Gabet and Mudd, 2010; Section 6.2) assert that the rate of soil production increases with increased tree throw so that a lower tree density (hence lower rate of tree throw) will result in a lower rate of

soil production, shallower soils and thus a lower soil water storage capacity. This will lead to an enhanced water limitation. Yetemen modelled only grass (since erosion is primarily a function of groundcover rather than tree cover), so it is unclear whether this will feed back into his asymmetry.

- Clay is a transformation product from rock weathering (Chapters 7, 8 and 10). Physical weathering breaks up the rock into fine particles with high specific surface area (with a low specific surface area, chemical reactions are very slow) and is primarily a function of cycles of wetting/drying. It is likely to be highest on those slopes with greatest soil moisture variability during the year. This variability is probably highest on the water-limited hillslope, but all of Yetemen's slopes have a high variability over the year (Figure 16.11), it is just that the EFSs dry out earlier in the summer. Chemical weathering requires sufficient water to facilitate the chemical reactions and is typically highest in wetter soils. Thus physical weathering is probably highest and chemical weathering lowest on the water-limited hillslopes. The net feedback effect is unclear and may be a function of other processes (e.g. microbiology, fungi).

16.1.3 Coupled Geodynamics and Landform Evolution Models

Coupled models of erosion and tectonics were one of the first uses of coupled landform evolution models. The first application was to couple a model for tectonics that included isostatic rebound and flexural rigidity to explore how the crust responded to flexural unloading as a result of erosion. The post-breakup dynamics of passive margins was an application for these types of coupled models of isostasy, crustal rigidity (the combination of isostasy and crustal rigidity is referred to as flexural isostasy) and surface erosion (Gilchrist et al., 1994; Kooi and Beaumont, 1994). The idea tested is illustrated in Figure 16.13. Initially after the breakup there is unloading on the coastal region as a result of the erosion on the steep coastal escarpment, which leads to isostatic rebound in the region surrounding the unloading. The area behind the escarpment up-warps, resulting in drainage on the other side of the coastal escarpment draining away from the escarpment rather than toward the escarpment. The importance of the crustal rigidity in distributing the uplift in the region around the unloading site is shown in Figures 12.6 and 12.7, both of which show the crustal deflection as a result of an imposed load (unloading is just the negative of loading).

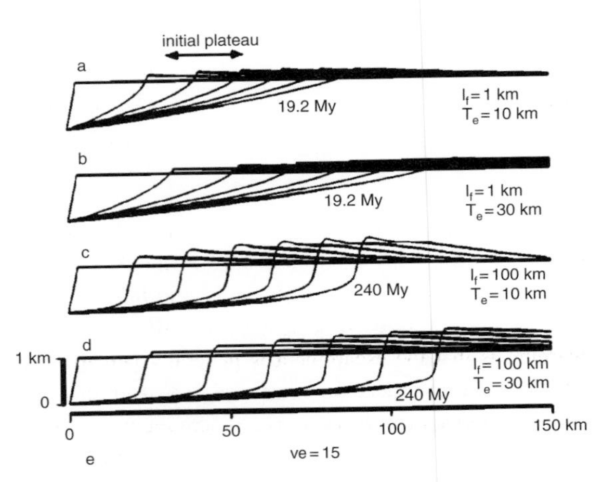

FIGURE 16.13: Profiles across an evolving passive margin escarpment for a variety of erosion rates and thickness of the crust T_e. The rigidity of the crust is linearly proportional to the thickness, so the stiffer crust has $T_e = 30$ km. The erosion rate is proportional to the travel distance for the sediment l_f so the highest erosion rate is for $l_f = 100$ km (from Kooi and Beaumont, 1994).

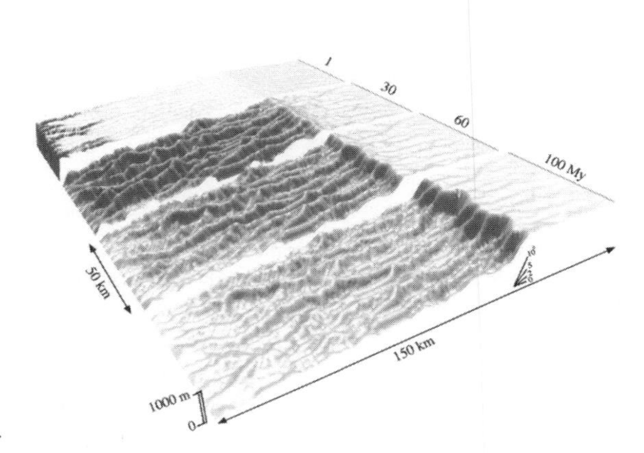

FIGURE 16.14: The evolution of a passive margin escarpment when flexural isostacy is coupled with erosion with the evolution proceeding from the initial condition at the back to the front with time (from Gilchrist et al., 1994).

Figure 16.14 shows the evolution of an escarpment using a model involving a transport-limited erosion, diffusion and flexural isostasy. Though initially all the flow is from right to left very quickly a catchment divide is created at the escarpment with the landform to the right of the escarpment flowing away from the escarpment.

One of the important insights from the works of Kooi, Gilchrist and colleagues was that a relative stable and 'permanent' escarpment was created even though a transport-limited erosion equation was used. It is commonly accepted that the detachment-limited transport equation based on stream power (i.e. Equation (4.29)) creates stable knickpoints and/or escarpments that travel upstream without dissipating, while the transport-limited equation (i.e. Equation (4.10)) tends to smooth out any knickpoints/escarpments.

The key insight is that the flexural isostasy creates a drainage divide that tends to stabilise the escarpment against dissipation even though a transport-limited transport equation was used. One consequence of this upwarping is that the escarpment is generally perceived to be stable because there is no process other than diffusion and mass movement to cause escarpment retreat, since no transport process is able to flow over the escarpment and erode it (e.g. van der Beek et al., 2002).

Subsequent work on passive margin evolution has been focussed on what additional or modified processes are required to control the rate of continued cliff retreat while maintaining the form of the escarpment (called 'parallel retreat'), including the initial topography and drainage network prior to passive margin creation. In the Kooi and Gilchrist modelling the parallel retreat observed in the model simulations (Figure 16.14) was a result of the diffusion process in the model, which increased transport at the escarpment as the escarpment steepened. No physical mechanism was postulated by the authors for what the diffusive process attempted to simulate, though a slope-dependent mass movement (e.g. landslide) would be reasonable. The representation of drainage divides in LEMs may play a part in the stabilisation of the drainage divide (Section 4.7.3).

Braun and van der Beek (2004) use GLUE (see Section 15.4) to explore two alternative hypotheses of evolution for the passive margin of southeast Australia: (1) slow parallel retreat of the escarpment or (2) sudden creation of the escarpment and subsequent erosion across the entire region rather than just at the escarpment. They did this by coupling a model of heat flow through the crust with a simplified version of their CASCADE landform evolution model and compared their model predictions with measurements of temperature across the escarpment from apatite fission track dating. They searched for the best fit to the data over the feasible range of parameters for the two different competing hypotheses, with the two parameters being the thickness of the crust and the heat flow through the crust. Unfortunately both models fit the data equally well, and the best fit parameters for both models gave thickness of the crust (~7 km) that were inconsistent with estimates derived from other sources (~17 km). While the GLUE analysis was unable to identify the best explanation for escarpment evolution, they were able to identify constraints on the geologic settings within which fission track dating could be used to distinguish competing hypotheses.

The other major application area for coupled models is at plate boundaries where mountain ranges are being built. The signature research sites for this are Taiwan and the South Island of New Zealand. The attraction of these two sites is the ability to couple an uplift model as a result of converging crustal plates, erosion model and a climate model where there is a time-invariant windward and leeward side to the mountain range so that rainfall (and thus runoff and erosion) was greater on the windward (western) side of the mountain range than the leeward side, creating asymmetry in the mountain range profile. Willet (1999) modelled the evolution of a one-dimensional cross section across a convergent orogen with a stream power–based detachment-limited fluvial erosion (Equation (4.29)). Using a viscous/plastic flow model, he modelled the converging plates (where one plate was subducting under another, and where the viscosity of the plate was dependent on temperature) and coupled this with the erosion model, which was used to estimate the unloading resulting from erosion of the topography. He then explored the spatial pattern of uplifting, the flow paths for the subsurface crustal material as the two plates converged, and the effect of orographic rainfall coupled with the evolving elevation (Section 3.1.1). Figure 16.15 shows the equilibrium result of experiments where the wind comes from different directions, so that the windward high rainfall occurs on different sides of the range. The grid shows the flow paths of the viscous/plastic flow of the crust, while the topographic surface is the result of the erosion of the uplifted crust. It is obvious from this figure that there has been considerable erosion of material because the height of the topographic divide (the highest point on the mountain range) is significantly lower than the maximum height of the exhumed crust. It is also clear that the location of the topographic divide is a function of the direction of the wind with the divide being skewed away from the side of the highest rainfall and toward the low rainfall side. Finally, the flow processes in the crust are different for the two cases. This shows that the crustal processes have responded to changes in the rainfall and the spatial pattern of rainfall, showing that a link between surface processes and plate dynamics at convergent orogens exists. Subsequent work by Willett and others (Goren et al., 2014) has improved the physical representation, and compared model predictions with a number of field sites, but the original conclusions remain broadly true.

The asymmetry resulting from the different rainfall on each side is consistent with the slope-area relationship in

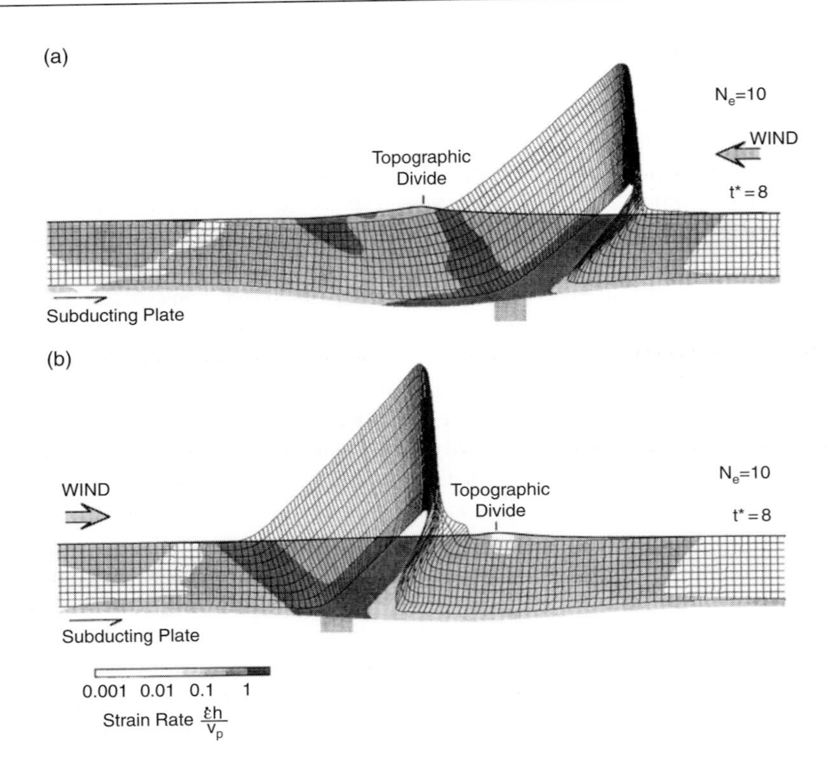

FIGURE 16.15: A coupled crustal convergence-erosion model with rainfall coming from (a) the left so rainfall is highest on the left of the range, (b) the right so rainfall is highest on the right of the range. The grid shows the path of the crustal material showing the pattern of exhumation of crustal material before erosion. The crustal plate from the left is subducting under the plate from the right. The topographic surface is the nearly flat line with a bump at the 'topographic divide' (from Willett, 1999).

Equation (3.21) (Figure 16.16). If everything except rainfall is the same on both sides of the divide, then on the windward, high rainfall, side the runoff is also higher (i.e. β_3) so that the slope is relatively lower on the windward side, and using Equation (3.21)

$$\frac{S_w}{S_l} = \left(\frac{\beta_{3l}}{\beta_{3w}}\right)^{\frac{m_2 m_3}{n_2}} \tag{16.1}$$

where the subscript w and l indicate the windward (high rainfall) and leeward side, respectively. Figure 16.16a shows Equation (16.1) where the erodibility is the same on both sides of the catchment divide. However, if vegetation responds to the rainfall, then the vegetation on the windward side will be denser than on the leeward side and provide greater erosion protection so that Equation (16.1) is then

$$\frac{S_w}{S_l} = \left(\frac{K_{2l}}{K_{2w}}\right)^{\frac{1}{n_2}} \left(\frac{\beta_{3l}}{\beta_{3w}}\right)^{\frac{m_2 m_3}{n_2}} \tag{16.2}$$

Figure 16.16b shows that as the effect of vegetation becomes increasingly dominant, the asymmetry due to rainfall disappears, and it is possible that the mountain range can become asymmetric in the opposite direction. The extent of this reversal in symmetry will depend on the strength of the vegetation response to rainfall changes. Table 14.1 shows that the erosion response to small changes in vegetation cover can be quite strong, so a reverse in the asymmetry is possible. Equations (16.1) and (16.2) are written using the detachment-limited transport model, which will be dominant only in the river network, but the vegetation will primarily determine the erosion from the hillslopes, which will be transport-limited (i.e. Equation (3.19)). The only difference is in the parameters but there are likely to be interactions between the hillslopes and channels not accounted for in Equations (16.1) and (16.2).

16.1.4 Coupled Soil Organic Matter, and Erosion and Deposition

At the time scales that organic matter changes occur (years to decades), the landform doesn't change greatly unless it is degrading quite badly. Thus to couple 'landform evolution' with organic matter, it is good enough to examine erosion and deposition on a fixed landform, and treat landform evolution as a second-order issue.

This was the approach adopted by Lacoste et al. (2015), who modelled soil organic carbon on a fixed landform (with a 2 m resolution DEM) at monthly resolution (which means that the surface drainage pattern did not change) but allowed the soil depth and elevation to change in response to the erosion and deposition at decadal resolution. They coupled the RothC soil organic carbon (SOC) model (see Section 10.3) with the LandSoil soil redistribution model (a simple transport-limited soil erosion and deposition as discussed in Section 4.2). Figure 16.17 shows a schematic of their model. They looked at a number of different agricultural sites with different cropping and tillage patterns to examine the implications of land management strategies and climate change on SOC storage and loss. The soil was modelled as eight layers down to 105 cm depth. Over their simulation period from 2010 to 2100, 22% of their 86 ha site had an elevation change due to erosion of more than 40 mm, and 19% had an elevation change due to deposition of more than 50 mm, while the net loss of soil (i.e. erosion deposition) for the site was about 0.5 mm. This is consistent with very slow landform evolution occurring during the study period.

Roth-C is a single layer compartment model for the top 30 cm of the soil. Lacoste extended Roth-C to have five SOC compartments (decomposable plant material, resistant plant material, microbial biomass, humified organic matter and inert organic matter), and the SOC dynamics of each of the eight layers was modelled independently. Organic matter was sourced from leaf litter at the surface and from decomposing roots within the profile, with the root depth distribution being specified for the agricultural crops being grown. For want of data, decomposition rates were assumed constant through the depth of the profile. Only two methods of movement of SOC were considered: (1) erosion and deposition that transported material spatially and (2) tillage that mixed the soil down to the depth of tillage (assumed to be 30 cm) but did not move soil spatially. No coupling between SOC and soil structure, and infiltration/runoff and SOC enrichment was considered.

Thus they ran two models in parallel; they described their models as 'lightly coupled'. LandSoil calculated the erosion and deposition on the DEM for every rainfall/erosion event. They summed up the erosion and deposition from every event at every node, and evolved the

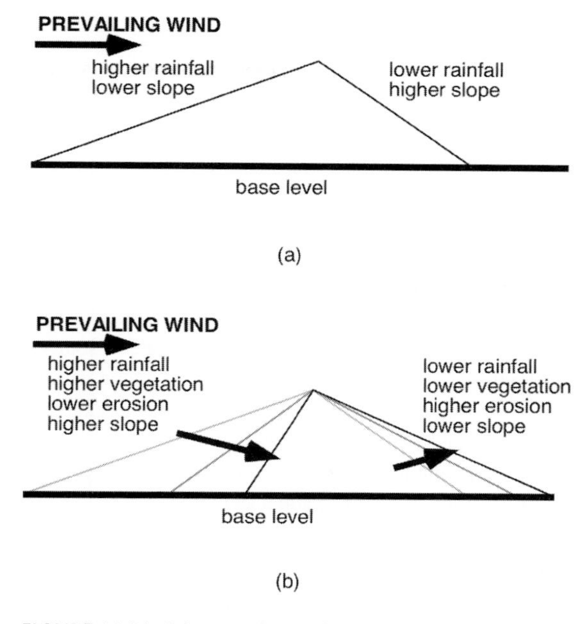

FIGURE 16.16: Schematic showing how asymmetry develops across a mountain range due to orographic rainfall (a) where the high rainfall has no impact on vegetation so erosion is higher on the left-hand side (i.e. K_2 in Equation (3.21) is the same on both sides), (b) where vegetation is stronger on the wetter side and thus reduces the erosion on the wet side, with the arrows indicating how the slopes on both sides would change for two cases with increasing strength of vegetation response to increased rainfall increases.

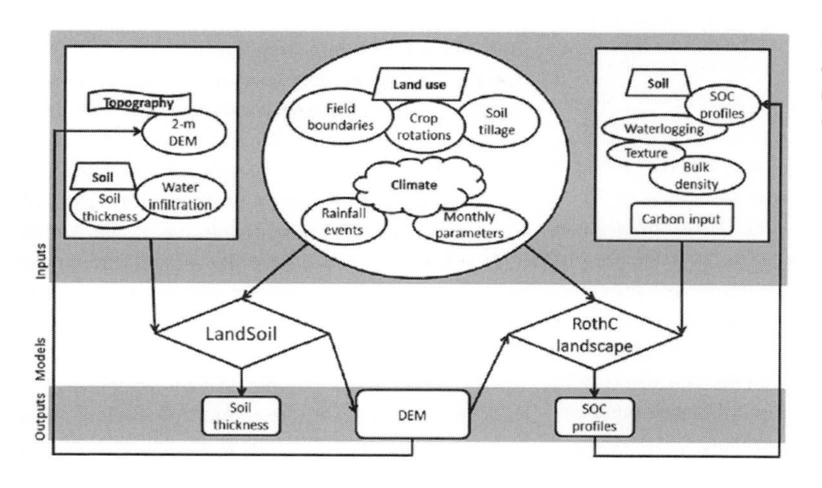

FIGURE 16.17: The model coupling approach for soil organic matter and erosion and deposition (from Lacoste et al., 2015).

landform and the soil layer boundaries every decade. Between decadal updates of the landform the elevations and the soil layer boundaries were considered unchanged. The erosion and deposition calculated for every event were also summed up monthly, and the SOM movement and decomposition calculated for every node monthly. They used one notable simplification to split the erosion from the SOC transport. They assumed that all soil that was deposited in each monthly timestep had the SOC content of the average of all the eroded areas in that same timestep. Thus they did not track soil and SOC from its erosion source to its deposition site. Rather, they simply assumed that all the eroded material had the same (on average) SOC content, irrespective of where it was sourced. It is unclear how important that assumption was because they did not plot any maps of the SOM over the catchment, only presenting catchment wide averages in their analysis. They made this simplification because LandSoil was unable to track sediment and SOC trajectories.

In their analysis they then compared SOC evolution for a number of climate change scenarios from IPCC projections and analysed the implications of the different climate projections for the different management practices across their field site. They considered their study a preliminary first of its kind assessment, and so they didn't provide definitive conclusions. However, they observed that management practices and changes in them were more significant than projected changes in climate to 2100. The different management practices also had different sensitivities to changes in climate, with the most intensively farmed being the most sensitive.

Given the number of assumptions and simplifications in the physics, it is unclear whether their conclusions are generally applicable. Rather, the reason for presenting this study here is as a proof-of-concept approach, with all its simplifications, for coupling SOC and landform evolution.

16.2 The 'Future'

The scope of this book has been to present and explain the mathematical principles used in the modelling of soilscape, landform and landscape evolution. While a wider and more comprehensive discussion of applications in this area was not possible, some discussion has been provided, as much as anything, to motivate why certain processes are modelled in the way they are, why certain approximations are both made and are justifiable, and to provide, using examples, some concrete applications of what can sometimes be rather esoteric abstractions.

That said, there are a range of exciting new applications for landscape evolution models, which are beginning to emerge out of practical needs and new science, and which I have made only passing reference to in this book. No doubt these new applications will lead to new physical approximations and insights.

16.2.1 Landscape Self-Organisation and Connectivity

One of the challenges of understanding the landscape as a functional system evolving in both space and time is the sheer range of active processes and the range of their spatial and temporal time scales and response times. This means to have a 'complete' model of the system, we might need to model the system at very fine time and space resolution, but then run these models over very long times and large areas to be able to capture the responses of the system. The models that we have discussed in the previous chapters all use considerable computer resources to simulate system response, so there is a compelling desire to simplify these models if underlying organising principles can be found.

At a personal level it was to understand the underlying organising principles of channel networks, based on the physics of channel network formation, that we developed the SIBERIA landform evolution model. This insight has been used to develop a deeper insight into why channel networks have the mathematical characteristics they do (e.g. the slope-area diagram, the cumulative area diagram; see Chapter 3) and has allowed us to make observations about channel response without needing to measure all the details of every individual channel.

In the same way, we can measure in fine detail, from remote sensing, the distribution of vegetation to include this into hydrology models. However, if we can understand the organising principles of vegetation distribution and function (and how they may coevolve with the landform and soils), we may be able to develop models of the response of a catchment without having to measure all the details of the vegetation. We have discussed several times the objective of developing subgrid-scale parameterisations of processes, so as to allow us to model at coarser resolution without having to resolve all the small-scale detail. Likewise we would like to be able to develop models of the organisation of soils in space and time so that we can then use a subgrid parameterisation of the soil, to use this to develop models of the catchment-scale integrated response of the hydrology to the soils.

One of the ways that these self-organisation principles can be studied is by developing models of the spatial and temporal coupling, and examining the equilibrium response and the evolutionary approach to that equilibrium, based on the principles in this book. On the basis of these simulations, we can then use these models to search for simple models that capture the behaviour shown by the complex models and in the field. This may seem to be an impossible task, but there is good reason to believe that this is an achievable objective. Field workers have found many empirical relationships in the field that seem to be true across many landscapes and catchments, across climates and across ecosystems. It is highly likely that these relationships result from the coevolution and approach to equilibrium of the processes and fluxes that have been discussed in this book. It is important to understand that when we discuss equilibrium, we don't mean a state where the system no longer changes with time, because there are many systems that oscillate in time, and this oscillation in time is in fact the equilibrium behaviour (this oscillation is the basis of the literature in the field of nonlinear systems and chaos).

One important aspect of self-organisation of these systems in space is the question of how the processes connect spatially. If we think of the patchy vegetation discussed in Chapter 14, then not only are we interested in the percentage cover of the hillslope by vegetation, but we are also interested in how those patches are arranged. Some arrangements of the patches are very effective at trapping runoff on the hillslope, and this better retention of water on the hillslope means that there is more water available for vegetation growth, which in an arid climate is a competitive advantage. This behaviour can be presented in terms of how well the flow paths of the hillslope are connected (a patch of vegetation interrupts the flow path because the water in that flow path infiltrates in that vegetation patch). If flow paths are well connected, then water is better able to flow off the hillslope and the vegetation function (which is limited by water availability) will be degraded. Thus flow connectivity is a measure of the ecological health of the hillslope. In addition, the connectivity of the flow paths drives the hydrologic response of the hillslope, and if we model large catchments, we would prefer not to be required to simulate every patch and the hillslopes between these patches. An 'effective' representation of the hydrology of the hillslope and the patches on it (and their dynamics to changing climate) that provides the integrated hydrology response of the hillslope would be a useful outcome. This effective model could be incorporated in the land surface schemes that are components of global climate models and that are now being used to predict the impact of climate change on hydrology and ecology.

But connectivity is about more than just water flow. When discussing the impacts of fire, we discussed the idea that some of the sediment postfire flows directly off the hillslope into the river, while some of it moves down the catchment only a short distance, to be moved at a later time by different postfire processes. Thus the fate of the latter sediment is a function of how well connected the transport trajectories are to the outlet of the catchment. A low connectivity to the catchment outlet will mean that the sediment will be more strongly retained within the catchment. An important question that can be asked is how the coevolution of the processes and the resulting self-organisation of the landscape impacts on the connectivity of the processes to the outlet of the catchment.

This question of connectivity and the principles that drive changes in it is a question that we have only the barest understanding of, and is a fertile area for future research.

16.2.2 Sustainable Landscapes

Sustainable landscapes is an application area that has been close to the author's heart and has motivated much of his work. It might be broadly described as 'quantitative sustainability assessment of landscapes'. If we wish to predict the trajectory of an environmental system into the future, then we need to be able to model the (typically nonlinear) transients, equilibria and feedbacks between the different parts of the landscape. This may involve complex coevolution problems that are difficult to untangle into their components, and where as a result of the feedbacks the trajectory may vary in nontrivial ways for even small changes in initial conditions and/or climate inputs. Many environmental feedbacks also take many years to exhibit themselves so that for a few years there may be little evidence of coevolution, but the final fate of the system may rely heavily upon them. One excellent example of this is erosion. Over the short term, agricultural erosion models applied on a fixed landform provide good estimates of the spatial distribution of soil erosion. However, over the long term as the landform/paddock/hillslope changes in response to erosion, these models begin to break down; gullies and rills form, they concentrate flow and locally increase erosion, while at the same time erosion is decreased outside the rills and gullies since the overland flow has concentrated into the gullies. The discussions of the previous chapter are replete with processes that when coupled have the potential to produce this long-term nonlinear behaviour.

We could rely upon natural analogues to estimate behaviour, but it is better if we can model this, because in many applications (1) we don't have natural analogues to these systems, (2) we wish to design an initial condition that provides the best long-term performance, and it is only with models that we can compare design alternatives or (3) the field observations required operate at time scales that are too long, or the impacts are too subtle, to be observable.

The author has pioneered applications of the LEMs in the area of constructed landforms where landforms are built to rehabilitate old mining sites, or as containment vessels for potentially hazardous wastes (mine tailings, chemical wastes and low-level nuclear waste repositories). In these cases a landform needs to be designed that will be stable over some design lifetime so as to contain the waste (typically 100 to 10,000 years, and over this time significant landform evolution typically occurs), to have minimal incremental impact on the surrounding landscape, be consistent with social and mine operational constraints and be cost-effective. It is rare that all these objectives can be met simultaneously, so a range of design alternatives need to be developed so that the positives and negatives can be explicitly subject to trade-off analysis. The Ranger Mine example discussed in Chapter 15 is just one published case where this tradeoff analysis has been ongoing for about 25 years. One final challenge is that the owner of the rehabilitated mine site typically cannot be held responsible for the site forever (what company will be around for the design lifetime of 10,000 years?), so at some stage government regulators must sign off on satisfactory performance and assume long-term responsibility. This requires a suite of defensible tests against which the field site can be assessed, and these tests should, ideally, be early indicators of longer-term performance, including all the effects of coevolution over the design lifetime. We would like these tests to be defensible in court because it is common for contentious designs to end up there under challenge by opponents of the particular development. In many western countries unrefereed computer model results are not admissible evidence in court proceedings. This remains a significant, and only partially resolved, challenge.

Finally, there is an intriguing science question that is at the heart of sustainable landscape design. What are the essential statistics to describe a landscape? If we know these, then we know what features to design into our man-made landscape. If we confine our thinking to just the landform for the moment, then if we know all the drainage lines (and thus the area draining through every point in the landform), we should be able to use the average slope-area relationship to construct the elevations of the landscape deterministically starting at the catchment outlet (this is the principle behind the QUEL model; Willgoose, 2001). However, we know that when we do this that the elevations, and the drainage paths derived from them, are commonly inconsistent with the original drainage pattern used to construct the elevations (Ibbitt et al., 1999), and an iteration is required to make the drainage lines, slope-area relationship and elevations consistent. The only thing that has been approximated has been the scatter around the mean slope-area relationship, indicating that there is something in this scatter that is an essential characteristic of the landform, and that it is not simply random scatter. It is likely that issues like this will arise in the use of the area-slope-d_{50} relationship (discussed in Chapter 11) for soil construction, and this is an area for future investigation.

16.2.3 Extraterrestrial Geomorphology

In the last two decades we have seen an explosion of mapping and high-resolution photography missions to other parts of the solar system. Baker (2008) makes the case that geomorphologists should be working on these applications, in addition to applications on Earth.

Other than sheer curiosity value there are several reasons why extraterrestrial applications might be important:

1. The parameterisations of many of the processes in this book are empirical in nature. They are thus dependent on the conditions in which they have been derived: earth's atmosphere, temperature and gravity as the main factors. We can reasonably ask whether the difference in gravity results in fundamental changes in the landform coevolution of the processes. No doubt many unknowns exist with respect to geology on the ground, but can we infer these other dependencies? There are the meandering rivers of liquid methane (i.e. different viscosity to water) on Titan (Burr et al., 2013), paleochannels from meandering rivers of water on Mars (Howard, 2009) and glaciers of solid nitrogen (i.e. different rheology) on Pluto (McKinnon et al., 2016). The question is whether the landforms that they generate are fundamentally different in some measurable way from those of Earth because of those differences.
2. Does the absence of biological activity on these planets result in fundamentally different processes? For instance, we know that biological activity is central to soil functioning, so the question is how does soil function in the absence of biological activity. We already have some information about this from colonisation studies of barren sites (volcanic eruptions) and

rehabilitated mine sites, but what are the long-term equilibria in the absence of biological activity (typically our barren sites recolonise long before they are approaching geomorphic equilibria, so we cannot be sure of what a nonbiological equilibrium looks like). If we can understand this question, we can then ask the question of whether we can remotely sense (i.e. either from satellite or from robots on the surface) the existence of life (Dietrich and Perron, 2006). For instance, given the recent recognition of the role of vegetation in governing channel geometry and river meandering, how did the meander scrolls on Mars and Pluto develop in the absence of vegetation (Howard, 2009)?
3. Can canyon formation on Earth (e.g. mega-floods at the end of the last ice age; Baker, 2009) be compared with canyon formation resulting from outburst floods from craters on Mars (e.g. Lapotre et al., 2016)? Can canyon geometry relationships change as a result of the difference in gravity, and can we infer Mars hydrology from the hydraulics of these canyons?
4. Can we infer past climate on the planets from the topography (Moore et al., 2014)?

16.2.4 Submarine Geomorphology

We conclude this discussion of new application areas a little closer to home, but to a large extent almost as unknown as other planets in the solar system: the submarine environment. Recent advances in multibeam sonar and gravity satellites mean that we can collect data about the sea floor and its changes, particularly before and after earthquakes, for example, the 2004 Sumatran earthquake (Vince, 2005) and the 2011 Tōhoku/Fukashima earthquake (Fujiwara et al., 2011). This also means that we can now observe the undersea topography much better than before. For example, the deep-sea search area in the Indian Ocean for Malaysian Airlines flight MH370 has now been imaged at 100 m resolution (Picard et al., 2017). From these new sensors we can see channels from the terrestrial rivers that travel considerable distances to the edge of the continental shelf (Gupta et al., 2007) and sometimes beyond. While some of this geomorphology no doubt results from terrestrial processes during low stands during ice ages when sea levels dropped (Section 1.3.1), there are visual differences between terrestrial canyons and these submarine canyons that circumstantially implicate other processes, for example, turbidity currents. We can also see landslide scars at the edge of the continental shelf. The practical importance of these areas follows from the considerable essential infrastructure (mostly fibre optic cables, but in some cases power cables, tunnels between islands and oil/gas production platforms) that are regularly damaged by undersea transport events. We have not discussed undersea and estuarine processes in this book, but terrestrial fluvial erosion has its parallel in the submarine environment: turbidity currents, flows downslope containing denser water (than the surrounding sea water) because of the entrained sediments. Gravity-driven mass movements driven by earthquakes also occur in the submarine environment, with the main difference being that air is no longer a factor, and the sediment is always saturated so that liquefaction is almost always a factor in both initiation and energy losses in the movements.

16.3 Conclusions

This book started in Chapter 1 with a rather optimistic presentation of a schematic for landscape evolution: 'The Model of Everything' (Figures 1.1 to 1.3). This schematic of The Model provided a holistic overview of the agenda of this book. The Model stressed the interconnected nature of all the processes. I am reminded of the quote from John Muir I put in the front of my PhD thesis in 1989: 'When we try and pick out anything by itself, we find it hitched to everything else in the Universe.' If anything, I feel it to be even more true today than I did back then. One could be demoralised by this observation and feel that the problem is so vast, interconnected and complicated that there is no hope of achieving anything. However, if the last 30 years of landform evolution modelling has proven anything, it is that by separating problems with different time and space scales, by using effective parameterisations for subgrid processes (so that the computer model runs in a feasible time) and, where necessary, making pragmatic simplifications based on field observation, we can chip away at the unknowns in The Model and develop an understanding of the underlying self-organisations in the system. The emergence of the Critical Zone research agenda (NAS, 2010) means that it is time to expand our focus away from solely landforms, and extend the approaches we've successfully used for landform evolution research to understand soilscape and ecosystem evolution, i.e. landscape evolution. That seems to me to be a worthwhile and achievable objective.

Having presented The Model, most of the book has subsequently been about filling in the (sometimes daunting) details in the schematic of The Model. I have tried not to avoid highlighting areas where our science or computational understanding is weak. Maybe a future PhD student or two will take the hint and discover that therein lies an interesting PhD topic. The last chapters have been more focussed on how to use a Landscape Evolution Model and presented some examples of recent uses of these models.

Before I finish, there are a number of topics that I think are worth reiterating.

Our current generation of models are computationally slow (a colleague calls them 'slowscape models' not 'landscape models'). Sometimes this slowness means that solution of problems at the scale that will provide science insight is impossible. For instance, for soilscape modelling our strictly physically based ARMOUR model (Willgoose and Sharmeen, 2006) was so slow that, while we could calibrate the model to field plot studies, we could barely simulate a hillslope over a few hundred years. However, once we had the insight that a state-state approximation was able to mimic the ARMOUR simulation results, the approach that eventually led to the mARM (Cohen et al., 2009, 2010) and SSSPAM (Welivitiya et al., 2016; Welivitiya, 2017) suite of models, and the theory further extending the state-state approach in Chapter 7, we were able to simulate (without a supercomputer) both spatially distributed soil profiles at the catchment scale, soil spatial organisation and coupled soilscape-landform evolution over periods up to a million years.

But there are other practical reasons that slow models are a problem. If we wish to quantitatively test our model predictions, then we need to become better at putting error bands on both our model predictions and field measurements. I see no other practical approach than Monte Carlo simulation (see Chapter 15) to do that. This will either require parallel computers that, at the moment, are out of the reach of all but the best-resourced research groups in the world, or we need faster models. These faster models will come from better solvers (e.g. can somebody please find a way to remove drainage direction algorithms from erosion solvers so we can take large timesteps), or approximations to more sophisticated physics that are easier and/or faster to solve (e.g. like our mARM and SSSPAM approximations to the ARMOUR model).

One emerging area is the use of landscape evolution models for teaching and professional practice. My own experience and experience of colleagues is that our existing models are probably too complex and difficult to use to be useful as routine tools for teaching geomorphology and mine rehabilitation principles. Simpler and better graphical interfaces are required. The command line is foreign to most undergraduates who have grown up with GUI-based software like Windows, Word and Excel. My experience is that even in the professional mine rehabilitation community a command line interface is unwelcome. And don't get me started on editing large text files that are too big or difficult to open with WordPad and the like. Finally, there is the plethora of different binary file formats (e.g. netCDF, HDF, geoTIFF, shape files), each of which require a different set of tools to manipulate unless you are proficient in coding with scripting languages like MATLAB or Python, or advanced GIS packages. Short of actually running a landscape evolution model, many colleagues have found the computer animations we regularly generate for research and commercial clients useful for teaching and seminars. Unfortunately a book is not the place to include animations, but I cannot overemphasise the amount of positive feedback that we have received from educators and clients as a result of freely distributing our animations.

Finally, I would like to return to a topic that has motivated much of our research at Newcastle for the last 30 years or so: the engineering application of landform evolution modelling to the clean-up and sustainability assessment of rehabilitated mines and low-level nuclear waste repositories. The first question we are always asked by new research partners and commercial clients is 'How do we know that the models are correct?' Rarely do they ask deep questions about the subtleties of the mathematics or the underlying science. Given that they are being asked or required to spend millions of dollars to (re)design their rehabilitation approach on the basis of the predictions from landscape evolution modelling studies, this is a perfectly reasonable question. Yet it is a devilishly difficult question to answer, and in ways that are not initially obvious. It is not simply that predictions are being made hundreds to thousands of years into the future (the facetious, but indefensible, answer is that we'll all be dead by then so who cares). It is, can we predict localised areas where enhanced protection might be required, against, for instance, gully erosion that might penetrate through a protective cap. This requires predictions where not only can we be sure that the gully/valley will occur in a specific location (see Chapter 15 for a deeper discussion of this), but also how deep the gully will be and how fast it will occur (e.g. some wastes become benign after a time). If we can do that (I'm not sure we can, as a general rule, do this everywhere), then can we give them some assessment (e.g. a probability) that the predictions are correct (e.g. with 50% confidence that these protection measures will work for 1,000 years).

At a recent meeting about a new project we had a discussion about how we believe that in general (due to all the randomness in the field) we cannot predict the exact location of gully occurrence unless very specific (and quite restrictive) landscape design principles are followed. One environmental manager asked the question 'How about we let the constructed landform evolve and see where gullies occur and then put protective measures (e.g. armouring) in that location. Will that work long term, or will we push the gully somewhere else over the long term?' This proposed management option opens a Pandora's box of issues, including (1) is short-term behaviour predictive of long-term behaviour, (2) the impact of localised armouring on the spatial distribution of long-term erosion and (3) do we have sufficient validation of our landform evolution tools to be able to defend predictions of unusual and untested engineering solutions like this. The answer we gave was that we don't know.

We now conclude with a topic near to my heart, the collection of field data. I'm sure, coming from me, that sentence surprised some of you. However, if we are to advance the science and practical applications of landscape evolution models we need stronger, more robust, testing of the models. The models discussed in this book (not just ours, but the models developed in other groups as well) make novel predictions, otherwise we would not develop them. Because these predictions are, well, novel, unless we are extremely lucky it is unlikely that field workers would have been motivated/funded to collect the data required to test those novel predictions. This new data collection task needs to be a collaboration between the model builders/users and the field researchers. The model builders need to find predictions that are testable in the field or the laboratory, while the field workers/experimentalists need to develop ways of collecting the new data in a way that can be used to perform these tests. This is a nontrivial task, not least because of the cultural differences between these two groups. Many field workers come from a purer science background whether that be geology, geography, agricultural and soil science or ecology, while the modellers typically come from an engineering, geophysics or mathematics background. Landscape evolution advances will come from being a broad church for all these workers, and for those workers of different backgrounds to be able to be deeply embedded together and understand each other's viewpoints, because it is unlikely in my view that any of us can do it on our own.

The development of the Critical Zone Observatories (CZOs) in the United States, is in my view, a step forward in that direction, even though the current ones do not, yet, in my view, fully encapsulate the broader range of topics that have been discussed in this book. Likewise there are a range of ongoing mini-CZOs around the world (but sadly without dedicated CZO-like funding) that are studying parts of the coupled landform-soilscape-ecosystem evolution system but not the whole (e.g. our SASMAS site, Rüdiger et al., 2007; *eriss*'s Tin Camp Creek site, Hancock et al., 2002; U. Cordoba's Santa Clotilde CZO, Román-Sánchez et al., 2017). No group is holistically addressing the whole. And the same is true of constructed landscapes, such as rehabilitated mines; there are no CZOs for mined land. But, as they say, baby steps.

Hopefully this discussion, and the book in general, will stimulate some creative juices (and hopefully trigger robust debate) among the book's readers, and generate ideas for topics worthy of intellectual attention. If I have achieved that, then writing this book has been worthwhile.

References

Abrahams, A. D. (1984), Channel networks: A geomorphological perspective, *Water Resources Research*, 20(2), 161–168.

Abrahams, A. D., and A. J. Parsons (1990), Determining the mean depth of overland flow in field studies of flow hydraulics, *Water Resources Research*, 26(3), 501–503.

Abrahams, A. D., and A. J. Parsons (1991a), Relation between sediment yield and gradient on debris covered hillslopes, Walnut Gulch, Arizona, *Bulletin of the Geological Society of America*, 103, 1109–1113.

Abrahams, A. D., and A. J. Parsons (1991b), Resistance to overland flow on desert pavements and its implications for sediment transport modeling, *Water Resources Research*, 27(8), 1827–1836.

Abrahams, A. D., and A. J. Parsons (1994), Hydraulics of interrill overland flow on stone-covered desert surfaces, *Catena*, 23(1–2), 111–140.

Adams, W. A. (1973), The effect of organic matter on the bulk and true densities of some uncultivated Podzolic soils, *European Journal of Soil Science*, 24(1), 10–17, doi:10.1111/j.1365-2389.1973.tb00737.x.

Ahad, T., T. A. Kanth, and S. Nabi (2015), Soil bulk density as related to texture, organic matter content and porosity in Kandi soils of District Kupwara (Kashmir Valley), India, *International Journal of Scientific Research*, 4(1), 198–200.

Ahnert, F. (1976), Brief description of a comprehensive three-dimensional process-response model for landform development, *Zeitschrift für Geomorphologie N.F. Supplement*, 25, 29–49.

Ahnert, F. (1977), Some comments on the quantitative formulation of geomorphological process in a theoretical model, *Earth Surface Processes*, 2, 191–201.

Ahr, S. W., L. C. Nordt, and S. L. Forman (2013), Soil genesis, optical dating, and geoarchaeological evaluation of two upland Alfisol pedons within the Tertiary Gulf Coastal Plain, *Geoderma*, 192, 211–226, doi:10.1016/j.geoderma.2012.08.016.

Albrecht, R. I., S. J. Goodman, D. E. Buechler, R. J. Blakeslee, and H. J. Christian (2016), Where are the lightning hotspots on Earth?, *Bulletin of the American Meteorological Society*, 97(11), 2015–2068, doi:10.1175/BAMS-D-14-00193.1.

Alley, R. B. (2014), *The two-mile time machine: Ice cores, abrupt climate change, and our future*, Princeton University Press, Princeton, NJ.

Alpert, P. (1986), Mesoscale indexing of the distribution of orographic precipitation over high mountains, *Journal of Climate and Applied Meteorology*, 25(4), 532–545, doi:10.1175/1520-0450(1986)025<0532:MIOTDO>2.0.CO;2.

Amundson, R. (1994), Towards the quantitative modeling of pedogenesis – A review – Comment – Functional vs mechanistic theories: The paradox of paradigms, *Geoderma*, 63(3–4), 299–302.

Amundson, R. (2001), The carbon budget in soils, *Annual Review of Earth and Planetary Sciences*, 29, 535–562, doi:10.1146/annurev.earth.29.1.535.

Ancey, C. (2007), Plasticity and geophysical flows: A review, *Journal of Non-Newtonian Fluid Mechanics*, 142, 4–35, doi:10.1016/j.jnnfm.2006.05.005.

Anders, A. M., and S. W. Nesbitt (2015), Altitudinal precipitation gradients in the tropics from Tropical Rainfall Measuring Mission (TRMM) precipitation radar, *Journal of Hydrometeorology*, 16(1), 441–448, doi:10.1175/JHM-D-14-0178.1.

Anderson, R. S. (2002), Modeling the tor-dotted crests, bedrock edges, and parabolic profiles of high alpine surfaces of the Wind River Range, Wyoming, *Geomorphology*, 46, 35–58.

Anderson, R. S. (2015), Particle trajectories on hillslopes: Implications for particle age and ^{10}Be structure, *Journal of Geophysical Research (Earth Surface)*, 120, 1626–1644, doi:10.1002/2015JF003479.

Anderson, R. S., and S. P. Anderson (2010), *Geomorphology: The mechanics and chemistry of landscapes*, Cambridge University Press, Cambridge.

Anderson, S. P., W. E. Dietrich, and G. H. Brimhall (2002), Weathering profiles, mass-balance analysis, and rates of solute loss: Linkages between weathering and erosion in a small, steep catchment, *Geological Society of America Bulletin*, 114(9), 1143–1158, doi:10.1130/0016-7606(2002)114<1143:WPMBAA>2.0.CO;2.

Andrews, D. J., and R. C. Bucknam (1987), Fitting degradation of shoreline scarps by a nonlinear diffusion model, *Journal of Geophysical Research (Solid Earth)*, 92(B12), 12857–12867, doi:10.1029/JB092iB12p12857.

Angers, D. A., et al. (1997), Impact of tillage practices on organic carbon and nitrogen storage in cool, humid soils of eastern Canada, *Soil and Tillage Research*, 41(3–4), 191–201, doi:10.1016/S0167-1987(96)01100-2.

Argus, D. F., and W. R. Peltier (2010), Constraining models of postglacial rebound using space geodesy: A detailed assessment of model ICE-5G (VM2) and its relatives, *Geophysical Journal International*, 181(2), 697–723, doi:10.1111/j.1365-246X.2010.04562.x.

Arshad, M. A., A. J. Franzluebbers, and R. H. Azooz (1999), Components of surface soil structure under conventional and no-tillage in northwestern Canada, *Soil Tillage Research*, 53(1), 41–47, doi:10.1016/S0167-1987(99)00075-6.

Asteriou, P., H. Saraglou, and G. Tsimbaos (2012), Geotechnical and kinematic parameters affecting the coefficients of restitution for rock fall analysis, *International Journal of Rock Mechanics & Mining Sciences*, 54, 103–113, doi:10.1016/j.ijrmms.2012.05.029.

Astete, C. E., W. D. Constant, L. J. Thibodeaux, R. K. Seals, and H. M. Delim (2015), Bioturbation-driven particle transport in surface soil: The biodiffusion coefficient mobility parameter, *Soil Science*, 180(1), 2–9, doi:10.1097/SS.0000000000000109.

Avirmed, O., I. C. Burke, M. L. Mobley, W. K. Lauenroth, and D. R. Schlaepfer (2014), Natural recovery of soil organic matter in 30–90-year-old abandoned oil and gas wells in sagebrush steppe, *Ecosphere*, 5(3), 1–13, doi:10.1890/ES13-00272.1.

Azooz, R. H., and M. A. Arshad (1996), Soil infiltration and hydraulic conductivity under long-term no-tillage and conventional tillage systems, *Canadian Journal of Soil Science*, 76(2), 143–152.

Baartman, J. E. M., A. J. A. M. Temme, J. M. Schoorl, M. H. Braakhekke, and T. Veldkamp (2012), Did tillage erosion play a role in millennial scale landscape development?, *Earth Surface Processes and Landforms*, 37(15), 1615–1626, doi:10.1002/esp.3262.

Baartman, J. E. M., A. J. A. M. Temme, T. Veldkamp, V. G. Jetten, and J. M. Schoorl (2013), Exploring the role of rainfall variability and extreme events in long-term landscape development, *Catena*, 109, 25–38, doi:10.1016/j.catena.2013.05.003.

Bagnold, R. A. (1936), The movement of desert sand, *Proceedings of the Royal Society of London. Series A, Mathematical and Physical Sciences* , 157(892), 594–620.

Bak, P., C. Tang, and K. Wiesenfeld (1988), Scale invariant spatial and temporal fluctuations in complex systems, in *Random fluctuations and pattern growth: Experiments and models*, edited by H. E. Stanley and N. Ostrowsky, pp. 329–335, Kluwer, Berlin.

Baker, V. R. (2008), Planetary landscape systems: A limitless frontier, *Earth Surface Processes and Landforms*, 33, 1341–1353, doi:10.1002/esp.1713.

Baker, V. R. (2009), The Channeled Scabland: A retrospective, *Annual Review of Earth and Planetary Sciences*, 37, 393–411, doi:10.1146/annurev.earth.061008.134726.

Baldwin, J. A., K. X. Whipple, and G. E. Tucker (2003), Implications of the shear stress river incision model for the timescale of postorogenic decay of topography, *Journal of Geophysical Research (Solid Earth)*, 108(B3), art. no. 2158, doi:10.1029/2001JB000550.

Balmforth, N. J., R. V. Craster, A. C. Rust, and R. Sassi (2007), Viscoplastic flow over an inclined surface, *Journal of Non-Newtonian Fluid Mechanics*, 142, 219–243, doi:10.1016/j.jnnfm.2006.07.013.

Barman, A. K., C. Varadachari, and K. Ghosh (1992), Weathering of silicate minerals by organic-acids. 1. Nature of cation solubilization, *Geoderma*, 53(1–2), 45–63.

Barreto, L., J. M. Schoorl, K. Kok, T. Veldkamp, and A. Hass (2013), Modelling potential landscape sediment delivery due to projected soybean expansion: A scenario study of the Balsas sub-basin, Cerrado, Maranhao state, Brazil, *Journal of Environmental Management*, 115, 270–277, doi:10.1016/j.jenvman.2012.11.017.

Barshad, I. (1959), Factors affecting clay formation, *Clays and Clay Minerals*, 6, 110–132.

Barzegar, A. R., A. Yousefi, and A. Daryashenas (2002), The effect of addition of different amounts and types of organic materials on soil physical properties and yield of wheat, *Plant and Soil*, 247(2), 295–301, doi:10.1023/A:1021561628045.

Bazin, L., et al. (2013), An optimized multi-proxy, multi-site Antarctic ice and gas orbital chronology (AICC2012): 120–800 ka, *Climate of the Past*, 9, 1715–1731, doi:10.5194/cp-9-1715-2013.

Beal, L. K., D. P. Huber, S. E. Godsey, S. K. Nawotniak, and K. A. Lohse (2016), Controls on ecohydrologic properties in desert ecosystems: Differences in soil age and volcanic morphology, *Geoderma*, 271, 32–41, doi:10.1016/j.geoderma.2016.01.030.

Bell, J. R. W., and G. R. Willgoose (1998), Monitoring of gully erosion at ERA Ranger Uranium Mine, Northern Territory, Australia, Internal Report 274, Environmental Research Institute of the Supervising Scientist, Jabiru, NT.

Bellugi, D., D. G. Milledge, W. E. Dietrich, J. McKean, J. T. Perron, E. B. Sudderth, and B. Kazian (2015a), A spectral clustering search algorithm for predicting shallow landslide size and location, *Journal of Geophysical Research (Earth Surface)*, 120, 300–324, doi:10.1002/2014JF003137.

Bellugi, D., D. G. Milledge, W. E. Dietrich, J. T. Perron, and J. McKean (2015b), Predicting shallow landslide size and location across a natural landscape: Application of a spectral clustering search algorithm, *Journal of Geophysical Research (Earth Surface)*, 120, 2552–2585, doi:10.1002/2015JF003520.

Bercovici, D. (2003), The generation of plate tectonics from mantle convection, *Earth and Planetary Science Letters*, 205(3–4), 107–121, doi:10.1016/S0012-821X(02)01009-9.

Bergeron, T. (1961), *Preliminary results of 'Project Pluvius'*, vol. 53, pp. 226–237. *International Association of Hydrological Sciences Publication*, Gentbrugge.

Bernoux, M., C. C. Cerri, C. Neill, and J. F. L. de Moraes (1998), The use of stable carbon isotopes for estimating soil organic matter turnover rates, *Geoderma*, 82(1–3), 43–58.

Beven, K. J. (1996), Equifinality and uncertainty in geomorphological modelling, in *The scientific nature of geomorphology: Proceedings of the 27th Binghampton Symposium in geomorphology, 27–29 September, 1996*, edited by B. L. Rhoads and C. E. Thorn, pp. 289–313, Wiley, Chichester, UK.

Beven, K. J. (2000), Uniqueness of place and process representations in hydrological modelling, *Hydrology and Earth System Sciences*, 4(2), 203–213.

Beven, K. J. (2012), *Rainfall-runoff modelling: The primer*, 2nd ed., Wiley-Blackwell, Chichester, UK.

Beven, K. J., and A. M. Binley (1992), The future of distributed models: Model calibration and uncertainty prediction, *Hydrological Processes*, 6, 279–298.

Beven, K. J., and K. Germann (1982), Macropores and water-flow in soils, *Water Resources Research*, 18(5), 1311–1325, doi:10.1029/WR018i005p01311.

Beven, K. J., and K. Germann (2013), Macropores and water flow in soils revisited, *Water Resources Research*, 49(6), 3071–3092, doi:10.1002/wrcr.20156.

Billings, S. A., R. W. Buddemeier, D. deB Richter, K. Van Oost, and G. Bohling (2010), A simple method for estimating the influence of eroding soil profiles on atmospheric CO_2, *Global Biogeochemal Cycles*, 24, GB2001, doi:10.1029/2009GB003560.

Birkeland, P. W. (1990), Soil-geomorphic research – A selective overview, *Geomorphology*, 3(3–4), 207–224, doi:10.1016/0169-555X(90)90004-A.

Bisdom, E. B. A., G. Stoops, J. Delvigne, P. Curmi, and H.-J. Altemuller (1982), Micromorphology of weathering Biotite and its secondary products, *Pedologie*, 32(2), 225–252.

Bishop, P. (2007), Long-term landscape evolution: Linking tectonics and surface processes, *Earth Surface Processes and Landforms*, 32, 329–365, doi:10.1002/esp.1493.

Boardman, J., A. J. Parsons, R. Holland, P. J. Holmes, and R. Washington (2003), Development of badlands and gullies in the Sneeuberg, Great Karoo, South Africa, *Catena*, 50(2–4), 165–184.

Bodí, M. B., D. A. Martin, V. N. Balfour, C. Santín, S. H. Doerr, P. Pereira, A. Cerdà, and J. Mataix-Sorda (2014), Wildland fire ash: Production, composition and eco-hydro-geomorphic effects, *Earth-Science Reviews*, 130, 103–127, doi:10.1016/j.earscirev.2013.12.007.

Boillat, J. L., and W. H. Graf (1982), Settling velocity of spherical particles in turbulent media, *Journal of Hydraulic Research*, 20, 395–413.

Bondeau, A., et al. (2007), Modelling the role of agriculture for the 20th century global terrestrial carbon balance, *Global Change Biology*, 13, 679–706, doi:10.1111/j.1365-2486.2006.01305.x.

Borga, M., G. D. Fontana, and F. Cazorzi (2002), Analysis of topographic and climatic control on rainfall-triggered shallow landsliding using a quasi-dynamic wetness index, *Journal of Hydrology*, 268, 56–71.

Bourrier, F. (2008), Modélisation de l'impact d'un bloc rocheux sur un terrain naturel, application à la trajectographie des chutes de blocs, PhD thesis, Institut National Polytechnique de Grenoble, Grenoble.

Bourrier, F., L. K. A. Dorren, F. Nicot, F. Berger, and F. Darve (2009), Toward objective rockfall trajectory simulation using a stochastic impact model, *Geomorphology*, 110, 68–79, doi:10.1016/j.geomorph.2009.03.017.

Bovy, B., J. Braun, and A. Demoulin (2016), A new numerical framework for simulating the control of weather and climate on the evolution of soil-mantled hillslopes, *Geomorphology*, 263, 99–112, doi:10.1016/j.geomorph.2016.03.016.

Bowman, D. M. J. S., G. S. Boggs, and L. D. Prior (2008), Fire maintains an Acacia aneura shrubland – Triodia grassland mosaic in central Australia, *Journal of Arid Environments*, 72, 34–47, doi:10.1016/j.jaridenv.2007.04.001.

Bradford, S. A., and S. Torkzaban (2008), Colloid transport and retention in unsaturated porous media: A review of interface-, collector-, and pore-scale processes and models, *Vadose Zone Journal*, 7(2), 667–681, doi:10.2136/vzj2007.0092.

Brantley, S. L., and M. I. Lebedeva (2011), Learning to read the chemistry of regolith to understand the Critical Zone, *Annual Review of Earth and Planetary Sciences*, 39, 387–416, doi:10.1146/annurev-earth-040809-152321.

Bras, R. L., and I. Rodriguez-Iturbe (1985), *Random functions and hydrology*, Addison-Wesley, New York.

Braun, J., A. M. Heimsath, and J. Chappell (2001), Sediment transport mechanisms on soil-mantled hillslopes, *Geology*, 29(8), 683–686.

Braun, J., J. Mercier, F. Guillocheau, and C. Robin (2016), A simple model for regolith formation by chemical weathering, *Journal of Geophysical Research (Earth Surface)*, 121, 2140–2171, doi:10.1002/2016JF003914.

Braun, J., and X. Robert (2005), Constraints on the rate of post-orogenic erosional decay from low-temperature thermochronological data: Application to the Dabie Shan, China, *Earth Surface Processes and Landforms*, 30, 1203–1225, doi:10.1002/esp.1271.

Braun, J., and M. Sambridge (1997), Modelling landscape evolution on geological time scales: A new method based on irregular spatial discretization, *Basin Research*, 9(1), 27–52.

Braun, J., and P. van der Beek (2004), Evolution of passive margin escarpments: What can we learn from low-temperature thermochronology?, *Journal of Geophysical Research (Earth Surface)*, 109(F4), F04009.

Brimhall, G. H., O. A. Chadwick, C. J. Lewis, W. Compston, I. S. Williams, K. J. Danti, W. E. Dietrich, M. E. Power, D. M. Hendricks, and J. Bratt (1992), Deformational mass transport and invasive processes in soil evolution, *Science*, 255(5045), 695–702.

Brimhall, G. H., and W. E. Dietrich (1987), Constitutive mass balance relations between chemical composition, volume, density, porosity, and strain in metasomatic hydrochemical systems: Results on weathering and pedogenesis, *Geochimica Cosmochimica Acta*, 51, 567–587.

Brimhall, G. H., C. J. Lewis, C. Ford, J. Bratt, G. Taylor, and O. Warin (1991), Quantitative geochemical approach to pedogenesis – Importance of parent material reduction, volumetric expansion, and Eolian influx in Lateritization, *Geoderma*, 51(1–4), 51–91.

Brocard, G. Y., J. K. Willenbring, T. E. Miller, and F. N. Scatena (2016), Relict landscape resistance to dissection by upstream migrating knickpoints, *Journal of Geophysical Research (Earth Surface)*, 121, 1182–1203, doi:10.1002/2015JF003678.

Brooks, R. J., M. A. Semenov, and P. D. Jamieson (2001), Simplifying Sirius: Sensitivity analysis and development of a meta-model for wheat yield prediction, *European Journal of Agronomics*, 14, 43–60.

Brunetti, M. T., F. Guzzetti, and M. Rossi (2009), Probability distributions of landslide volumes, *Nonlinear Processes in Geophysics*, 16(2), 179–188.

Buck, W. R. (1993), Effect of lithospheric thickness on the formation of high- and low-angle normal faults, *Geology*, 21(10), 933–936, doi:10.1130/0091-7613(1993) 021<0933:EOLTOT> 2.3.CO;2.

Bufe, A., C. Paola, and D. W. Burbank (2016), Fluvial bevelling of topography controlled by lateral channel mobility and uplift rate, *Nature Geoscience*, 9(9), 706–710, doi:10.1038/ngeo2773.

Bull, L. J., and M. J. Kirkby (1997), Gully processes and modelling, *Progress in Physical Geography*, 21(3), 354–374, doi:10.1177/030913339702100302.

Bull, L. J., and M. J. Kirkby (2002), Channel heads and channel extension, in *Dryland rivers: Hydrology and geomorphology of semi-arid channels*, edited by L. J. Bull and M. J. Kirkby, pp. 263–298, Wiley, Chichester, UK.

Burke, B. C., A. M. Heimsath, and A. F. White (2007), Coupling chemical weathering with soil production across soil-mantled landscapes, *Earth Surface Processes and Landforms*, 32, 853–873, doi:10.1002/esp.1443.

Burnett, S. A., J. A. Hattey, J. E. Johnson, A. L. Swann, D. I. Moore, and S. L. Collins (2012), Effects of fire on belowground biomass in Chihuahuan desert grassland, *Ecosphere*, 3(11), 107, doi:10.1890/ES12-00248.1.

Burr, D. M., et al. (2013), Fluvial features on Titan: Insights from morphology and modeling, *Geological Society of America Bulletin*, 125(3–4), 299–321, doi:10.1130/B30612.1.

Burroughs, E. R., and B. R. Thomas (1977), Declining root strength in Douglas-Fir after felling as a factor in slope stability, Rep. INT-190, US Department of Agriculture, Ogden, Utah.

Buss, H. L., P. B. Sak, S. M. Webb, and S. L. Brantley (2008), Weathering of the Rio Blanco quartz diorite, Luquillo Mountains, Puerto Rico: Coupling oxidation, dissolution, and fracturing, *Geochimica Cosmochimica Acta*, 72, 4488–4507, doi:10.1016/j.gca.2008.06.020.

Buzzi, O., A. Giacomini, and M. Spadari (2012), Laboratory investigation of high values of restitution coefficients, *Rock Mechanics and Rock Engineering*, 45, 35–43, doi:10.1007/s00603-011-0183-0.

Byrne, K. A., and G. Kiely (2008), *Evaluation of models (PaSim, RothC, CENTURY and DNDC) for simulation of grassland carbon cycling at plot, field and regional scale*, Environment Protection Agency, Wexford, Ireland.

Cagnoli, B., and M. Manga (2003), Pumice-pumice collisions and the effect of the impact angle, *Geophysical Research Letters*, 30(12), 1636, doi:10.1029/2003GL017421.

Campbell, C. S., P. W. Cleary, and M. Hopkins (1995), Large-scale landslide simulations: Global deformation, velocities and basal friction, *Journal of Geophysical Research (Solid Earth)*, 100(B5), 8267–8283, doi:10.1029/94JB00937.

Camporeale, C., E. Perucca, L. Ridolfi, and A. M. Gurnell (2013), Modelling the interactions between river morphodynamics and riparian vegetation, *Reviews of Geophysics*, 51, 379–414, doi:10.1002/rog.20014.

Canadell, J. G., R. B. Jackson, J. R. Ehleringer, H. A. Mooney, O. E. Sala, and E. D. Schulze (1996), Maximum rooting depth of vegetation types at the global scale, *Oecologia*, 108(4), 583–595, doi:10.1007/BF00329030.

Canales, J. B., G. Ito, R. S. Detrick, and J. Sinton (2002), Crustal thickness along the western Galapagos Spreading Center and the compensation of the Galapagos hotspot swell, *Earth and Planetary Science Letters*, 203, 311–327.

Carson, M. A., and M. J. Kirkby (1972), *Hillslope form and process*, Cambridge University Press, London.

Cawson, J. G., G. J. Sheridan, H. G. Smith, and P. N. J. Lane (2012), Surface runoff and erosion after prescribed burning and the effect of different fire regimes in forests and shrublands: A review, *International Journal of Wildland Fire*, 21, 857–872, doi:10.1071/WF11160.

Celia, M. A., and W. G. Gray (1991), *Numerical methods for differential equations: Fundamental concepts for scientific & engineering applications*, Prentice-Hall, New York.

Cerdà, A., and S. H. Doerr (2005), Influence of vegetation recovery on soil hydrology and erodibility following fire: An 11-year investigation, *International Journal of Wildland Fire*, 14, 423–437, doi:10.1071/WF05044.

Certini, G. (2005), Effects of fire on properties of forest soils: A review, *Oecologia*, 143, 1–10, doi:10.1007/S00442-0041788-8.

Certini, G., R. Scalenghe, and W. I. Woods (2013), The impact of warfare on the soil environment, *Earth-Science Reviews*, 127, 1–15, doi:10.1016/j.earscirev.2013.08.009.

Chadwick, O. A., G. H. Brimhall, and D. M. Hendricks (1990), From a black to a gray box: A mass balance interpretation of pedogenesis, *Geomorphology*, 3, 369–390.

Chase, C. G. (1992), Fluvial landsculpting and the fractal dimension of topography, *Geomorphology*, 5(1/2), 39–57, doi:10.1016/0169-555X(92)90057-U.

Chatanantavet, P., and G. Parker (2009), Physically based modeling of bedrock incision by abrasion, plucking, and macroabrasion, *Journal of Geophysical Research (Earth Surface)*, 114, F04018, doi:10.1029/2008JF001044.

Chatanantavet, P., and G. Parker (2011), Quantitative testing of model of bedrock channel incision by plucking and macroabrasion, *Journal of Hydraulic Division – ASCE*, 137(11), 1311–1317, doi:10.1061/(ASCE)HY.1943-7900.0000421.

Chau, K. T., R. H. C. Wong, and J. J. Wu (2002), Coefficient of restitution and rotational motions of rockfall impacts, *International Journal of Rock Mechanics & Mining Sciences*, 39, 69–77, doi:10.1016/S1365-1609(02)00016–3.

Chen, A., J. Darbon, and J.-M. Morel (2014), Landscape evolution models: A review of their fundamental equations, *Geomorphology*, 219, 68–86, doi:10.1016/j.geomorph.2014.04.037.

Chen, M., G. R. Willgoose, and P. M. Saco (2015), Evaluation of the hydrology of the IBIS land surface model in a semi-arid catchment, *Hydrological Processes*, 29, 653–670, doi:10.1002/hyp.10156.

Cheng, D.-L., and K. J. Niklas (2007), Above- and below-ground biomass relationships across 1534 forested communities, *Annals of Botany*, 99, 95–102, doi:10.1093/aob/mcl206.

Chia, E. K., M. Bassett, D. G. Nimmo, S. W. J. Leonard, E. G. Ritchie, M. F. Clarke, and A. F. Bennett (2015), Fire severity and fire-induced landscape heterogeneity affect arboreal mammals in fire-prone forests, *Ecosphere*, 6(10), 190.

Chien-Yuan, C., Y. Fan-Chieh, L. Sheng-Chi, and C. Kei-Wai (2007), Discussion of landslide self-organized criticality and the initiation of debris flow, *Earth Surface Processes and Landforms*, 32(2), 197–209, doi:10.1002/esp.1400.

Chigara, M., and T. Oyama (1999), Mechanism and effect of chemical weathering of sedimentary rocks, *Engineering Geology*, 55(1), 3–14, doi:10.1016/S0013-7952(99)00102-7.

Chowdhury, A. F. M. K. (2017), Development and evaluation of stochastic rainfall models for urban water security assessment, PhD Thesis, University of Newcastle, Callaghan, Australia.

Chowdhury, A. F. M. K., N. Lockart, G. R. Willgoose, G. Kuczera, A. S. Kiem, and N. Parana Manage (2018), Development and evaluation of a stochastic daily rainfall model with long term variability, *Hydrology and Earth System Sciences*, doi:10.5194/hess-2017-84.

Claessens, L., G. B. M. Heuvelink, J. M. Schoorl, and A. Veldkamp (2005), DEM resolution effects on shallow landslide hazard and soil redistribution modelling, *Earth Surface Processes and Landforms*, 30, 461–477, doi:10.1002/esp.1155.

Clarke, L. E. (2015), Experimental alluvial fans: Advances in understanding of fan dynamics and processes, *Geomorphology*, 244, 135–145, doi:10.1016/j.geomorph.2015.04.013.

Clarke, R. H. (1979), Reservoir properties of conglomerates and conglomerate sandstones, *American Association of Petroleum Geologists Bulletin*, 63, 799–809.

Cohen, S. (2010), Spatial description of soil properties through landscape-pedogenesis modelling, PhD thesis, University of Newcastle, Callaghan, Australia.

Cohen, S., G. R. Willgoose, and G. R. Hancock (2008), A methodology for calculating the spatial distribution of the area-slope equation and the hypsometric integral within a catchment, *Journal of Geophysical Research (Earth Surface)*, 113, F03027, doi:10.1029/2007JF000820.

Cohen, S., G. R. Willgoose, and G. R. Hancock (2009), The mARM spatially distributed soil evolution model: A computationally efficient modeling framework and analysis of hillslope soil surface organization, *Journal of Geophysical Research (Earth Surface)*, 114, F03001, doi:10.1029/2008JF001214.

Cohen, S., G. R. Willgoose, and G. R. Hancock (2010), The mARM3D spatially distributed soil evolution model: Three-dimensional model framework and analysis of hillslope and landform responses, *Journal of Geophysical Research (Earth Surface)*, 115, F04013, doi:10.1029/2009JF001536.

Cohen, S., G. R. Willgoose, and G. R. Hancock (2013), Soil response to late-Quaternary climatic oscillations, new insights based on numerical simulations, *Quaternary Research*, 79(3), 452–457, doi:10.1016/j.yqres.2013.01.001.

Cohen, S., T. Svoray, S. Sela, G. R. Hancock, and G. R. Willgoose (2015), The effect of sediment-transport, weathering and aeolian mechanisms on soil evolution, *Journal of Geophysical Research (Earth Surface)*, 120(2), 260–274, doi:10.1002/2014JF003186.

Cohen, S., T. Svoray, S. Sela, G. R. Hancock, and G. R. Willgoose (2016), Soilscape evolution of aeolian-dominated hillslopes during the Holocene: Investigation of sediment transport mechanisms and climatic-anthropogenic drivers, *Earth Surface Dynamics*, 5, 101–112, doi:10.5194/esurf-5-101-2017.

Coleman, K., and D. S. Jenkinson (2014), RothC – A model for the turnover of carbon in soil. Model description and users guide, Rothamsted Research, UK.

Coleman, K., D. S. Jenkinson, G. J. Crocker, P. R. Grace, J. Klir, M. Korschens, P. R. Poulton, and D. D. Richter (1997), Simulating trends in soil organic carbon in long-term experiments using RothC-26.3, *Geoderma*, 81(1–2), 29–44.

Collins, D. B. G., and R. L. Bras (2008), Climatic control of sediment yield in dry lands following climate and land cover change, *Water Resources Research*, 44, W10405, doi:10.1029/2007WR006474.

Collins, D. B. G., and R. L. Bras (2010), Climatic and ecological controls of equilibrium drainage density, relief, and channel concavity in dry lands, *Water Resources Research*, 46, W04508, doi:10.1029/2009WR008615.

Collins, D. B. G., R. L. Bras, and G. E. Tucker (2004), Modeling the effects of vegetation-erosion coupling on landscape evolution, *Journal of Geophysical Research (Earth Surface)*, 109(F3), F03004.

Collins, K. M., O. F. Price, and T. D. Penman (2015), Spatial patterns of wildfire ignitions in south-eastern Australia, *International Journal of Wildland Fire*, 24, 1098–1108, doi:10.1071/WF15054.

Conrad, C. P., and C. Lithgow-Bertelloni (2002), How mantle slabs drive plate tectonics, *Science*, 298(5591), 207–209.

Corti, G., A. Agnelli, G. Certini, and F. C. Ugolini (2001), The soil skeleton as a tool for disentangling pedogenetic history: A case study in Tuscany, central Italy, *Quaternary International*, 78, 33–44, doi:10.1016/S1040-6182(00)00113-0.

Coulthard, T. J. (2001), Landscape evolution models: A software review, *Hydrological Processes*, 15(1), 165–173.

Coulthard, T. J., D. M. Hicks, and M. J. Van De Wiel (2007), Cellular modelling of river catchments and reaches: Advantages, limitations and prospects, *Geomorphology*, 90, 192–207, doi:10.1016/j.geomorph.2006.10.030.

Coulthard, T. J., M. G. Macklin, and M. J. Kirkby (2002), A cellular model of Holocene upland river basin and alluvial fan evolution, *Earth Surface Processes and Landforms*, 27(3), 269–288.

Coulthard, T. J., and C. J. Skinner (2016), The sensitivity of landscape evolution models to spatial and temporal rainfall resolution, *Earth Surface Dynamics*, 4, 757–771, doi:10.5194/esurf-4-757-2016.

Coventry, R. J., A. J. Moss, and E. Verster (1988), Thin surface soil layers attributable to rain-flow transportation on low-angle slopes: An example from semi-arid tropical Queensland, *Australia, Earth Surface Processes and Landforms*, 13, 421–430.

Cox, N. R. (1980), On the relationship between bedrock lowering and regolith thickness, *Earth Surface Processes*, 5(3), 271–274, doi:10.1002/esp.3760050305.

Cramer, W., and C. B. Field (1999), Comparing global models of terrestrial net primary productivity (NPP): Introduction, *Global Change Biology*, 5(Suppl. 1), 3–4.

Crave, A., and P. Davy (2001), A stochastic ''precipiton'' model for simulating erosion/sedimentation dynamics, *Computers & Geosciences*, 27(7), 815–827, doi:10.1016/S0098-3004(00)00167-9.

Culling, W. E. H. (1960), Analytical theory of erosion, *Journal of Geology*, 68(3), 336–344.

Culling, W. E. H. (1963), Soil creep and the development of hillside slopes, *Journal of Geology*, 71(2), 127–161.

Cundall, P. A., and O. D. L. Strack (1979), A discrete numerical model for granular assemblies, *Geotechnique*, 29(1), 47–65, doi:10.1680/geot.1979.29.1.47.

Dagbovie, A. S., and J. A. Sheratt (2014), Pattern selection and hysteresis in the Rietkerk model for banded vegetation in semi-arid environments, *Journal of the Royal Society Interface*, 11, 20140465, doi:10.1098/rsif.2014.0465.

Dahlen, F. A. (1990), Critical taper model of fold-and-thrust belts and accretionary wedges, *Annual Review of Earth and Planetary Sciences*, 18, 55–99, doi:10.1146/annurev.earth.18.1.55.

Daily, J. W., and D. R. F. Harleman (1966), *Fluid dynamics*, Addison-Wesley, Reading, MA.

Davidson, E. A., E. Belk, and R. D. Boone (1998), Soil water content and temperature as independent or confounded factors controlling soil respiration in a temperate mixed hardwood forest, *Global Change Biology*, 4(2), 217–227, doi:10.1046/j.1365-2486.1998.00128.x.

Davidson, E. A., and I. A. Janssens (2006), Temperature sensitivity of soil carbon decomposition and feedbacks to climate change, *Nature*, 440 (7081), 165–173, doi:10.1038/nature04514.

Davis, W. M. (1899), The geographical cycle, *Geographical Journal*, 14, 481–504.

Davy, P., and D. Lague (2009), Fluvial erosion/transport equation of landscape evolution models revisited, *Journal of Geophysical Research (Earth Surface)*, 114, F03007, doi:10.1029/2008JF001146.

DeBano, L. F. (2000), The role of fire and soil heating on water repellency in wildland environments: A review, *Journal of Hydrology*, 231–232, 195–206.

De Marco, A., A. E. Gentile, C. Arena, and A. V. De Santo (2005), Organic matter, nutrient content, and biological activity in burned and unburned soils of a Mediterranean maquis area of southern Italy, *International Journal of Wildland Fire*, 14, 365–377, doi:10.1071/WF05030.

Dennison, P. E., S. C. Brewer, J. D. Arnold, and M. A. Moritz (2014), Large wildfire trends in the western United States, 1984–2011, *Geophysical Research Letters*, 41, 2928–2933, doi:10.1002/2014GL059576.

DeNovio, N. M., J. E. Saiers, and J. N. Ryan (2004), Colloid movement in unsaturated porous media: Recent advances and future directions, *Vadose Zone Journal*, 3(2), 338–251, doi:10.2113/3.2.338.

Dessert, C., B. Dupre, J. Gaillardet, L. M. François, and C. J. Allegre (2003), Basalt weathering laws and the impact of basalt weathering on the global carbon cycle, *Chemical Geology*, 202, 257–273, doi:10.1016/j.chemgeo.2002.10.001.

de Vries, H., T. Becker, and B. Eckhardt (1994), Power law distribution of discharge in ideal networks, *Water Resources Research*, 30(12), 3541–3543.

de Vries, W., J. J. M. van Grinsven, N. van Breemen, E. E. J. M. Leeters, and P. C. Jansen (1995), Impacts of acid deposition on concentrations and fluxes of solutes in acid sandy forest soils in the Netherlands, *Geoderma*, 67(1–2), 17–43, doi:10.1016/0016-7061(94)00056-G.

Dewitte, O., M. Daoudi, C. Bosco, and M. Van Der Eeckhaut (2015), Predicting the susceptibility to gully initiation in data-poor regions, *Geomorphology*, 228, 101–115, doi:10.1016/j.geomorph.2014.08.010.

Dialynas, Y. G., S. Bastola, R. L. Bras, S. A. Billings, D. Markewitz, and D. D. Richter (2016), Topographic variability and the influence of soil erosion on the carbon cycle, *Global Biogeochemical Cycle*, 30, doi:10.1002/2015GB005302.

Dietrich, W. E. (1982), Settling velocity of natural particles, *Water Resources Research*, 18(6), 1615–1626, doi:10.1029/WR018i006p01615.

Dietrich, W. E., and D. R. Montgomery (1998), SHALSTAB: A digital terrain model for mapping shallow landslide potential, National Council for Air and Stream Improvement.

Dietrich, W. E., and J. T. Perron (2006), The search for a topographic signature of life, *Nature*, 439(7075), 411–418.

Dietrich, W. E., R. Reiss, M. L. Hsu, and D. R. Montgomery (1995), A process based model for colluvial soil depth and shallow landsliding using digital elevation data, *Hydrological Processes*, 9(3/4), 383–400, doi:10.1002/hyp.3360090311.

Dietrich, W. E., and P. Whiting (1989), Boundary shear stress and sediment transport in river meanders of sand and gravel, in *River Meandering*, edited by S. Ikeda and G. Parker, pp. 1–40, American Geophysical Union, Washington, DC.

Dietrich, W. E., C. J. Wilson, D. R. Montgomery, J. McKean, and R. A. Bauer (1992), Erosion thresholds and land surface morphology, *Geology*, 20, 675–679.

Dietrich, W. E., C. J. Wilson, D. R. Montgomery, and J. McKean (1993), Analysis of erosion thresholds, channel networks, and landscape morphology using a digital terrain model, *Journal of Geology*, 101(2), 259–278.

Dijkstra, J. J., J. C. L. Meeussen, and R. N. J. Comans (2004), Leaching of heavy metals from contaminated soils: An experimental and modeling study, *Environmental Science & Technology*, 38(16), 4390–4395, doi:10.1021/es049885v.

Doetterl, S., A. A. Berhe, E. Nadeu, Z. Wang, M. Sommer, and P. Fiener (2016), Erosion, deposition and soil carbon: A review of process-level controls, experimental tools and models to address C cycling in dynamic landscapes, *Earth-Science Reviews*, 154, 102–122, doi:10.1016/j.earscirev.2015.12.005.

Dorren, L. K. A. (2003), A review of rockfall mechanics and modelling approaches, *Progress in Physical Geography*, 27(1), 69–87.

Dorren, L. K. A., F. Berger, C. le Hir, E. Mermin, and P. Tardif (2005), Mechanisms, effects and management implications of rockfall in forests, *Forest Ecology and Management*, 215, 183–195, doi:10.1016/j.foreco.2005.05.012.

Dorren, L. K. A., F. Berger, and U. S. Putters (2006), Real-size experiments and 3-D simulation of rockfall on forested and non-forested slopes, *Natural Hazards and Earth System Sciences*, 6, 145–153.

Driscoll, N. W., and G. D. Karner (1994), Flexural deformation due to Amazon fan loading: A feedback mechanism affecting sediment delivery to margins, *Geology*, 22, 1015–1018, doi:10.1130/0091-7613(1994)022<1015:FDDTAF> 2.3.CO;2.

Dunai, T. J. (2010), *Cosmogenic nuclides: Principles, concepts and applications in the earth surface sciences*, Cambridge University Press, Cambridge.

Dunkerley, D. L. (1997), Banded vegetation: Development under uniform rainfall from a simple cellular automaton model, *Plant Ecology*, 129, 103–111.

Dunkerley, D. L. (2000), Hydrologic effects of dryland shrubs: Defining the spatial extent of modified soil water uptake rates at an Australian desert site, *Journal of Arid Environments*, 45, 159–172, doi:10.1006/jare.2000.0636.

Dunkerley, D. L. (2002), Infiltration rates and soil moisture in a groved mulga community near Alice Springs, arid central Australia: Evidence for complex internal rainwater redistribution in a runoff–runon landscape, *Journal of Arid Environments*, 51(2), 199–219, doi:10.1006/jare.2001.0941.

Dunkerley, D. L., and K. J. Brown (1995), Runoff and runon areas in a patterned chenopod shrubland, arid western New South Wales, Australia: Characteristics and origin, *Journal of Arid Environments*, 30(1), 41–55.

Dunkerley, D. L., and K. J. Brown (1999), Banded vegetation near Broken Hill, Australia: Significance of surface roughness and soil physical properties, *Catena*, 37(1–2), 75–88.

Dunne, T., D. V. Malmon, and K. B. J. Dunne (2016), Limits on the morphogenetic role of rain splash transport in hillslope evolution, *Journal of Geophysical Research (Earth Surface)*, 121, doi:10.1002/2015JF003737.

Dunne, T., D. V. Malmon, and S. M. Mudd (2010), A rain splash transport equation assimilating field and laboratory measurements, *Journal of Geophysical Research (Earth Surface)*, 115, F01001, doi:10.1029/2009JF001302.

Durda, D. D., W. F. Bottke, B. L. Enke, W. J. Merline, E. Asphaug, D. C. Richardson, and Z. M. Leinhardt (2004), The formation of asteroid satellites in large impacts: Results from numerical simulations, *Icarus*, 170(1), 243–257.

Durda, D. D., W. F. Bottke, D. Nesvorny, B. L. Enke, W. J. Merline, E. Asphaug, and D. C. Richardson (2007), Size–frequency distributions of fragments from SPH/N-body simulations of asteroid impacts: Comparison with observed asteroid families, *Icarus*, 186(2), 498–516.

Durda, D. D., N. Movshovitz, D. C. Richardson, E. Asphaug, A. Morgan, A. R. Rawlings, and C. Vest (2011), Experimental determination of the coefficient of restitution for meter-scale granite spheres, *Icarus*, 211(1), 849–855.

Eagleson, P. S. (2002), *Ecohydrology: Darwinian expression of vegetation form and function*, Cambridge University Press, Cambridge.

Einstein, H. A. (1950), The bed-load function for sediment transportation in open channel flows, Technical Bulletin 1026, US Department of Agriculture, Washington, DC.

Elias, E. A., R. Cichota, H. H. Torriani, and Q. de Jong van Lier (2004), Analytical soil-temperature model: Correction for temporal variation of daily amplitude, *Soil Science Society of America Journal*, 68, 784–788, doi:10.2136/sssaj2004.7840.

Enquist, B. J., and K. J. Niklas (2002), Global allocation rules for patterns of biomass partitioning in seed plants, *Science*, 295(5559), 1517–1520.

Eppes, M. C., and D. Griffing (2010), Granular disintegration of marble in nature: A thermal-mechanical origin for a grus and corestone landscape, *Geomorphology*, 117, 170–180, doi:10.1016/j.geomorph.2009.11.028.

Espinoza, J. C., S. Chavez, J. Ronchail, C. Junquas, K. Takahashi, and W. Lavado (2015), Rainfall hotspots over the southern tropical Andes: Spatial distribution, rainfall intensity, and relations with large-scale atmospheric circulation, *Water Resources Research*, 51, 3459–3475, doi:10.1002/2014WR016273.

Espirito-Santo, F. D. B., et al. (2014), Size and frequency of natural forest disturbances and the Amazon forest carbon balance, *Nature Communications*, 5, 1–6, doi:10.1038/ncomms4434.

Evans, K. G., and R. J. Loch (1996), Using the RUSLE to identify factors controlling erosion rates of mine spoils, *Land Degradation and Development*, 7, 267–277.

Evans, K. G., M. J. Saynor, and G. R. Willgoose (1996), The effect of vegetation on waste rock erosion, Ranger Uranium Mine, Northern Territory, *Bulletin of the Australian Institute of Mining and Metallurgy*, 6, 21–23.

Evans, K. G., M. J. Saynor, G. R. Willgoose, and S. J. Riley (2000), Post-mining landform evolution modelling. I. Derivation of sediment transport model and rainfall-runoff model parameters, *Earth Surface Processes and Landforms*, 25(7), 743–763.

Evans, K. G., and G. R. Willgoose (2000), Post-mining landform evolution modelling. II. Effects of vegetation and surface ripping, *Earth Surface Processes and Landforms*, 25(8), 803–823.

Fagherazzi, S., A. D. Howard, and P. L. Wiberg (2002), An implicit finite difference method for drainage basin evolution, *Water Resources Research*, 38(7), art. no.-1116.

FAO (2009), *Harmonized World Soil Database (version 1.1)*, FAO, Rome, Italy, and IIASA, Laxenburg, Austria.

Fatichi, S., C. Pappas, and V. Y. Ivanov (2016), Modeling plant–water interactions: an ecohydrological overview from the cell to the global scale, *Wiley Interdisciplinary Reviews – Water*, 3(3), 327–368, doi:10.1002/wat2.1125.

Field, J. B., and G. R. Anderson (2003), Biological agents in regolith processes: Case study on the Southern Highlands, NSW, paper presented at Advances in Regolith: Proceedings of the CRC LEME Regional Regolith Symposia, 2003, Cooperative Research Centre for Landscape Environments and Mineral Exploration (CRC LEME), Canberra, www.crcleme.org.au/Pubs/Advancesinregolith/Field_Anderson.pdf.

Finke, P. A. (2012), Modeling the genesis of luvisols as a function of topographic position in loess parent material, *Quaternary International*, 265, 3–17, doi:10.1016/j.quaint.2011.10.016.

Finke, P. A., A. Samouelian, M. Sourez-Bonnet, B. Laroche, and S. S. Cornu (2015), Assessing the usage potential of SoilGen2 to predict clay translocation under forest and agricultural land uses, *European Journal of Soil Science*, 66, 194–205, doi:10.1111/ejss.12190.

Finke, P. A., T. Vanwalleghem, E. Opolot, J. Poesen, and J. Deckers (2013), Estimating the effect of tree uprooting on variation of soil horizon depth by confronting pedogenetic simulations to measurements in a Belgian loess area, *Journal of Geophysical Research (Earth Surface)*, 118, 2124–2139, doi:10.1002/jgrf.20153.

Finney, M. A. (1999), Mechanistic modeling of landscape fire patterns, in *Spatial modeling of forest landscape change: Approaches and applications*, edited by D. J. Mladenoff and W. L. Baker, pp. 186–209, Cambridge University Press, Cambridge.

Finney, M. A. (2004), FARSITE: Fire Area Simulator – Model Development and Evaluation, Rep. *RMRS-RP-4*, USDA Forest Service Rocky Mountain Research Station.

Fleming, R. W., and A. M. Johnson (1975), Rates of seasonal creep of silty clay soil, *Quarterly Journal of Engineering Geology and Hydrogeology*, 8 (1), 1–29, doi:10.1144/GSL.QJEG.1975.008.01.01.

Fleskens, L., M. J. Kirkby, and B. J. Irvine (2016), The PESERA-DESMICE modeling framework for spatial assessment of the physical impact and economic viability of land degradation mitigation technologies, *Frontiers in Environmental Science*, 4, 31, doi:10.3389/fenvs.2016.00031.

Fletcher, R. C., and S. L. Brantley (2010), Reduction of bedrock blocks as corestones in the weathering profile: Observations and models, *American Journal of Science*, 310, 131–164, doi:10.2475/03.2010.01].

Fletcher, R. C., H. L. Buss, and S. L. Brantley (2006), A spheroidal weathering model coupling porewater chemistry to soil thicknesses during steady-state denudation, *Earth and Planetary Science Letters*, 244, 444–457, doi:10.1016/j.epsl.2006.01.055.

Fontúrbel, M. T., A. Barreiro, J. A. Vega, A. Martín, E. Jiménez, T. Carballas, C. Fernández, and M. Díaz-Raviña (2012), Effects of an experimental fire and post-fire stabilization treatments on soil microbial communities, *Geoderma*, 191, 51–60, doi:10.1016/j.geoderma.2012.01.037.

Fordham, A. W. (1990), Weathering of biotite into dioctahedral clay minerals, *Clay Minerals*, 25, 51–63.

Foster, G. R. (1982), Modelling the erosion process, in *Hydrologic Modelling of Small Watersheds*, edited by C. T. Haan, pp. 295–380, American Society of Agricultural Engineers, St Joseph, Missouri.

Foster, G. R., D. C. Flanagan, M. A. Nearing, L. J. Lane, L. M. Risse, and S. C. Finkner (1995), Chapter 11: Hillslope erosion component, in *USDA-Water Erosion Prediction Project Hillslope Profile and Watershed Model Documentation, NSERL Report #10*, edited by D. C. Flanagan and M. A. Nearing, US Department of Agriculture ARS, West Lafayette, Indiana.

Fox, M., L. Goren, D. A. May, and S. D. Willett (2014), Inversion of fluvial channels for paleorock uplift rates in Taiwan, *Journal of Geophysical Research (Earth Surface)*, 119, 1853–1875, doi:10.1002/2014JF003196.

Fox, R. W., and A. T. McDonald (1998), *Introduction to Fluid Mechanics*, 5th ed., Wiley, Chichester, UK.

Fraser, H. J. (1935), Experimental study of the porosity and permeability of clastic sediments, *Journal of Geology*, 83(8), 910–1010.

Freer, J., J. J. McDonnell, K. J. Beven, N. E. Peters, D. A. Burns, R. P. Hooper, B. Aulenbach, and C. Kendall (2002), The role of bedrock topography on subsurface storm flow, *Water Resources Research*, 38(12), art. no.-1269.

Freeze, R. A., and J. A. Cherry (1979), *Groundwater*, Prentice Hall, Englewood Cliffs, NJ.

Friedlingstein, P., et al. (2006), Climate-carbon cycling feedback analysis results from the C4MIP model intercomparison, *Journal of Climate*, 19 (14), 3337–3353.

Fujiwara, T., S. Kodaira, T. No, Y. Kaiho, N. Takahashi, and Y. Kaneda (2011), The 2011 Tohoku-Oki earthquake: Displacement reaching the trench axis, *Science*, 334(6060), 1240, doi:10. 1126/science. 1211554.

Furbish, D. J., A. E. Ball, and M. Schmeeckle (2012a), A probabilistic description of the bed load sediment flux: 4. Fickian diffusion at low transport rates, *Journal of Geophysical Research (Earth Surface)*, 117, F03034, doi:10.1029/2012JF002356.

Furbish, D. J., E. M. Childs, P. K. Haff, and M. W. Schmeeckle (2009), Rain splash of soil grains as a stochastic advection-dispersion process, with implications for desert plant-soil interactions and land-surface evolution, *Journal of Geophysical Research (Earth Surface)*, 114, F00A03, doi:10.1029/2009JF001265.

Furbish, D. J., and S. Fagherazzi (2001), Stability of creeping soil and implications for hillslope evolution, *Water Resources Research*, 37(10), 2607–2618.

Furbish, D. J., P. K. Haff, J. C. Roseberry, and M. Schmeeckle (2012b), A probabilistic description of the bed load sediment flux: 1. Theory, *Journal of Geophysical Research (Earth Surface)*, 117, F03031, doi:10.1029/2012JF002352.

Furbish, D. J., K. K. Hamner, M. Schmeeckle, M. N. Borosund, and S. M. Mudd (2007), Rain splash of dry sand revealed by high-speed imaging and sticky paper splash targets, *Journal of Geophysical Research (Earth Surface)*, 112, F01001, doi:10.1029/2006JF000498.

Furbish, D. J., J. C. Roseberry, and M. Schmeeckle (2012c), A probabilistic description of the bed load sediment flux: 3. The particle velocity distribution and the diffusive flux, *Journal of Geophysical Research (Earth Surface)*, 117, F03033, doi:10.1029/2012JF002355.

Gabet, E. J. (2000), Gopher bioturbation: Field evidence for nonlinear hillslope diffusion, *Earth Surface Processes and Landforms*, 25(13), 1419–1428.

Gabet, E. J., and T. Dunne (2003), Sediment detachment by rain power, *Water Resources Research*, 39(1), ESG1, doi:10.1029/2001WR000656.

Gabet, E. J., and S. M. Mudd (2010), Bedrock erosion by root fracture and tree throw: A coupled biogeomorphic model to explore the humped soil production function and the persistence of hillslope soils, *Journal of Geophysical Research (Earth Surface)*, 115, F04005, doi:10.1029/2009JF001526.

Gabet, E. J., O. J. Reichman, and E. W. Seabloom (2003), The effects of bioturbation on soil processes and sediment transport, *Annual Review of Earth and Planetary Sciences*, 31, 249–273.

Garcia-Corona, R., E. Benito, E. de Blas, and M. E. Varela (2004), Effects of heating on some soil physical properties related to its hydrological behaviour in two north-western Spanish soils, *International Journal of Wildland Fire*, 13, 195–199, doi:10.1071/WF03068.

Garreaud, R. D., M. Vuille, R. Compagnucci, and J. Marengo (2009), Present-day South American climate, *Paleogeography Paleoclimatology Paleoecology*, 281, 180–195, doi:10.1016/j.palaeo.2007.10.032.

Gasparini, N. M., G. E. Tucker, and R. L. Bras (1999), Downstream fining through selective particle sorting in an equilibrium drainage network, *Geology*, 27(12), 1079–1082.

Gasparini, N. M., G. E. Tucker, and R. L. Bras (2004), Network-scale dynamics of grain-size sorting: Implications for downstream fining, stream-profile concavity, and drainage basin morphology, *Earth Surface Processes and Landforms*, 29(4), 401–421.

Gasparini, N. M., K. X. Whipple, and R. L. Bras (2007), Predictions of steady state and transient landscape morphology using sediment-flux-dependent river incision models, *Journal of Geophysical Research (Earth Surface)*, 112, F03S09, doi:10.1029/2006JF000567.

Géminard, J.-C., and W. Losert (2002), Frictional properties of biodisperse granular matter: Effect of mixing ratio, *Physical Review E*, 65(4), D041301, doi:10.1103/PhysRevE.65.041301.

Géminard, J.-C., W. Losert, and J. P. Gollub (1999), Frictional mechanics of wet granular material, *Physical Review E*, 59(5), D5881, doi:10.1103/PhysRevE.59.5881.

Gercek, H. (2007), Poisson's ratio values for rocks, *International Journal of Rock Mechanics & Mining Sciences*, 44(1), 1–13, doi:10.1016/j.ijrmms.2006.04.011.

Gerten, D. (2013), A vital link: Water and vegetation in the Anthropocene, *Hydrology and Earth System Sciences*, 17, 3841–3852, doi:10.5194/hess-17-3841-2013.

Gerya, T. (2010), *Introduction to numerical geodynamic modelling*, Cambridge University Press, Cambridge.

Ghannoum, O. (2009), C_4 photosynthesis and water stress, *Annals of Botany*, 103, 6350644, doi:10.1093/aob/mcn093.

Ghezzehei, T. A., and D. Or (2001), Rheological properties of wet soils and clays under steady and oscillatory stresses, *Soil Science Society of America*, 65, 624–637.

Giacomini, A., O. Buzzi, B. Renard, and G. P. Giani (2009), Experimental studies on fragmentation of rockfalls on impact with rock surfaces, *International Journal of Rock Mechanics & Mining Sciences*, 46, 708–715, doi:10.1016/j.ijrmms.2008.09.007.

Gignoux, J., J. I. House, D. Hall, D. Masse, H. B. Nacro, and L. Abbadie (2001), Design and test of a generic cohort model of soil organic matter decomposition: The SOMKO model, *Global Ecology and Biogeography*, 10(6), 639–660.

Gilbert, G. (1909), The convexity of hillslopes, *Journal of Geology*, 17, 344–350.

Gilchrist, A. R., H. Kooi, and C. Beaumont (1994), Post-Gondwana geomorphic evolution of south-western Africa: Implications for the controls on landscape development from observations and numerical experiments, *Journal of Geophysical Research (Solid Earth)*, 99(B6), 12211–12228.

Gläser, G., H. Wernli, A. Kerkweg, and F. Teubler (2015), The transatlantic dust transport from North Africa to the Americas – Its characteristics and source regions, *Journal of Geophysical Research (Atmospheres)*, 120 (121), 11231–11252, doi:10.1002/2015JD023792.

Gómez-Villar, A., and J. M. García-Ruiz (2000), Surface sediment characteristics and present dynamics in alluvial fans of the central Spanish Pyrenees, *Geomorphology*, 34(3–4), 127–144, doi:10.1016/S0169-555X(99)00116-6.

Goodfellow, B. W., G. E. Hilley, S. M. Webb, L. S. Sklar, S. Moon, and C. A. Olson (2016), The chemical, mechanical, and hydrological evolution of weathering granitoid, *Journal of Geophysical Research (Earth Surface)*, 121, doi:10.1002/2016JF003822.

Goodwin, I. D., T. D. van Ommen, M. A. J. Curran, and P. A. Mayewski (2004), Mid latitude winter climate variability in the South Indian and southwest Pacific regions since 1300 AD, *Climate Dynamics*, 22, 783–794, doi:10.1007/s00382-004-0403-3.

Goren, L., S. D. Willett, F. Herman, and J. Braun (2014), Coupled numerical–analytical approach to landscape evolution modeling, *Earth Surface Processes and Landforms*, 39(4), 522–545, doi:10.1002/esp.3514.

Gorsevski, P. V., P. E. Gessler, J. Boll, W. J. Elliot, and R. B. Foltz (2006), Spatially and temporally distributed modeling of landslide susceptibility, *Geomorphology*, 80, 178–198, doi:10.1016/j.geomorph.2006.02.011.

Graf, W. H. (1984), *Hydraulics of sediment transport*, Water Resources Publications, Highlands Ranch, CO.

Graham, R. C., A. M. Rossi, and K. R. Hubbert (2010), Rock to regolith conversion: Producing hospitable substrates for terrestrial ecosystems, *GSA Today*, 20 (2), 4–9, doi:10.1130/GSAT57A.1.

Greene, R. S. B., C. J. Chartres, and K. A. Hodgkinson (1990), The effects of fire on the soil in a degraded semi-arid woodland. I. Cryptogam cover and physical and micromorphological properties, *Australian Journal of Soil Research*, 28, 755–777.

Gregorich, E. G., K. J. Greer, D. W. Anderson, and B. C. Liang (1998), Carbon distribution and losses: Erosion and deposition effects, *Soil Tillage Research*, 47(3–4), 291–302, doi:10.1016/S0167-1987(98)00117-2.

Griffiths, D. V., J. Huang, and G. F. deWolfe (2011), Numerical and analytical observations on long and infinite slopes, *International Journal for Numerical and Analytical Methods in Geomechanics*, 35, 569–585, doi:10.1002/nag.909.

Grimm, V., et al. (2006), A standard protocol for describing individual-based and agent-based models, *Ecological Modelling*, 198, 115–126, doi:10.1016/j.ecolmodel.2006.04.023.

Guerit, L., F. Metiver, O. Devauchelle, E. Lajeunesse, and L. Barrier (2014), Laboratory alluvial fans in one dimension, *Physical Review E*, 90, 022203, doi:10.1103/PhysRevE.90.022203.

Güneralp, I., and R. A. Marston (2012), Process–form linkages in meander morphodynamics: Bridging theoretical modeling and real world complexity, *Progress in Physical Geography*, 36(6), 718–746, doi:10.1177/0309133312451989.

Gupta, S., J. S. Collier, A. Palmer-Fengate, and G. Potter (2007), Catastrophic flooding origin of shelf valley systems in the English Channel, *Nature*, 448(19 July), 342–346, doi:doi:10.1038/nature06018.

Gyasi-Agyei, Y., G. R. Willgoose, and F. P. de Troch (1995), Effects of vertical resolution and map scale of digital elevation maps on geomorphologic parameters used in hydrology, *Hydrological Processes*, 9(3/4), 121–140.

Hack, J. T. (1957), Studies of longitudinal stream profiles in Virginia and Maryland, USGS Professional Papers 294-B, USGS.

Hairsine, P. B., and C. W. Rose (1992a), Modelling water erosion due to overland flow using physical principles: 1 Sheet flow, *Water Resources Research*, 28(1), 237–243.

Hairsine, P. B., and C. W. Rose (1992b), Modelling water erosion due to overland flow using physical principles: 2 Rill flow, *Water Resources Research*, 28(1), 245–250.

Hairsine, P. B., and G. C. Sander (2009), Comment on 'A transport-distance based approach to scaling erosion rates': Parts 1, 2 and 3 by Wainwright et al., *Earth Surface Processes and Landforms*, 34, 882–885, doi:10.1002/esp.1782.

Hallet, P. D., S. Caul, T. J. Daniell, P. Barre, and E. Paterson (2010), The rheology of rhizosphere formation by root exudates and soil microbes, paper presented at 19th World Congress of Soil Science, 1–6 August 2010, Brisbane, Australia.

Hammer, P. T. C., R. M. Clowes, F. A. Cook, K. Vasudevan, and A. J. van der Velden (2013), The big picture: A lithospheric cross section of the North American continent, *GSA Today*, 21(6), 4–10, doi:10.1130/GSATG95A.1.

Han, S. C., J. Sauber, S. B. Luthcke, C. Ji, and F. F. Pollitz (2008), Implications of postseismic gravity change following the great 2004 Sumatra-Andaman earthquake from the regional harmonic analysis of GRACE intersatellite tracking data, *Journal of Geophysical Research (Solid Earth)*, 113(B11), B11413, doi:10.1029/2008JB005705.

Hancock, G. R. (2003), Effect of catchment aspect ratio on geomorphological descriptors, in *Prediction in Geomorphology*, edited by P. R. Wilcock and R. M. Iverson, pp. 217–230, American Geophysical Union, Washington, DC.

Hancock, G. R. (2005), The use of digital elevation models in the identification and characterisation of catchments, *Hydrological Processes*, 19, 1727–1749.

Hancock, G. R., T. J. Coulthard, and J. B. C. Lowry (2016), Long-term landscape trajectory – Can we make predictions about landscape form and function for post-mining landforms?, *Geomorphology*, 266, 121–132, doi:10.1016/j.geomorph.2016.05.014.

Hancock, G. R., T. J. Coulthard, C. Martinez, and J. D. Kalma (2011), An evaluation of landscape evolution models to simulate decadal and centennial scale soil erosion in grassland catchments, *Journal of Hydrology*, 398 (3–4), 171–183, doi:10.1016/j.jhydrol.2010.12.002.

Hancock, G. R., K. G. Evans, J. J. McDonnell, and L. Hopp (2012), Ecohydrological controls on soil erosion and landscape evolution, *Ecohydrology*, 5(4), 478–490, doi:10.1002/eco.241.

Hancock, G. R., K. G. Evans, G. R. Willgoose, D. R. Moliere, M. J. Saynor, and R. J. Loch (2000), Medium term erosion simulation of an abandoned mine site using the SIBERIA landscape evolution model, *Australian Journal of Soil Research*, 38, 249–263.

Hancock, G. R., J. Hugo, A. Webb, and L. Turner (2017a), Sediment transport in steep forested catchments – An assessment of scale and disturbance, *Journal of Hydrology*, 547, 613–622, doi:10.1016/j.jhydrol.2017.02.022.

Hancock, G. R., J. B. C. Lowry, and T. J. Coulthard (2015a), Catchment reconstruction – Erosional stability at millennial time scales using landscape evolution models, *Geomorphology*, 231, 15–27, doi:10.1016/j.geomorph.2014.10.034.

Hancock, G. R., J. B. C. Lowry, T. J. Coulthard, K. G. Evans, and D. R. Moliere (2010), A catchment scale evaluation of the SIBERIA and CAESAR landscape evolution models, *Earth Surface Processes and Landforms*, 35(8), 863–875, doi:10.1002/esp.1863.

Hancock, G. R., J. B. C. Lowry, and C. Dever (2017b), Surface disturbance and erosion by pigs: A medium term assessment for the Monsoonal tropics, *Land Degradation and Development*, 28(1), 255–264, doi:10.1002/ldr.2636.

Hancock, G. R., J. B. C. Lowry, C. Dever, and M. Braggins (2015b), Does introduced fauna influence soil erosion? A field and modelling assessment, *Science of the Total Environment*, 518–519, 189–200, doi:10.1016/j.scitotenv.2015.02.086.

Hancock, G. R., J. B. C. Lowry, D. R. Moliere, and K. G. Evans (2008), An evaluation of an enhanced soil erosion and landscape evolution model: A case study assessment of the former Nabarlek uranium mine, Northern Territory, Australia, *Earth Surface Processes and Landforms*, 33(13), 2045–2063, doi:10.1002/esp.1653.

Hancock, G. R., J. B. C. Lowry, and M. J. Saynor (2017c), Surface armour and erosion – Impacts on long-term landscape evolution, *Land Degradation and Development*, doi:10.1002/ldr.2738.

Hancock G. R., C. Martinez, K. G. Evans, and D. R. Moliere. (2006). A comparison of SRTM and high-resolution digital elevation models and their use in catchment geomorphology and hydrology – Australian examples, *Earth Surface Processes and Landforms*, 31, 809–824.

Hancock, G. R., and G. R. Willgoose (2001), The use of a landscape simulator in the validation of the SIBERIA catchment evolution model: Declining equilibrium landforms, *Water Resources Research*, 37(7), 1981–1992.

Hancock, G. R., and G. R. Willgoose (2002), The use of a landscape simulator in the validation of the SIBERIA landscape evolution model: Transient landforms, *Earth Surface Processes and Landforms*, 27(12), 1321–1334.

Hancock, G. R., and G. R. Willgoose (2004), An experimental and computer simulation study of erosion on a mine tailings dam wall, *Earth Surface Processes and Landforms*, 29(4), 457–475.

Hancock, G. R., G. R. Willgoose, and K. G. Evans (2002), Testing of the SIBERIA landscape evolution model using the Tin Camp Creek, Northern Territory, Australia, field catchment, *Earth Surface Processes and Landforms*, 27(2), 125–143.

Hantson, S., S. Pueyo, and E. Chuvieco (2016), Global fire size distribution: From power law to log-normal, *International Journal of Wildland Fire*, 25, 403–412, doi:10.1071/WF15108.

Hart, S. A., and N. J. Luckai (2014), Charcoal carbon pool in North American boreal forests, *Ecosphere*, 5(8), 99, doi:10.1890/ES13-00086.1.

Hasbargen, L. E., and C. Paola (2000), Landscape instability in an experimental drainage basin, *Geology*, 28(12), 1067–1070.

Hasegawa, K. (1977), Computer simulation of the gradual migration of meandering channels, in *Proceedings of the Hokkaido Branch, Japan Society of Civil Engineering*, pp. 197–202 (in Japanese).

Heimsath, A. M., J. Chappell, N. A. Spooner, and D. G. Questiaux (2003), Creeping soil, *Geology*, 30(2), 111–114.

Heimsath, A. M., W. E. Dietrich, K. Nishiizummi, and R. C. Finkel (1997), The soil production function and landscape equilibrium, *Nature*, 388(6640), 358–361.

Heimsath, A. M., W. E. Dietrich, K. Nishiizummi, and R. C. Finkel (1999), Cosmogenic nuclides, topography, and the spatial variation of soil depth, *Geomorphology*, 27(1–2), 151–172.

Heimsath, A. M., D. J. Furbish, and W. E. Dietrich (2005), The illusion of diffusion: Field evidence for depth-dependent sediment transport, *Geology*, 33(12), 949–952, doi:10.1130/G21868.1.

Heimsath, A. M., G. R. Hancock, and D. Fink (2009), The 'humped' soil production function: Eroding Arnhem Land, Australia, *Earth Surface Processes and Landforms*, 34(12), 1674–1684.

Heister, K. (2016), How accessible is the specific surface area of minerals? A comparative study with Al-containing minerals as model substances, *Geoderma*, 263, 8–15, doi:10.1016/j.geoderma.2015.09.001.

Henderson, F. M. (1966), *Open channel flow*, Macmillan, New York.

Hénin, S., and M. Dupuis (1945), Essai de bilan de la matière organique des sols, *Annales agronomiques*, 15, 161–172.

Hergarten, S. (2003), Landslides, sandpiles, and self-organized criticality, *Natural Hazards and Earth System Sciences*, 3, 505–514.

Hilinski, T. E. (2001), Implementation of exponential depth distribution of organic carbon in the CENTURY model, Colorado State University, Fort Collins. www.nrel.colostate.edu/projects/century5/reference/html/Century/exp-c-distrib.htm.

HilleRisLambers, R., M. Rietkerk, F. van Den Bosch, H. H. T. Prins, and H. de Kroon (2001), Vegetation pattern formation in semi-arid grazing systems, *Ecology*, 82(1), 50–61, doi:10.1890/0012-9658(2001)082[0050:VPFISA]2.0.CO;2.

Hilton, J. E., C. Miller, A. L. Sullivan, and C. Rucinski (2015), Effects of spatial and temporal variation in environmental conditions on simulation of wildfire spread, *Environmental Modelling Software*, 67, 118–127, doi:10.1016/j.envsoft.2015.01.015.

Ho, J.-Y., K. T. Lee, T.-C. Chang, Z.-Y. Wang, and Y.-H. Liao (2012), Influences of spatial distribution of soil thickness on shallow landslide prediction, *Engineering Geology*, 124, 38–46, doi:10.1016/j.enggeo.2011.09.013.

Ho, M., D. C. Verdon-Kidd, A. S. Kiem, and R. N. Drysdale (2014), Broadening the spatial applicability of paleoclimate information – A case study for the Murray-Darling Basin, Australia, *Journal of Climate*, 27(7), 2477–2495, doi:10.1175/JCLI-D-13-00071.1.

Hobley, D. E. J., J. M. Adams, S. S. Nudurupati, E. W. H. Hutton, N. M. Gasparini, E. Istanbulluoglu, and G. E. Tucker (2017), Creative computing with Landlab: An open-source toolkit for building, coupling, and exploring two-dimensional numerical models of Earth-surface dynamics, *Earth Surface Dynamics*, 5, 21–46, doi:10.5194/esurf-5-21-2017.

Hobley, E. U., G. R. Willgoose, S. Frisia, and G. E. Jacobsen (2013), Environmental and site factors controlling the vertical distribution and radiocarbon ages of organic carbon in a sandy soil, *Biology and Fertility of Soils*, 49(8), 1015–1026, doi:10.1007/s00374-013-0800-z.

Hobley, E. U., G. R. Willgoose, S. Frisia, and G. E. Jacobsen (2014), Vertical distribution of charcoal in a sandy soil: Evidence from DRIFT spectra, SEM and radiocarbon dating, *European Journal of Soil Science*, 65(7), 751–762, doi:10.1111/ejss.12171.

Hobley, E. U., and B. Wilson (2016), The depth distribution of organic carbon in the soils of eastern Australia, *Ecosphere*, 7(1), 1–21, doi:e01214.10.1002/ecs2.1214.

Hobley, E., B. Wilson, A. Wilkie, J. Gray, and T. Koen (2015), Drivers of soil organic carbon storage and vertical distribution in Eastern Australia, *Plant and Soil*, 390(1–2), 111–127, doi:10.1007/s11104-015-2380-1.

Hodson, M. E. (2006), Does reactive surface area depend on grain size? Results from pH 3, 25 °C far-from-equilibrium flow-through dissolution experiments on anorthite and biotite, *Geochimica et Cosmochimica Acta*, 70(7), 1655–1667, doi:10.1016/j.gca.2006.01.001.

Hole, F. D. (1981), Effects of animals on soils, *Geoderma*, 25(1–2), 75–112, doi:10.1016/0016-7061(81)90008-2.

Holmes, K. W., R. D. A, S. Sweeney, I. Numata, E. Matricardi, T. W. Biggs, G. Batista, and O. A. Chadwick (2004), Soil databases and the problem of establishing regional biogeochemical trends, *Global Change Biology*, 10(5), 796–814, doi:10.1111/j.1529-8817.2003.00753.x.

Holzworth, D. P., V. Snow, S. Janssen, I. N. Athanasiadis, M. Donatelli, G. Hoogenboom, J. W. White, and P. Thorburn (2015), Agricultural production systems modelling and software: Current status and future prospects, *Environmental Modelling Software*, 72, 276–286, doi:10.1016/j.envsoft.2014.12.013.

Hoosbeek, M. R. (1994), Towards the quantitative modeling of pedogenesis: A review – Reply – Pedology beyond the soil landscape paradigm: Pedodynamics and the connection to other sciences, *Geoderma*, 63(3–4), 303–307.

Hoosbeek, M. R., and R. B. Bryant (1992), Towards the quantitative modeling of pedogenesis – A review, *Geoderma*, 55(3–4), 183–210.

Horwath, W. (2007), Carbon cycling and formation of soil organic matter, in *Soil Microbiology, Ecology, and Biochemistry*, edited by E. A. Paul, pp. 303–340, Academic Press, Amsterdam.

Hosseini, M., J. J. Keizer, O. G. Pelayo, S. A. Prats, C. J. Ritsema, and V. Geissen (2016), Effect of fire frequency on runoff, soil erosion, and loss of organic matter at the micro-plot scale in north-central Portugal, *Geoderma*, 269, 126–137, doi:10.1016/j.geoderma.2016.02.004.

Houghton, R. A. (2007), Balancing the global carbon budget, *Annual Review of Earth and Planetary Sciences*, 35, 313–347, doi:10.1146/annurev.earth.35.031306.140057.

Hovius, N., C. P. Stark, and P. A. Allen (1997), Sediment flux from a mountain belt derived by landslide mapping, *Geology*, 25(3), 231–234, doi:10.1130/0091-7613(1997)025<0231:SFFAMB>2.3.CO;2.

Howard, A. D. (1971), Simulation of stream networks by headward growth and branching, *Geographical Analysis*, 3, 29–50.

Howard, A. D. (1980), Thresholds in river regimes, in *Thresholds in Geomorphology, 9th Binghampton Geomorphology Symposium, 1978*, edited by D. R. Coates and J. D. Vitek, pp. 227–258, Allen and Unwin, Boston.

Howard, A. D. (1992), Modelling channel migration and floodplain sedimentation in meandering streams, in *Lowland floodplain rivers*, edited by P. Carling and G. E. Petts, pp. 1–42, Wiley, Chichester, UK.

Howard, A. D. (1994), A detachment-limited model of drainage-basin evolution, *Water Resources Research*, 30(7), 2261–2285.

Howard, A. D. (1996), Modelling channel evolution and floodplain morphology, in *Floodplain processes*, edited by M. G. Anderson, D. E. Walling, and P. D. Bates, pp. 15–62, Wiley, Chichester, UK.

Howard, A. D. (1998), Long profile development of bedrock channels: Interaction of weathering, mass wasting, bed erosion, and sediment transport, in *Rivers over rocks: Fluvial processes in bedrock channels*, edited by K. J. Tinkler and E. E. Wohl, pp. 297–319, America Geophysical Union, Washington, DC.

Howard, A. D. (2009), How to make a meandering river, *Proceedings of the National Academy of Sciences USA*, 106(41), 17246–17246.

Howard, A. D., W. E. Dietrich, and M. A. Seidl (1994), Modeling fluvial erosion on regional to continental scales, *Journal of Geophysical Research (Solid Earth)*, 99(B7), 13971–13986.

Howard, A. D., and G. Kerby (1983), Channel changes in badlands, *Geological Society of America Bulletin*, 94(6), 739–752.

Howard, A. D., and T. R. Knutson (1984), Sufficient conditions for river meandering – A simulation approach, *Water Resources Research*, 20(11), 1659–1667.

Huang, H. Q., and G. R. Willgoose (1993), Some scale dependent properties of distributed rainfall-runoff models, paper presented at Towards the 21st Century, Hydrology and Water Resources Symposium, Institution of Engineers (Aust.), Newcastle.

Huang, X., and J. D. Niemann (2006), An evaluation of the geomorphically effective event for fluvial processes over long periods, *Journal of Geophysical Research (Earth Surface)*, 111, doi:10.1029/2006JF000477.

Huang, X., and J. D. Niemann (2008), How do streamflow generation mechanisms affect watershed hypsometry?, *Earth Surface Processes and Landforms*, 33, 751–772, doi:10.1002/esp.1573.

Hudson, B. D. (1994), Soil organic matter and available water capacity, *Journal of Soil and Water Conservation*, 49(2), 189–194.

Huggett, R. J. (1975), Soil landscape systems: A model for soil genesis, *Geoderma*, 13, 1–22.

Huntly, N., and R. Inouye (1988), Pocket gophers in ecosystems: Patterns and mechanisms, *Bioscience*, 38(1), 786–793.

Hupy, J. P., and R. J. Schaetzl (2006), Introducing 'Bombturbation', a singular type of soil disturbance and mixing, *Soil Science*, 171(11), 823–836, doi:10.1097/01.ss.0000228053.08087.19.

Hupy, J. P., and R. J. Schaetzl (2008), Soil development on the WWI battlefield of Verdun, France, *Geoderma*, 145, 37–49, doi:10.1016/j.geoderma.2008.01.024.

Hutchinson, M. F. (1995), Interpolating mean rainfall using thin plate smoothing splines, *International Journal of Geographical Information Systems*, 9(4), 385–403, doi:10.1080/02693799508902045.

Hutchinson, M. F. (1998), Interpolation of rainfall data with thin plate smoothing splines – Part II: Analysis of Topographic Dependence, *Journal of Geographic Information and Decision Analysis*, 2(2), 152–167.

Ibbitt, R. P., G. R. Willgoose, and M. J. Duncan (1999), Channel network simulation models compared with data from the Ashley River, New Zealand, *Water Resources Research*, 35(12), 3875–3890.

Ikeda, H., G. Parker, and K. Sawai (1981), Bend theory of river meanders: Part I, Linear development, *Journal of Fluid Mechanics*, 112, 363–377.

IPCC (2007), *Climate change 2007: The physical science basis. Contribution of Working Group 1 to the Fourth Assessment report of the Intergovernmental Panel on Climate Change*, edited by S. Solomon, D. Qin, M. Manning, Z. Chen, M. Marquis, K. B. Averyt, M. Tignor, and H. L. Miller, p. 996, Cambridge University Press, Cambridge.

Istanbulluoglu, E., and R. L. Bras (2005), Vegetation-modulated landscape evolution: Effects of vegetation on landscape processes, drainage density, and topography, *Journal of Geophysical Research (Earth Surface)*, 110, F02012, doi:10.1029/2004JF000249.

Istanbulluoglu, E., D. G. Tarboton, R. T. Pack, and C. H. Luce (2004), Modeling of the interactions between forest vegetation, disturbances, and sediment yields, *Journal of Geophysical Research (Earth Surface)*, 109, F01009, doi:10.1029/2003JF000041.

Iverson, R. M. (1990), Groundwater flow fields in infinite slopes, *Geotechnique*, 40(1), 139–143.

Iverson, R. M. (1997), The physics of debris flows, *Reviews of Geophysics*, 35(3), 245–296.

Iverson, R. M., and R. P. Denlinger (2001), Flow of variably fluidized granular masses across three-dimensional terrain 1. Coulomb mixture theory, *Journal of Geophysical Research (Solid Earth)*, 106(B1), 537–552.

Iverson, R. M., and D. L. George (2015), A depth-averaged debris-flow model that includes the effects of evolving dilatancy. I. Physical basis, *Proceedings of the Royal Society of London. Series A, Mathematical and Physical Sciences*, 470, 20130819, doi:10.1098/rspa.2013.0819.

Iverson, R. M., M. Logan, and R. P. Denlinger (2004), Granular avalanches across irregular three-dimensional terrain: 2. Experimental tests, *Journal of Geophysical Research (Earth Surface)*, 109(F1), F01015, doi:10.1029/2003JF000084.

Iverson, R. M., and J. W. Vallance (2001), New views of granular mass flows, *Geology*, 29(2), 115–118, doi:10.1130/0091-7613(2001)029<0115:NVOGMF>2.0.CO;2.

Ivins, E. R., and D. Wolf (2008), Glacial isostatic adjustment: New developments from advanced observing systems and modeling, *Journal of Geodynamics*, 46, 69–77, doi:10.1016/j.jog.2008.06.002.

Ijjasz-Vasquez, E. J., R. L. Bras, I. Rodriguez-Iturbe, R. Rigon, and A. Rinaldo (1993), Are river basins Optimal Channel Networks, *Advances in Water Resources*, 16(1), 69–79.

Jackson, R. B., J. G. Canadell, J. R. Ehleringer, H. A. Mooney, O. E. Sala, and E. D. Schulze (1996), A global analysis of root distributions for terrestrial biomes, *Oecologia*, 108(3), 389–411, doi:10.1007/BF00333714.

Jagercikova, M., O. Evrard, J. Balesdent, I. Lefevre, and S. S. Cornu (2014), Modeling the migration of fallout radionuclides to quantify the contemporary transfer of fine particles in Luvisol profiles under different land uses and farming practices, *Soil Tillage Research*, 140, 82–97, doi:10.1016/j.still.2014.02.013.

James, A. L., J. J. McDonnell, I. Tromp-van Meerveld, and N. E. Peters (2010), Gypsies in the palace: Experimentalist's view on the use of 3-D physics-based simulation of hillslope hydrological response, *Hydrological Processes*, 24(26), 3878–3893, doi:10.1002/hyp.7819.

James, T. S., and W. J. Morgan (1990), Horizontal motions due to post-glacial rebound, *Geophysical Research Letters*, 17(7), 957–960.

Jarvis, N. J., K. G. Villholth, and B. Ulen (1999), Modelling particle mobilization and leaching in macroporous soil, *European Journal of Soil Science*, 50(4), 621–632.

Jenkinson, D. S., and K. Coleman (2008), The turnover of organic carbon in subsoils. Part 2. Modelling carbon turnover, *European Journal of Soil Science*, 59, 400–413, doi:10.1111/j.1365-2389.2008.01026.x.

Jenkinson, D. S., and J. H. Rayner (1977), The turnover of soil organic matter in some of the Rothamsted classical experiments, *Soil Science*, 123 (5), 298–305, doi:10.1097/00010694-197705000-00005.

Jenny, H. (1941), *Factors of soil formation – A system of quantitative pedology*, McGraw-Hill, New York.

Jenny, H. (1961), Derivation of state factor equations of soils and ecosystems, *Proceedings of the Soil Science Society of America*, 25, 385–388.

Jobbágy, E. G., and R. B. Jackson (2000), The vertical distribution of soil organic carbon and its relation to climate and vegetation, *Ecological Applications*, 10(2), 423–436, doi:10.2307/2641104.

Johansen, M. P., T. E. Hakonson, and D. D. Breshears (2001), Post-fire runoff and erosion from rainfall simulation: Contrasting forests with shrublands and grasslands, *Hydrological Processes*, 15(15), 2953–2965.

Johnson, B. C., C. S. Campbell, and H. J. Melosh (2016), The reduction of friction in long runout landslides as an emergent phenomenon, *Journal of Geophysical Research (Earth Surface)*, 121, 881–889, doi:10.1002/2015JF003751.

Johnson, C. G., B. P. Kokelaar, R. M. Iverson, M. Logan, R. G. LaHusen, and J. M. N. T. Gray (2012), Grain-size segregation and levee formation in geophysical mass flows, *Journal of Geophysical Research (Earth Surface)*, 117, F01032, doi:10.1029/2011JF002185.

Johnson, I. R. (2008), Biophysical pasture model documentation: Model documentation for DairyMod. EcoMod and the SGS Pasture Model, http://imj.com.au/wp-content/uploads/2014/08/GrazeMod.pdf.

Johnson, J. P., and K. X. Whipple (2007), Feedbacks between erosion and sediment transport in experimental bedrock channels, *Earth Surface Processes and Landforms*, 32, 1048–1062, doi:10.1002/esp.1471.

Johnston, C. A., et al. (2004), Carbon cycling in soil, *Frontiers in Ecology and the Environment*, 2(10), 522–528, doi:10.1890/1540-9295(2004)002[0522:CCIS]2.0.CO;2.

Johnstone, S. A., and G. E. Hilley (2015), Lithologic control on the form of soil-mantled hillslopes, *Geology*, 43(1), 83–86, doi:10.1130/G36052.1.

Julien, P. Y. (2014), Downstream hydraulic geometry of alluvial rivers, in *Sediment dynamics from the summit to the sea*, IAHS Publ. 367, IAHS, New Orleans, doi:10.5194/piahs-367-3-2015.

Julien, P. Y., and J. Wargadalam (1995), Alluvial channel geometry – Theory and applications, *Journal of Hydraulic Division – ASCE*, 121(4), 312–325, doi:10.1061/(ASCE)0733–9429(1995)121:4(312).

Kaiser, M., D. P. Zederer, R. H. Ellerbrook, M. Sommer, and B. Ludwig (2016), Effects of mineral characteristics on content, composition, and stability of organic matter fractions separated from seven forest topsoils of different pedogenesis, *Geoderma*, 263, 1–7, doi:10.1016/j.geoderma.2015.08.029.

Kalma, J. D., T. R. McVicar, and M. F. McCabe (2008), Estimating land surface evaporation: A review of methods using remotely sensed surface temperature data, *Surveys in Geophysics*, 29, 421–469, doi:10.1007/s10712-008-9037-z.

Katagis, T., I. Z. Gitas, P. Toukiloglou, S. Veraverbeke, and R. Goosens (2014), Trend analysis of medium- and coarse-resolution time series image data for burned area mapping in a Mediterranean ecosystem, *International Journal of Wildland Fire*, 23, 668–677, doi:10.1071/WF12055.

Kavetski, D., G. Kuczera, and S. W. Franks (2006), Bayesian analysis of input uncertainty in hydrological modeling: 1. Theory, *Water Resources Research*, 42(3), W03407.

Keating, B. A., et al. (2003), An overview of APSIM, a model designed for farming systems simulation, *European Journal of Agronomy*, 18, 267–288.

Keeley, J. E. (2009), Fire intensity, fire severity and burn severity: A brief review and suggested usage, *International Journal of Wildland Fire*, 18, 116–126, doi:10.1071/WF07049.

Kelemen, P. B., and M. D. Behn (2016), Formation of lower continental crust by relamination of buoyant arc lavas and plutons, *Nature Geoscience*, 9(3), 197–205, doi:10.1038/ngeo2662.

Kinnell, P. I. A. (2010), Event soil loss, runoff and the Universal Soil Loss Equation family of models: A review, *Journal of Hydrology*, 385, 384–397, doi:10.1016/j.jhydrol.2010.01.024.

Kirchner, J. W. (1993), Statistical inevitability of Horton's laws and the apparent randomness of stream channel networks, *Geology*, 21, 591–594.

Kirkby, M. J. (1967), Measurement and theory of soil creep, *Journal of Geology*, 75(4), 359–378.

Kirkby, M. J. (1971), Hillslope process-response models based on the continuity equation, in *Slopes: Form and process*, pp. 15–30, Institute of British Geographers, London.

Kirkby, M. J. (1976), Tests of the random network model and its application to basin hydrology, *Earth Surface Processes*, 1, 197–212, doi:10.1002/esp.3290010302.

Kirkby, M. J. (1977), Soil development models as a component of slope models, *Earth Surface Processes*, 2, 203–230, doi:10.1002/esp.3290020212.

Kirkby, M. J. (1985), A basis for soil profile modelling in a geomorphic context, *European Journal of Soil Science*, 36(1), 97–121, doi:10.1111/j.1365-2389.1985.tb00316.x.

Kirkby, M. J. (1989), A model to estimate the impact of climatic-change on hillslope and regolith form, *Catena*, 16(4–5), 321–341, doi:10.1016/0341-8162(89)90018-0.

Kirkby, M. J. (2000), Limits to modelling in the Earth and environmental sciences, in *Geocomputation*, edited by R. J. Abrahart, S. Openshaw, and L. M. See, pp. 374–386, Taylor and Francis, London.

Kirkby, M. J., L. J. Bull, J. Poesen, J. Nachtergaele, and L. Vandekerckhove (2003), Observed and modelled distributions of channel and gully heads – with examples from SE Spain and Belgium, *Catena*, 50(2–4), 415–434.

Kirkby, M. J., and I. Statham (1975), Surface stone movement and scree formation, *Journal of Geology*, 83(3), 349–362.

Kirkby, M. J., and G. R. Willgoose (2005), Weathering limited or transport limited removal in bedrock channel systems, paper presented at IAG, 7–11 September, Zaragoza, Spain.

Klein, F. W. (2016), Lithospheric flexure under the Hawaiian volcanic load: Internal stresses and a broken plate revealed by earthquakes, *Journal of Geophysical Research (Solid Earth)*, 121, 2400–2428, doi:10.1002/2015JB012746.

Klemann, V., Z. Martinec, and E. R. Ivins (2008), Glacial isostasy and plate motion, *Journal of Geodynamics*, 46, 95–103, doi:10.1016/j.jog.2008.04.005.

Klepeis, K. A., G. L. Clarke, and T. Rushmer (2003), Magma transport and coupling between deformation and magmatism in the continental lithosphere, *GSA Today*.

Knisel, W. (1980), CREAMS: A field-scale model for chemicals, runoff, and erosion from Agricultural Management Systems, Conservation Research Report No. 26, US Department of Agriculture, Washington, DC.

Knorr, W., and V. Lakshmi (2001), Assimilation of fAPAR and surface temperature into a land surface and vegetation model, in *Land surface hydrology, meteorology and climate: Observations and modeling*, edited by V. Lakshmi, J. Albertson, and J. Schaake, pp. 177–200, American Geophysical Union, Washington, DC.

Koltermann, C. E., and S. M. Gorelick (1992), Paleoclimatic signature in terrestrial flood deposits, *Science*, 256, 1775–1782.

Koltermann, C. E., and S. M. Gorelick (1995), Fractional packing model for hydraulic conductivity derived from sediment mixtures, *Water Resources Research*, 31(12), 3283–3297.

Konno, K. (2016), A general parameterized mathematical food web model that predicts a stable green world in the terrestrial ecosystem, *Ecological Monographs*, 86(2), 190–214, doi:10.1890/15-1420.

Kooi, H., and C. Beaumont (1994), Escarpment evolution on high-elevation rifted margins: Insights derived from a surface process model that combines diffusion, advection and reaction, *Journal of Geophysical Research (Solid Earth)*, 99(B6), 12191–12209.

Kotroni, V., and K. Lagouvardos (2008), Lightning occurrence in relation with elevation, terrain slope, and vegetation cover in the Mediterranean, *Journal of Geophysical Research (Atmospheres)*, D21118, doi:10.1029/2008JD010605.

Kuczera, G. (1987), Prediction of water yield reductions following a bushfire in ash-mixed species eucalypt forest, *Journal of Hydrology*, 94 (3–4), 215–236.

Kuczera, G. (1989), An application of Bayesian nonlinear-regression to hydrologic models, *Advances in Engineering Software and Workstations*, 11(3), 149–155.

Kuehl, S. A., et al. (2016), A source-to-sink perspective of the Waipaoa River margin, *Earth-Science Reviews*, 153(301–334), doi:10.1016/j.earscirev.2015.10.001.

Kump, L. R., S. L. Brantley, and M. A. Arthur (2000), Chemical weathering, atmospheric CO_2, and climate, *Annual Review of Earth and Planetary Sciences* , 28, 611–667, doi:10.1146/annurev.earth.28.1.611.

Kyriakidis, P. C., J. Kim, and N. L. Miller (2001), Geostatistical mapping of precipitation from rain gauge data using atmospheric and terrain characteristics, *Journal of Applied Meteorology*, 40, 1855–1877.

Lacoste, M., V. Viaud, D. Michot, and C. Walter (2015), Landscape-scale modelling of erosion processes and soil carbon dynamics under land-use and climate change in agroecosystems, *European Journal of Soil Science*, 66(4), 780–791, doi:10.1111/ejss.12267.

Laflen, J. M., L. J. Lane, and G. R. Foster (1991), WEPP: A new generation of erosion prediction technology, *Journal of Soil and Water Conservation*, 46(1), 34–38.

Lague, D. (2014), The stream power river incision model: Evidence, theory and beyond, *Earth Surface Processes and Landforms*, 39, 38–61, doi:10.1002/esp.3462.

Lague, D., A. Crave, and P. Davy (2003), Laboratory experiments simulating the geomorphic response to tectonic uplift, *Journal of Geophysical Research (Solid Earth)*, 108(B1), 2008, doi:10.1029/2002JB001785.

Lague, D., N. Hovius, and P. Davy (2005), Discharge, discharge variability, and the bedrock channel profile, *Journal of Geophysical Research (Earth Surface)*, 110, F04006, doi:10.1029/2004JF000259.

Lal, R. (2003), Soil erosion and the global carbon budget, *Environment International*, 29, 437–450, doi:10.1016/S0160-4120(02)00192-7.

Lamb, M. P., W. E. Dietrich, and L. S. Sklar (2008), A model for fluvial bedrock incision by impacting suspended and bed load sediment, *Journal of Geophysical Research (Earth Surface)*, 113, F03025, doi:10.1029/2007JF000915.

Lamb, M. P., N. J. Finnegan, J. S. Scheingross, and L. S. Sklar (2016), New insights into the mechanics of fluvial bedrock erosion through flume experiments and theory, *Geomorphology*, 244, 33–55, doi:10.1016/j.geomorph.2015.03.003.

Lambeck, K., and P. Johnston (1998), The viscosity of the Mantle: Evidence from analyses of glacial-rebound phenomena, in *The Earth's Mantle: Composition, structure and evolution*, edited by I. Jackson, pp. 461–502, Cambridge University Press, Cambridge.

Lambeck, K., H. Rouby, A. Purcell, Y. Sun, and M. Sambridge (2014), Sea level and global ice volumes from the Last Glacial Maximum to the Holocene, *Proceedings of the National Academy of Sciences U.S.A*, 111 (43), 15296–15303, doi:10.1073/pnas.1411762111.

Lambers, H., F. S. Chapin, and T. L. Pons (2008), *Plant physiological ecology*, 2nd ed., Springer, New York.

Lancaster, S. T., and R. L. Bras (2002), A simple model of river meandering and its comparison to natural channels, *Hydrological Processes*, 16(1), 1–26.

Lane, E. W. (1955), Design of stable channels, *Transactions of the American Society of Civil Engineers*, 120(1), 1234–1260.

Lane, L. J., E. D. Shirley, and V. P. Singh (1988), Modelling erosion on hillslopes, in *Modelling geomorphological systems*, edited by M. G. Anderson, pp. 287–308, John Wiley & Sons, New York.

Lane, P. N. J., G. J. Sheridan, and P. J. Noske (2006), Changes in sediment loads and discharge from small mountain catchments following wildfire in south eastern Australia, *Journal of Hydrology*, 331, 495–510, doi:10.1016/j.jhydrol.2006.05.035.

Langbein, W. B., and S. A. Schumm (1958), Yield of sediment in relation to mean annual precipitation, *Transactions of the American Geophysical Union*, 30(6), 1076–1084, doi:10.1029/ TR039i006p01076.

Langhans, C., H. G. Smith, D. M. O. Chong, P. Nyman, P. N. J. Lane, and G. J. Sheridan (2016), A model for assessing water quality risk in catchments prone to wildfire, *Journal of Hydrology*, 534, 407–426, doi:10.1016/j.jhydrol.2015.12.048.

Lapotre, M. G. A., M. P. Lamb, and R. M. E. Williams (2016), Canyon formation constraints on the discharge of catastrophic outburst floods of Earth and Mars, *Journal of Geophysical Research (Planets)*, 121, doi:10.1002/2016JE005061.

Larcher, W. (2003), *Physiological plant ecology: Ecophysiology and stress physiology of functional groups*, Springer-Verlag, Berlin.

Larsen, I. J., D. R. Montgomery, and O. Korup (2010), Landslide erosion controlled by hillslope material, *Nature Geoscience*, 3, 247–251, doi:10.1038/NGEO776.

Larsen, L. G., M. B. Eppinga, P. Passalacqua, W. M. Getz, K. A. Rose, and M. Liang (2016), Appropriate complexity landscape modeling, *Earth-Science Reviews*, 160, 111–130, doi:10.1016/j.earscirev.2016.06.016.

Lasaga, A. C. (1984), Chemical kinetics of water-rock interactions, *Journal of Geophysical Research*, 89(B6), 4009–4025, doi:10.1029/JB089iB06p04009.

Lauer, J. W., and G. Parker (2004), Modeling channel-floodplain co-evolution in sand-bed streams, in ASCE World Water and Environmental Resources 2004 Congress, 27 June–1 July, p. 10, ASCE, Salt Lake City.

Lavier, L. L., and W. R. Buck (2002), Half graben versus large-offset low-angle normal fault: Importance of keeping cool during normal faulting, *Journal of Geophysical Research (Solid Earth)*, 107(B6), *ETG* 8–1, doi:10.1029/2001JB000513.

Lavier, L. L., W. R. Buck, and A. N. B. Poliakov (2000), Factors controlling normal fault offset in an ideal brittle layer, *Journal of Geophysical Research (Solid Earth)*, 105(B10), 23431–23442.

LeB. Hooke, R. (2005), *Principles of glacier mechanics*, Cambridge University Press, Cambridge.

Lebedeva, M. I., R. C. Fletcher, V. N. Balashov, and S. L. Brantley (2007), A reactive diffusion model describing transformation of bedrock to saprolite, *Chemical Geology*, 244, 624–645, doi:10.1016/j.chemgeo.2007.07.008.

Legates, D. R., R. Mahmood, D. F. Levia, T. L. DeLiberty, S. M. Quiring, C. Houser, and F. E. Nelson (2010), Soil moisture: A central and unifying theme in physical geography, *Progress in Physical Geography*, 35(1), 65–86, doi:10.1177/0309133310386514.

Legros, J. P., and G. Pedro (1985), The causes of particle-size distribution in soil profiles derived from crystalline rocks, France, *Geoderma*, 36(1), 15–25.

Lehsten, V., A. Arneth, A. Spessa, K. Thonicke, and A. Moustakas (2016), The effect of fire on tree–grass coexistence in savannas: A simulation study, *International Journal of Wildland Fire*, 25, 137–146, doi:10.1071/WF14205.

Leithold, E. L., N. E. Blair, and K. W. Wegmann (2016), Source-to-sink sedimentary systems and global carbon burial: A river runs through it, *Earth-Science Reviews*, 153, 30–42, doi:10.1016/j.earscirev.2015.10.011.

Leopold, L. B., M. G. Wolman, and J. P. Miller (1964), *Fluvial processes in geomorphology*, Freeman, London.

Li, F., C. Dyt, and C. Griffiths (2004), 3D modelling of flexural isostatic deformation, *Computers & Geosciences*, 30, 1105–1115, doi:10.1016/j.cageo.2004.08.005.

Li, Z., L. Liu, J. Chen, and H. H. Teng (2016), Cellular dissolution at hypha- and spore-mineral interfaces revealing unrecognized mechanisms and scales of fungal weathering, *Geology*, 44(4), 319–322, doi:10.1130/G37561.1.

Little, D. A., J. B. Field, and S. A. Welch (2005), Metal Dissolution from Rhizosphere and non-Rhizosphere soils using low molecular weight organic acids, paper presented at Regolith 2005: Ten Years of CRC LEME, Cooperative Research Centre for Landscape Environments and Mineral Exploration (CRC LEME), Canberra.

Llovet, J., M. Ruiz-Valera, R. Josa, and V. R. Vallejo (2009), Soil responses to fire in Mediterranean forest landscapes in relation to the previous stage of land abandonment, *International Journal of Wildland Fire*, 18, 222–232, doi:10.1071/WF07089.

Lobry de Bruyn, L. A., and A. J. Conacher (1990), The role of termites and ants in soil modification: A review, *Australian Journal of Soil Research*, 28(1), 55–93.

Lobry de Bruyn, L. A., and A. J. Conacher (1994), The bioturbation activity of ants in agricultural and naturally vegetated habitats in semi-arid environments, *Australian Journal of Soil Research*, 32, 555–570, doi:10.1071/SR9940555.

Lockart, N., G. R. Willgoose, G. Kuczera, A. S. Kiem, A. F. M. K. Chowdhury, N. P. Manage, L. Zhang, and C. Twomey (2016), Case study on the use of dynamically downscaled GCM data for assessing water security on coastal NSW, *Journal of Southern Hemisphere Earth Systems Science*, 66(2), 177–202.

Loewenherz-Lawrence, D. S. (1994), Theoretical constraints on the development of surface rills: Mode shapes, amplitude limitations and implications for nonlinear evolution, in *Process Models and Theoretical Geomorphology*, edited by M. J. Kirkby, pp. 315–334, Wiley, Chichester, UK.

Lopéz, F., and M. García (1998), Open-channel flow through simulated vegetation: Suspended sediment transport modeling, *Water Resources Research*, 34(9), 2341–2352, doi:10.1029/98WR01922.

Lucas, Y. (2001), The role of plants in controlling rates and products of weathering: Importance of biological pumping, *Annual Review of Earth and Planetary Sciences* , 29, 135–163.

Luchi, R., J. M. Hooke, G. Zolezzi, and W. Bertoldi (2010), Width variations and mid-channel bar inception in meanders: River Bollin (UK), *Geomorphology*, 119, 1–8, doi:10.1016/j.geomorph.2010.01.010.

Ludwig, J. A., D. J. Tongway, and S. G. Marsden (1999), Stripes, strands or stipples: Modelling the influence of three landscape banding patterns on resource capture and productivity in semi-arid woodlands, Australia, *Catena*, 37(1–2), 257–273.

Ludwig, J. A., B. P. Wilcox, D. D. Breshears, D. J. Tongway, and A. C. Imeson (2005), Vegetation patches and runoff-erosion as interacting eco-hydrological processes in semiarid landscapes, *Ecology*, 86(2), 288–297.

Lugato, E., F. Bampa, P. Panagos, L. Montanarella, and A. Jones (2015), Potential carbon sequestration of European arable soils estimated by modelling a comprehensive set of management practices, *Global Change Biology*, 20, 3557–3567, doi:10.1111/gcb.12551.

Lugato, E., P. Panagos, F. Bampa, A. Jones, and L. Montanarella (2014), A new baseline of organic carbon stock in European agricultural soils using a modelling approach, *Global Change Biology*, 20, 313–316, doi:10.1111/gcb.12292.

Lugato, E., K. Paustian, P. Panagos, A. Jones, and P. Borrelli (2016), Quantifying the erosion effect on current carbon budget of European agricultural soils at high spatial resolution, *Global Change Biology*, 22, 1976–1984, doi:10.1111/gcb.13198.

Lynch, A. H., J. Beringer, P. Kershaw, A. Marshall, S. Mooney, N. Tapper, C. Turney, and S. Van Der Kaars (2007), Using the paleorecord to evaluate climate and fire interactions in Australia, *Annual Review of Earth and Planetary Sciences*, 35, 215–239, doi:10.1146/annurev.earth.35.092006.145055.

McBratney, A. B., I. O. A. Odeh, T. F. A. Bishop, M. S. Dunbar, and T. M. Shatar (2000), An overview of pedometric techniques for use in soil survey, *Geoderma*, 97(3–4), 293–327.

McBratney, A. B., M. L. M. Santos, and B. Minasny (2003), On digital soil mapping, *Geoderma*, 117(1–2), 3–52, doi:10.1016/S0016-7061(03)00223-4.

McCarthy, M. A., A. M. Gill, and R. A. Bradstock (2001), Theoretical fire-interval distributions, *International Journal of Wildland Fire*, 10(1), 73–77, doi:10.1071/wf01013.

McFadden, L. D., and P. L. K. Knuepfer (1990), Soil geomorphology: The linkage of pedology and surficial processes, *Geomorphology*, 3(3/4), 197–205, doi:10.1016/0169-555X(90)90003-9.

McGuire, L. A., J. D. Pelletier, J. A. Gomez, and M. A. Nearing (2013), Controls on the spacing and geometry of rill networks on hillslopes: Rain splash detachment, initial hillslope roughness, and the competition between fluvial and colluvial transport, *Journal of Geophysical Research (Earth Surface)*, 118, 241–256, doi:10.1002/jgrf.20028.

McKenzie, B. M., and A. R. Dexter (1993), Size and orientation of burrows made by the earthworms Aporrectodea-Rosea and a-Caliginosa, *Geoderma*, 56(1–4), 233–241.

McKinnon, W. B., et al. (2016), Convection in a volatile nitrogen-ice-rich layer drives Pluto's geological vigour, *Nature*, 534(7605), 82–85, doi:10.1038/nature18289.

McVicar, T. R., T. G. Van Niel, L. T. Li, M. F. Hutchinson, X. M. Mu, and Z. H. Liu (2007), Spatially distributing monthly reference evapotranspiration and pan evaporation considering topographic influences, *Journal of Hydrology*, 338, 196–220, doi:10.1016/j.jhydrol.2007.02.018.

Maestre, F. T., et al. (2015), Increasing aridity reduces soil microbial diversity and abundance in global drylands, *Proceedings of the National Academy of Sciences U.S.A*, 112(51), 15684–15689, doi:10.1073/pnas.1516684112.

Maher, K. (2010), The dependence of chemical weathering rates on fluid residence time, *Earth and Planetary Science Letters*, 294(1–2), 101–110. doi:10.1016/j.epsl.2010.03.010.

Maher, K., C. I. Steefel, A. F. White, and D. A. Stonestrom (2009), The role of reaction affinity and secondary minerals in regulating chemical weathering rates at the Santa Cruz Soil Chronosequence, California, *Geochimica et Cosmochimica Acta*, 73, 2804–2831, doi:10.1016/j.gca.2009.01.030.

Major, J. J., and R. M. Iverson (1999), Debris-flow deposition: Effects of pore-fluid pressure and friction concentrated at flow margins, *Geological Society of America Bulletin*, 111(10), 1424–1434.

Malamoud, K., A. B. McBratney, B. Minasny, and D. J. Field (2009), Modelling how carbon affects soil structure, *Geoderma*, 149(1–2), 19–26, doi:10.1016/j.geoderma.2008.10.018.

Malamud, B. D., J. D. A. Millington, and G. L. W. Perry (2005), Characterizing wildfire regimes in the Unites States, *Proceedings of the National Academy of Sciences U.S.A*, 102(13), 4694–4699, doi:10.1073/pnas.0500880102.

Malamud, B. D., G. Morein, and D. L. Turcotte (1998), Forest fires: An example of self-organized critical behavior, *Science*, 281(5384), 1840–1842.

Malamud, B. D., D. L. Turcotte, F. Guzzetti, and P. Reichenback (2004), Landslide inventories and their statistical properties, *Earth Surface Processes and Landforms*, 29, 687–711, doi:10.1002/esp.1064.

Mann, M. E., R. S. Bradley, and M. K. Hughes (1999), Northern hemisphere temperatures during the past millennium: Inferences, uncertainties, and limitations, *Geophysical Research Letters*, 26(6), 759–762, doi:10.1029/1999GL900070.

Mao, L., J. R. Cooper, and L. E. Frostick (2011), Grain size and topographical differences between static and mobile armour layers, *Earth Surface Processes*, 36, 1321–1334, doi:10.1002/esp.2156.

Markgraf, W., C. W. Watts, W. R. Whalley, T. Hrkac, and R. Horn (2012), Influence of organic matter on rheological properties of soil, *Applied Clay Science*, 64, 25–33, doi:10.1016/j.clay.2011.04.009.

Marshall, J. S., and W. M. Palmer (1948), The distribution of raindrops with size, *Journal of Meteorology*, 5, 165–166.

Martin, Y. E. (2007), Wildfire disturbance and shallow landsliding in coastal British Columbia over millennial time scales: A numerical modelling study, *Catena*, 69(3), 206–219, doi:10.1016/j.catena.2006.05.006.

Martinez-Casasnovas, J. A. (2003), A spatial information technology approach for the mapping and quantification of gully erosion, *Catena*, 50 (2–4), 293–308.

Mataix-Solera, J., I. Gómez, J. Navarro-Pedreño, C. Guerrero, and R. Moral (2002), Soil organic matter and aggregates affected by wildfire in a Pinus halepensis forest in a Mediterranean environment, *International Journal of Wildland Fire*, 11, 107–114, doi:10.1071/WF02020.

Matsuoka, N., and K. Moriwaki (1992), Frost heave and creep in the Sør Rondane Mountains, Antarctica, *Arctic and Alpine Research*, 24(4), 271–280.

Mead, S. R., and P. W. Cleary (2015), Validation of DEM prediction for granular avalanches on irregular terrain, *Journal of Geophysical Research (Earth Surface)*, 120, 1724–1742, doi:10.1002/2014JF003331.

Mein, R. G., E. M. Laurenson, and T. A. McMahon (1976), Simple nonlinear model for flood estimation, *Journal of Hydraulic Division – ASCE* 100(NHY11), 1507–1518.

Meisina, C., and S. Scarabelli (2007), A comparative analysis of terrain stability models for predicting shallow landslides in colluvial soils, *Geomorphology*, 87, 207–223, doi:10.1016/j.geomorph.2006.03.039.

Melini, D., P. Gegout, G. Spada, and M. A. King (2015), REAR: A Regional Elastic Rebound calculator, GitHub.

Mesa, O. J. (1986), Analysis of channel networks parameterized by elevation, PhD thesis, University of Mississippi.

Metherell, A. K., L. A. Harding, C. V. Cole, and W. J. Parton (1994), CENTURY Soil organic matter environment, Technical Documentation Agrosystem, Version 4.0, Great Plains System Research Unit Technical Report No. 4, USDA-ARS, Fort Collins, CO.

Michaelides, K., and G. J. Martin (2012), Sediment transport by runoff on debris-mantled dryland hillslopes, *Journal of Geophysical Research (Earth Surface)*, 117, F03014, doi:10.1029/2012JF002415.

Michaelides, K., and M. B. Singer (2014), Impact of coarse sediment supply from hillslopes to the channel in runoff-dominated, dryland fluvial systems, *Journal of Geophysical Research (Earth Surface)*, 119, 1205–1221, doi:10.1002/2013JF002959.

Migon, P., and M. F. Thomas (2002), Grus weathering mantles – Problems of interpretation, *Catena*, 49(1–2), 5–24.

Millar, R. G. (2005), Theoretical regime equations for mobile gravel-bed rivers with stable banks, *Geomorphology*, 64(3–4), 207–220, doi:10.1016/j.geomorph.2004.07.001.

Milledge, D. G., D. Bellugi, J. A. McKean, A. L. Densmore, and W. E. Dietrich (2014), A multidimensional stability model for predicting shallow landslide size and shape across landscapes, *Journal of Geophysical Research (Earth Surface)*, 119, 2481–2504, doi:10.1002/2014JF003135.

Minasny, B., P. Finke, U. Stockmann, T. Vanwalleghem, and A. B. McBratney (2015), Resolving the integral connection between pedogenesis and landscape evolution, *Earth-Science Reviews*, 150, 102–120, doi:10.1016/j.earscirev.2015.07.004.

Minasny, B., and A. B. McBratney (1999), A rudimentary mechanistic model for soil production and landscape development, *Geoderma*, 90(1–2), 3–21, doi:10.1016/S0016-7061(98)00115-3.

Minasny, B., and A. B. McBratney (2001), A rudimentary mechanistic model for soil formation and landscape development II. A two-dimensional model incorporating chemical weathering, *Geoderma*, 103 (1–2), 161–179, doi:10.1016/S0016-7061(01)00075-1.

Minasny, B., and A. B. McBratney (2006), Mechanistic soil–landscape modelling as an approach to developing pedogenetic classifications, *Geoderma*, 133(1–2), 138–149, doi:10.1016/j.geoderma.2006.03.042.

Minasny, B., A. B. McBratney, and S. Salvador-Blanes (2008), Quantitative models for pedogenesis – A review, *Geoderma*, 144(1–2), 140–157, doi:10.1016/j.geoderma.2007.12.013.

Meriam, J. L., L. G. Kraige, and J. N. Bolton (2012), *Engineering mechanics: Statics*, Wiley, Chichester, UK.

Mitchell, A., and O. Hungr (2017), Theory and calibration of the Pierre 2 stochastic rock fall dynamics simulation program, *Canadian Geotechnical Journal*, 54(1), 18–30, doi:10.1139/cgj-2016-0039.

Mitchell, P. B. (1985), Some aspects of the role of bioturbation in soil formation in south-eastern Australia, PhD thesis, Macquarie University.

Mitchell, S. G., and E. E. Humphries (2015), Glacial cirques and the relationship between equilibrium line altitudes and mountain range height, *Geology*, 43(1), 35–38, doi:10.1130/G36180.1.

Mitrovica, J. X., and A. M. Forte (2004), A new inference of mantle viscosity based upon joint inversion of convection and glacial isostatic adjustment data, *Earth and Planetary Science Letters*, 225(1–2), 177–189, doi:10.1016/j.epsl.2004.06.005.

Moglen, G. E., and R. L. Bras (1995), The importance of spatially heterogeneous erosivity and the Cumulative Area Distribution within a basin evolution model, *Geomorphology*, 12(3), 173–185.

Moglen, G. E., E. A. B. Eltahir, and R. L. Bras (1998), On the sensitivity of drainage density to climate change, *Water Resources Research*, 34(4), 855–862.

Montelli, R., G. Noilet, F. A. Dahlen, G. Masters, E. R. Engdahl, and S. H. Hung (2004), Finite-frequency tomography reveals a variety of plumes in the mantle, *Science*, 303(5656), 338–343, doi:10.1126/science.1092485.

Montgomery, D. R., and W. E. Dietrich (1988), Where do channels begin?, *Nature*, 336, 232–234.

Montgomery, D. R., and W. E. Dietrich (1989), Source areas, drainage density and channel initiation, *Water Resources Research*, 25(8), 1907–1918, doi:10.1029/WR025i008p01907.

Montgomery, D. R., and W. E. Dietrich (1994), A physically based model for the topographic control on shallow landsliding, *Water Resources Research*, 30(4), 1153–1171.

Montgomery, D. R., and E. Foufoula-Georgiou (1993), Channel network source representation using digital elevation models, *Water Resources Research*, 29(12), 3925–3935, doi:10.1029/93WR02463.

Montgomery, D. R., K. M. Schmidt, W. E. Dietrich, and J. McKean (2009), Instrumental record of debris flow initiation during natural rainfall: Implications for modeling slope stability, *Journal of Geophysical Research (Earth Surface)*, 114, F01031, doi:10.1029/2008JF001078.

Moody, J. A., R. A. Shakesby, P. R. Robichaud, S. H. Cannon, and D. A. Martin (2013), Current research issues related to post-wildfire runoff and erosion processes, *Earth-Science Reviews*, 122, 10–37, doi:10.1016/j.earscirev.2013.03.004.

Moore, J. G. (1987), Subsidence of the Hawaiian Ridge, in Volcanism in Hawaii, edited by R. W. Decker, T. L. Wright, and P. H. Stauffer, USGS Professional Paper 1350.

Moore, J. M., A. D. Howard, and A. M. Morgan (2014), The landscape of Titan as witness to its climate evolution, *Journal of Geophysical Research (Planets)*, 119(9), 2060–2077, doi:10.1002/2014JE004608.

Moreno de Las Heras, M., P. M. Saco, and G. R. Willgoose (2012), A comparison of SRTM V4 and ASTER GDEM for hydrological applications in low relief terrain, *Photogrammetric Engineering and Remote Sensing*, 78(7), 757–766.

Moreno de Las Heras, M., P. M. Saco, G. R. Willgoose, and D. J. Tongway (2012), Variations in hydrological connectivity of Australian semiarid landscapes indicate abrupt changes in ecosystem functionality, *Journal of Geophysical Research (Biogeosciences)*, 117, G03009, doi:10.1029/2011JG001839.

Morin, R. H. (2005), Negative correlation between porosity and hydraulic conductivity in sand-and-gravel aquifers at Cape Cod, Massachusetts, USA, *Journal of Hydrology*, 316, 43–52, doi:10.1016/j.jhydrol.2005.04.013.

Mudd, S. M., and D. J. Furbish (2004), Influence of chemical denudation on hillslope morphology, *Journal of Geophysical Research (Earth Surface)*, 109, F02001, doi:10.1029/2003JF000087.

Munson, B. R., D. F. Young, and T. H. Okiishi (1998), *Fundamentals of fluid mechanics*, Wiley, Chichester, UK.

Murphy, B., J. Russell-Smith, and L. Prior (2010), Frequent fires reduce tree growth in northern Australian savannas: Implications for tree demography and carbon sequestration, *Global Change Biology*, 16(1), 331–343, doi:10.1111/j.1365-2486.2009.01933.x.

Murray, A. B., and C. Paola (2003), Modelling the effect of vegetation on channel pattern in bedload rivers, *Earth Surface Processes and Landforms*, 28(2), 131–143.

National Academy of Science (NAS) (2010), *Landscapes on the edge: New horizons for research of earth's surface*, National Academies Press, Washington, DC.

Navarre-Stichler, A. K., D. R. Cole, G. Rother, L. Jin, H. L. Buss, and S. L. Brantley (2013), Porosity and surface area evolution during weathering of two igneous rocks, *Geochimica et Cosmochimica Acta*, 109, 400–413, doi:10.1016/j.gca.2013.02.012.

Nayyar, H., and D. Gupta (2006), Differential sensitivity of C_3 and C_4 plants to water deficit stress: Association with oxidative stress and antioxidants, *Environmental and Experimental Botany*, 58(1–3), 106–113, doi:10.1016/j.envexpbot.2005.06.021.

Nearing, M. A., G. R. Foster, L. J. Lane, and S. C. Finkner (1989), A process-based soil erosion model for USDA-Water Erosion Prediction Project Technology, *American Society of Agricultural Engineers*, 32(5), 1587–1593.

Nelson, P. A., J. G. Vendetti, W. E. Dietrich, J. W. Kirchner, H. Ikeda, F. Iseya, and L. S. Sklar (2009), Response of bed surface patchiness to reductions in sediment supply, *Journal of Geophysical Research (Earth Surface)*, 114, F02005, doi:10.1029/2008JF001144.

Nepf, H. (2012), Hydrodynamics of vegetated channels, *Journal of Hydraulic Research*, 50(3), 262–279, doi:10.1080/00221686.2012.696559.

Nesbitt, S. W., and A. M. Anders (2009), Very high resolution precipitation climatologies from the Tropical Rainfall Measuring Mission precipitation radar, *Geophysical Research Letters*, 36, L15815, doi:10.1029/2009GL038026.

Nicholas, A. P., and T. A. Quine (2010), Quantitative assessment of landform equifinality and palaeoenvironmental reconstruction using geomorphic models, *Geomorphology*, 121, 167–183, doi:10.1016/j.geomorph.2010.04.004.

Niklasson, M., and A. Granström (2000), Numbers and sizes of fires: Long-term spatially explicit fire history in a Swedish boreal landscape, *Ecology*, 81(6), 1484–1499.

Nimmo, J. R., and K. S. Perkins (2002), Aggregate stability and size distribution, in *Methods of soil analysis, Part 4 – Physical methods*, edited by J. H. Dane and G. C. Topp, pp. 317–328, Soil Science Society of America, Madison, WI.

Nolan, R. H., P. N. J. Lane, R. G. Benyon, R. A. Bradstock, and P. J. Mitchell (2014a), Changes in evapotranspiration following wildfire in resprouting eucalypt forests, *Ecohydrology*, 7, 1363–1377, doi:10.1002/eco.1463.

Nolan, R. H., P. N. J. Lane, R. G. Benyon, R. A. Bradstock, and P. J. Mitchell (2015), Trends in evapotranspiration and streamflow following wildfire in resprouting eucalypt forests, *Journal of Hydrology*, 524, 614–624, doi:10.1016/j.jhydrol.2015.02.045.

Nolan, R. H., P. J. Mitchell, R. A. Bradstock, and P. N. J. Lane (2014b), Structural adjustments in resprouting trees drive differences in post-fire transpiration, *Tree Physiology*, 34(2), 123–136, doi:10.1093/treephys/tpt125.

North Greenland Ice Core Project (NGRIP) (2004a), High-resolution record of Northern Hemisphere climate extending into the last interglacial period, *Nature*, 431(7005), 147–151.

North Greenland Ice Core Project (NGRIP) (2004b), North Greenland Ice Core Project Oxygen Isotope Data, NOAA/NGDC Paleoclimatology Program, Boulder CO, USA, IGBP PAGES/World Data Center for Paleoclimatology Data Contribution Series no. 2004-059.

Noy-Meir, I. (1973), Desert ecosystems: Environment and producers, *Annual Review of Ecology and Systematics*, 4, 25–51.

NRCCA (2016), Northeast Region Certified Crop Advisor (NRCCA) Study Resources, Cornell University, Ithaca, NY, http://nrcca.cals.cornell.edu/soil/CA2/CA0212.1–3.php.

Nyman, P., G. J. Sheridan, and P. N. J. Lane (2013), Hydro-geomorphic response models for burned areas and their applications in land management, *Progress in Physical Geography*, 37(6), 787–812, doi:10.1177/0309133313508802.

Nyman, P., G. J. Sheridan, J. A. Moody, H. G. Smith, P. J. Noske, and P. N. J. Lane (2013), Sediment availability on burned hillslopes, *Journal of Geophysical Research (Earth Surface)*, 118, 2451–2467, doi:10.1002/jgrf.20152.

Nyman, P., G. J. Sheridan, H. G. Smith, and P. N. J. Lane (2011), Evidence of debris flow occurrence after wildfire in upland catchments of south-east Australia, *Geomorphology*, 125, 383–401, doi:10.1016/j.geomorph.2010.10.016.

O'Brien, B. J., and J. D. Stout (1978), Movement and turnover of soil organic matter as indicated by carbon isotope measurements, *Soil Biology & Biochemistry*, 10(4), 309–317, doi:10.1016/0038-0717(78)90028-7.

O'Callaghan, J. F., and D. M. Mark (1984), The extraction of drainage networks from digital elevation data, *Computer Vision, Graphics and Image Processing*, 28(3), 323–344, doi: 10.1016/S0734-189X(84)80011-0.

O'Donnell, A. J., M. M. Boer, W. L. McCaw, and P. F. Grierson (2011), Climatic anomalies drive wildfire occurrence and extent in semi-arid shrublands and woodlands of southwest Australia, *Ecosphere*, 2(11), 127, doi:10.1890/ES11-00189.1.

O'Donnell, A. J., M. M. Boer, W. L. McCaw, and P. F. Grierson (2014), Scale-dependent thresholds in the dominant controls of wildfire size in semi-arid southwest Australia, *Ecosphere*, 5(7), 93, doi:10.1890/ES14-00145.1.

Ogawa, M. (2008), Mantle convection: A review, *Fluid Dynamics Research*, 40, 379–398, doi:10.1016/j.fluiddyn.2007.09.001.

Olive, J.-A., M. D. Behn, and L. C. Malatesta (2014), Modes of extensional faulting controlled by surface processes, *Geophysical Research Letters*, 41, 6725–6733, doi:10.1002/2014GL061507.

Ollier, C. D., and C. F. Pain (1996), *Regolith, soils and landforms*, Wiley, Chichester, UK.

O'Reilly, S. Y., W. L. Griffin, Y. H. Poudjom Djomani, and P. Morgan (2001), Are lithospheres forever? Tracking changes in subcontinental lithospheric mantle through time, *GSA Today*, 11(4), 4–10.

Ouimet, R. (2008), Using compositional change within soil profiles for modelling base cation transport and chemical weathering, *Geoderma*, 145 (3–4), 410–418, doi:10.1016/j.geoderma.2008.01.007.

Oyama, T., and M. Chigara (1999), Weathering rate of mudstone and tuff on old unlined tunnel walls, *Engineering Geology*, 55(1), 15–27, doi:10.1016/S0013-7952(99)00103-9.

Pack, R. T., D. G. Tarboton, and C. N. Goodwin (1998), The SINMAP approach to terrain stability mapping, in *Eighth International Congress of the International Association for Engineering Geology and the Environment*, edited by D. Moore and O. Hungr, pp. 1157–1165, Vancouver, 21–25 September.

Panagos, P., P. Borrelli, K. Meusburger, C. Alewell, E. Lugato, and L. Montanarella (2015), Estimating the soil erosion cover-management factor at the European scale, *Land Use Policy*, 48, 38–50, doi:10.1016/j.landusepol.2015.05.021.

Pandey, S., and H. Rajaram (2016), Modeling the influence of preferential flow on the spatial variability and time-dependence of mineral weathering rates, *Water Resources Research*, 52, doi:10.1002/2016WR019026.

Parana Manage, N. (2017), Testing the hydrologic validity of downscaled rainfall data for water security assessment, PhD thesis, University of Newcastle, Callaghan, Australia.

Pardini, G. (2003), Fractal scaling of surface roughness in artificially weathered smectite-rich soil regoliths, *Geoderma*, 117(1–2), 157–167.

Parizek, J. R., and G. H. Girty (2014), Assessing volumetric strains and mass balance relationships resulting from biotite-controlled weathering: Implications for the isovolumetric weathering of the Boulder Creek Granodiorite, Boulder County, Colorado, USA, *Catena*, 120, 29–45, doi:10.1016/j.catena.2014.03.019.

Parker, G. (1990), Surface-based bedload transport relation for gravel rivers, *Journal of Hydraulic Research*, 28(4), 417–436.

Parker, G., and P. C. Klingeman (1982), On why gravel beds streams are paved, *Water Resources Research*, 18(5), 1409–1423, doi:10.1029/WR018i005p01409.

Parker, G., K. Sawai, and S. Ikeda (1982), Bend theory of river meanders: Part II, *Nonlinear deformation of finite amplitude bends,* Journal of Fluid Mechanics, 115, 303–314.

Parker, G., Y. Shimizu, G. V. Wilkerson, E. C. Eke, E. C. Abad, J. D. Lauer, C. Paola, W. E. Dietrich, and V. R. Voller (2011), A new framework for modeling the migration of meandering rivers, *Earth Surface Processes and Landforms*, 36, 70–86, doi:10.1002/esp.2113.

Parker, G., P. R. Wilcock, C. Paola, W. E. Dietrich, and J. Pitlick (2007), Physical basis for quasi-universal relations describing bankfull hydraulic geometry of single-thread gravel bed rivers, *Journal of Geophysical Research (Earth Surface)*, 112, F04005, doi:10.1029/2006JF000549.

Parsons, A. J., A. D. Abrahams, and S. H. Luk (1990), Hydraulics of interrill overland flow on a semi-arid hillslope, southern Arizona, *Journal of Hydrology*, 117, 255–273.

Parsons, A. J., A. D. Abrahams, and J. R. Simanton (1992), Microtopography and soil-surface materials on semi-arid piedmont hillslopes, southern Arizona, *Journal of Arid Environments*, 22, 107–115.

Parsons, A. J., J. Wainwright, A. D. Abrahams, and J. R. Simanton (1997), Distributed dynamic modelling of interrill overland flow, *Hydrological Processes*, 11(14), 1833–1859.

Parton, W. J., J. W. B. Stewart, and C. V. Cole (1988), Dynamics of C, N, P and S in grassland soils: A model, *Biogeochemistry*, 4(1), 109–131, doi:10.1007/BF02180320.

Paton, T. R., G. S. Humphreys, and P. B. Mitchell (1995), *Soils: A new global view*, CRC Press, Boca Raton, FL.

Paulson, A., S. Zhong, and J. Wahr (2007), Inference of mantle viscosity from GRACE and relative sea level data, *Geophysical Journal International*, 171, 497–508, doi:10.1111/j.1365-246X.2007.03556.x.

Peakall, J., and E. J. Sumner (2015), Submarine channel flow processes and deposits: A process-product perspective, *Geomorphology*, 244, 95–120, doi:10.1016/j.geomorph.2015.03.005.

Peckham, S. D., E. W. H. Hutton, and B. Norris (2013), A component-based approach to integrated modeling in the geosciences: The design of CSDMS, *Computers & Geosciences*, 53, 3–12, doi:10.1016/j.cageo.2012.04.002.

Peltier, W. R. (1998), Postglacial variations in the level of the sea: Implications for climate dynamics and soil-earth geophysics, *Reviews of Geophysics*, 36(4), 441–500.

Peltier, W. R. (1999), Global sea level rise and glacial isostatic adjustment, *Global Planetary Change*, 20, 93–123.

Peltier, W. R. (2004), Global glacial isostasy and the surface of the ice-age Earth: The ICE-5G (VM2) model and GRACE, *Annual Reviews of Earth and Planetary Sciences*, 32, 111–149, doi:10.1146/annurev.earth.32.082503.144359.

Peng, J., L. Dan, and M. Huang (2014), Sensitivity of global and regional terrestrial carbon storage to the direct CO_2 effect and climate change based on the CMIP5 model intercomparison, *PLoS ONE*, 9(4), e95282, doi:10.1371/journal.pone.0095282.

Penman, T. D., R. A. Bradstock, and O. Price (2013), Modelling the determinants of ignition in the Sydney Basin, Australia: Implications for future management, *International Journal of Wildland Fire*, 22(4), 469–478, doi:10.1071/wf12027.

Perera, H. J., and G. R. Willgoose (1998), A physical explanation of the cumulative area diagram, *Water Resources Research*, 34(5), 1335–1345.

Perez, L., and S. Dragicevic (2012), Landscape-level simulation of forest insect disturbance: Coupling swarm intelligent agents with GIS-based cellular automata model, *Ecological Modelling*, 231, 53–64, doi:10.1016/j.ecolmodel.2012.01.020.

Perron, J. T. (2011), Numerical methods for nonlinear hillslope transport laws, *Journal of Geophysical Research (Earth Surface)*, 116, F02021, doi:10.1029/2010JF001801.

Perron, J. T., and L. H. Royden (2013), An integral approach to bedrock river profile analysis, *Earth Surface Processes and Landforms*, 38, 570–576, doi:10.1002/esp.3302.

Peterman, W., and D. Bachelet (2012), Climate change and forest dynamics: A soils perspective, in *Soils and food security*, edited by R. E. Hester and R. M. Harrison, pp. 158–182, Royal Society of Chemistry, London.

Peterman, W., D. Bachelet, K. Ferschweiler, and T. Sheehan (2014), Soil depth affects simulated carbon and water in the MC2 dynamic global vegetation model, *Ecological Modelling*, 294, 84–93, doi:10.1016/j.ecolmodel.2014.09.025.

Petersen, J. F., D. Sack, and R. E. Gabler (2012), *Physical geography*, 10th ed., Brooks/Cole, Belmont, CA.

Petit, J. R., et al. (1999a), Climate and atmospheric history of the past 420,000 years from the Vostok ice core, Antarctica, *Nature*, 399(6735), 429–436, doi:10.1038/20859.

Petit, J. R. (1999b), Vostok Ice Core Data for 420,000 Years, NOAA/NGDC Paleoclimatology Program, Boulder CO, IGBP PAGES/World Data Center for Paleoclimatology Data Contribution Series #2001-076.

Pfister, L., and J. W. Kirchner (2017), Debates – Hypothesis testing in hydrology: Theory and practice, *Water Resources Research*, 53, doi:10.1002/2016WR020116.

Philip, J. R. (1991), Hillslope infiltration, *Water Resources Research*, 27(1), 109–117.

Phillips, J. D. (1993), Stability implications of the state factor model of soils as a nonlinear dynamical system, *Geoderma*, 58(1–2), 1–15.

Picard, K., B. Brooke, and M. F. Coffin (2017), Geological insights from Malaysia Airlines flight MH370 search, *EOS*, 98, doi:10.1029/2017EO069015.

Pierson, F. B., D. H. Carlson, and K. E. Spaeth (2002), Impacts of wildfire on soil hydrological properties of steep sagebrush-steppe rangeland, *International Journal of Wildland Fire*, 11, 145–151, doi:10.1071/WF02037.

Pilgrim, D. H. (1976), Travel times and nonlinearity of flood runoff from tracer measurements on a small watershed, *Water Resources Research*, 12(3), 487–496.

Pilgrim, D. H. (1977), Isochrones of travel time and distribution of flood storage from a tracer study on a small watershed, *Water Resources Research*, 13(3), 587–595.

Placzkowska, E., M. Gornik, E. Mocior, B. Peek, P. Potoniec, B. Rzonca, and J. Siwek (2015), Spatial distribution of channel heads in the Polish Flysch Carpathians, *Catena*, 127, 240–249, doi:10.1016/j.catena.2014.12.033.

Plante, A. F., and W. J. Parton (2007), The dynamics of soil organic matter and nutrient cycling, in *Soil Microbiology, Ecology, and Biochemistry*, edited by E. A. Paul, pp. 433–470, Academic Press, Amsterdam.

Poesen, J., J. Nachtergaele, G. Verstraeten, and C. Valentin (2003), Gully erosion and environmental change: Importance and research needs, *Catena*, 50(2–4), 91–133.

Pollen-Bankhead, N., and A. Simon (2010), Hydrologic and hydraulic effects of riparian root networks on streambank stability: Is mechanical root-reinforcement the whole story?, *Geomorphology*, 116, 353–362, doi:10.1016/j.geomorph.2009.11.013.

Polyakov, V. O., and R. Lal (2004), Modeling soil organic matter dynamics as affected by soil water erosion, *Environment International*, 30, 547–556, doi:10.1016/j.envint.2003.10.011.

Polychronaki, A., I. Z. Gitas, and A. Minchella (2014), Monitoring post-fire vegetation recovery in the Mediterranean using SPOT and ERS imagery, *International Journal of Wildland Fire*, 23, 631–642, doi:10.1071/WF12058.

Popper, K. R. (1959), *The Logic of Scientific Discovery*, Hutchinson, London.

Portenga, E. W., and P. R. Bierman (2011), Understanding Earth's eroding surface with ^{10}Be, *GSA Today*, 21(8), 4–10, doi:10.1130/G111A.1.

Prentice, I. C., W. Cramer, S. P. Harrison, R. Leemans, R. A. Monserud, and A. M. Solomon (1992), A global biome model based on plant physiology and dominance, soil properties and climate, *Journal of Biogeography*, 19(2), 117–134.

Price, F., and R. A. Bradstock (2011), Quantifying the influence of fuel age and weather on the annual extent of unplanned fires in the Sydney region of Australia, *International Journal of Wildland Fire*, 20(1), 142–151, doi:10.1071/wf10016.

Price, O. E., R. A. Bradstock, J. E. Keeley, and A. D. Syphard (2012), The impact of antecedent fire area on burned area in southern California coastal ecosystems, *Journal of Environmental Management*, 113, 301–307, doi:10.1016/j.jenvman.2012.08.042.

Proffitt, G. T., and A. J. Sutherland (1983), Transport of non-uniform sediments, *Journal of Hydraulic Research*, 21, 33–43.

Prosser, I. P., and B. Abernethy (1996), Predicting the topographic limits to a gully network using a digital terrain model and process thresholds, *Water Resources Research*, 32(7), 2289–2298.

Prosser, I. P., W. E. Dietrich, and J. Stevenson (1995), Flow resistance and sediment transport by concentrated overland flow in a grassland, *Geomorphology*, 13, 71–86, doi:10.1016/0169-555X(95)00020-6.

Pumpanen, J., H. Ilvesniemi, and P. Hari (2003), A process-based model for predicting soil carbon dioxide efflux and concentration, *Soil Science Society of America*, 67(2), 402–413, doi:10.2136/sssaj2003.4020.

Quillet, A., C. Peng, and M. Garneau (2010), Toward dynamic global vegetation models for simulating vegetation–climate interactions and feedbacks: Recent developments, limitations, and future challenges, *Environmental Reviews*, 18, 333–353, doi:10.1139/A10-016.

Radoane, M., I. Ichim, and N. Radoane (1995), Gully distribution and development in Moldova, Romania, *Catena*, 24(2), 127–146, doi:10.1016/0341-8162(95)00023-L.

Rajaram, H., and M. Arshadi (2016), A similarity solution for reaction front propagation in a fracture–matrix system, *Philosophical Transactions of the Royal Society of London A*, 374, 20150424, doi:10.1098/rsta.2015.0424.

Ramankutty, P., M. Ryan, R. Lawes, J. Speijers, and M. Renton (2013), Statistical emulators of a plant growth simulation model, *Climate Research*, 55, 253–265, doi:10.3354/cr01138.

Rasmussen, E., R. A. Dahlgren, and R. J. Southard (2010), Basalt weathering and pedogenesis across an environmental gradient in the southern Cascade Range California, USA, *Geoderma*, 154, 473–485, doi:10.1016/j.geoderma.2009.05.019.

Rawls, W. J., Y. A. Pachepsky, J. C. Ritchie, T. M. Sobecki, and H. Bloodworth (2003), Effect of soil organic carbon on soil water retention, *Geoderma*, 116(1–2), 61–76.

Recking, A. (2010), A comparison between flume and field bed load transport data and consequences for surface-based bed load transport prediction, *Water Resources Research*, 46, W03518, doi:10.1029/2009WR008007.

Refice, A., E. Giachetta, and D. Capolongo (2012), SIGNUM: A Matlab, TIN-based landscape evolution model, *Computers & Geosciences*, 45, 293–303, doi:10.1016/j.cageo.2011.11.013.

Reneau, S. L., D. Katzman, G. A. Kuyumjian, A. Lavine, and D. V. Malmon (2007), Sediment delivery after a wildfire, *Geology*, 35(2), 151–154, doi:10.1130/G23288A.1.

Rengers, F. K., and G. E. Tucker (2014), Analysis and modeling of gully headcut dynamics, North American high plains, *Journal of Geophysical Research (Earth Surface)*, 119, 983–1003, doi:10.1002/2013JF002962.

Rethemeyer, J., C. Kramer, G. Gleixner, B. John, T. Yamashita, H. Flessa, N. Andersen, M. J. Nadeau, and P. M. Grootes (2005), Transformation of organic matter in agricultural soils: Radiocarbon concentration versus soil depth, *Geoderma*, 128(1–2), 94–105.

Richter, D. D., and D. Markewitz (1995), How deep is soil?, *Bioscience*, 45(9), 600–609, doi:10.2307/1312764.

Richter, D. D., D. Markewitz, S. Trumbore, and C. G. Wells (1999), Rapid accumulation and turnover of soil carbon in a re-establishing forest, *Nature*, 400(6739), 56–58, doi:10.1038/21867.

Riebe, C. S., J. W. Kirchner, and R. C. Finkel (2003), Long-term rates of chemical weathering and physical erosion from cosmogenic nuclides and geochemical mass balance, *Geochimica et Cosmochimica Acta*, 67(22), 4411–4427, doi:10.1016/S0016-7037(03)00382-X.

Riebe, C. S., J. W. Kirchner, and R. C. Finkel (2004), Erosional and climatic effects on long-term chemical weathering rates in granitic landscapes spanning diverse climate regimes, *Earth and Planetary Science Letters*, 224(3–4), 547–562, doi:10.1016/j.epsl.2004.05.019.

Rieke-Zapp, D., J. Poesen, and M. A. Nearing (2007), Effects of rock fragments incorporated in the soil matrix on concentrated flow hydraulics and erosion, *Earth Surface Processes and Landforms*, 32, 1063–1076, doi:10.1002/esp.1469.

Rietkerk, M., M. C. Boerlijst, F. van Langevelde, R. HilleRisLambers, J. van de Koppel, L. Kumar, H. H. T. Prins, and A. M. de Roos (2002), Self-organization of vegetation in arid ecosystems, *American Naturalist*, 160 (4), 524–530, doi:10.1086/342078.

Rigon, R., A. Rinaldo, I. Rodriguez-Iturbe, R. L. Bras, and E. Ijjasz-Vasquez (1993), Optimal Channel Networks – A framework for the study of river basin morphology, *Water Resources Research*, 29(6), 1635–1646.

Riley, S. J., B. Gardiner, F. Hancock, and C. Uren (1991), Concentrated flow simulation experiments, waste rock dumps, Ranger Uranium Mine, 1990 results, Internal Report 49, Supervising Scientist for the Alligator Rivers Region.

Rinaldo, A., G. K. Vogel, R. Rigon, and I. Rodriguez-Iturbe (1995), Can one gauge the shape of a basin?, *Water Resources Research*, 31(4), 1119–1127.

Ritchie, J. C., G. W. McCarty, E. R. Venteris, and T. C. Kaspar (2007), Soil and soil organic carbon redistribution on the landscape, *Geomorphology*, 89, 163–171, doi:10.1016/j.geomorph.2006.07.021.

Robertson, A., E. Githumbi, and D. Colombaroli (2016), Paleofires and models illuminate future fire scenarios, *EOS*, 97, 11, doi:10.1029/2016EO049933.

Rodrigues, M., P. Ibarra, M. Echeverría, F. Pérez-Cabello, and J. de la Riva (2014), A method for regional-scale assessment of vegetation recovery time after high-severity wildfires: Case study of Spain, *Progress in Physical Geography*, 38(5), 556–575, doi:10.1177/0309133314542956.

Rodriguez-Iturbe, I., E. J. Ijjasz-Vasquez, R. L. Bras, and D. G. Tarboton (1992a), Power law distributions of discharge mass and energy in river basins, *Water Resources Research*, 28(4), 1089–1093.

Rodriguez-Iturbe, I., and J. M. Mejia (1974), On the transformation of point rainfall to areal rainfall, *Water Resources Research*, 10(4), 729–735, doi:10.1029/WR010i004p00729.

Rodriguez-Iturbe, I., and A. Porporato (2005), *Ecohydrology of water-controlled ecosystems: Soil moisture and plant dynamics*, Cambridge University Press, Cambridge.

Rodriguez-Iturbe, I., and A. Rinaldo (2001), *Fractal river basins: Chance and self-organization*, Cambridge University Press, Cambridge.

Rodriguez-Iturbe, I., A. Rinaldo, R. Rigon, R. L. Bras, E. Ijjasz-Vasquez, and A. Marani (1992b), Fractal structures as least energy patterns – The case of river networks, *Geophysical Research Letters*, 19(9), 889–892.

Rodriguez-Iturbe, I., and J. B. Valdes (1979), The geomorphologic structure of hydrologic response, *Water Resources Research*, 15(6), 1409–1420.

Roe, G. H. (2005), Orographic precipitation, *Annual Review of Earth and Planetary Sciences*, 33, 645–671, doi:10.1146/annurev.earth.33.092203.122541.

Roe, G. H., and M. T. Brandon (2011), Critical form and feedbacks in mountain-belt dynamics: Role of rheology as a tectonic governor, *Journal of Geophysical Research (Solid Earth)*, 116, B02101, doi:10.1029/2009JB006571.

Roe, G. H., D. R. Montgomery, and B. Hallet (2002), Effects of orographic precipitation variations on the concavity of steady-state river profiles, *Geology*, 30(2), 143–146.

Roe, G. H., D. R. Montgomery, and B. Hallet (2003), Orographic precipitation and the relief of mountain ranges, *Journal of Geophysical Research (Solid Earth)*, 108(B6), ETG15, doi:10.1029/2001JB001521.

Roe, G. H., K. X. Whipple, and J. K. Fletcher (2008), Feedbacks among climate, erosion, and tectonics in a critical wedge orogen, *American Journal of Science*, 308, 815–842, doi:10.2475/07.2008.01.

Roering, J. J. (2004), Soil creep and convex-upward velocity profiles: Theoretical and experimental investigation of disturbance-driven sediment transport on hillslopes, *Earth Surface Processes and Landforms*, 29, 1597–1612, doi:10.1002/esp.1112.

Roering, J. J. (2008), How well can hillslope evolution models 'explain' topography? Simulating soil transport and production with high-resolution topographic data, *Geological Society of America Bulletin*, 120(9/10), 1248–1262, doi:10.1130/B26283.1.

Roering, J. J., J. W. Kirchner, and W. E. Dietrich (1999), Evidence for nonlinear, diffusive sediment transport on hillslopes and implications for landscape morphology, *Water Resources Research*, 35(3), 853–870.

Roering, J. J., J. W. Kirchner, L. S. Sklar, and W. E. Dietrich (2001), Hillslope evolution by nonlinear creep and landsliding: An experimental study, *Geology*, 29(2), 143–146, doi:10.1130/0091-7613(2001)029<0143:HEBNCA>2.0.CO;2.

Roering, J. J., J. T. Perron, and J. W. Kirchner (2007), Functional relationships between denudation and hillslope form and relief, *Earth and Planetary Science Letters*, 264, 245–258, doi:10.1016/j.epsl.2007.09.035.

Roering, J. J., K. M. Schmidt, J. D. Stock, W. E. Dietrich, and D. R. Montgomery (2003), Shallow landsliding, root reinforcement, and the spatial distribution of trees in the Oregon Coast Range, *Canadian Geotechnical Journal*, 40, 237–253, doi:10.1139/T02-113.

Rogers, N. (2007), An introduction to the structure and composition of the Earth, in *An Introduction to Our Dynamic Planet*, edited by N. Rogers, pp. 1–46, Cambridge University Press, Cambridge.

Román-Sánchez, A., T. Vanwalleghem, A. Peña, A. Laguna, and J. V. Giráldez (2017), Controls on soil carbon storage from topography and vegetation in a rocky, semi-arid landscapes, *Geoderma*, doi:10.1016/j.geoderma.2016.10.013.

Roseberry, J. C., D. J. Furbish, and M. Schmeeckle (2012), A probabilistic description of the bed load sediment flux: 2. Particle activity and motions, *Journal of Geophysical Research (Earth Surface)*, 117, F03032, doi:10.1029/2012JF002353.

Rothermel, R. C. (1972), A mathematical model for predicting fire spread in wildland fire, Rep. INT-115, Intermountain Forest and Range Experiment Station, USDA Forest Service, Ogden, UT.

Rothermel, R. C. (1991), Predicting behavior and size of crown fires in the northern Rocky Mountains, Rep. INT-438, Intermountain Research Station, USDA Forest Service, Ogden, UT.

Roy, S. G., P. O. Koons, P. Upton, and G. E. Tucker (2015), The influence of crustal strength fields on the patterns and rates of fluvial incision, *Journal of Geophysical Research (Earth Surface)*, 120, 275–299, doi:10.1002/2014JF003281.

Royden, L. H., and J. T. Perron (2013), Solutions of the stream power equation and application to the evolution of river longitudinal profiles, *Journal of Geophysical Research (Earth Surface)*, 118(2), 497–518, doi:10.1002/jgrf.20031.

Ruddiman, W. F. (2013), *Earth's climate: Past and future*, 3rd ed., Freeman, New York.

Rüdiger, C., G. R. Hancock, H. M. Hemakumura, B. Jacobs, J. D. Kalma, C. Martinez, M. Thyer, J. P. Walker, T. Wells, and G. R. Willgoose (2007), Goulburn River experimental catchment data set, *Water Resources Research*, 43(10), W10403.

Ruimy, A., L. Kergoat, A. Bondeau, and Postdam NPP Model Intercomparison Team (1999), Comparing global models of terrestrial net primary productivity (NPP): Analysis of differences in light absorption and light-use efficiency, *Global Change Biology*, 5(Suppl. 1), 56–64.

Rumpel, C., V. Chaplot, O. Planchon, J. Bernadou, C. Valentin, and A. Mariotti (2006), Preferential erosion of black carbon on steep slopes with slash and burn agriculture, *Catena*, 65, 30–40, doi:10.1016/j.catena.2005.09.005.

Rumpel, C., A. Ba, F. Darboux, V. Vhaplot, and O. Planchon (2009), Erosion budget and process selectivity of black carbon at meter scale, *Geoderma*, 154, 131–137, doi:10.1016/j.geoderma.2009.10.006.

Rumpel, C., J. Leifield, C. Santin, and S. Doerr (2015), Movement of biochar in the environment, in *Biochar for environmental management: Science, technology and implementation*, edited by J. Lehmann and S. Joseph, pp. 283–300, Routledge, Oxford.

Russell-Smith, J., P. G. Ryan, and R. Durieu (1997), A LANDSAT-MSS derived fire history of Kakadu National Park, monsoonal northern Australia, 1980–94: seasonal extent, frequency and patchiness, *Journal of Applied Ecology*, 34, 748–766.

Ryan, S. A., K. A. Dwire, and M. K. Dixon (2011), Impacts of wildfire on runoff and sediment loads at Little Granite Creek, western Wyoming, *Geomorphology*, 129, 113–130, doi:10.1016/j.geomorph.2011.01.017.

Sachau, T., D. Koehn, and C. Passchier (2013), Mountain building under extension, *American Journal of Science*, 313, 326–344, doi:10.2475/04.2013.03.

Saco, P. M., and M. Moreno de Las Heras (2013), Ecogeomorphic coevolution of semiarid hillslopes: Emergence of banded and striped vegetation patterns through interaction of biotic and abiotic processes, *Water Resources Research*, 49(1), 115–126, doi:10.1029/2012WR012001.

Saco, P. M., G. R. Willgoose, and G. R. Hancock (2006), Spatial organization of soil depths using a landform evolution model, *Journal of Geophysical Research (Earth Surface)*, 111, F02016, doi:02010.01029/02005JF000351.

Saco, P. M., G. R. Willgoose, and G. R. Hancock (2007), Eco-geomorphology and banded vegetation patterns in arid and semi-arid regions, *Hydrology and Earth System Sciences*, 11(6), 1717–1730, doi:10.5194/hess-11-1717-2007.

Saft, M., M. C. Peel, A. W. Western, J. M. Perraud, and L. Zhang (2016a), Bias in streamflow projections due to climate-induced shifts in catchment response, *Geophysical Research Letters*, 43(4), 1574–1581, doi:10.1002/2015GL067326.

Saft, M., M. C. Peel, A. W. Western, and L. Zhang (2016b), Predicting shifts in rainfall-runoff partitioning during multiyear drought: Roles of dry period and catchment characteristics, *Water Resources Research*, 52, doi:10.1002/2016WR019525.

Saft, M., A. W. Western, L. Zhang, M. C. Peel, and N. J. Potter (2015), The influence of multiyear drought on the annual rainfall-runoff relationship: An Australian perspective, *Water Resources Research*, 51, 2444–2463, doi:10.1002/2014WR015348.

Saito, K., and T. Oguchi (2005), Slope of alluvial fans in humid regions of Japan, Taiwan and the Philippines, *Geomorphology*, 70, 147–162, doi:10.1016/j.geomorph.2005.04.006.

Sala, O. E., W. J. Parton, L. Joyce, and W. K. Laurenrath (1988), Primary production of the central grassland region of the United States, *Ecology*, 69(1), 40–45.

Sallares, V., and A. Calahorrano (2007), Geophysical characterization of mantle melting anomalies: A crustal view, in *Plates, plumes and planetary processes*, Geological Society of America Special Paper 430, edited by G. R. Foulger and D. M. Jurdy, pp. 507–524, Geological Society of America, Boulder, CO, doi:10.1130/2007.2430(25).

Salvador-Blanes, S., B. Minasny, and A. B. McBratney (2007), Modelling long-term in situ soil profile evolution: Application to the genesis of soil profiles containing stone layers, *European Journal of Soil Science*, 58, 1535–1548.

Samonil, P., K. Kral, and L. Hort (2010), The role of tree uprooting in soil formation: A critical literature review, *Geoderma*, 157, 65–79, doi:10.1016/j.geoderma.2010.03.018.

Sanchidrian, J. A., F. Ouchterlony, P. Segarra, and P. Moser (2014), Size distribution functions for rock fragments, *International Journal of Rock Mechanics & Mining Sciences*, 71, 381–394, doi:10.1016/j.ijrmms.2014.08.007.

Sankaran, M., et al. (2005), Determinants of woody cover in African savannas, *Nature*, 438(7069), 846–849, doi:10.1038/NATURE04070.

Santi, P. M., V. G. deWolfe, J. D. Higgins, S. H. Cannon, and J. E. Gartner (2008), Sources of debris flow material in burned areas, *Geomorphology*, 96, 310–321, doi:10.1016/j.geomorph.2007.02.022.

Sauer, D., P. Finke, R. Sorensen, R. Sperstad, I. Schulli-Maurer, H. Hoeg, and K. Stahr (2012), Testing a soil development model against southern Norway soil chronosequences, *Quaternary International*, 265, 18–31, doi:10.1016/j.quaint.2011.12.018.

Saxton, K. E., and W. J. Rawls (2006), Soil water characteristic estimates by texture and organic matter for hydrologic solutions, *Soil Science Society of America*, 70, 1569–1578, doi:10.2136/sssaj2005.0117.

Schaetzl, R. J. (2014), Professor Donald Johnson's list of Landmark Papers (1878–1998) in geomorphology and soil geomorphology – An appreciation, *Progress in Physical Geography*, 28(1), 129–137, doi:10.1177/0309133314522284.

Schaetzl, R. J., and M. L. Thompson (2015), *Soils genesis and geomorphology*, 2nd ed., Cambridge University Press, Cambridge.

Schmidt, K. M., J. J. Roering, J. D. Stock, W. E. Dietrich, D. R. Montgomery, and T. Schaub (2001), The variability of root cohesion as an influence on shallow landslide susceptibility in the Oregon Coast Range, *Canadian Geotechnical Journal*, 38, 995–1024, doi:10.1139/cgj-38–5-995.

Schmidt, M. W. I., et al. (2011), Persistence of soil organic matter as an ecosystem property, *Nature*, 478, 49–56, doi:10.1038/nature10386.

Schimel, D. S., B. H. Braswell, E. A. Holland, R. McKeown, D. S. Ojima, T. H. Painter, W. J. Parton, and A. R. Townsend (1994), Climatic, edaphic, and biotic controls over storage and turnover of carbon in soils, *Global Biogeochemical Cycles*, 8(3), 279–293.

Schönbrodt, S., P. Saumer, T. Behrens, C. Seeber, and T. Scholten (2010), Assessing the USLE crop and management factor C for soil erosion modeling in a large mountainous watershed in Central China, *Journal of Earth Science*, 21(6), 835–845, doi:10.1007/s12583-010-0135-8.

Schoorl, J. M., A. J. A. M. Temme, and T. Veldkamp (2014), Modelling centennial sediment waves in an eroding landscape – Catchment complexity, *Earth Surface Processes and Landforms*, 39(11), 1526–1537, doi:10.1002/esp.3605.

Schoorl, J. M., and A. Veldkamp (2001), Linking land use and landscape process modelling: A case study for the Alora region (south Spain), *Agriculture, Ecosystems & Environment*, 85(1–3), 281–292, doi:10.1016/s0167-8809(01)00194-3.

Scott, K. M., and C. F. Pain (2009), *Regolith science*, Springer, Dordrecht, the Netherlands.

Sella, G. F., S. Stein, T. H. Dixon, M. Craymer, T. S. James, S. Mazzotti, and R. K. Dokka (2007), Observation of glacial isostatic adjustment in 'stable' North America with GPS, *Geophysical Research Letters*, 34(2), L02306, doi:10.1029/2006GL027081.

Shakesby, R. A. (2011), Post-wildfire soil erosion in the Mediterranean: Review and future research directions, *Earth-Science Reviews*, 105(3–4), 71–100, doi:10.1016/j.earscirev.2011.01.001.

Shakesby, R. A., and S. H. Doerr (2006), Wildfire as a hydrological and geomorphological agent, *Earth-Science Reviews*, 74, 269–307, doi:10.1016/j.earscirev.2005.10.006.

Sharmeen, S., and G. R. Willgoose (2006), The interaction between armouring and particle weathering for eroding landscapes, *Earth Surface Processes and Landforms*, 31(10), 1195–1210, doi:10.1002/esp.1397.

Sharmeen, S., and G. R. Willgoose (2007), A one-dimensional model for simulating armouring and erosion on hillslopes. 2. Long-term erosion and armouring predictions for two contrasting mine spoils, *Earth Surface Processes and Landforms*, 32(10), 1437–1453, doi:10.1002/esp.1482.

Sharples, J. J., G. J. Cary, P. Fox-Hughes, S. Mooney, J. P. Evans, M.-S. Fletcher, M. Fromm, P. F. Grierson, R. McRae, and P. Baker (2016), Natural hazards in Australia: Extreme bushfire, *Climate Change*, 139, 85–99, doi:10.1007/s10584-016-1811-1.

Sheridan, G. J., P. N. J. Lane, and P. J. Noske (2007), Quantification of hillslope runoff and erosion processes before and after wildfire in a wet eucalyptus forest, *Journal of Hydrology*, 343, 12–28, doi:10.1016/j.jhydrol.2007.06.005.

Shreve, R. L. (1967), Infinite topologically random channel networks, *Journal of Geology*, 75, 178–186.

Shuin, Y., N. Hotta, M. Suzuki, and K. Ogawa (2012), Estimating the effects of heavy rainfall conditions on shallow landslides using a distributed landslide conceptual model, *Physics and Chemistry of the Earth*, 49, 44–51, doi:10.1016/j.pce.2011.06.002.

Shull, D. H. (2001), Transition-matrix model of bioturbation and radionuclide diagenesis, *Limnology and Oceanography*, 46(4), 905–916.

Sidle, R. C. (1992), A theoretical model of the effects of timber harvesting on slope stability, *Water Resources Research*, 28(7), 1897–1910.

Sidman, G., D. P. Guertin, D. C. Goodrich, D. Thoma, D. Falk, and I. S. Burns (2016), A coupled modelling approach to assess the effect of fuel treatments on post-wildfire runoff and erosion, *International Journal of Wildland Fire*, 25, 351–362, doi:10.1071/WF14058.

Silva, J. S., F. C. Rego, and S. Mazzoleni (2006), Soil water dynamics after fire in a Portuguese shrubland, *International Journal of Wildland Fire*, 15, 99–111, doi:10.1071/WF04057.

Simonett, D. S. (1967), Landslide distribution and earthquakes in the Bewani and Torricelli Mountains, New Guinea, in *Landform studies from Australia and New Guinea*, edited by J. N. Jennings and J. A. Mabbutt, pp. 64–84, Cambridge University Press, Cambridge.

Simunek, J., M. Senja, H. Saito, M. Sakai, and M. T. van Genuchten (2008), The HYDRUS-1D software package for simulating the one-dimensional movement of water, heat, and multiple solutes in variably-saturated media, version 4.0, 315 pp, University of California.

Sitch, S., et al. (2003), Evaluation of ecosystem dynamics, plant geography and terrestrial carbon cycling in the LPJ dynamic global vegetation model, *Global Change Biology*, 9(2), 161–185.

Six, J., H. Bossuyt, S. Degryze, and K. Denef (2004), Review: A history of research on the link between (micro)aggregates, soil biota, and soil organic matter dynamics, *Soil Tillage Research*, 79, 7–31, doi:10.1016/j.still.2004.03.008.

Sklar, L. S., and W. E. Dietrich (1998), River longitudinal profiles and bedrock incision models: Stream power and the influence of sediment supply, in *Rivers over rock: Fluvial processes in bedrock channels*, edited by K. J. Tinkler and E. E. Wohl, AGU, Washington, DC.

Sklar, L. S., and W. E. Dietrich (2001), Sediment and rock strength controls on river incision into bedrock, *Geology*, 29(12), 1087–1090, doi:10.1130/0091-7613(2001)029<1087:SARSCO>2.0.CO;2.

Sklar, L. S., and W. E. Dietrich (2004), A mechanistic model for river incision into bedrock by saltating bedrock, *Water Resources Research*, 40(6), W06301.

Sklar, L. S., and W. E. Dietrich (2008), Implications of the saltation–abrasion bedrock incision model for steady-state river longitudinal profile relief and concavity, *Earth Surface Processes and Landforms*, 33(7), 1129–1151, doi:10.1002/esp.1689.

Sklar, L. S., and W. E. Dietrich (2012), Correction to A mechanistic model for river incision into bedrock by saltating bed load, *Water Resources Research*, 48, W06902, doi:10.1029/2012WR012267.

Sklar, L. S., and J. A. Marshall (2016), The problem of predicting the size distribution of sediment supplied by hillslopes to rivers, *Geomorphology*, 277, 31–49, doi:10.1016/j.geomorph.2016.05.005.

Sloan, S. W. (1987), A fast algorithm for constructing Delaunay triangulation in the plane, *Advances in Engineering Software*, 9(1), 34–55.

Smith, H. G., G. J. Sheridan, P. N. J. Lane, P. Nyman, and S. Haydon (2011), Wildfire effects on water quality in forest catchments: A review with implications for water supply, *Journal of Hydrology*, 396, 170–192, doi:10.1016/j.jhydrol.2010.10.043.

Smith, P., et al. (1997), A comparison of the performance of nine soil organic matter models using datasets from seven long-term experiments, *Geoderma*, 81(1–2), 153–225.

Smith, T. R., and F. P. Bretherton (1972), Stability and the conservation of mass in drainage basin evolution, *Water Resources Research*, 8(6), 1506–1529.

Solyom, P. B., and G. E. Tucker (2004), Effect of limited storm duration on landscape evolution, drainage basin geometry, and hydrograph shapes, *Journal of Geophysical Research (Earth Surface)*, 109, F03012.

Solyom, P. B., and G. E. Tucker (2007), The importance of the catchment area–length relationship in governing non-steady state hydrology, optimal junction angles and drainage network pattern, *Geomorphology*, 88, 84–108, doi:10.1016/j.geomorph.2006.10.014.

Sørensen, M. (2004), On the rate of aeolian sand transport, *Geomorphology*, 59(1–4), doi:10.1016/j.geomorph.2003.09.005.

Southard, J. B. (2006), 12.090 Introduction to fluid motions, sediment transport, and current-generated sedimentary structures, MIT OpenCourseWare, Massachusetts Institute of Technology, Cambridge, MA.

Spada, G., G. Ruggieri, L. S. Sorenson, K. Nielson, D. Melini, and F. Colleoni (2012), Greenland uplift and regional sea level changes from ICESat observations and GIA modelling, *Geophysical Journal International*, 189, 1457–1474, doi:10.1111/j.1365-246X.2012.05443.x.

Stallard, R. F. (1998), Terrestrial sedimentation and the carbon cycle: Coupling weathering and erosion to carbon burial, *Global Biogeochemical Cycles*, 12(2), 231–257.

Stark, C. P., and F. Guzzetti (2009), Landslide rupture and the probability distribution of mobilized debris volumes, *Journal of Geophysical Research (Earth Surface)*, 114, F00A02, doi:10.1029/2008JF001008.

Stark, C. P., and N. Hovius (2001), The characterization of landslide size distributions, *Geophysical Research Letters*, 28(6), 1091–1094, doi:10.1029/2000GL008527.

Steefel, C. I. (2009), *CrunchFlow: Software for modelling multicomponent reactive flow and transport, user manual*, Earth Sciences Division, Lawrence Berkeley National Laboratory, Berkeley, CA, www.csteefel.com/CrunchFlowManual.pdf.

Steel, Z. L., H. D. Safford, and J. H. Viers (2015), The fire frequency-severity relationship and the legacy of fire suppression in California forests, *Ecosphere*, 6(1), 8, doi:10.1890/ES14-00224.1.

Stewart, V. I., W. A. Adams, and H. H. Abdulla (1970), Quantitative pedological studies on soils derived from Silurian mudstones. II. The relationship between stone content and apparent density of the fine earth, *Journal of Soil Science*, 21, 248–255.

Stocker, T. F., D. Qin, G.-K. Plattner, M. M. B. Tignor, S. K. Allen, J. Boschung, A. Nauels, Y. Xia, V. Bex, and P. M. Midgley (Eds.) (2013), *Working Group I Contribution to the Fifth Assessment Report of the Intergovernmental Panel on Climate Change*, Cambridge University Press, Cambridge.

Stockmann, U., B. Minasny, and A. B. McBratney (2014), How fast does soil grow?, *Geoderma*, 216, 48–61, doi:10.1016/j.geoderma.2013.10.007.

Stockmann, U., et al. (2013), The knowns, known unknowns and unknowns of sequestration of soil organic carbon, *Agriculture, Ecosystems & Environment*, 164, 80–99, doi:10.1016/j.agee.2012.10.001.

Strahler, A. N. (1952), Dynamic basis of geomorphology, *Bulletin of the Geological Society of America*, 63, 923–938.

Strahler, A. N. (1952), Hypsometric (area-altitude) analysis of erosional topography, *Bulletin of the Geological Society of America*, 63, 1117–1142.

Sullivan, A. L. (2009a), Wildland surface fire spread modelling, 1990–2007. 1: Physical and quasi-physical models, *International Journal of Wildland Fire*, 18, 349–368, doi:10.1071/WF06143.

Sullivan, A. L. (2009b), Wildland surface fire spread modelling, 1990–2007. 2: Empirical and quasi-empirical models, *International Journal of Wildland Fire*, 18, 369–386, doi:10.1071/WF06142.

Sullivan, A. L. (2009c), Wildland surface fire spread modelling, 1990–2007. 3: Simulation and mathematical analogue models, *International Journal of Wildland Fire*, 18, 387–403, doi:10.1071/WF06144.

Surkan, A. J. (1969), Synthetic hydrographs: Effects of network geometry, *Water Resources Research*, 5(1), 112–128.

Sweeney, K. E., J. J. Roering, and C. Ellis (2015), Experimental evidence for hillslope control of landscape scale, *Science*, 349(6243), 51–53, doi:10.1126/science.aab0017.

Syvitski, J. P. M., S. D. Peckham, R. Hilberman, and T. Mulder (2003), Predicting the terrestrial flux of sediment to the global ocean: A planetary perspective, *Sedimentary Geology*, 162, 5–24, doi:10.1016/S0037-0738(03)00232-X.

Tarboton, D. G. (1997), A new method for the determination of flow directions and upslope areas in grid digital elevation models, *Water Resources Research*, 33(2), 309–319.

Tarboton, D. G., R. L. Bras, and I. Rodriguez-Iturbe (1989), Scaling and elevation in river networks, *Water Resources Research*, 25(9), 2037–2052.

Tarolli, P. (2014), High-resolution topography for understanding Earth surface processes: Opportunities and challenges, *Geomorphology*, 216, 295–312, doi:10.1016/j.geomorph.2014.03.008.

Taylor, S. H., B. S. Ripley, F. I. Woodward, and C. P. Osborne (2011), Drought limitation of photosynthesis differs between C_3 and C_4 grass species in a comparative experiment, *Plant, Cell and Environment*, 34, 65–75, doi:10.1111/j.1365-3040.2010.02226.x.

Teles, V., G. de Marsily, and É. Perrier (1998), Sur une nouvelle approche de modélisation de la mise en place des sédiments dans une plaine alluviale pour en représenter l'hétérogénéité, *Comptes Rendus de l'Academie des Sciences Serie II Fascicule A – Sciences de la terre et des planètes*, 327(9), 597–606, doi:10.1016/S1251-8050(99)80113-X.

Temme, A. J. A. M., L. Claessens, A. Veldkamp, and J. M. Schoorl (2011), Evaluating choices in multi-process landscape evolution models, *Geomorphology*, 125, doi:10.1016/j.geomorph.2010.10.007.

Temme, A. J. A. M., I. Peeters, E. Buis, A. Veldkamp, and G. Govers (2011), Comparing landscape evolution models with quantitative field data at the millennial time scale in the Belgian loess belt, *Earth Surface Processes and Landforms*, 36(10), 1300–1312, doi:10.1002/esp.2152.

Temme, A. J. A. M., and T. Vanwalleghem (2016), LORICA – A new model for linking landscape and soil profile evolution: Development and sensitivity analysis, *Computers & Geosciences* 90, 131–143, doi:10.1016/j.cageo.2015.08.004.

Temme, A. J. A. M., and A. Veldkamp (2009), Multi-process Late Quaternary landscape evolution modelling reveals lags in climate response over small spatial scales, *Earth Surface Processes and Landforms*, 34(4), 573–589, doi:10.1002/esp.1758.

Tesemma, Z. K., Y. Wei, M. C. Peel, and A. W. Western (2015), The effect of year-to-year variability of leaf area index on Variable Infiltration Capacity model performance and simulation of runoff, *Advances in Water Resources*, 83, 310–322, doi:10.1016/j.advwatres.2015.07.002.

Tessler, N., L. Wittenberg, and N. Greenbaum (2013), Soil water repellency persistence after recurrent forest fires on Mount Carmel, Israel, *International Journal of Wildland Fire*, 22, 515–526, doi:10.1071/WF12063.

Thomas, I. A., P. Jordan, O. Shine, O. Fenton, P. E. Mellander, P. Dunlop, and P. N. C. Murphy (2017), Defining optimal DEM resolutions and point densities for modelling hydrologically sensitive areas in agricultural catchments dominated by microtopography, *International Journal of Applied Earth Observation and Geoinformation*, 54, 38–52, doi:10.1016/j.jag.2016.08.012.

Thompson, S. E., C. J. Harman, P. Heine, and G. G. Katul (2010), Vegetation infiltration relationships across climatic and soil type gradients, *Journal of Geophysical Research (Biogeosciences)*, 115, G02023, doi:10.1029/2009JG001134.

Thonicke, K., A. Spessa, I. C. Prentice, S. P. Harrison, L. Dong, and C. Carmona-Moreno (2010), The influence of vegetation, fire spread and fire behaviour on biomass burning and trace gas emissions: Results from a process-based model, *Biogeosciences*, 7, 1991–2011, doi:10.5194/bg-7-1991-2010.

Tilman, D. (1994), Competition and biodiversity in spatially structured habitats, *Ecology*, 75(1), 2–16, doi:10.2307/1939377.

Ting, I. P., and W. E. Loomis (1965), Further studies concerning stomatal diffusion, *Plant Physiology*, 40(2), 220–228, doi:10.1104/pp.40.2.220.

Tisdall, J. M. (1996), Formation of soil aggregates and accumulation of soil organic matter, in *Structure and organic matter storage in agricultural soils*, edited by M. R. Carter and B. A. Stewart, pp. 57–96, CRC Press, Boca Raton, FL.

Tisdall, J. M., and J. M. Oades (1982), Organic matter and water stable aggregates in soils, *European Journal of Soil Science*, 33(2), 141–163.

Tomkin, J. H., M. T. Brandon, F. J. Pazzaglia, J. R. Barbour, and S. D. Willett (2003), Quantitative testing of bedrock incision models for the Clearwater River, NW Washington State, *Journal of Geophysical Research (Solid Earth)*, 108(B6), ETG 10, doi:10.1029/2001JB000862.

Tongway, D. J., and J. A. Ludwig (1990), Vegetation and soil patterning in semi-arid mulga lands of Eastern Australia, *Australian Journal of Ecology*, 15, 23–34.

Torn, M. S., S. E. Trumbore, O. A. Chadwick, P. M. Vitousek, and D. M. Hendricks (1997), Mineral control of soil organic carbon storage and turnover, *Nature*, 389(6647), 170–173, doi:10.1038/38260.

Torri, D., and J. Poesen (2014), A review of topographic threshold conditions for gully head development in different environments, *Earth-Science Reviews*, 130, 73–85, doi:10.1016/j.earscirev.2013.12.006.

Tranter, G., B. Minasny, A. B. McBratney, B. Murphy, N. J. McKenzie, M. Grundy, and D. M. Brough (2007), Building and testing conceptual and empirical models for predicting soil bulk density, *Soil Use and Management*, 23(4), 437–443, doi:0.1111/j.1475-2743.2007.00092.x.

Trumbore, S. (2000), Age of soil organic matter and soil respiration: Radiocarbon constraints on belowground C dynamics, *Ecological Applications*, 10(2), 399–411, doi:10.1890/1051-0761(2000)010[0399:AOSOMA]2.0.CO;2.

Trumbore, S. (2009), Radiocarbon and soil carbon dynamics, *Annual Review of Earth and Planetary Sciences*, 37, 47–66, doi:10.1146/annurev.earth.36.031207.124300.

Tucker, G., S. Lancaster, N. Gasparini, and R. Bras (2001a), The channel-hillslope integrated landscape development model (CHILD), in *Landscape erosion and evolution modelling*, edited by R. S. Harmon and W. W. Doe, pp. 349–388, Kluwer, New York.

Tucker, G. E., and R. L. Bras (2000), A stochastic approach to modeling the role of rainfall variability in drainage basin evolution, *Water Resources Research*, 36(7), 1953–1964.

Tucker, G. E., and G. R. Hancock (2010), Modelling landscape evolution, *Earth Surface Processes and Landforms*, 35(1), 28–50, doi:10.1002/esp.1952.

Tucker, G. E., S. T. Lancaster, N. M. Gasparini, R. L. Bras, and S. M. Rybarczyk (2001b), An object-oriented framework for distributed hydrologic and geomorphic modeling using triangulated irregular networks, *Computers & Geosciences*, 27(8), 959–973.

Tucker, G. E., and R. L. Slingerland (1994), Erosional dynamics, flexural isostasy, and long-lived escarpments – A numerical modeling study, *Journal of Geophysical Research (Solid Earth)*, 99(B6), 12229–12243, doi:10.1029/94JB00320.

Tucker, G. E., and K. X. Whipple (2002), Topographic outcomes predicted by stream erosion models: Sensitivity analysis and intermodel comparison, *Journal of Geophysical Research (Solid Earth)*, 107(B9), art. no.-2179.

Turcotte, D. L., and G. Schubert (2002), *Geodynamics*, Cambridge University Press, Cambridge.

Turowski, J. M., and D. Rickenmann (2009), Tools and cover effects in bedload transport observations in the Pitzbach, Austria, *Earth Surface Processes and Landforms*, 34, 26–37, doi:10.1002/esp.1686.

Tushingham, A. M., and W. R. Peltier (1991), ICE-3G: A new global model of late Pleistocene deglaciation based upon geophysical predictions of post-glacial relative sea level change, *Journal of Geophysical Research (Solid Earth)*, 96(B3), 4497–4523.

Uijlenhoet, R. (2001), Raindrop size distributions and radar reflectivity–rain rate relationships for radar hydrology, *Hydrology and Earth System Sciences*, 5(4), 615–627.

Unger, P. W., and O. R. Jones (1998), Long-term tillage and cropping systems affect bulk density and penetration resistance of soil cropped to dryland wheat and grain sorghum, *Soil Tillage Research*, 45(1–2), 39–57, doi:10.1016/S0167-1987(97)00068-8.

USDA-ARS (1999), Soil quality test kit guide, USDA-ARS, Soil Quality Institute.

USDA-ARS (2008), *User's reference guide: Revised universal soil loss equation Version 2 (RUSLE2)*, USDA-ARS, Washington, DC.

USGS (2015), PHREEQC (Version 3) – A computer program for speciation, batch-reaction, one-dimensional transport, and inverse geochemical calculations, http://wwwbrr.cr.usgs.gov/projects/GWC_coupled/phreeqc/.

Valentin, C., J. Poesen, and Y. Li (2005), Gully erosion: Impacts, factors and control, *Catena*, 63(2–3), 132–153, doi:10.1016/j.catena.2005.06.001.

van Ash, T. J. W., and P. M. B. van Genuchten (1990), A comparison between theoretical and measured creep profiles of landslides, *Geomorphology*, 3(1), 45–55, doi:10.1016/0169-555X(90)90031-K.

Vandaele, K., and J. Poesen (1995), Spatial and temporal patterns of soil-erosion rates in an agricultural catchment, Central Belgium, *Catena*, 25(1–4), 213–226.

Vandekerckhove, L., J. Poesen, and G. Govers (2003), Medium-term gully headcut retreat rates in Southeast Spain determined from aerial photographs and ground measurements, *Catena*, 50(2–4), 329–352, doi:10.1016/S0341-8162(02)00132–7.

Vandekerckhove, L., J. Poesen, D. O. Wijdenes, and G. Gyssels (2001), Short-term bank gully retreat rates in Mediterranean environments, *Catena*, 44(2), 133–161, doi:10.1016/S0341-8162(00)00152-1.

VandenBygaart, A. J., E. G. Gregorich, and B. L. Helgason (2015), Cropland C erosion and burial: Is buried soil organic matter biodegradable?, *Geoderma*, 239–240, 240–249, doi:10.1016/j.geoderma.2014.10.011.

van der Beek, P., M. A. Summerfield, J. Braun, R. W. Brown, and A. Fleming (2002), Modeling postbreakup landscape development and denudational history across the southeast African (Drakensberg Escarpment) margin, *Journal of Geophysical Research (Solid Earth)*, 107(B12), art. no.-2351.

Van De Wiel, M. J., T. J. Coulthard, M. G. Macklin, and J. Lewin (2007), Embedding reach-scale fluvial dynamics within the CAESAR cellular automaton landscape evolution model, *Geomorphology*, 90, 283–301, doi:10.1016/j.geomorph.2006.10.024.

van Dijk, A. I. J. M., L. A. Bruijnzeel, and C. J. Rosewell (2002a), Rainfall intensity-kinetic energy relationships: A critical literature appraisal, *Journal of Hydrology*, 261(1–4), 1–23, doi:10.1016/S0022-1694(02)00020-3.

van Dijk, A. I. J. M., A. G. C. A. Meesters, and L. A. Bruijnzeel (2002b), Exponential distribution theory and the interpretation of splash detachment and transport experiments, *Soil Science Society of America Journal*, 66, 1466–1474, doi:10.2136/sssaj2002.1466.

Van Eck, C. M., J. P. Nunes, D. C. S. Vieira, S. Keestra, and J. J. Keizer (2016), Physically-based modelling of the post-fire runoff response of a forest catchment in central Portugal: Using field versus remote sensing based estimates of vegetation recovery, *Land Degradation and Development*, 27, 1535–1544, doi:10.1002/ldr.2507.

van Grinsven, J. J. M., and W. H. van Riemsdijk (1992), Evaluation of batch and column techniques to measure weathering rates in soils, *Geoderma*, 52(1–2), 41–57.

van Hooff, P. (1983), Earthworm activity as a cause of splash erosion in a Luxembourg forest, *Geoderma*, 31(3), 195–204.

Vanmaercke, M., et al. (2016), How fast do gully headcuts retreat?, *Earth-Science Reviews*, 154, 336–355, doi:10.1016/j.earscirev.2016.01.009.

van Mantgem, P. J., et al. (2009), Widespread increase of tree mortality in the western United States, *Science*, 323(5913), 521–524, doi:10.1126/science.1165000.

Vanoni, V. A. (1975), *Sedimentation engineering*, ASCE, New York.

Van Oost, K., and M. M. Bakker (2012), Soil productivity and erosion, in *Soil ecology and ecosystem services*, edited by D. H. Wall, R. D. Bardgett, V. Behan-Pelletier, J. E. Herrick, T. H. Jones, K. Ritz, J. Six, D. R. Strong, and W. H. van der Putten, pp. 301–314, Oxford University Press, Oxford.

van Veen, J. A., and E. A. Paul (1981), Organic carbon dynamics in grassland soils. 1. Background information and computer simulation, *Canadian Journal of Soil Science*, 61(2), 185–201.

Vanwalleghem, T., J. Poesen, A. B. McBratney, and J. Deckers (2010), Spatial variability of soil horizon depth in natural loess-derived soils, *Geoderma*, 157, 37–45, doi:10.1016/j.geoderma.2010.03.013.

Vanwalleghem, T., H. Saito, Y. Hayakawa, and T. Oguchi (2013), Interaction between soil formation and landslide occurrence, paper presented at EGU General Assembly 2013, Geophysical Research Abstracts, EGU2013-8665-1, European Geophysical Union, Vienna.

Vanwalleghem, T., U. Stockmann, B. Minasny, and A. B. McBratney (2013), A quantitative model for integrating landscape evolution and soil formation, *Journal of Geophysical Research (Earth Surface)*, 118, 1–17, doi:10.1029/2011JF002296.

Varadachari, C., A. K. Barman, and K. Ghosh (1994), Weathering of silicate minerals by organic-acids. 2. Nature of residual products, *Geoderma*, 61(3–4), 251–268.

Veenstra, J. J., and C. L. Burras (2015), Soil profile transformation after 50 years of agricultural land use, *Soil Science Society of America*, 79, 1154–1162, doi:10.2136/sssaj2015.01.0027.

Venevsky, S., K. Thonicke, S. Sitch, and W. Cramer (2002), Simulating fire regimes in human-dominated ecosystems: Iberian Peninsula case study, *Global Change Biology*, 8(10), 984–998.

Ventsel, E., and T. Krauthammer (2001), *Thin plates and shells: Theory: analysis, and applications*, Marcel Dekker, New York.

Verdon, D. C., A. S. Kiem, and S. W. Franks (2004), Multi-decadal variability of forest fire risk – Eastern Australia, *International Journal of Wildland Fire*, 13(2), 165–171.

Vereecken, H., et al. (2016), Modelling soil processes: Review, key challenges and new perspectives, *Vadose Zone Journal*, 15(5), doi:10.2136/vzj2015.09.0131.

Vico, G., and A. Porporato (2008), Modelling C_3 and C_4 photosynthesis under water-stressed conditions, *Plant and Soil*, 313, 187–203, doi:10.1007/s11104-008-9691-4.

Viles, H. A., L. A. Naylor, N. E. A. Carter, and D. Chaput (2008), Biogeomorphological disturbance regimes: Progress in linking ecological and geomorphological systems, *Earth Surface Processes and Landforms*, 33, 1419–1435, doi:10.1002/esp.1717.

Vince, G. (2005), Tsunami seabed shows massive disruption, *New Scientist*, www.newscientist.com/article/dn7465-tsunami-seabed-shows-massive-disruption/.

Vincent, K. R., and O. A. Chadwick (1994), Synthesizing bulk-density for soils with abundant rock fragments, *Soil Science Society of America Journal*, 58, 455–464.

Volkwein, A., K. Schllenberg, V. Labiouse, F. Agliardi, F. Berger, F. Bourrier, L. K. A. Dorren, W. Gerber, and M. Jaboyedoff (2011), Rockfall characterisation and structural protection – A review, *Natural Hazards and Earth System Sciences*, 11, 2617–2651, doi:10.5194/nhess-11-2617-2011.

Voroney, R. P. (2007), The soil habitat, in *Soil microbiology, ecology, and biochemistry*, edited by E. A. Paul, pp. 25–52, Academic Press, Amsterdam.

Vrieling, A. (2006), Satellite remote sensing for water erosion assessment: A review, *Catena*, 65, 2–18, doi:10.1016/j.catena.2005.10.005.

Wainwright, J. (2008), Can modelling enable us to understand the role of humans in landscape evolution?, *Geoforum*, 39(2), 659–674, doi:10.1016/j.geoforum.2006.09.011.

Wainwright, J., and J. D. A. Millington (2010), Mind, the gap in landscape-evolution modelling, *Earth Surface Processes and Landforms*, 35, 842–855, doi:10.1002/esp.2008.

Wainwright, J., A. J. Parsons, J. R. Cooper, P. Gao, J. A. Gillies, L. Mao, J. D. Orford, and P. G. Knight (2015), The concept of transport capacity in geomorphology, *Reviews of Geophysics*, 53, 115–1202, doi:10.1002/2014RG000474.

Wainwright, J., A. J. Parsons, E. N. Müller, R. E. Brazier, and D. M. Powell (2009), Response to Hairsine's and Sander's 'Comment on "A transport-distance based approach to scaling erosion rates"': Parts 1, 2 and 3 by Wainwright et al., *Earth Surface Processes and Landforms*, 34(6), 886–890.

Wainwright, J., A. J. Parsons, E. N. Müller, R. E. Brazier, D. M. Powell, and B. Fenti (2008a), A transport-distance approach to scaling erosion rates: 1. Background and model development, *Earth Surface Processes and Landforms*, 33, 813–826, doi:10.1002/esp.1624.

Wainwright, J., A. J. Parsons, E. N. Müller, R. E. Brazier, D. M. Powell, and B. Fenti (2008b), A transport-distance approach to scaling erosion rates: 2. Sensitivity and evaluation of MAHLERAN, *Earth Surface Processes and Landforms*, 33, 962–984, doi:10.1002/esp.1623.

Wainwright, J., A. J. Parsons, E. N. Müller, R. E. Brazier, D. M. Powell, and B. Fenti (2008c), A transport-distance approach to scaling erosion rates: 3. Evaluating scaling characteristics of MAHLERAN, *Earth Surface Processes and Landforms*, 33, 1113–1128, doi:10.1002/esp.1622.

Walker, J. P., and G. R. Willgoose (1999), On the effect of digital elevation model accuracy on hydrology and geomorphology, *Water Resources Research*, 35(7), 2259–2268.

Walker, J. P., G. R. Willgoose, and J. D. Kalma (2001), One-dimensional soil moisture profile retrieval by assimilation of near-surface measurements: A simplified soil moisture model and field application, *Journal of Hydrometeorology*, 2(4), 356–373.

Wang, E., M. J. Robertson, G. L. Hammer, P. S. Carberry, D. Holzworth, H. Meinke, S. C. Chapman, J. N. G. Hargreaves, N. I. Huth, and G. McLean (2002), Development of a generic crop model template in the cropping system model APSIM, *European Journal of Agronomy*, 18, 121–140.

Wang, Y., and F. Tonon (2011), Dynamic validation of a discrete element code in modeling rock fragmentation, *International Journal of Rock Mechanics & Mining Sciences*, 48, 535–545, doi:10.1016/j.ijrmms.2011.02.003.

Warburton, J., D. G. Milledge, and R. Johnson (2008), Assessment of shallow landslide activity following the January 2005 storm, northern Cumbria, *Proceedings of the Cumberland Geological Society*, 7, 263–283.

Wardle, D. A. (1992), A comparative-assessment of factors which influence microbial biomass carbon and nitrogen levels in soil, *Biological Reviews of the Cambridge Philosophical Society*, 67(3), 321–358, doi:10.1111/j.1469-185X.1992.tb00728.x.

Wasson, R. J., R. K. Mazari, B. Starr, and G. Clifton (1998), The recent history of erosion and sedimentation on the Southern Tablelands of southeastern Australia: Sediment flux dominated by channel incision, *Geomorphology*, 24(4), 263–372, doi:10.1016/S0169-555X(98)00019-1.

Watson, F. G. R., R. A. Vertessy, T. A. McMahon, B. G. Rhodes, and I. S. Watson (1999), The hydrologic impacts of forestry on the Maroondah catchments Rep. 99/1, Cooperative Research Centre for Catchment Hydrology, Melbourne.

Watts, A. B. (2001), *Isostasy and flexure of the lithosphere*, Cambridge University Press, Cambridge.

Wechsler, S. P. (2007), Uncertainties associated with digital elevation models for hydrologic applications: A review, *Hydrology and Earth System Sciences* 11, 1481–1500.

Weissel, J. K., and M. A. Seidl (1997), Influence of rock strength properties on escarpment retreat across passive continental margins, *Geology*, 25 (7), 631–635, doi:10.1130/0091-7613(1997) 025<0631:IORSPO> 2.3.CO;2.

Welivitiya, W. D. D. P. (2017), A next generation spatially distributed model for soil profile dynamics and pedogenesis, PhD thesis, University of Newcastle, Callaghan, Australia.

Welivitiya, W. D. D. P., G. R. Willgoose, G. R. Hancock, and S. Cohen (2016), Exploring the sensitivity on a soil area-slope-grading relationship to changes in process parameters using a pedogenesis model, *Earth System Dynamics*, 4, 607–625, doi:10.5194/esurf-4-607-2016.

Wells, T., P. Binning, and G. R. Willgoose (2005), The role of moisture cycling in the weathering of a quartz chlorite schist in a tropical environment: Findings of a laboratory simulation, *Earth Surface Processes and Landforms*, 30(4), 413–428, doi:10.1002/esp.1149.

Wells, T., P. Binning, G. R. Willgoose, and G. R. Hancock (2006), Laboratory simulation of the salt weathering of schist: I. Weathering of schist blocks in a seasonally wet tropical environment, *Earth Surface Processes and Landforms*, 31(3), 339–354, doi:10.1002/esp.1248.

Wells, T., G. R. Hancock, C. Dever, and C. Martinez (2013), Application of RothPC-1 to soil carbon profiles in cracking soils under minimal till cultivation, *Geoderma*, 207–208, 144–153, doi:10.1016/j.geoderma.2013.05.018.

Wells, T., G. R. Hancock, C. Dever, and D. Murphy (2012), Prediction of vertical soil organic carbon profiles using soil properties and environmental tracer data at an untilled site, *Geoderma*, 170, 337–346, doi:10.1016/j.geoderma.2011.11.006.

Wells, T., G. R. Willgoose, and P. Binning (2007), Laboratory simulation of the salt weathering of schist: II. Fragmentation of fine schist particles, *Earth Surface Processes and Landforms*, 32(5), 687–697, doi:10.1002/esp.1450.

Wells, T., G. R. Willgoose, and G. R. Hancock (2008), Modelling weathering pathways and processes for salt induced fragmentation of quartz-chlorite schist, *Journal of Geophysical Research (Earth Surface)*, 113, F01014, doi:10.1029/2006JF000714.

West, A. J. (2012), Thickness of the chemical weathering zone and implications for erosional and climatic drivers of weathering and for carbon-cycle feedbacks, *Geology*, 40(9), 811–814, doi:10.1130/G33041.1.

West, N., E. Kirby, P. Bierman, and B. A. Clarke (2014), Aspect-dependent variations in regolith creep revealed by meteoric ^{10}Be, *Geology*, 42(6), 507–510, doi:10.1130/G35357.1.

Western, A. W., R. B. Grayson, G. Blöschl, G. R. Willgoose, and T. A. McMahon (1999), Observed spatial organization of soil moisture and its relation to terrain indices, *Water Resources Research*, 35(3), 797–810.

Whipple, K. X. (1997), Open-channel flow of Bingham fluids: Applications in debris-flow research, *Journal of Geology*, 105(2), 243–262, doi:10.1086/515916.

Whipple, K. X. (2009), The influence of climate on the tectonic evolution of mountain belts, *Nature Geoscience*, 2, 97–104, doi:10.1038/ngeo413.

Whipple, K. X., G. S. Hancock, and R. S. Anderson (2000), River incision into bedrock: Mechanics and relative efficacy of plucking, abrasion, and cavitation, *Geological Society of America Bulletin*, 112(3), 490–503.

Whipple, K. X., and B. J. Meade (2004), Controls on the strength of coupling among climate, erosion, and deformation in two sided, frictional orogenic wedges at steady state, *Journal of Geophysical Research (Earth Surface)*, 109(F1), F01011, doi:10.1029/2003JF000019.

Whipple, K. X., and G. E. Tucker (1999), Dynamics of the stream-power river incision model: Implications for height limits of mountain ranges, landscape response timescales, and research needs, *Journal of Geophysical Research (Solid Earth)*, 104(B8), 17661–17674.

Whipple, K. X., and G. E. Tucker (2002), Implications of sediment-flux-dependent river incision models for landscape evolution, *Journal of Geophysical Research (Solid Earth)*, 107(B2), art. no.-2039.

White, A. F., and S. L. Brantley (2003), The effect of time on the weathering of silicate minerals: Why do weathering rates differ in the laboratory and field?, *Chemical Geology*, 202(3–4), 479–506, doi:10.1016/j.chemgeo.2003.03.001.

White, A. F., A. E. Blum, M. S. Schulz, T. D. Bullen, J. W. Harden, and M. L. Peterson (1996), Chemical weathering rates of a soil chronosequence on granitic alluvium: I. Quantification of mineralogical and surface area changes and calculation of primary silicate reaction rates, *Geochimica et Cosmochimica* Acta, 60(14), 2533–2550.

White, A. F., M. S. Schulz, D. A. Stonestrom, D. V. Vivit, J. Fitzpatrick, T. D. Bullen, K. Maher, and A. E. Blum (2009), Chemical weathering of a marine terrace chronosequence, Santa Cruz, California. Part II: Solute profiles, gradients and the comparisons of contemporary and long-term weathering rates, *Geochimica et Cosmochimica Acta*, 73, 2769–2803, doi:10.1016/j.gca.2009.01.029.

White, A. F., M. S. Schulz, D. V. Vivit, A. E. Blum, D. A. Stonestrom, and S. P. Anderson (2008), Chemical weathering of a marine terrace chronosequence, Santa Cruz, California I: Interpreting rates and controls based on soil concentration–depth profiles, *Geochimica et Cosmochimica Acta*, 72(1), 36–68, doi:10.1016/j.gca.2007.08.029.

White, A. F., M. S. Schulz, D. V. Vivit, A. E. Blum, D. A. Stonestrom, and J. W. Harden (2005), Chemical weathering rates of a soil chronosequence on granitic alluvium: III. Hydrochemical evolution and contemporary solute fluxes and rates, *Geochimica et Cosmochimica Acta*, 69(8), 1975–1996, doi:10.1016/j.gca.2004.10.003.

Wickens, G. E., and F. W. Collier (1971), Some vegetation patterns in the Republic of Sudan, *Geoderma*, 6, 43–59, doi:10.1016/0016-7061(71)90050-4.

Wickert, A. D. (2015), Open-source modular solutions for flexural isostasy: gFlex v1.0, *Geoscientific Model Development Discussions*, 8, 4245–4292, doi:10.5194/gmdd-8-4245-2015.

Wilcock, P. R., and J. C. Crowe (2003), Surface-based transport model for mixed-size sediment, *Journal of Hydraulic Engineering*, 129(2), 120–128, doi:10.1061/ASCE 0733-9429(2003)129:2(120).

Wilkinson, M. T., G. S. Humphreys, J. Chappell, K. Fifield, and B. Smith (2003), Estimates of soil production in the Blue Mountains, Australia, using cosmogenic 10Be, Advances in Regolith: Proceedings of the CRC LEME Regional Regolith Symposia, 2003, Cooperative Research Centre for Landscape Environments and Mineral Exploration (CRC LEME), Canberra, http://crcleme.org.au/Pubs/Advancesinregolith/Wilkinson_et_al.pdf.

Wilkinson, M. T., P. J. Richards, and G. S. Humphreys (2009), Breaking ground: Pedological, geological, and ecological implications of soil bioturbation, *Earth-Science Reviews*, 97, 257–272, doi:10.1016/j.earscirev.2009.09.005.

Willett, S. (1999), Orogeny and orography: The effects of erosion on the structure of mountain belts, *Journal of Geophysical Research (Solid Earth)*, 104(B12), 28957–28981.

Willett, S. D., R. Slingerland, and N. Hovius (2001), Uplift, shortening, and steady-state topography in active mountain belts, *American Journal of Science*, 301(4–5), 455–485.

Willgoose, G. R. (1989), A physically based channel network and catchment evolution model, PhD thesis, Department of Civil Engineering, Massachusetts Institute of Technology, Cambridge, MA. Also published as Willgoose, G. R., R. L. Bras, and I. Rodriguez-Iturbe (1989), A physically based channel network and catchment evolution model, Tech. Report 322, Ralph M. Parsons Laboratory, Dept. of Civil Engineering, MIT, Cambridge, MA.

Willgoose, G. R. (1994a), A physical explanation for an observed area-slope-elevation relationship for declining catchments, *Water Resources Research*, 30(2), 151–159.

Willgoose, G. R. (1994b), A statistic for testing the elevation characteristics of landscape simulation models, *Journal of Geophysical Research (Solid Earth)*, 99(B7), 13987–13996.

Willgoose, G. R. (1995), A preliminary assessment of the effect of vegetation on the long-term stability of the proposed above-grade rehabilitation strategy at Ranger Uranium Mine, Open File Report 119, Environmental Research Institute of the Supervising Scientist, Jabiru, Australia.

Willgoose, G. R. (1997), A hydrodynamic particle tracking algorithm for simulating settling of sediment, *Mathematics and Computers in Simulation*, 43, 343–349.

Willgoose, G. R. (2001), Erosion processes, catchment elevations and landform evolution modelling, in *Gravel Bed Rivers 2000*, edited by P. Mosley, pp. 507–530, Hydrology Society, Christchurch, New Zealand.

Willgoose, G. R. (2005a), Mathematical modeling of whole-landscape evolution, *Annual Review of Earth and Planetary Sciences*, 33, 443–459.

Willgoose, G. R. (2005b), User manual for SIBERIA (Version 8.30), www.telluricresearch.com/siberia_8.30_manual.pdf.

Willgoose, G. R. (2009), TELLUSIM: A Python plug-in based computational framework for spatially distributed environmental and earth sciences modelling, paper presented at 18th World IMACS/MODSIM Congress, Cairns, Australia, 13–17 July 2009.

Willgoose, G. R. (2010), Assessment of the erosional stability of encapsulation caps and covers at the millennial timescale: Current capabilities, research issues and operational needs, paper presented at Proceedings of the Workshop on Engineered Barrier Performance Related to Low-level Radioactive Waste, Decommissioning, and Uranium Mill Tailings facilities, NUREG/CP-0195, Office of Nuclear Regulatory Research, Rockville, Maryland, 3–5 August 2010.

Willgoose, G. R. (2011), Modelling bushfire impact on hydrology: The implications of the fire modelling approach on the climate change impact, in *MSSANZ MODSIM 2011*, pp. 3664–3670, Modelling and Simulation Society of Australia and New Zealand, Perth.

Willgoose, G. R. (2015), The interactions between evolving soils and landforms, and the importance of relative response times, in *EGU General Meeting*, pp. EGU2015–8220, EGU, Vienna.

Willgoose, G. R., R. L. Bras, and I. Rodriguez-Iturbe (1990), A model of river basin evolution, *EOS*, 71(47), 1806–1807.

Willgoose, G. R., R. L. Bras, and I. Rodriguez-Iturbe (1991a), A coupled channel network growth and hillslope evolution model. 1. Theory, *Water Resources Research*, 27(7), 1671–1684.

Willgoose, G. R., R. L. Bras, and I. Rodriguez-Iturbe (1991b), A coupled channel network growth and hillslope evolution model. 2. Nondimensionalization and applications, *Water Resources Research*, 27(7), 1685–1696.

Willgoose, G. R., R. L. Bras, and I. Rodriguez-Iturbe (1991c), A physical explanation of an observed link area-slope relationship, *Water Resources Research*, 27(7), 1697–1702.

Willgoose, G. R., and Y. Gyasi-Agyei (1995), New technology in hydrology and erosion modeling for mine rehabilitation, paper presented at APCOM XXV Application of Computers and Operations Research in the Mineral Industries, Australian Institute of Mining and Metallurgy, Brisbane.

Willgoose, G. R., and G. R. Hancock (1998), Revisiting the hypsometric curve as an indicator of form and process in transport-limited catchments, *Earth Surface Processes and Landforms*, 23(7), 611–623.

Willgoose, G. R., and G. R. Hancock (2010), Applications of long-term erosion and landscape evolution models, in *Handbook of erosion modelling*, edited by R. P. C. Morgan and M. A. Nearing, pp. 339–359, Wiley-Blackwell, Oxford.

Willgoose, G. R., G. R. Hancock, and G. A. Kuczera (2003), A framework for the quantitative testing of landform evolution models, in *Predictions in geomorphology*, edited by P. R. Wilcock and R. M. Iverson, pp. 195–216, American Geophysical Union, Washington, DC.

Willgoose, G., and G. Kuczera (1995), Estimation of subgrid scale kinematic wave parameters for hillslopes, *Hydrological Processes*, 9(3–4), 469–482.

Willgoose, G. R., G. A. Kuczera, and B. J. Williams (1995), DISTFW-NLFIT: Rainfall-runoff and erosion model calibration and model uncertainty assessment suite, Research Report 108.03.1995, Department of Civil Engineering and Surveying, University of Newcastle, Australia, Callaghan.

Willgoose, G. R., and H. J. Perera (2001), A simple model for saturation excess runoff generation based on geomorphology, steady state soil moisture, *Water Resources Research*, 37(1), 147–156.

Willgoose, G. R., and S. J. Riley (1998a), Application of a catchment evolution model to the prediction of long term erosion on the spoil heap at Ranger Uranium Mines: Initial analysis, Supervising Scientist Report 132, Australian Government Publishing Service, Canberra. Originally published as Willgoose, G. R., and S. J. Riley (1993), Application of a catchment evolution model to the prediction of long-term erosion on the spoil heap at Ranger Uranium Mine, Open File Report 107, Office of the Supervising Scientist, Jabiru.

Willgoose, G. R., and S. J. Riley (1998b), An assessment of the long-term erosional stability of a proposed mine rehabilitation, *Earth Surface Processes and Landforms*, 23, 237–259.

Willgoose, G. R., and S. Sharmeen (2006), A one-dimensional model for simulating armouring and erosion on hillslopes. 1. Model development and event-scale dynamics, *Earth Surface Processes and Landforms*, 31(8), 970–991, doi:10.1002/esp.1398.

Williams, B. J. (2006), *Hydrobiological modelling: Processes, numerical methods and applications*, Lulu Press, Newcastle, Australia.

Wilson, C. J., K. J. Crowell, and L. J. Lane (2006), Surface erosion modelling for the repository waste cover at Los Alamos National Laboratory Technical Area 54, Material Disposal Area G, Rep. LA-UR-05-7771, Los Alamos.

Wischmeier, W. H., and D. D. Smith (1978), *Predicting rainfall erosion losses – A guide to conservation planning, USDA Agriculture Handbook 537*, US Government Printing Office, Washington, DC. http://naldc.nal.usda.gov/download/CAT79706928/PDF.

Witten, T. A., and L. M. Sander (1981), Diffusion-limited aggregation, a kinetic critical phenomenon, *Physical Review Letters*, 47(19), 1400–1403.

Wohl, E. E., R. O. Hall, K. B. Lininger, N. A. Sutfin, and D. M. Walters (2017), Carbon dynamics of river corridors and the effects of human alterations, *Ecological Monographs*, doi:10.1002/ecm.1261.

Woods, S. W., and V. N. Balfour (2008), The effect of ash on runoff and erosion after a severe forest wildfire, Montana, USA, *International Journal of Wildland Fire*, 17, 535–548, doi:10.1071/WF07040.

Wu, S., R. L. Bras, and A. P. Barros (2006), Sensitivity of channel profiles to precipitation properties in mountain ranges, *Journal of Geophysical Research (Earth Surface)*, 111, F01024, doi:10.1029/2004JF000164.

Wu, W., and R. C. Sidle (1995), A distributed slope stability model for steep forested basins, *Water Resources Research*, 31(8), 2097–2110.

Wyllie, D. C. (2014), Calibration of rockfall modeling parameters, *International Journal of Rock Mechanics & Mining Sciences*, 67, 170–180, doi:10.1016/j.ijrmms.2013.10.002.

Yang, C. T. (1973), Incipient motion and sediment transport, *Journal of Hydraulic Division – ASCE*, 99(HY10), 1679–1704.

Yeteman, O., E. Istanbulluoglu, and A. R. Duvall (2015a), Solar radiation as a global driver of hillslope asymmetry: Insights from an ecogeomorphic landscape evolution model, *Water Resources Research*, 51, 9843–9861, doi:10.1002/2015WR017103.

Yetemen, O., E. Istanbulluoglu, J. Flores-Cervantes, E. R. Vivoni, and R. L. Bras (2015b), Ecohydrologic role of solar radiation on landscape evolution, *Water Resources Research*, 51, 1127–1157, doi:10.1002/2014WR016169.

Yoo, K., R. Amundson, A. M. Heimsath, and W. E. Dietrich (2005), Erosion of upland hillslope soil organic carbon: Coupling field measurements with a sediment transport model, *Global Biogeochemical Cycles*, 19, GB3003, doi:10.1029/2004GB002271.

Yoo, K., R. Amundson, A. M. Heimsath, and W. E. Dietrich (2006), Spatial patterns of soil organic carbon on hillslopes: Integrating geomorphic processes and the biological C cycle, *Geoderma*, 130(1–2), 47–65.

Yoo, K., and S. M. Mudd (2008), Toward process-based modeling of geochemical soil formation across diverse landforms: A new mathematical framework, *Geoderma*, 146, 248–260, doi:10.1016/j.geoderma.2008.05.029.

Yu, C.-K., and L.-W. Cheng (2013), Distribution and mechanisms of orographic precipitation associated with Typhoon Morakot, *Journal of Atmospheric Sciences*, 70(9), 2894–2915, doi:10.1175/JAS-D-12-0340.1.

Yu, N., B. Boudevillian, G. Delrieu, and R. Uijlenhoet (2012), Estimation of rain kinetic energy from radar reflectivity and/or rain rate based on a scaling formulation of the raindrop size distribution, *Water Resources Research*, 48, W04505, doi:10.1029/2011WR011437.

Yu, Y. Y., P. A. Finke, H. B. Wu, and Z. T. Guo (2013), Sensitivity analysis and calibration of a soil carbon model (SoilGen2) in two contrasting loess forest soils, *Geoscientific Model Development*, 6, 29–44, doi:10.5194/gmd-6-29-2013.

Zaitlin, B., and M. Hayashi (2012), Interactions between soil biota and the effects on geomorphological features, *Geomorphology*, 157–158, 142–152, doi:10.1016/j.geomorph.2011.07.029.

Zhang, J. J., and L. R. Bentley (2005), Factors determining Poisson's ratio, Research Report 62, University of Calgary, CREWES, Calgary.

Zhang, K., J. S. Kimball, and S. W. Running (2016), A review of remote sensing based actual evapotranspiration estimation, *Wiley Interdisciplinary Reviews – Water*, 3(6), 834–853, doi:10.1002/wat2.1168.

Zhang, W., and D. R. Montgomery (1994), Digital elevation model grid size, landscape representation and hydrologic simulations, *Water Resources Research*, 30(4), 1019–1028.

Zhang, W., J. Niu, V. L. Morales, X. Chen, A. G. Hay, J. Lehmann, and T. S. Steenhuis (2010), Transport and retention of biochar particles in porous media: Effect of pH, ionic strength, and particle size, *Ecohydrology*, 3, 497–508, doi:10.1002/eco.160.

Zhang, Z. F., A. L. Ward, and J. M. Keller (2009), Determining the porosity and saturated hydraulic conductivity of binary mixtures, Pacific Northwest National Laboratory, Oak Ridge.

Zhong, S., A. Paulson, and J. Wahr (2003), Three-dimensional finite-element modelling of Earth's viscoelastic deformation: Effects of lateral variations in lithospheric thickness, *Geophysical Journal International*, 155(2), 679–695, doi:10.1046/j.1365-246X.2003.02084.x.

Zhou, X., E. Istanbulluoglu, and E. R. Vivoni (2013), Modeling the ecohydrological role of aspect-controlled radiation on tree-grass-shrub coexistence in a semiarid climate, *Water Resources Research*, 49, 2872–2895, doi:10.1002/wrcr.20259.

Zolezzi, G., R. Luchi, and M. Tubino (2012), Modelling morphodynamic processes in meandering rivers with spatial width variations, *Reviews of Geophysics*, 50, RG4005, doi:10.1029/2012RG000392.

Index